CORROSION PREDICTION AND PREVENTION IN MOTOR VEHICLES

ELLIS HORWOOD SERIES IN CORROSION AND ITS PREVENTION

Series Editors:
A. D. MERCER, Principal Scientific Officer, National Physical Laboratory and Scientific Secretary, European Federation of Corrosion
D. R. HOLMES, Manager/Consultant, High Temperature Corrosion Projects, National Corrosion Service, National Physical Laboratory
D. H. SHARPE, O.B.E., lately General Secretary, Society of Chemical Industry; formerly General Secretary, Institution of Chemical Engineers; and former Technical Director, Confederation of British Industry.

This new series provides up-to-date compelling texts on the theory and practice of corrosion science and technology. There is special emphasis on cost-effective preventative measures and on the most recent developments in new materials and techniques for the study of corrosion products. The books will usually be at the post-graduate research or practising corrosion engineer/technologist level, but will also be suitable for informing advanced undergraduate and master courses of the most recent developments. Together, they will provide substantial coverage of the latest developments in corrosion phenomena and protective measures.

PROTECTION OF METALS FROM CORROSION IN STORAGE AND TRANSIT
P. D. DONOVAN, Principal Scientific Officer, Ministry of Defence, Royal Armament Research and Development Establishment
HIGH ALLOY STAINLESS STEELS FOR CRITICAL SEAWATER APPLICATIONS
Editors: J. W. OLDFIELD and P. GUHA
BASIC CORROSION AND OXIDATION, Second Edition
J. M. WEST, Department of Metallurgy, University of Sheffield
CORROSION OF REINFORCEMENT IN CONCRETE CONSTRUCTION
A. P. CRANE, Pall Process Filtration, Hampshire
CORROSION PREDICTION AND PREVENTION IN MOTOR VEHICLES
H. McARTHUR, Faculty of Technology and Construction, Leicester Polytechnic
CORROSION AND MARINE GROWTH ON OFFSHORE STRUCTURES
J. R. LEWIS, Senior Lecturer in Chemistry, University of Aberdeen, and A. D. MERCER, Principal Scientific Officer, National Physical Laboratory, Teddington.

Books in Corrosion Science Published for the Institution of Corrosion Science and Technology

CATHODIC PROTECTION: Theory and Practice
Editors: V. ASHWORTH, Global Corrosion Consultants, Telford, and C. J. L. BOOKER, City of London Polytechnic
DEWPOINT CORROSION
Editor: D. R. HOLMES, Manager/Consultant, High Temperature Corrosion Projects, National Corrosion Service, National Physical Laboratory
COATINGS AND SURFACE TREATMENT FOR CORROSION AND WEAR RESISTANCE
Editors: K. N. STRAFFORD and P. K. DATTA, School of Materials Engineering, Newcaste upon Tyne Polytechnic, and C. G. GOOGAN, Global Corrosion Consultants Limited, Telford
PLANT CORROSION: Prediction of Materials Performance
Editors: J. E. STRUTT and J. R. NICHOLLS, School of Industrial Science, Cranfield Institute of Technology, Bedford

CORROSION PREDICTION AND PREVENTION IN MOTOR VEHICLES

HUGH McARTHUR, B.A., M.A., Ph.D.
Faculty of Technology and Construction
Leicester Polytechnic

ELLIS HORWOOD LIMITED
Publishers · Chichester

Halsted Press: a division of
JOHN WILEY & SONS
New York · Chichester · Brisbane · Toronto

First published in 1988 by
ELLIS HORWOOD LIMITED
Market Cross House, Cooper Street,
Chichester, West Sussex, PO19 1EB, England
The publisher's colophon is reproduced from James Gillison's drawing of the ancient Market Cross, Chichester.

Distributors:
Australia and New Zealand:
JACARANDA WILEY LIMITED
GPO Box 859, Brisbane, Queensland 4001, Australia
Canada:
JOHN WILEY & SONS CANADA LIMITED
22 Worcester Road, Rexdale, Ontario, Canada
Europe and Africa:
JOHN WILEY & SONS LIMITED
Baffins Lane, Chichester, West Sussex, England
North and South America and the rest of the world:
Halsted Press: a division of
JOHN WILEY & SONS
605 Third Avenue, New York, NY 10158, USA
South-East Asia
JOHN WILEY & SONS (SEA) PTE LIMITED
37 Jalan Pemimpin # 05–04
Block B, Union Industrial Building, Singapore 2057
Indian Subcontinent
WILEY EASTERN LIMITED
4835/24 Ansari Road
Daryaganj, New Delhi 110002, India

British Library Cataloguing in Publication Data
McArthur, Hugh
Corrosion prediction and prevention in motor vehicles. —
(Ellis Horwood series in mechanical engineering).
1. Motor Vehicles — Corrosion
I. Title
629.2′32 TL154

Library of Congress Card No. 87–26089

ISBN 0–7458–0357–1 (Ellis Horwood Limited)
ISBN 0–470–20995–X (Halsted Press)

Phototypeset in Times by Ellis Horwood Limited
Printed in Great Britain by Butler & Tanner, Frome, Somerset

Table of contents

Preface

In writing a book of this kind, one accumulates knowledge from, and indebtedness to, a wide range of people, not least to the participants and authors of published papers at corrosion conferences and in a corrosion text-books. I particularly wish to acknowledge the help given by this formal interchange, which has sharpened up the theory which has been developed here in the body of this book.

Most engineers in the automotive industries know the results of corrosion, but very little about the mechanism or corrosion theory. However, they feel on very safe ground because neither do their listeners or the purchasers of their products. A section is included in the book concerning some of the 'mechanisms' of corrosion given to the author during lecture tours of the United States, Canada, Japan and England. A private motorist expects an automotive engineer to know about motor vehicle degradation and so accepts the 'myths' about corrosion which are often expounded with such authority. If you don't know better, you can't verbally compete in such a situation. It is human nature to remain silent or to feel intimidated, and acceptance of the 'respected engineer's views', however wrong, is the result. If this book rectifies this human characteristic about the subject of corrosion, then it will have achieved one of its objects.

The book is dedicated to the scientists and technologists in the production industries, to those students studying engineering, materials science or motor vehicle courses, and to designers, engineers and those of the general public whose primary function is to combat the premature ageing of their most valuable asset — the motor car.

Since corrosion involves chemical change in the metal, it is obvious that the reader will have to be familiar with the fundamental principles of chemistry and chemical reactions. Most corrosion reactions are determined by structure and chemical composition of the metal and so some elements of metallurgy are also included. Chaper 6 onwards includes the elementary modelling of the material world. These models help the scientist in his interpretation of the outside world.

These 'models' allow for a discussion of the laws of nature and perform the same role as the architect's model which portrays the building from the plans. These plans would be difficult for the layman to interpret. Hence it is the deliberate policy of the author to keep these scientific elements to a level where they can be easily interpreted by an enthusiastic quester of information and thus mathematical equations, unless absolutely necessary, are not included. For example, the Nernst equation is not included, but the elementary Ohm law is where it is required to explain the electro-coating process.

Each scientific 'package' is well illustrated and cross-referenced for easier understanding. A bibliography is given to which reference may be made for further reading. The exception to this rule is in dealing with dislocations, where the concept is perhaps unique.

This book is based on the careful examination, interpretation and experimentation of the corrosion reaction by the author. Hence, if there are any misinterpretations, they are the author's and no-one else's. The information is advisory to the corrosion of the motor vehicle in general, and the application to a specific model and make is solely the risk of the user.

1

Setting the scene

1.1 INTRODUCTION

Acquisition of proficiency in scientific understanding in the narrow field of motor vehicle corrosion requires a grasp of a number of fundamental concepts and an elementary knowledge of engineering, physics and chemistry. It also requires the ability to apply some fundamental concepts to both familiar and unfamiliar problems. It is interdisciplinary and therefore presents a problem to an author in collecting the many scientific principles together and arranging them in some logical order. The book has been written so as to give an understanding of the corrosion mechanism without a detailed knowledge of any one science. The detailed knowledge of science has been arranged in Chapters 6–12, which follow the main corrosion text so as not to spoil, for the layman, the theory and derivation of the build-up of the aggressive micro-climate which becomes a free-rider in the out-of-sight areas within the vehicle. These chapters may therefore be omitted by the reader without impairing the comprehension of the subject of automobile corrosion. The complete book will develop the knowledge of the reader in corrosion and materials technology. It is suggested that the references to Chapters 6–12 are left for those who wish to specialize in some aspect of science or to be a specialist in motor vehicle corrosion, and until the general mechanism of the micro-climate build-up is understood. The layman should approach Chapters 6–12 with the certain knowledge that science is learned in small 'packets of knowledge', and then only by repetitive reading. The amount learned from each technical subsection on one reading will be perhaps 5–45%, and so each such subsection will require several readings and discussion. If all the technical subsections are read, slowly and painstakingly, with discussion or thought over a period of time for ease of assimilation, the book will assist laymen, garage personnel or the student studying engineering courses to gain a deeper insight into corrosion processes. Science, and its associated modelling, is not learned overnight and so it is recognized that this approach will be conceptually difficult to the person either without a scientific base upon which to study or without the discussion from interested colleagues. The author hopes that, from this book, even non-scientists should be able to look at their future motor cars as regards longevity and good design in the corrosion sense and to interpret the external manifestations of corrosion with a new insight. Remember (see Section 5.2) that **good corrosion design is free, bad corrosion design is expensive**. The residual second-hand value resides in the body and not in the mechanical power train. Corrosion is localised into relatively small areas (about 1–5% of the vehicle is adversely affected). Corrosion is usually internally nucleated on internal panels of the body. Corrosion starts in crevices and in 'out-of-sight' areas and beneath dirt deposits or mud accumulations.

It would be a good design goal if the body could outlast one if not two power trains and the internal fabric and physical age of the vehicle governed its economic life, rather than the relatively small areas of the vehicle body which are so badly affected by corrosion.

1.2 CORROSION AS A CONSUMER

Corrosion has been aptly described as 'an insidious consumer of our stocks of raw materials, a squanderer of our productive capacity and a dissipater of the fruits of our labours' (Rogers, 1967). The actual 'raw cost' of the materials making up the motor vehicle is a small fraction of its added value cost in labour and other non-corroding materials. A small family car weighing 1000 kg is made up of the materials shown in Table 1.1.

Table 1.1 — Material content of a motor vehicle

360 kg of sheet steel, and it is the body which limits the life of a car and not the mechanical parts
250 kg of wrought steel
140 kg of cast iron (mainly engine)
30 kg of light alloy
15 kg of copper wire (does not corrode)
35 kg of plastics (does not corrode) (Table 11.4)
50 kg of glass (does not corrode) (Section 6.45.1)
60 kg rubber (tyres) which wear out

remainder is oil, water, carpets etc.

About 38–45% of the purchase price of a new car is tax, delivery and dealer margins; VAT and special car taxes account for 21% (see Table 1.3).
From B. B. Hundy, Resources Policy, September 1976, p. 180, Table 2.

The consequences and study of corrosion have become in recent years an integral part of life since the Hoar Committee (Hoar, 1971) estimated in 1971 that corrosion costs amounted to about 4% of the gross national product (GNP) (see Table 5.1). Put crudely, this means that a third of the annual steel output in one year goes back to rust within that year. With energy not being a renewable resource, and energy costs increasing, protection against corrosion will become more desirable, if not fashionable, owing to the fact that energy costs are a major factor in the production of the corrodible metallic components. Often the damage and the repair costs of this corrosion take other non-corrodible and unworn but high cost attachments to the scrap yard. This is certainly true of the motor vehicle, where personal transport is the most expensive commodity a person ever acquires. Normally, one would argue that a house is the most expensive purchase one undertakes: indeed the total monetary outlay is initially very large but at the end of the purchase period there is a 'return' on the capital outlay, because house prices tend to keep pace with inflation (Fig. 1.1).

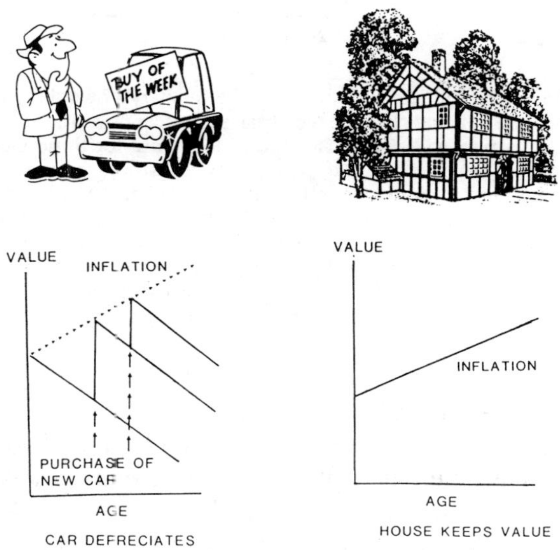

Fig. 1.1 — The comparison of the value of a house with that of a car. A house is normally given by most analysts as the most expensive item a person ever purchases. A house appreciates annually as the paper money decreases in value owing to inflation with only a small loss, if any, for wear and tear. This should be compared with the value of a motor vehicle which depreciates with age and 'wear and tear'. This heavy depreciation is mostly due to the corrosion of about 5% of the total body-work. 95% of the body-work remains uncorroded.

This is not so with motor vehicles. Consider the following example: a new car, costing say £8000 for the next 40 years. A depreciation of roughly 22% of the initial purchase price (22% × £8000 = £1760) is the first year's depreciation (Fig. 1.2). This depreciation cost soon mounts up. For example, a person buying a new car each year would lose £70 000 over the 40 years of personal motoring (40 × £1760). This sum lost does not include any interest on capital or servicing. Servicing, however could be absorbed in the concept that there would be no need for a service if the car was replaced every year.

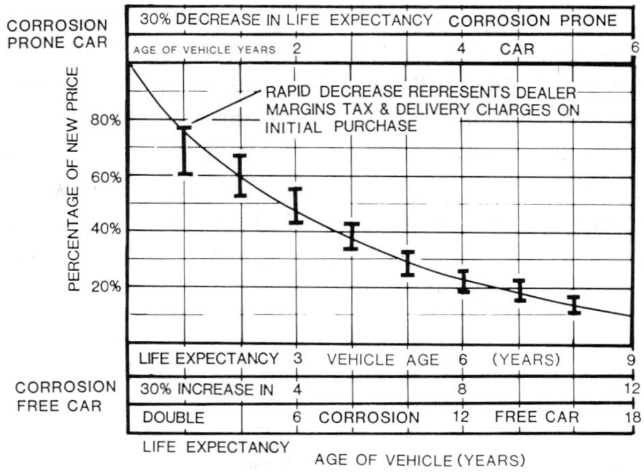

Fig. 1.2 — If any motorists' price guide, or even the industry's own 'blue book' prices, are taken and expressed as a percentage of the replacement new car price and plotted against age of vehicle a curve of this type will result. The figures taken are for those models which have remained basically unchanged so as not to represent a breakthrough in technology. The new replacement price is taken so as to take in the inflation element. The horizontal axis can be altered to represent a longer-life vehicle of 30% and 100% increase in life. This increase in life is obtained solely by good corrosion design and therefore is free. The scatter represents the second-hand value ascribed to the vehicle by the second-hand owners who are the 'true owners' of the models produced by the manufacturers. The vehicle makes with the greater depreciation loss are usually those which suffer the most corrosion.

1.3 MARKET DISTORTIONS

The purchase of a motor vehicle is not the same as the purchase of a house or other utility, say a washing machine. These utility items are usually paid for by the user and keeper and are often not resold until the useful life has expired. A motor vehicle on the other hand is more often purchased or subsidized by a company or agent with the specific intention of reselling it while there is some useful life remaining. It is estimated that over 80% of the new vehicles are purchased as 'company perks' and that 79% of the cars travelling into London are not driven by the purchaser and are sold within three years of purchase. This means that the life of a car with its second-hand owner(s) far exceeds the life with its first owner by a factor of four or five. The second-hand owner(s) therefore have more to lose from corrosion than do the original purchasers of the motor vehicle. It is the second-hand owner who has least direct effect on the manufacturers concerning corrosion because there is always a time delay in the effect of these second-hand consumers on the trade-in prices.

1.4 THE FIRST OWNER PROBLEM

The problem therefore can best be summarised as follows:

> for a motor vehicle to have better corrosion protection, the first purchaser must be prepared to pay a higher price for a commodity from which he will not derive any benefit.

The advantages which accrue from this higher initial purchase price are to the benefit of those owners in the latter years of the vehicle's life. Any private owner of a well kept vehicle will always claim that a well kept vehicle does not reflect a higher trade-in price when it is exchanged for a new car. Fig. 1.2 shows a curve which indicates the depreciation, as a percentage of the new purchase price, for six popular vehicles with age of car along the horizontal axis. By altering the scale of this axis, the trade-in price for a more durable 'long-life' car can be calculated with several life scales. If it were not for the market distortions discussed above, and in subsequent sections, these percentage figures would most probably apply.

1.5 THE LIFE OF A CAR

The average life of an average car is about 10–12 years, although there is considerable scatter. One manufacturer claims that 50% of its cars are still being re-registered after 19 years (Table 1.2). The average life has increased in recent years, but motor vehicle purchase still does not represent a good investment. The areas which corrode within a car are those relatively small areas which are badly designed (see Section 2.5.1) with regards to corrosion, as will be shown in Chapter 5. These areas could be

Table 1.2 — The percentage of cars remaining

Car make	Percentage remaining of year model		
	1962 (about 16 years old)	1966 (about 12 years old)	1970 (about 8 years old)
All passenger cars	27	62	94
BMC	16	37	88
BMW	13	49	94
Citroen	16	49	94
DKW/Audi	10	42	91
Fiat	12	38	82
Ford	16	47	91
Mercedes-Benz	30	75	97
Opel	16	53	92
Peugeot	19	42	85
Renault	13	40	90
SAAB	17	40	93
Simca	4	34	88
Toyota	—[a]	54	88
Vauxhall	15	39	86
Volkswagen	26	67	93
Volvo	56	88	97

The percentages of the originally registered passenger cars of the major car makes, year models 1962, 1966 and 1970, which remain at the end of 1978.

[a]The car make was not represented in Sweden at this time.

Data from I. E. Svenson, Swedish Corrosion Institute, and *Cars go Rusty*, AB Svensk Bilprovning.

considerably improved by better design, and greater attention to detail, and therefore would represent a minimal cost to the first and subsequent owners. However, greater longevity could mean that the increase in the number of cars on the road could affect the sale of new vehicles, which if taken to a logical situation would mean roads becoming choked with ageing motor vehicles, such that the most affluent motorists took to the air. Thus corrosion protection does not follow the normal free enterprise marketplace rules, as the driving force behind offering a better standard of corrosion protection increases the cost of the product, which gives a competitive financial edge to those manufacturers who have not offered this additional 'attractive' durability. Secondly, the first-time owner of a company-purchased car is only interested in getting the best internal specification for the sum allowed. Durability, in which the driver has no financial stake, is not considered in these cases. Therefore it is understandable that the motor manufacturers were hesitant in offering additional corrosion durability packages which only benefitted second and subsequent owners and which could act to disadvantage them in the first new car registrations. Therefore some legislation was necessary to bring this contradiction to a halt. This legislation was proposed by the Canadian Government and is discussed in Section 1.12.

1.6 HIGH FIRST COSTS: STAINLESS STEEL EXHAUSTS

Most of the consumable products (e.g. exhausts (Section 8.4)) on a motor vehicle could be, and have been, improved by using better materials which are more resistant to corrosion. An example of this is the exhaust system, which can be obtained at various costs. The cheaper exhausts are usually manufactured from mild steel (plain carbon steel) and last 18 months whereas the more expensive ones (see Table 8.4), lasting three years or more, are of special aluminized steel. Stainless steel exhausts are advertized with various guarantees, usually five years, with instant replacement if there is any failure. Stainless steel is considerably more expensive than aluminized steel. The raw costs of aluminized steel are about 2% more than those of mild steel. When one considers the economics of a two-year exhaust life, loss of use, expensive labour costs and difficulty of fitting a replacement, it makes sense to use the more expensive steel. However, only the after-market trade dealing with replacements still uses the inferior mild steel, although stainless steel is offered as an alternative.

1.7 MANUFACTURING AND MARKET SUBSIDIES

It is estimated that the present market subsidy of the company car syndrome amounts to some £2,000 million per annum (Section 1.8). This figure is more than the UK Government spends on subsidizing the whole of the combined public transport by bus and rail networks. In order to substantiate this figure, certain tax matters will have to be discussed. Most people find tax matters unbelievably complicated. Rather than get bogged down in the derivation of this subsidy, accept the figure and proceed to Section 1.9. Tax rates and regulations vary annually and so this may be altered in subsequent tax years.

In the purchase and use of a new 'company' car we have five interested parties. These are: (1) the Government, through the Inland Revenue, and Customs and

Excise, (2) the motor vehicle manufacturer, (3) the prospective purchaser, (the company), (4) the prospective user-driver, who wants the best vehicle specification which the money allocation will purchase, and (5) the dealer, who also represents the Customs and Excise.

Eight out of ten cars are purchased as company cars either directly or by leasing arrangements, while the remainder are purchased by the average motorist, who will be called Mr PAYE (Mr Pay As You Earn). When Mr PAYE buys a car there are only three interested parties and these are (see Table 1.3) (1) the motor vehicle manufacturer, (2) the prospective purchaser, who is also the prospective user and driver, and (3) the dealer, who also represents the Customs and Excise,

The Government, through the Inland Revenue, has now dropped out because, for every £1,000 earned, Mr PAYE will have paid £300 income tax (assuming that tax remains at 30%), leaving Mr PAYE with £700 to spend. For the various rates of tax see Table 1.8. In this text, a 30% rate will be used to show how the figures are obtained. Thus, for every £1,000 to spend on the vehicle, Mr PAYE will have to earn £1,430.00 (£1,428.57). Now, if one takes into account value added tax (VAT) and special car tax (SCT), the figures become frightening. To spend £1,000 on a car attracts nearly £130 VAT and £83.33 SCT. The figures are shown in detail in Table 1.3.

Table 1.3 — Amount of car purchased from gross pay in the United Kingdom

Tax rate	Mr PAYE earned (gross pay)	Inland Revenue takes	Leaving Mr PAYE to spend (ex-showroom)	15% VAT	SCT	Amount of car purchased (ex-factory)
30%	£1428.50	£428.50	£1000	£130.43	£83.33	£786.24
29%	£1408	£408	£1000	£130.43	£83.33	£786.24
27%	£1370	£370	£1000	£130.43	£83.33	£786.24
25%	£1333	£333	£1000	£130.43	£83.33	£786.24

Hence for a £7,862 motor car (1.6 litre engine ex-factory), bought outright (costs **£10,000** ex-showroom) with no trade-in, Mr PAYE will have to have earned £14,300 (30% tax) and contributed **£4,300** to the Inland Revenue as income tax. This figure is derived here to compare it with the Inland Revenue's tax assessment made as 'car benefit' in the UK (see Tables 1.7 , 1.10, 1.11 and 1.12).

Assuming this vehicle is traded in after three years' use, its trade-in price will be 50% of the new purchase price (Fig. 1.2), and therefore £5,000 is lost in depreciation over the three years. This is equivalent to **£1,666.60** per annum. In order to spend this amount it would require earnings of £7,150 (gross pay) over three years, or £2,383 per annum in gross pay.

In order to run this vehicle further expenditure will be required as shown in Table 1.4. This gives a total expenditure, excluding servicing, of **£1,020.00**, which again is after having paid tax. To pay out this amount Mr PAYE has to earn (gross)

Table 1.4 — Added costs

Road tax	£100.00
Insurance	£200.00
Petrol (12,000 miles, 30 miles per gallon, 1 gallon costs £1.80)	£720.00
Total expenditure excluding servicing	£1,020.00

£1,458.60 per annum. Combining both depreciation and running costs (petrol etc.), we can find out how much Mr. PAYE has to earn to pay this:

$$N^{a} \times (\text{Depreciation} + \text{petrol}) = \text{total gross pay} \tag{1.1}$$

Total cost per annum before paying tax is £2,383.00 + £1,458.60 = £3,841.60. This amount of money (plus the initial purchase cost) could have been gaining interest at between 7% and 9% (income tax paid) at a building society. The total amount which could be invested is shown in Table 1.5. The total cost of vehicle ownership which is paid out of earned income and after paying tax is shown in Table 1.6.

Table 1.5 — Possible investment amount: year 1

Initial cost (purchase price of car)	£10,000
Running costs	£1,020
Total	£11,020
Interest at 7½% (tax paid) on £11,020[b]=£826.50	£827
Subtotal	£11,847[c]
This subtotal is after tax (29%) has been paid	
Equivalent Cost = Sum before tax ×	
1.41[a]=£11,847×141[a]	£16,686
Total income tax paid to achieve this transation	£4,838[d]

These costs are worked out at 1986–1987 rates. New rates have now been announced which reduce income tax from 29% to 27%. *a* For a tax rate of x%, the multiplying factor N is N=100/(100−x). A tax rate of 29% N≈1.41; 40% N≈1.67; 45% N≈1.81; 50% N=2, etc. *b* Inland Revenue believes this 7½% is too low. *c* Compare this interest figure with the car benefit rates in Table 1.7. *d* Compare this to couple B taxation on 'fuel and car benefit', Tables 1.10 and 1.12.

Table 1.6 — True costs of motoring

After tax	
Initial purchase cost	Excluded because depreciation has been included
Running costs	£1,020
Depreciation	£1,660
Total	£2,680
Equivalent gross pay	
£2,680 is pay after deduction of tax	
Gross pay = £2,660 × 1.41 (29% rate)	**£3,774**
Total income tax paid £3,774−£2,680=	**£1,094**

In contrast, if this vehicle was purchased by a company for an executive Mr Perk and given to him as part of his 'duties and pay', the Inland Revenue will deem this to be worth only £480.00 (gross pay) to Mr Perk, as shown in Table 1.7. (Mr Perk is also taxed under PAYE. Mr Perk is different from Mr PAYE in that Mr Perk has a vehicle provided by the company, while Mr PAYE has to provide his transport out of earned income and after tax). The Inland Revenue will therefore claw back from Mr Perk £480 x 29% = £139.20 for the car and £480 (gross pay) for the free petrol, making a total of tax liability of gross pay of £960 and therefore a tax demand of £960 x 29% = £278.40 (correct for the year 1986-1987 budget). The 'correct' benefit in kind should therefore be in the first year of usage some £16,686.00 (at 29% tax rates), as shown in Table 1.5, and in the second year £3,774 Table 1.6. Mr Perk never has to make the initial purchase, nor find the depreciation, or money for the general running costs, and can invest this 'extra capital (£10,000)' in very shrewd ways which will give him an extra £5,000–6,000 in capital gains allowances in any one year. (Note: this cannot be entered into here but the first £6,300 for the tax year 1986/1987 is usually free of capital gains tax.) The actual car benefit scale is given in Table 1.7.

Table 1.7 — Car benefit rates

Original market value of car	Car benefit for age of car at April 5, 1986		Fuel benefit
	1–4 years	Over 4 years	
Up to £16,000			
Engine capacity			
Up to 1300 cc	£375	£250	£375
1301–1800 cc	**£480**	£320	£480
Over 1800 cc	£750	£500	£750
£16,000–£24,000	£1,100	£740	£750
Over £24,000	£1,725	£1,150	£750

1986–1987 rates applied.

If the economics of car use in this very large sector is further examined, this is not only a meritorious 'perk', it is also Government subsidy to a section of the community who are also fairly well paid, albeit heavily taxed on income, as shown in Table 1.8. The 'car benefit' is widely used in advertisements (Fig. 1.3) to promote the advancement of personnel without contributing to the 'public purse' in the form of tax on the *true benefit in kind*. It will remain a 'perk' until the government, through the Inland Revenue puts the 'levy' of £480 on Mr Perk up to the colossal figure of (N *times Depreciation*) for the 'car benefit', and N×£1,020 for the 'fuel, road tax and insurance benefit', and some figure (such as **£826.50** (Table 1.5)) on the 'capital asset' which can accrue from the income derived in not having to purchase the vehicle. Only at this point will the company executive Mr Perk be paying the true

Table 1.8 — UK tax bands

Rates of tax	1985–1986 band	1986–1987 band	1987–1988 band
27%	—	—	£1–17,900
30%	£1–16,200	—	—
40%	£16,201–19,200	£17,201–20,200	£17,901–20,400
45%	£19,201–24,400	£20,201–25,400	£20,401–25,400
50%	£24,401–32,300	£25,401–33,300	£25,401–33,300
55%	£32,301–40,200	£33,301–41,200	£33,301–41,200
60%	Over £40,200	Over £41,200	Over £41,200
Personal allowances			
Single person	£2,205	£2,335	£2,425
Married man	£3,455	£3,655	£3,795
Wife's earned income	£2,205	£2,335	£2,425

Fig. 1.3 — Newspaper advertisements for managers with the added benefit of a company car. At the time of printing the average industrial wage is £6,500

Inland Revenue tax rate for the benefit of the vehicle as will be shown by considering two identical couples A and B (Section 1.7.1). As shown in Table 1.8, the company car perk is more beneficial as the tax rates rise. In the figures given above, the 'benefit in kind' has been worked out at the standard rate of 29% (27% and due to fall to 25% in the UK). (See footnote (a) in Table 1.6.)

1.7.1 Comparison of two couples A and B

The cases of two couples both earning the same amount with a combined income of £30,000 are compared. Couple A have to buy their own car (Table 1.9) and couple B have a company car on which both car and fuel benefit are paid.

Table 1.9 — Couple A: PAYE, own car

	Tax rate applied	
	1986–1987	1987–1988
Annual salary	£30,000	£30,000
Less married man's allowance	£3,655	£3,795
Pays tax on[a]	£26,345	£26,205
27% on first £17,900	—	£4,833
29% on first £17,200	£4,640	—
40% on next £2,999 (£2,499)	£1,200	£999.60
45% on next £5,199 (£4,999)	£2,340	£2,249.55
50% on next £945 (£804)	£470	£402
Total tax paid	£8,650	£8,484.15
Residual income before car expenses	**£21,350**†	£21,515.85 (an increase of £165.85, i.e. 0.7%)
Cost of 1600 cc car in year 1 (£10,000 + running costs (£1,020)		£11,020
Cost in year 2: depreciation (£1,666) + running costs (£1,020)		£2,686
Available spending income over two years (£42,700 – £13,706)		**£28,994**†
Residual income in year 3: depreciation + running costs (£2,686) included; interest, repairs etc. not included (£21,350–£2,686)=		£18,664

a Where two values are given, the value in parentheses refers to 1987–1988.
† Figures used in Table 1.10.

Couple B, who have the unlimited use of the company car for social and domestic pleasure, will have a levy made on them by the Inland Revenue for the benefit in kind. This is shown in Table 1.10. Couple B will also be better off to the tune of **£12,736** over the two year period (£6,368 each year of **spendable** income). After purchase the amount by which couple B are better off than couple A is £41,730÷2 −£18,664=£20,865−£18,664=£2,201 showing how the initial purchase price of the vehicle distorts the picture.

Table 1.10 — Couple B with company car: income tax liability of Mr Perk

Car benefit[a] (£480) + fuel tax benefit[a] (£480)	£960[a]
This amount is charged at the higher rate of 50%	
Tax at 50% × £960 for year 1	£480
For year 2 it is the same	£480
Paid to the Inland Revenue over two-year period for the use of a car	£960[d]
Residual income over two year period is £42,700[c] less extra tax of £960	**£41,730**
Difference between couple A and couple B in residual income (£41,730 – £28,994) (From Table 1.9)	£12,736
If added interest is included to acknowledge the fact that couple B have £10,000 to invest safely at 7.5% × £10,000 = £750 per annum added income[b]=£1,500 over 2 years	£14,236

a These benefits were increased in the Budget. As the 1986–1987 rates are being applied, the car and fuel benefits for 1986–1987 will remain unaltered for the purposes of this argument.
b The Inland Revenue believes this to be too low. A value of 10% should be used. With an added interest of £1,000 the difference between the incomes of couples A and B is £13,736.
c Table 1.9 twice £21,350.
d Compare this figure to the figure of £4,838 (Table 1.5).

1.8 GOVERNMENT SUBSIDY?

Now if we look at the company's books for the running of this car we can see where the subsidy comes in. The cost of this car is £7,862 (ex-works) or £10,000 ex-showroom. There is no refund on the VAT element, or the SCT, but the company is allowed to write down 25% of the purchase price of the new car (provided for Mr Perk) against the liability of company tax (corporation tax). Therefore in the company books the following transaction would appear. (As there are special cases involved, the total purchase price will be considered as £7,862; the exceptions will be considered in case B.)

Writing down allowances (W.D.A): Case A

In the first year the writing down allowance is 25% of £7,862 which is £1,965.50. The car is now on the books at £5,896.50 at the start of the second year and so for the second year the writing down allowance is 25% of £5,896.50 which is £1,474.13. The car is now on the books at £4,422.38 and so for the third year the writing down allowance is 25% of £4,422.38 which is £1,105.59.

The car at the end of the third year is now on the books at £3,316.79 and so can be sold at this figure. Note that the true trade-in price derived from Fig. 1.2 is £3,921 (50% of £7,862). If this price (£3,921) were received for the vehicle, then this additional sum (£3,921 – £3,316 = £605) would be taken off the writing down allowance. This will be considered later.

The total writing down allowance is £4,545.

Writing down allowances (W.D.A.): Case B

In the first year the writing down allowance is 25% of £10,000 which is £2,500. The

Inland Revenue will only allow a maximum of £2,000 to be written down. The car is now on the books at £8,000 at the start of the second year and so for the second year the writing down allowance is 25% of £8,000 which is £2,000. This full amount can be written down. The car is now on the books at £6,000 and so for the third year the writing down allowance is 25% of £6,000 which is £1,500

The total writing down allowance is £5,500.

The car at the end of the third year is now on the books at £4,500 and so can be sold at this figure. Note that the true trade-in price is £5,000 (50% of the purchase price of £10,000). If this price (£5,000) were received for the vehicle, then this additional sum (£5,000 − £4,500) = £500) would be taken off the writing down allowance. This will be considered later.

Consider case A with a Mr Perk whose combined income stays within the basic rate of 29%. This would be a combined income of £20,000. The total amount that has been set aside against the profits of this company is £7,862.00 − £3,316.79 = £4,545.20

Company tax is complicated in that it is 29% for small companies whose annual profit is less than £100,000 and for larger companies with increased profits the rate of tax is 35%. For the purposes of illustration and explanation, a tax rate of 30% will be used. Therefore this writing down allowance of £4,545.20 has saved the company a tax bill of 30% × £4,545 = **£1,363.50** and by the expenditure of £4,545 it has been able to give a very good perk at the fraction of the true cost as judged by our tax laws.

If the higher trade-in price had been received this extra £605 would be subtracted from the writing down allowance above, making £3,940 to set against profits and thus making a company tax bill of 30% × £3,940 = **£1,182.00**. Thus the company has only gained £181.50 to put into dividends although it did get £605 more for the trade-in. Hence it is often better to allow the vehicle to be purchased by the company executive at the written down figure of £3,316 and to pass the saving of £605.00 on to him. (This has now been closed by the Finance Act, which allows the company employee purchaser to be taxed on this added deficiency (£605) at the personal tax rates which can rise up to 60% (Table 1.8). It would be more equitable in society to have the initial benefits also raised to reflect a true benefit rather than a token benefit as at present.)

The Government has now 'lost' income derived from corporation tax as a result of the provision of the company vehicle through the writing down allowances as shown in the following table (Table 1.11). It has also lost the private petrol amount used by Mr Perk. The private running costs were given in Table 1.4. as £720.00 (petrol) and other associated costs making a total of £1,020.00, where the vehicle is used privately and *not* on company business, *but* only £278.00 per annum has been recouped from Mr Perk (the company executive paying 29% tax) for the car and fuel benefit. This means that the Government has subsidized the vehicle provision for this employee by £628[†] per annum. This is money 'lost' to the Government's purse for every company-provided vehicle where private use is allowed.

If the running costs are now examined in greater detail, the 'system' becomes even stranger. There is no VAT on the insurance and road tax element for the

† [Max writing down allowance £2,000 + Petrol £1,020]×30%=£906. £906−claw back £278=£628.

Table 1.11 — Inland Revenue losses due to corporation tax loss on depreciation over a three-year period

Provision of company car, Case A W.D.A. of £4,554×30%	= £1,363.20
Provision of company car, Case B W.D.A. of £5,500×30%	= £1,650.00
Petrol One year at £720−£93.91 VAT (VAT is claimed back by the company)=£626.09×30%	= £187.83 per annum
Petrol. Loss over 3 years = 3×£187.83 per annum	= £563
Total cost of fuel and writing down allowance (W.D.A.) to the Inland Revenue over 3 years=£563+£1,363 up to £1,650	= £2,011 up to £2,298
Average **annual** loss to Government = £2,100÷3 (av)	= **£700**[a]
Inland Revenue Claw back (CB) from Mr Perk through the PAYE (assuming maximum rate of personal tax is 29%) is:	
Car and fuel benefit (Table 1.7) £960×29% = **'Claw Back'**	= £278.00
Government loss from corporation tax = effective subsidy	of £700.00
Government subsidy £700.00 + VAT reclaimed £93.91=£793.91	= **£794.00**
Net loss = £794.00 − Claw back of £278.00	= **£516**[a](to **Table 1.12**)

a Note this figure is less than the subsidy derived of page 13 which used the maximum writing allowance of £2,000.

vehicle. There is VAT on the petrol bill of £720.00 which is £93.91, which can be claimed back by the company from the Customs and Excise. Hence the Government is losing £93.91 in VAT plus that which it lost in the petrol's not being available for Corporation tax (£187 in Table 1.11) (Total, £280.50) and claiming back only £139 as the fuel benefit. Thus on the 'petrol benefit' the Government has lost £93.91 VAT and corporation tax of £188.00, making a total of £282.00, and has only clawed back as benefit in kind £139.00: the Government subsidy of fuel benefit is therefore £143 per annum.

The amount clawed back depends on the tax bracket which the salary attracts and the price and size of the vehicle (Table 1.7). In Table 1.12 the standard 1600 cc engine is used as demonstrating the point. From Table 1.12, it can be seen that the Government never recoups on the 'claw back' on both fuel and car benefit what it loses on corporation tax and reclaimed VAT. When the company employee Mr Perk gets into the 60% tax bracket, it is very unlikely that he will stay with a 1600 cc vehicle costing £10,000. If Mr Perk has a more expensive vehicle, calculations show that the Government never claws back from the beneficiary anything like the subsidy it has made in writing down allowances and VAT on petrol, etc.

The Government therefore loses and thereby subsidizes this motoring section of the community as the fuel benefit ought to be £720×N for petrol and the car benefit should be N times a percentage of the initial purchase price in the first year so that PAYE purchasers are not disadvantaged as shown in Tables 1.9 to 1.12 as typified by couples A and B (N being 100/(100−R) where R is the highest band of tax pertaining to the individual). As shown by this section, the residual value of a company-owned vehicle is not of paramount importance. What is important is the added extras which enable couple B to enjoy the fruits of their labour without the disadvantage of high

Table 1.12 — Government claw back (CB): Couple B's (Mr Perk) payment to the Inland Revenue compared to couple A's (Mr PAYE) at different tax rates

Maximum tax band of couple	29%	40%	45%	50%	60%
Inland Revenue claw back on					
Fuel Benefit £480 (**a**)	£139	£192	£216	£240	£288
Car Benefit £480 (**b**)	£139	£192	£216	£240	£288
Couple B's tax payment to the Inland Revenue = Claw back (a)+(b) (**c**)	£278	£384	£432	£480	£576
Estimated Government subsidy = £794 (Table 1.11)−(claw back)	**£516**	£410	£362	£314	£218
Equivalent couple A's payment to the Inland Revenue to cover depreciation and petrol of £2,680 (Table 1.6) (**d**)	**£1,094**	£1,787	£2,193	£2,680	£4,020
Couple A's extra payment (over couple B's) to the Inland Revenue for the *same vehicle facility* (**d–c**)	£816	£1,403	£1,761	£2,200	£3,444

tax premiums which falls on the same salaried couple A. This form of discrimination *must* be divisive and it is wrong to subsidize those in our society who are well paid. More importantly it completely distorts the 'new' car market and makes certain corrosion protection schemes uneconomic and results in penny-pinching systems, the absence of which would benefit the motoring public who have to purchase their transport requirements out of taxable income. Note: The balancing charge also brings a further subsidy from the Government. Enough has been written to set the scene.

1.9 PREMATURE AGEING: IS IT A MEANS OF INCREASING SALES?

One is constantly asked whether the corrosion of the vehicle (some may call it the premature ageing of a car) is designed in and used as a means of increasing new car sales. The evidence is that it is *not*. If it were, one would expect the Western countries to produce an inferior vehicle with regard to corrosion than, say, the Eastern Communist countries. It seems that this is not the case. The sites of corrosion are the same whatever the manufacturer. Some manufacturer's models are worse than others. The general sites are shown in Fig. 1.4.

The book will attempt to look at the reasons why these characteristic locations corrode in all makes of vehicles, and through a unified theory for the build-up of the micro-climate and corrosion mechanism any engineer, or even the layman, will be able to predict the areas which are going to show early signs of corrosion. The reasons for the consistent areas of corrosion in each car model will be outlined in the book with the general scientific principles involved.

There is a constant battle to improve the market share of the manufacturer, and it is important that the big leasing companies can get rid of their used cars at a reasonable price because the same tax advantages do not apply to them.

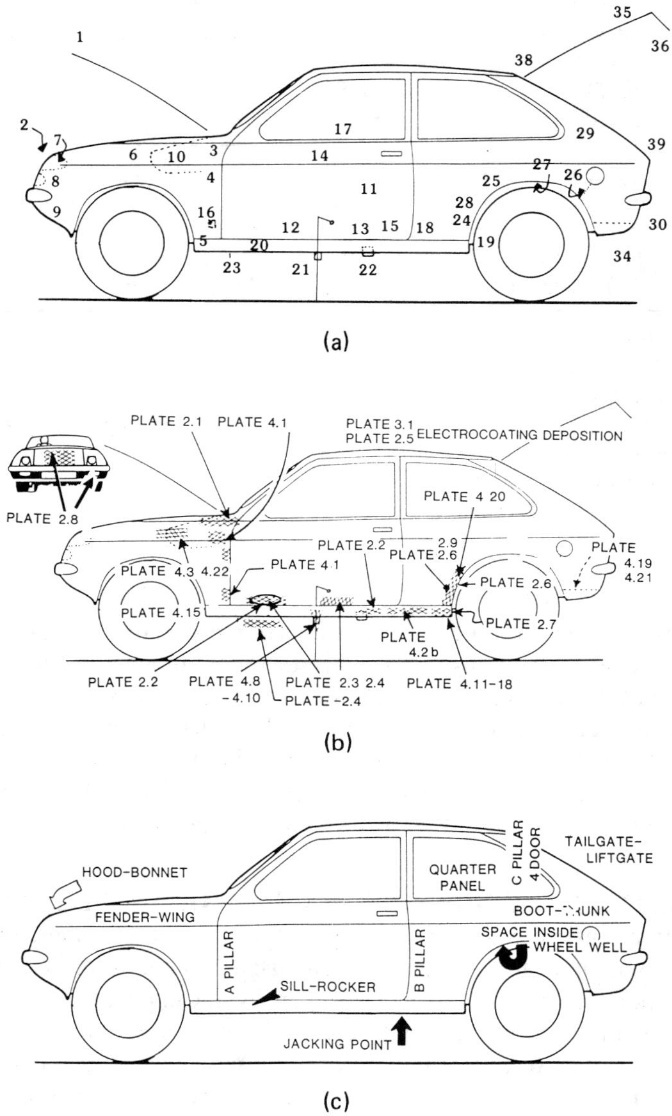

Fig. 1.4 — A diagrammatic view of the areas of corrosion on a vehicle depicted by (a) reference number and (b) plate number; (c) some of the terminology used in the motor trade of the areas of the vehicle in different countries. The plate numbers in (b) refer to the photographs of the internal panels of a stripped-down low-mileage vehicle shown in Chapter 4. The areas shown can be broken down into two categories. There are those areas where corrosion is due to the accumulation of mud deposits which are not washed away by the owner (owner faults). The other areas are where the corrosion is internally nucleated on internal surfaces of the vehicle and **cannot** be washed away by the owner. These are categorized as (design faults). These aspects will be discussed throughout the text.

1.10 INTRODUCTION — ANNUAL SALT USAGE

Because the second-hand owner is almost certainly a Mr PAYE, and for every £1,000 of depreciation — £1,430 has to be earned, most private motorists are concerned with the rapid deterioration of their motor vehicle, resulting from internally initiated corrosion during the winter de-icing programme. The vehicle manufacturers, have in the past, laid the blame for the rapid corrosion upon the road de-icing salts used during the winter period (Fig. 1.5). Other factors such as coastal environments, high humidity and industrial pollutants (Figs. 1.6(a) and 1.6(b), e.g. acid rain, contribute to the corrosive environment, but it is the road de-icing salts which are the most destructive (Fig. 1.6(c)). The winter environment of the motor vehicle is not unlike the marine splash zone which is known to be the the most corrosive.

The reason for this will be discussed in Chapters 8 (Section 8.12) and 10 (Section 10.14). It is suggested that, for a first reading, readers less conversant with corrosion and the motor vehicle should read slowly through the chapters without reference to the later technical chapters or plates as these will propagate the reader through the text at too fast a rate. At a second or third reading due notice can be taken of the technical aspects.

A careful examination of low mileage accident-damaged vehicles has revealed the mechanism of corrosion and has shown up many design faults in the vehicle manufacture. These design faults give rise to the conclusion that manufacturers only have themselves to blame for the increased incidence of corrosion with salt usage. This results (Section 10.22) because the motor manufacturers have not investigated the corrosion process–mechanism properly, in order to postulate a mechanism for predicting the sites of corrosion. Hence some of the ad hoc rectification processes which were an attempt to reduce the incidence of corrosion in the sill–rockers actually accelerated the corrosion (see Section 4.5 and Plates 4.16 and 4.17(a)). Some of these design faults have been eliminated now that the vehicle manufacturer

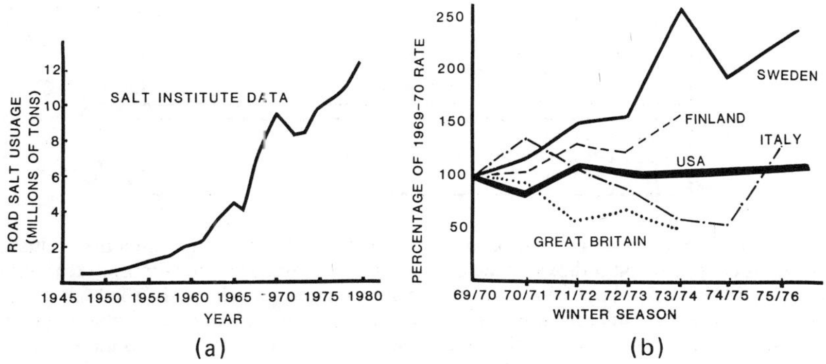

Fig. 1.5 — (a) The annual usage of salt required to keep our roads free of snow and black ice. The increased usage does not reflect a return to colder weather or another ice age. The increased usage reflects society's increased expectation of trouble-free motoring into work. Perhaps this is another subsidy to the affluent motorist. (b) Salt usage for different countries affected by snow and low overnight temperatures. (After Designing for Automotive Corrosion Prevention, Society of Automotive Engineers Report, November 1978, p. 78. Vehicle Corrosion caused by de-icing salts, American Public Works Association, Special Report 34, September 1978. Reported in R. Baboian, Automotive Corrosion by De-icing Salts, NACE, 1981. Redrawn and reproduced with thanks.)

(a)

Metal	Atmosphere		
	Industrial	Marine	Rural
Aluminium	0.81	0.71	0.025
Copper	1.19	1.32	0.58
Lead	0.43	1.41	0.48
Zinc	5.13	1.60	0.86
Mild steel	13.72	6.35	5.08
Weathering steel	2.54	3.81	1.27

Penetration rates of metals in different atmospheric conditions (μm /year after 10 years' exposure).

(b)

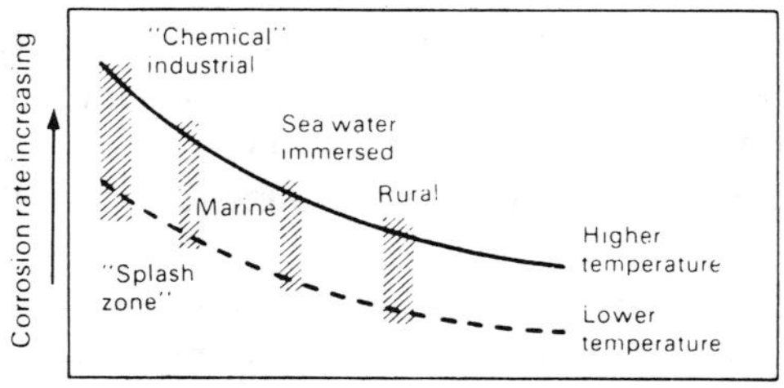

Corrosion rates and influence of aggressive environments.

(c)

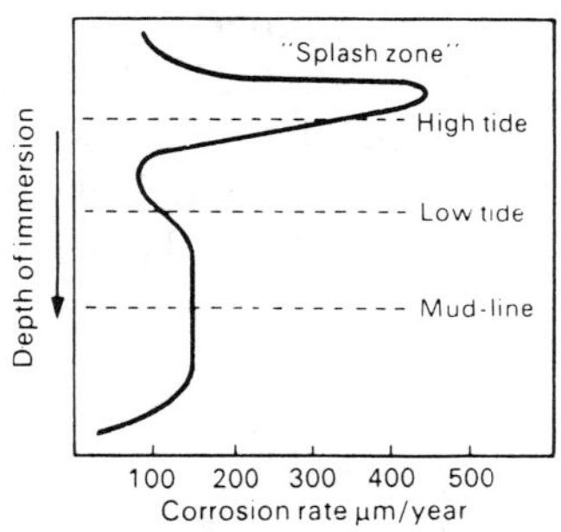

Corrosion rates at different positions for steel partly immersed in sea water.

Rule of thumb
A 30°C temperature change will result in a 10 times change in corrosion rate.

Fig. 1.6 — (a) The corrosion rates of various metals in three different environments. Industrial atmospheres contain sulphur dioxide (see also Fig. 10.5) whereas marine atmospheres contain sodium chloride (to a first approximation chemically the same material as the road salts), while a rural atmosphere is considered as one devoid of these previously mentioned contaminants and is the least corrosive atmosphere to the metals listed. (b) The corrosion rate is increased with increase of temperature. Usually a 10°C rise in temperature will double the rate of corrosion. However, in the case of the motor vehicle this is not quite true as the higher temperature will also cause the 'micro-climate environment' to dry out. The extent to which the environment dries out depends on the relative humidity (Section 10.6). However, remember that when a liquid dries out the dissolved salts are concentrated into the areas which are the last to dry (Section 3.6): compare Figs. 3.7 and 3.8. (c) Sea water is generally considered to be a very corrosive liquid. Shown here are the corrosion rates for various depths of immersion in sea water. The splash zone is the most corrosive area. The automotive vehicle operates in this 'splash zone' (from wheel spray) during the winter period. (Source of information: Department of Industry.)

is giving a six-year warranty. It is well known, in the industry, that road salts only accelerate the corrosion reaction and quickly expose those areas where, for one reason or another, the road dirt has been allowed to accumulate over the vehicle's life. Hence the increased use of salt on the roads (Figs. 1.5(a) and 1.5(b)), during the winter only exposes the manufacturer's design faults, which often arise as a result of the unrepresentative accelerated corrosion tests used at the prototype stage (Sections 2.10 and 10.22). These accelerated tests do not reflect the combination of the motor vehicle corrosion mechanism and the science behind the compaction of road dirt and mud accumulations (see Figs. 6.10, 6.11 and 10.10). In order to describe how the internally initiated corrosion arises, several related parameters have to be studied. It is the intention of this book to give the reader a thorough understanding of the corrosion process, without too much technical jargon in Chapters 1–5.

1.11 MANUFACTURER'S WARRANTIES

It is probably worth looking at the reason why the motor vehicle manufacturers started offering their own warranties and to consider why many of these design faults have now disappeared.

Historically throughout the conferences on motor vehicle corrosion, there were two arguments.

(1) Corrosion was a form of 'built-in obsolescence', which meant that for every car scrapped a new one could be built and profitably sold. It was thought that, if cars lasted longer, the market would become saturated and restrict profitable production.

(2) Most first car purchases were by companies and given as a 'perk' to people who had no monetary outlay in the longevity of the vehicle, and it was considered that this first-time buyer would not want to spend extra on corrosion protection systems from which they would not benefit. The after-market rust-proofing stations proved that this was not the case.

Some proposed legislation in Canada concerning the Corrosion of the vehicle made the manufacturers very conscious of corrosion as an unacceptable form of premature ageing.

1.12 WHAT IS THIS PROPOSED LEGISLATION?

The Canadian government threatened to legislate on consumer rights and to bring in some 'anti-corrosion' laws which would make the manufacturer responsible for the rectification costs of the following.

(1) Cosmetic corrosion for three years. This gave rise to the first black plastic attachments on the leading outward flared rear wheel arches of the certain vehicle makes (Section 4.6, item 28, and Plate 4.20).

(2) Penetration corrosion for five years, from internally initiated corrosion on the inside of panels which then rusted through to the outside (Sunday polish surface) (Section 2.7 *et seq.*, Plate 2.6 and Chapters 4 and 5).

(3) Structural corrosion for eight years. Structural corrosion was considered to be corrosion on those parts of the vehicle which were the force-transmitting members, e.g. suspension struts, areas of spring hangers etc.

Rather than bring in legislation, it was argued that a voluntary code would suffice. This voluntary code meant that the manufacturers had to pay more attention to design to counteract corrosion. To be successful, corrosion protection systems must be associated with good anti-corrosion design. This aspect will be brought out in Chapters 4 and 5 of this book.

The corrosion rate for steel is approximately 0.033 mm/year and with vehicle panels becoming thinner, in order to save weight (Fig. 5.2) and to result in better petrol consumption (Fig. 5.1), there is little room to 'allow' for corrosion on vehicle panels with this tight 'legislation'.

One unacceptable issue that has arisen as a result of the manufacturer's warranty has been that this warranty ties the vehicle owner in to the manufacturer's dealer network for both spare panels and parts, as only 'authorized' parts and repair procedures could be used to keep the warranty 'alive'.

1.13 DOES CORROSION CAUSE ACCIDENTS?

One of the big red herrings about corrosion is that it causes accidents. There are probably many sightings of a vehicle slumped on the road owing to the ultimate failure of the car body at the top of the MacPherson struts. Onlookers then imagine what would have been the outcome if this collapse had taken place at high speed. Statistically most of these collapses take place at low speeds, either when the vehicle is braking or negotiating at low speeds (e.g. parking). At low speeds the loading torques on the suspension are very much higher than they are at high speeds (bumps and other irregularities in the road excepted).

1.14 WHAT ABOUT THE CORROSION OF BRAKE PIPES?

If suspensions show corrosion failure only in high loads at low speed, then surely one would argue that the pitting corrosion of the hydraulic brake pipes will lead to accidents. Corrosion could cause the hydraulic pipes to burst under severe breaking which could take place at high speed (Figs. 8.10 and 8.11). Once again statistically only 0.02% of the accidents are caused by brake line corrosion which results in a total loss of pressure. The corrosion of brake pipes and the methods available to overcome this problem are considered again in Section 8.6.

1.15 WHAT CAUSES MOST ACCIDENTS?

The cause of most accidents is driver carelessness (often called driving without due care and attention). In the next section this cause of accidents only will be considered.

The function of the car designer is to produce a rigid passenger cell or cage, which is protected by a collapsible front and rear structure (Plate 1.1) so that, in the event of an accident, these structures collapse progressively with accident time. In this way they absorb some of the kinetic energy of the two colliding vehicles. Some manufacturers produce designs which incorporate 'crumple zones' which 'tell' the structure where and how to collapse. An important question is: at what age does the vehicle body lose this progressive collapse with accident time as a result of corrosion?.

Plate 1.1 — An advertisement by one manufacturer proposing the philosophy of the passenger safety cell. Whilst this may be true when the vehicle is new, do the same principles still apply when the vehicle is 10 years old, corrosion having made its mark on the structural energy absorbing members.

1.16 CAR SAFETY DURING ACCIDENTS — IS A CAR 'CRASH-SAFE' AT ANY AGE?

If the function of the car designer is to produce a rigid passenger cage, and a collapsible front and rear structure which collapses progressively with accident time, then surely this concept will be affected by corrosion, which is so manifest on the wings, engine valance and 'A' pillar. (The A pillar (Fig. 1.4(c)) is the pillar upon which the front doors and windscreen are hung and fixed). A new car has to go through certain crash tests before it can be marketed. An older car is *never* tested in a crash test.

The important questions relating corrosion and accidents are as follows (in this exercise the cause of the accident is NOT considered).

(1) Does corrosion negate this philosophy of energy absorption during accident time?

(2) At what age does the structure lose this ability to collapse and progressively absorb energy as a result of corrosion.

(3) Is there any more danger to the occupants of an older vehicle than to those of a new one?

The number of people killed or seriously injured in a particular make and model of car in the UK is a closely guarded secret, available only to manufacturers and Government officials. Figures taken from those countries (e.g. the United States) where such information is available and published, are graphically presented with respect to the age of car in Fig. 1.7, curve B. On this curve is superimposed (curve A)

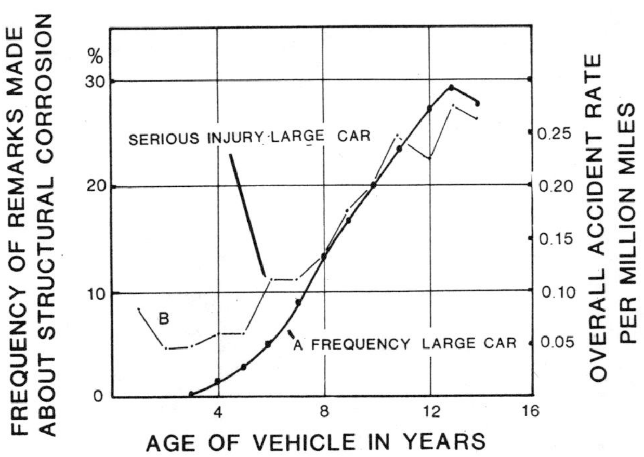

Fig. 1.7 — Relative frequency of remarks concerning frame and structural corrosion of the motor vehicle superimposed upon the crash rates per million miles with regard to the age of vehicle. The curve is discussed in the text.

The sources of data used are as follows:
curve A, G. E. Svenson, Swedish Corrosion Institute (personal communication), also published in 'Cars go Rusty', AB Svensk Bilprovning;
curve B, A. M. Dutt and D. A. Reinfurt, Accident Involvement and Crash Injury Rates by Make, Model, and Year of Car, Highway Safety Research Centre, University of North Carolina, DOT-HS-4-00897, D. A. Reinfurt, personal communication.

the frequency of remarks made about structural corrosion where annual government safety tests (from Sweden) are required to licence the vehicle for road use.

Fig. 1.7 shows that there may be a correlation between the number of accident victims with the age of the car in which the occupants are carried (curve B). The curves are plotted with respect to the age of the vehicle. It is important to emphasize that corrosion is not the cause of the accident but rather the reason why the occupants were admitted to hospital owing to the injuries received as a result of the non-progressive energy absorption with accident time (Fig. 1.8). Several explanations are available for this trend.

(1) The automotive designer is achieving his goal in that the younger or newer vehicles are more crashworthy (Fig. 1.8; Plate 1.1).

(2) The older vehicle, being cheaper, is run by young 18-25 year olds who are more accident prone than the older person who can afford to run a new car.

(3) Corrosion is affecting the ability of the steel to absorb energy during the 'accident time' and so the rapid deceleration when the engine comes into contact with the bulkhead causes very high 'g' forces which impart some further injury to the occupants. This effect is shown diagramatically in Fig. 1.8.

Interpretation (1), one hopes, is obviously true. Interpretation (2), that the older vehicle is used by a younger more inexperienced driver than the newer car, is more difficult to analyse. However, this correlation has been disproved by an investigation carried out by the Transport Road Research Laboratory (TRRL) into the accidents with pedestrians injured and vehicle age. This shows that the ratio of pedestrians to occupants injured by a four-year-old car is higher than the ratio for a 12-year-old car.

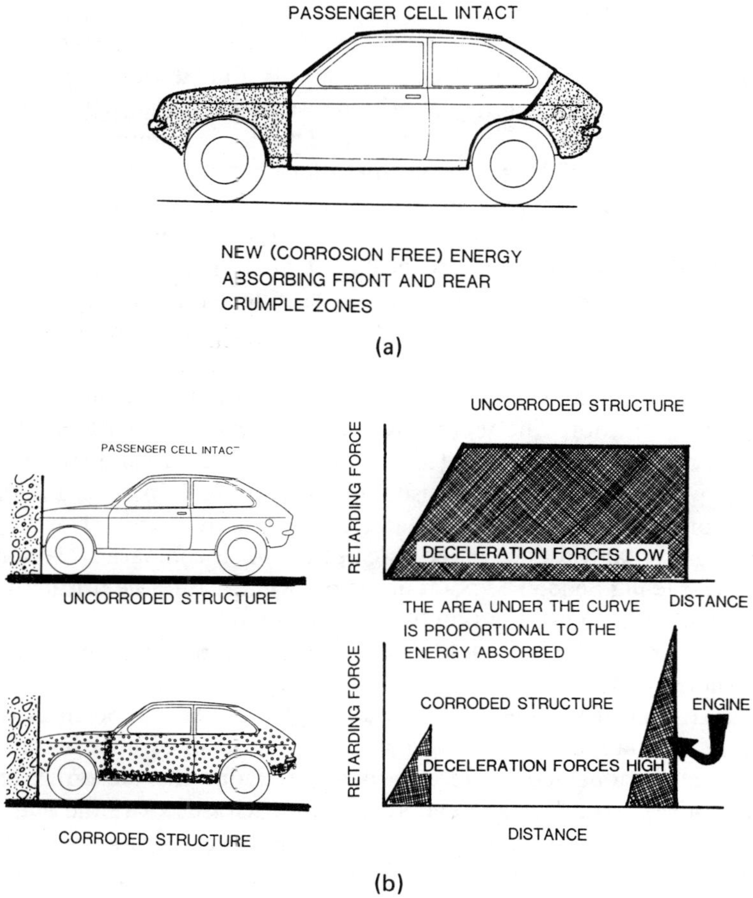

Fig. 1.8 — (a) Areas of corrosion in energy-absorbing sections with respect to the passenger cell (Plate 1.1). (b) Comparison of corroded and non-corroded vehicles crashing into a concrete wall with their energy absorptions with accident time (note: the corroded vehicle's energy absorption is not uniform with distance or crash time and so produces large 'g' forces).

Thus it cannot be the reckless young driver responsible for the increase in the curves shown in Fig. 1.7. (The sources of data used to prepare this statement are as follows: curve A, G. E. Svenson, Swedish Corrosion Institute (personal communication) (also published in *Cars go Rusty*, AB Svensk Bilprovning; curve B, A. M. Dutt and D. A. Reinfurt, Accident involvement and crash injury rates by make, model, and year of car, Highway Safety Research Centre, University of North Carolina, DOT-HS-4-00897, and D. A. Reinfurt, (personal communication); ratio of pedestrians to occupants injured, I. D. Neilson, Head of Vehicle Safety Division, Transport and Road Research Laboratory, Crowthorne, (personal communication) (the source of the data used to prepare this comment is The National Police Collection of Road Accident Injury Data (Stats 19)).)

1.17 TENSILE TESTS

In order to understand this section of the chapter we must understand a concept which engineers use to compare the 'carrying capacity' of materials. Chewing gum is soft and pliable and can be deformed easily between the teeth. When stretched it deforms readily and 'thins' at one position. This local thinning or waisting is technically called 'necking'. However chewing gum is quite capable of supporting the weight of a person when dropped and trodden on the floor. It would be no good as chewing gum if the jaws had to do this amount of effort and yet it is the same material. The reason why the chewing gum supports the weight of a person is that it can deform and spread out until the load-carrying area is large enough for it to stop deforming or spreading. Therefore engineers consider the deformation load divided by the load carring area, which is called stress.

Now, the chewing gum deforms at the same stress whether it is in between teeth or on the pavement floor. The teeth give a lower cross-sectional area and so the yielding stress can be achieved with a lower jaw force. This principle will be used in Section 2.15, where a drawing pin is used on 'paint blisters' in order to determine whether the corrosion is internally nucleated or due to a stone chip. The sharp point of the drawing pin can produce a very high stress for a comparatively small load applied to its large flat head. The point of the pin will fracture the brittle oxide, but the flat head will not damage the skin tissue of the thumb.

Engineers therefore deform some materials in compression, e.g. bricks and concrete, whereas other materials are deformed in tension, e.g. metals. The difference arises because some materials are 'notch sensitive' (Sections 6.42 and 6.51). Most readers are familiar with the breaking of glass or ceramic bathroom tiles by producing a diamond score line and then supporting the tile on match-sticks placed beneath the score line and evenly loading each side. The tile will fracture along the score line. (If it does not, it is because there were stress-raising cracks of smaller size going off at right-angles to the main score line. Stress-raising notches are dealt with in Section 6.51.) Materials which are tested in tension are not sensitive to stress-raising notches, e.g. metals. Metals tend to 'blunt' the notch by concentrating all the deformation at these areas and so can be loaded up to fracture in tension. This concentration is rather like the waisting of the stretched chewing gum which will stretch under its own weight when the cross-sectional area becomes small enough. Engineers call the relationship between the stress required to produce a certain extension per unit length a **stress–strain** curve. Such stress–strain curves are reproduced in Fig. 1.9. It stands to reason that, if the metal extends under a tensile stress, then its cross-sectional area carrying the load becomes smaller because the metal has not increased in total volume. Hence the **strain** is always defined as the increase in length divided by the original length. However, there are *two* stresses. There is the **true** stress and the **conventional** stress as defined in Fig. 1.9. With metals it is not very important to distinguish between the true and conventional stresses as there is not very much much difference until the metal starts to neck. Necking is the point where all the deformation is concentrated within the deforming area (as with the chewing gum). However, necking in a metal does not take place as early as it does with chewing gum. Therefore it is important to consider the reduced cross-sectional area where the metal is thinned in cross-sectional area by the phenomena of corrosion, as this causes early or premature 'necking'. Hence the reduced stress-

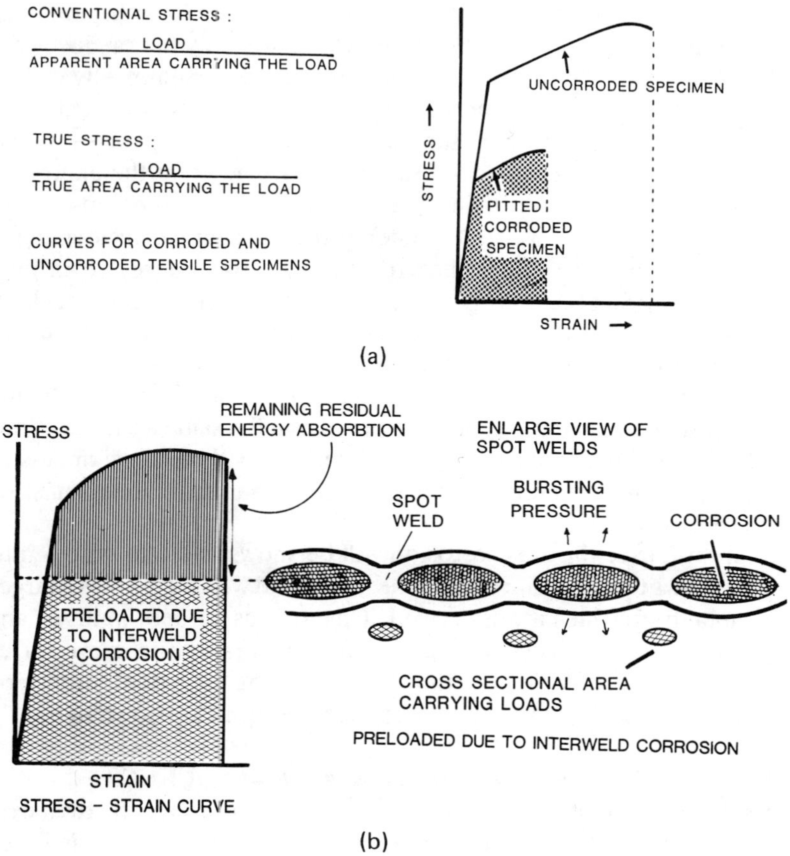

Fig. 1.9 — (a) Corrosion affects the ability of a steel specimen to withstand high stresses as shown in the conventional stress–strain curve. The area beneath the stress–strain curve is proportional to the energy absorbed by the sample. Corrosion-pitted samples have lower stress–strain curves than non-corroded samples. (b) Another way of looking at the problem is to consider that the corrosion in the inter-welds will pre-load the inter-weld by a process of 'oxide jacking' caused by the increase in volume on oxidation (Chapter 8, Fig. 8.9). The energy of absorption is now much less because of this pre-loading which shifts the origin vertically upwards.

–strain curve for the corroded steel is compared with that of the uncorroded sample. Fig. 1.9 shows the effect of comparing a corroded steel specimen with a non-corroded sample. The tensile stength of the corroded sample is reduced considerably by the following factors.

(1) The cross sectional area is reduced by the corrosion and therefore higher stresses are involved as a result of corrosion.

(2) The localized corrosion (called pitting) (Figs. 7.14 and 2.11 and Section 7.11) acts as a stress raiser and concentrates the stresses and deforms prematurely at the pitted area. The deformation is all concentrated within this region and the metal 'prematurely necks' within this corroded area. Thus this corrosion acts in a similar way to the scribe mark placed on a kitchen tile to make it break in a straight line.

Hence corrosion affects the ability of a steel specimen to withstand high stresses as shown by the conventional stress–strain curve. Usually the engineer requires to know the energy absorbed by the deformation process. Scientifically this is given by the area beneath the stress–strain curve which is directly proportional to the energy absorbed by the sample. As steel samples affected by pitting corrosion have lower stress–strain curves than uncorroded samples, it follows that the energy of absorption in a corroded steel is less than the energy absorbed in the non-corroded steel. Hence an older car will absorb less energy than a new car in a similar accident situation. This is due to the areas of corrosion in the front energy-absorbing sections of the car. Therefore if the retardation forces are compared during accident time for a corroded and an uncorroded vehicle crashing into, for example, a concrete wall (Fig. 1.8) with their respective energy absorptions with accident time, the corroded vehicle behaves less favourably. Hence occupants in an older car will suffer more severe injury than their counterparts in a new car in the same accident. Note that the corroded vehicle's energy absorption is not uniform with distance or crash time and so the corroded structure produces large '*g*' forces when the engine is pushed into the bulkhead (or fire-wall).

In Chapter 2 (Section 2.5) corrosion will be shown to occur within inter-welds. Fig. 1.9(b) shows the corrosion within the inter-weld which occupies a larger volume than the metal from which it was derived. This stresses the spot welds by an 'internal bursting pressure' which is superimposed on the stress–strain curve for externally applied (accident) forces. This has the effect of raising the zero reference point such that the energy of absorption is now very much reduced as shown in Fig. 1.9(b).

1.18 ENERGY, THE LONG-LIFE CAR AND RESOURCES

Energy is the fundamental unit of cost. This energy represents the total expenditure in energy (man's own 'energy' and capital investment are not included) required to get the product (e.g. a metal) to the marketplace as coil or strip to be used by the end-user, in this case the motor vehicle manufacturer. For steel, this cost will reflect the amont of energy required to extract the material from its mineral existance within the Earth's crust, its chemical conversion into the metal. The loss of metal as 'scrap' is high with steel, or any other metal which requires forming between two mould presses. This arises because the sheet metal must be clamped around the outer perimeter to prevent buckling between the presses. This clamped area is then cut off and scrapped. The example given in Fig. 1.10 is for steel panels of a car. Each process

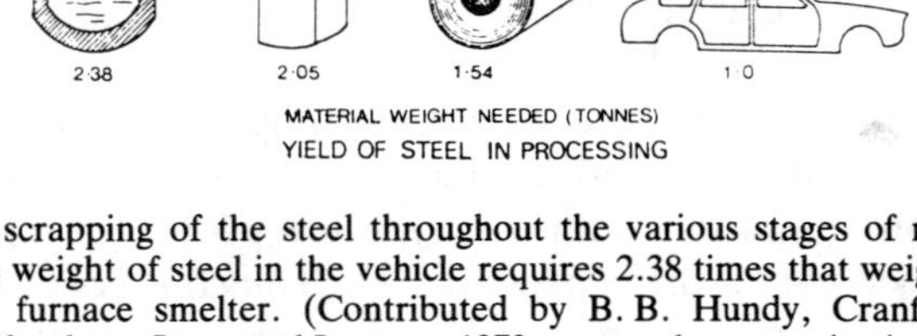

Fig. 1.10 — The scrapping of the steel throughout the various stages of manufacture and usage. Hence the weight of steel in the vehicle requires 2.38 times that weight of steel to be produced at the furnace smelter. (Contributed by B. B. Hundy, Cranfield Institute of Technology, Inaugural Lecture, 1979, personal communication.)

from the furnace to the ingot, to hot and cold rolling (Fig. 6.41), involves some wastage. The highest wastage is in the manufacture of the pressed panels as certain defects can occur within the steel which make a panel show up the deformation markings (Section 6.35). These will give poor paint reflectance (Fig. 6.18). Other materials, e.g. plastics, can be produced to shape with much less wastage. This will involve a high capital investment in the injection moulding equipment. The wastage is the plastic which 'solidifies' within the mould. A good example of this is on model aeroplanes where the whole injection moulding is produced in order to keep the separate parts together. A longer life car can be produced (Section 4.1 and Chapter 5), but this costs energy as different materials require different energy inputs as shown in Table 1.13. This increase in energy is reflected in higher purchase costs of the raw materials. There are several ways of denoting energy usage. These are the joule (J) or kilowatt hour (kW h). The kilowatt hour is the unit of energy with which most people will be familiar since it is the method used by the electricity boards in calculating the electrical consumption of domestic premises. It is the amount of energy consumed by a one kilowatt electric fire in one hour. The kilowatt hour is a 'modern' universal unit (unlike the weights and volumes which differ depending on the country of origin). (A brief historical introduction to the fundamental unit of length is given in Section 7.1. For our present purposes the relationships will be used as follows:

Table 1.12b

1 J is equivalent to 2.77778×10^{-7} kWh;
1 kWh is the energy consumed by a one-bar electric fire in 1 hour;
1 kWh is equivalent to 3.6×10^{6} J;
1 tonne is 1000 kg and is equivalent to 0.984 UK ton;
1 US ton (often called a short ton) is equivalent to 907.19 kg;
1 UK ton (often called a long ton) is equivalent to 1016 kg;
the modern unit of volume is the litre;
the US gallon is 3.7851 litres whilst the UK gallon is 4.5461 litres;
UK measurements will be used throughout unless otherwise specified.

The parameters which influence the designer's choice of material (Fig. 8.1) are a combination of the materials properties and their cost. The energy required to extract the metals used in automotive production is given in Table 1.13. However, the designer is also concerned with the associated mechanical properties (stiffness, tensile strength etc.) and the weight of the material for a given mechanical strength. All materials and metals can be made stiffer by correct design. A metal can be stiffened by producing reverse curvatures (Plate 2.5) and right-angle bends within the sheet metal. Fig. 1.11 shows the cost of various materials in their energy per specific tensile strength and the energy cost per specific stiffness. Both of these are 'performance terms' for the materials to be considered. Wood was one of the first materials used in the motor vehicle construction and is very cost effective in these terms. However, the skill and labour costs of the 'production system' made wood uneconomic.

Table 1.13 — Costs and usages of materials in cars

Material	Energy to produce material (kW h/tonne)	UK usage (tons)	World-wide usage (tons)
Steel	8000	1500000	43000000
Cast iron	4000	180000	5000000
Aluminium	65000	38000	1600000
Zinc	15000	9000	500000
Copper	20000	36000	36000
Lead	13000	12000	350000
Glass	6000	66000	1500000
Plastics	25000	110000	2750000

Contributed by B. B. Hundy, Cranfield Institute of Technology, Inaugural lecture, 1979 (personal communication).
At the end of the vehicle life, the scrap produced is contaminated with these elements, which cause poor mechanical properties in steel (Section 6.34).

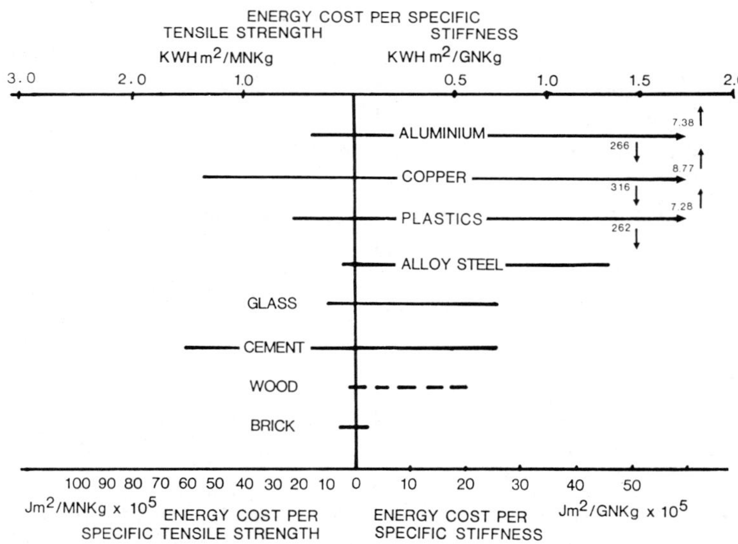

Fig. 1.11 — A useful performance parameter for the use of a material is the energy cost per mechanical property. This can be Young's modulus, density or as shown here the energy cost per specific tensile strength and stiffness. This compares the materials with those other common units such as brick and cement. Cement can of course only be used in compression (Section 12.2). (Redrawn from information supplied by courtesy of Pilkington Glass, Ormskirk (D. S. Oliver and G. T. Copley, Glass for Construction Purposes) and acknowledged with thanks.)

Most private motorists are most concerned with energy when it comes to the fuel (petrol or gasoline) bills and the consumption. A calculation, made in 1979, showed that if the vehicle body were made out of aluminium it would not save any energy over the lifetime of the vehicle. This comes about because it would take more energy to made the aluminium initially for the parts of the vehicle and this would not be recouped by the saving in fuel costs. The calculations are reproduced in Table 1.14.

Table 1.14 — Steel and aluminium energy costs

	Steel body	Aluminium body
Mass (lb)	500	300
Energy required (kW h)	1785	8705
Difference (kW h)	6920	
Difference allowing for scrap during manufacture and offcuts (kW h) (from Fig. 1.10)	10500	
Assume: overall mass saving 400 lbs (double the actual savings above); Fuel saving is 0.00001 gallons per mile per lb; 9000 miles per annum for life of 11 years;		
Fuel saving (400 gallons) in energy terms (kW h)	5600	
Difference in energy to build the lighter vehicle and fuel saving (kW h) 10500−5600	4900	
Extra energy required for the lighter aluminium-bodied vehicle	4900	

Calculations based upon making a vehicle lighter by using aluminium instead of steel for some panels and therefore saving on fuel. See Plate 6.1.

Contributed by B. B. Hundy, Cranfield Institute of Technology, Inaugural Lecture, 1979 (personal communication).

However, aluminium-bodied commercial vehicles are a more economical proposition as more load can be carried for a given volume of enclosure, and therefore fewer vehicles would be required. Commercial vehicle operators like having the lightest possible body in order to increase the payload. There is legislation for the axle weights of commercial vehicles and so what is carried as 'body' cannot be carried as a payload.

Therefore, to save energy in global energy terms, certain requirements have to be met.

(1) Pay attention to total energy costs in producing a material ready for the market place (Fig. 1.11, Table 1.13 and Table 6.12(d)).
(2) Increase the usage of the material so that there is less wastage.
(3) Design to consume less fuel but not at the expense of (1) (see Figs. 5.1–5.3).

(4) Make the car last longer, by better anticorrosion design (Section 5.1) and by reducing the subsidy for its use (Section 1.7). This would automatically increase trade-in prices as shown in Fig. 1.2. Manufacturers would be competitive on structural longevity and depreciation.
(5) Avoid material wastage and off-cuts. Some newer steels will cause problems in scrap in the future (Section 6.34).
(6) Ensure that the vehicle can be scrapped more effectively and the high-energy-cost materials, e.g. copper, aluminium and plastics, are fully realized. (Plastics are the least problem as their density is low and so their actual weight is small compared with that of the other metals cited.)

A longer vehicle life car would cause severe problems in later years for the motor industry. The manufacturers exist to make and sell new cars. The heavy subsidy on the provision of a 'company transport system' means that the more affluent companies will provide air flights either by small privately owned aeroplanes or helicopters if the roads become choked with ageing motor vehicles. Hundy (B. B. Hundy, Cranfield Institute of Technology, Inaugural Lecture, 1979 (personal communication)) estimated that if a long-life car was marketed in 1980 it would lead to bad times within the motor industry and perhaps to unemployment. The longer-life vehicle would affect new vehicle registrations in the years after 1987 as shown in Fig. 1.12.

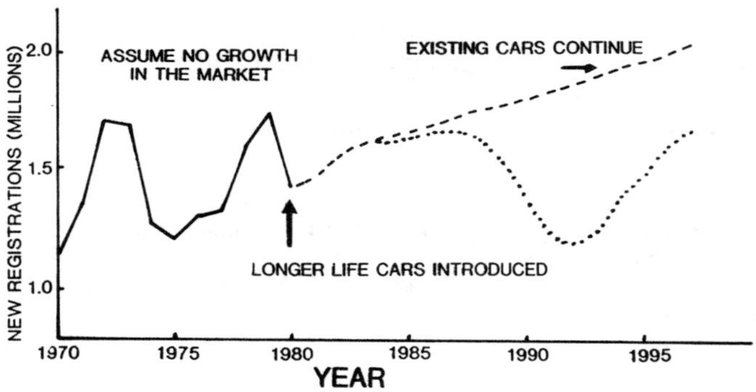

Fig. 1.12 — The new registrations in the UK are represented here with the projected view if a longer-life vehicle were to be introduced in 1980. The assumption made is that there is no growth in the market. (Contributed by B. B. Hundy, Cranfield Institute of Technology, Inaugural Lecture, 1979, personal communication.)

REFERENCES

Hoar, T. P. (1971). Report of the Committee on Corrosion and Protection, HMSO.
Svenson, G. E. (1973). *Cars Go Rusty*, AB Svensk Bilprovning.
Svenson, G. E. (1978). *Weak Points of Cars*, AB Svensk Bilprovning.
Rogers, T. H. (1967). *Marine Corrosion*, Newnes (p. viii).

2

Corrosion mechanism and build-up of the micro-climate

2.1 WHAT IS CORROSION? — MISCONCEPTIONS OF THE CORROSION REACTION

The word 'corrosion' is derived from the Latin word *corrodere* which means to gnaw away, or to wear away. Indeed the dictionary definition is 'a process in which a solid, especially a metal, is eaten away and changed by a chemical action, in the presence of water by an electrolytic process'. Corrosion is not an organic pest, fungus or like dry rot in timber — the dry rot of steel. It cannot be killed nor can the metal be 'regenerated' from the rust by any liquid potent. Once the metal is converted into its oxide, it must be heated up to the melting point with carbon (as the reducing agent) to reproduce the metallic state. The oxide is very brittle and has little energy-absorbing properties. Rust also contains water (Section 6.14, point (10)) and harbours road salts, thus aggravating the problem of corrosion.

A metal is not porous (Fig. 2.1(a)), nor does it 'sweat' water. A metal is in fact a very dense substance and there is no room for water in its structure. Water appears on the surface of a metal because atmospheric water vapour condenses on the cold surface. Road salts on the metal increase this condensation of water (section 10.5). Road salts do not cause corrosion; they only accelerate it. Hence the water sometimes found in paint blisters is not sweated from the metal (see Sections 2.15 and 8.12—8.14). The corrosion of the motor vehicle is not due to the manufacturer's using 'recycled scrap' from rust-prone models, nor is it a 'latent defect' carried within the steel being used (Sections 6.34.4).

Corrosion is not caused by the car cutting the earth's magnetic field so that the car body acts like a generator (Fig. 2.2).

Corrosion is not due to the direct combination of oxygen and steel and is not accelerated by parking a vehicle over grass where more oxygen is generated from the chlorophyll in the green grass or trees (Fig. 2.3). Cars parked continuously under trees will be subjected to a higher humidity and this affects the corrosion rate (see Fig. 10.5).

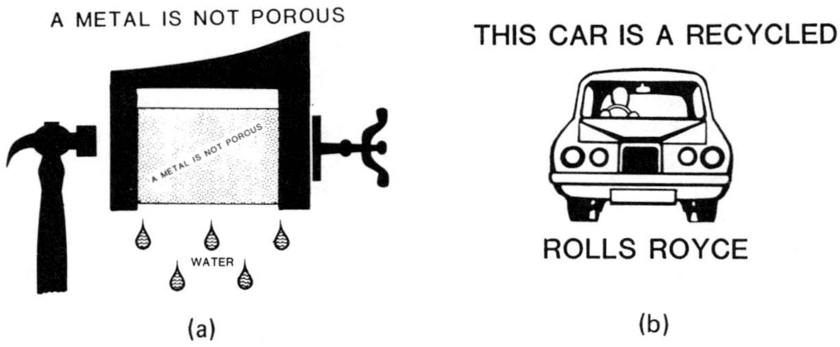

Fig. 2.1 — (a) A metal is not porous, nor does it sweat water. (b) Corrosion is not due to 'residual rot' from previously recycled corroded vehicles. Although impurities in the steel may cause corrosion in laboratory experiments, this is not considered the most important parameter in the corrosion of the motor vehicle.

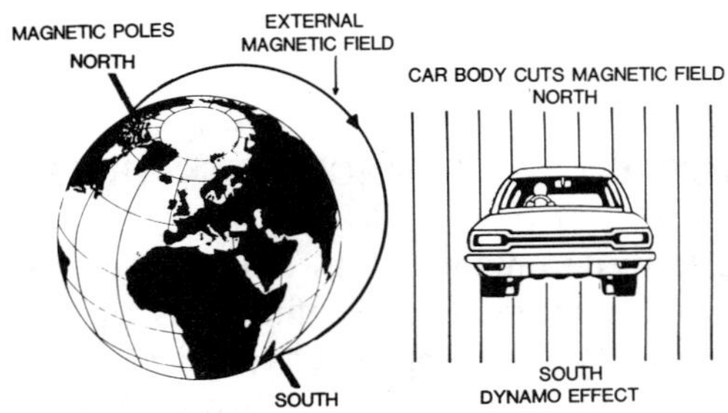

Fig. 2.2 — The Earth contains magnetic North and South poles, which do not coincide with the geographic North and South poles. Some engineers postulate that it is the car cutting this magnetic field which acts as a small generator or dynamo. The electric current generated is considered as the cause of corrosion.

Fig. 2.3 — When plants photosynthesize sugars from carbon dioxide and water, oxygen is given off. This oxygen does not accelerate the corrosion rate of the car when it is parked over grass or under trees. The trees and grass will produce a higher humidity which affects drying and allows the winter de-icing salts to absorb water (Section 10.5).

Corrosion is not solely 'friction' or 'abrasion' of road debris on the underside of the vehicle (Sections 10.12 and 10.13). This does lead to corrosion by the formation of a local micro-climate at the non-adherent paint–metal interface (Figs. 2.6(f) and 8.18; Sections 8.12 and 8.13).

2.2 SO WHAT IS CORROSION? THE CORROSION REACTION

Corrosion is an electrical phenomenon and is associated with a very small 'battery potential' (Fig. 2.4) rather like the 1.5 V of the dry cell battery which powers so many transistor radios, toys or flash-lights (Fig. 2.5). These 'proper' or commercial batteries (Section 7.16) turn chemical energy, by the corrosion of another metal called zinc, into electrical energy, the current which drives the toys. Corrosion therefore is accompanied by the flow of an electrical current. Stop the electrical current and you stop corrosion. Increase the electrical current by producing a good conductor of electricity by providing road de-icing salts and the corrosion is increased. An increase in the current increases the corrosion with the corresponding loss of metal from the structure (Sections 7.19 and 7.20). Hence one can only reduce corrosion by providing a 'suitable environment' in which the current cannot flow in the liquid environment (this is usually performed by a paint film). A suitable analogy is, reduce or throw away the electrolyte in your 12 V car battery and you destroy the 12 V supply.

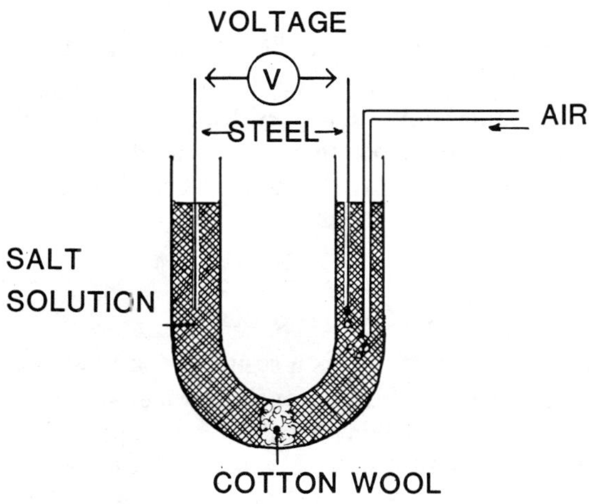

Fig. 2.4 — Two pieces of mild steel connected to a DVM, as shown, and immersed in a saline electrolyte which is oxygenated differentially will show a potential difference. The potential difference developed rises with increased oxygenation of one arm with respect to the adjacent arm. The tissue paper or cotton wool is there merely to prevent the oxygenated liquids in one of the two arms from mixing with the deoxygenated electrolyte by the stirring action. Potential reversals up to ± 90 mV may be obtained, depending on which arm is oxygenated. If the two arms are oxygenated to the same extent there is no potential difference and in this state can be likened to a 'run down' battery. For comparison, a car battery produces 12 V = 12000 mV. The oxygenated arm is the cathode or positive terminal of the 'corrosion battery'. The deoxygenated arm and the area which dissolves is the anode or negative terminal of the corrosion battery.

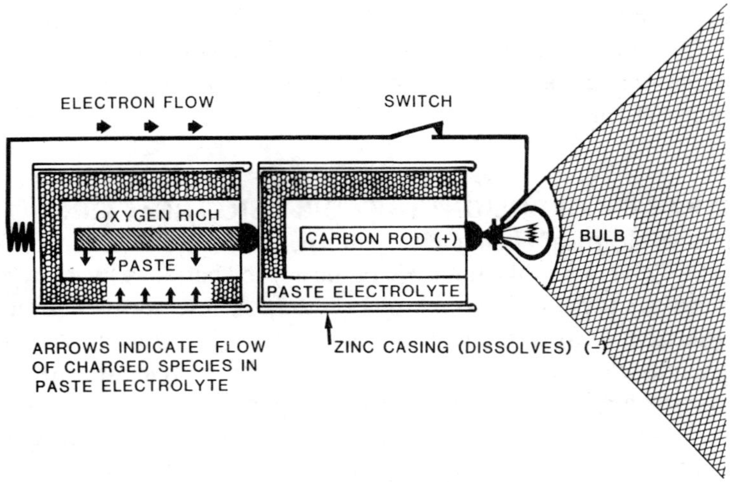

Fig. 2.5 — A cross-section through a typical torch or flash-light. The electrons are shown flowing in the external metal circuit, and the charged species in the 'paste electrolyte' to provide the complete electrical circuit. The electrical energy is produced by the dissolution of the zinc case. The electrons are consumed at the carbon rod, which in corrosion technology is termed the cathode, and is the positive terminal of the battery. The wasting away of the zinc case can be stopped by switching off the flow of electrons in the outer metallic circuit. The zinc is termed the anode in corrosion technology and is the negative terminal of the battery. The dry cell battery is discussed in greater detail in Section 7.16.

The term **corrosion** is therefore applied to **unwanted** or **unexpected** metal degradation processes and is something to be avoided, but we are quite happy to use, and to make better, the same chemical process for the supply of electrical energy and to call it a battery. A battery has a positive and a negative terminal (or electrode), at which specific chemical reactions take place (see Chapter 7, in particular Sections 7.2–7.4 and 7.16).

2.3 CORROSION IN THE MOTOR CAR

Corrosion of the motor car (Fig. 1.4) is a result of the flow of electricity from one region of the metal (of the car) which is not well oxygenated (which acts as negative electrode) to another region which is plentifully supplied with oxygen (which is the positive electrode) and in the presence of water or an electrolyte (Section 10.1). In corrosion the 'battery' terminology (negative and positive) terminals are not used. These terminals are called the **anode** and **cathode** respectively. It is the oxygen-denuded region which corrodes.

2.4 DEMONSTRATION OF THIS BATTERY PROCESS — THE DIFFERENTIAL AERATION CELL

This potential difference or voltage of the 'equivalent battery' due to the difference in the oxygen availability can be shown by taking two pieces of steel and immersing them in a salt solution as shown in Fig. 2.4. The two pieces of steel are connected to a potential-measuring instrument — a digital voltmeter (DVM). One side of the U-tube is oxygenated by blowing air through the liquid, while the other side remains relatively de-oxygenated. The cotton wool prevents the two liquids from mixing, while at the same time allowing the corrosion process to take place between the two liquids in either arm (see also Figs. 7.16 and 7.17, Section 7.11 and Section 6.14). Potential reversals or anodes changing to cathodes and vice versa may be obtained depending on which arm is oxygenated. For this reason the 'cell' or battery is called the **differential aeration cell.**

2.5 WHERE ARE THESE CELLS AND THEIR ELECTRODES (EQUIVALENT BATTERIES)?

These cells form short-circuited batteries because the car body is one integral electrical conductor, which is unlike the two pieces of steel used above. These 'equivalent batteries' occur at all places where there is a difference in the availability of oxygen (Fig. 2.6). The equivalent batteries can be formed at the following locations.

(1) Areas of bad design as they harbour road dirt which restricts natural drying. (Corrosion can only proceed in the presence of a moist environment. **Bad design, in the corrosion sense, is any design which allows the accumulation of a corrosive micro-climate, the presence of which the 'lay public' are unaware of but the motor manufacturers should be aware of in view of their expertise in the subject matter.)**
(2) Beneath road dust 'dirt' (Fig. 2.6(a)) or other surface deposits.
(3) In inter-welds or clinch flanges or hemming folds. These also form stagnant liquid environments which present a small liquid surface to the atmosphere and so drying by evaporation is severely restricted. More importantly, oxygen replenishment at the metal surface which is furthest from the liquid surface is restricted or denied.
(4) Beneath mud poultices or layers
(5) At the corrosion front or 'pit' (Fig. 7.14) which will explain the localized 'pitting' nature and speed of 'rust through' corrosion. This is often termed 'inside-out' corrosion owing to the fact that it is initiated by the deposition of salts on the inner panels of box members which then corrode through by the localized pitting to the outer surface.
(6) Beneath non-adherent but coherent surface coatings (Figs. 8.18–8.21).
(7) At threaded screw sections (Fig. 2.10). Corrosion here means that very high torques have to be applied to loosen corroded studs.

It is suggested that the subsequent chapters are not read until after the contents of the main body of the book have been understood, as the more technical chapters are intended for further reading to bring the reader up to a greater understanding of motor vehicle corrosion in conjunction with the associated chemical and physical sciences. The remainder of this section poses some questions and their answers are given in Sections 7.13 *et seq*.

Motor vehicle corrosion is the application of the differential aeration cell to the micro-climate which forms during the life of the vehicle. This micro-climate becomes

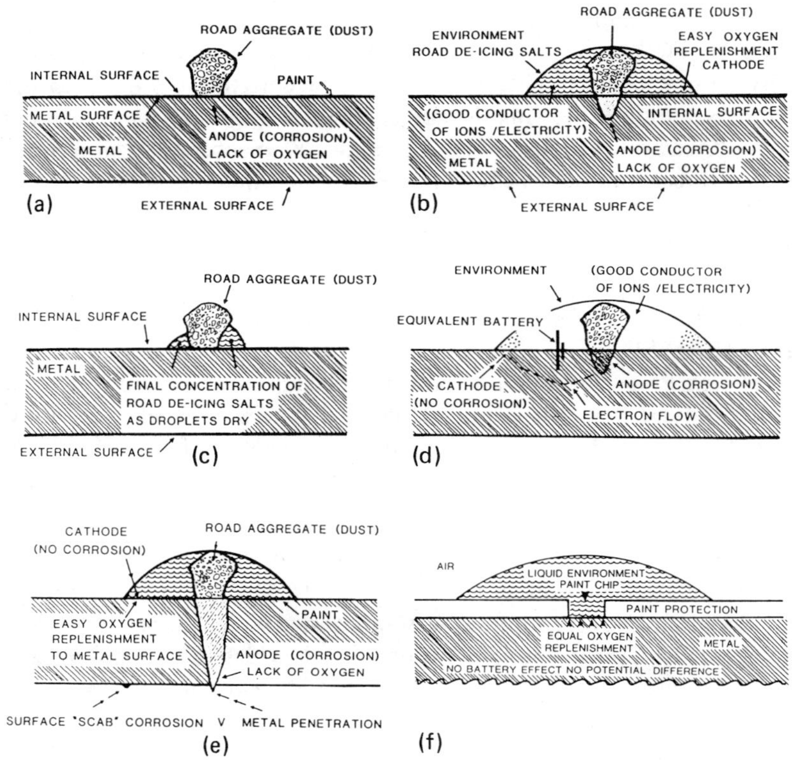

ANODE REACTION

STEEL

$$Fe_{(METAL)} \longrightarrow Fe^{2+}_{(LIQUID)\ WATER} + 2(e)_{(ELECTRONS)\ METAL}$$

CATHODE REACTION

$$O_{2\,(AIR)} + 2H_2O_{(WATER)} + 4e_{\ METAL} \longrightarrow 4OH^-_{\ LIQUID}$$

PRODUCT OF ANODE + PRODUCT OF CATHODE ⟶ RUST

$$Fe^{2+} + 2\,OH^- \longrightarrow Fe_2O_3$$

(g)

Fig. 2.6 — (a) Schematic view of the road de-icing salts on the surface of the car body. (b) Schematic view of the road de-icing salts absorbing water from the atmosphere to form the differential aeration cell and (c) concentrating at the base of the road aggregate on drying, which helps to fix the aggressive environment upon vertical surfaces. (d) The short circuited equivalent battery is generated by the difference in oxygen availability. The oxygenated region is the cathode or positive terminal of the 'corrosion battery'. The deoxygenated region and the area which dissolves is the anode or negative terminal of the corrosion battery. Note that the cathodic electrode is much bigger than the anodic region. Compare this with the paint chip (f). (e) The difference between internally initiated corrosion (metal penetration by pitting) owing to the lack of oxygen replenishment beneath these surface deposits and surface-initiated corrosion or scab corrosion from stone chips. (f) A stone chip does not cause very severe corrosion in the early stages of life. Both anodes and cathodes are equally oxygenated and dry rapidly. Anodic and cathodic areas do not enter into the problem as this is effectively a 'run-down' corrosion cell. (g) The chemical reactions pertaining to the differential aeration cell (these are perhaps best left to a second and third reading, and after reading Sections 6.13 and 6.14). If the two areas (anode and cathode) are of equal size and oxygenated to the same extent there is no corrosion because there is no potential difference. Severe pitting corrosion results when large cathodes (perimeter of the micro-droplet) are in contact with small anodes (area beneath the dust particle) (Sections 7.11 and 7.12).

a free-rider and the subsequent sections will develop a theory which will enable the layman to understand and even to predict the areas of corrosion. In order to test and concentrate the reader's mind the following questions are concerned with Fig. 2.6.

Fig. 2.6(b) shows the road de-icing salt adsorbing atmospheric water.

Fig. 2.6(c) gives a schematic view of the road de-icing salts concentrating at the base of the road aggregate, which helps to fix the aggressive saline environment upon vertical surfaces. The liquid envelope which is produced by the uptake of water on humid days (Section 10.5) is also shown. As an exercise, try to indicate which is the region or area which has most oxygen replenishment. Answers are given in Section 7.13.

Fig. 2.6(d) shows the short circuit which is developed by this form of corrosion. There is no way of breaking this electrical circuit as there was in the case of the torch battery and circuit (compare Figs. 2.6(d) and 2.5). Figure 2.6(e) shows diagrammatically the difference between internally initiated pitting corrosion, causing metal penetration, and scab corrosion (surface corrosion beneath a painted surface). Indicate which are the areas of oxygen denudation on Fig. 2.6(e). The answer is given in Section 7.13.

Figure 2.4(e) also shows diagrammatically the effect of a stone chip on the paintwork of a vehicle. Such paint chipping is relatively harmless. Why? This is because the metal surface is subjected to an equal amount of oxygen and so the corrosion battery gives *zero potential difference,* i.e. it is like a run-down battery — lifeless. What will happen if the corrosion process starts to 'undercut' the paint film as shown in Figs. 8.18 and 8.21? There will be a voltage generated as the area beneath the paint film will be denuded of oxygen.

2.6 HOW CAN THIS CORROSION BE STOPPED?

The battery in the toy or torch (Fig. 2.5) is switched off to stop the flow of electricity and so prevents the unnecessary run-down of the battery. This breaks the metallic part of the circuit. This method of switching off the corrosion process is not available in the case of the motor car, because to use it we would have to cut the car up. One aspect of corrosion (Section 7.11) which is needed to explain the corrosion phenomenon is the analogy of electrical charge to an escalator: whatever happens on the upwards or tread side (passenger side) must happen in reverse beneath the treads on the underside (Fig. 7.17). In corrosion the flow of electricity by electrical charges (called electrons, and analogous to the under-side escalator steps) in the metal must be balanced by the flow of charged entities in the electrolyte or liquid environment (the salt droplet). Therefore, if there is no water environment outside the metal, there can be no flow of these charged entities and so there is no corrosion. This gives another method (like the switch) on the torch of stopping the flow of corrosion current in the metal. This method is also available in the case of the car battery; throw away the electrolyte of the 12 V car battery and one has lost the 12 V. Hence, if one uses the same concept in the case of the motor car, we must isolate the metal surface from the agent causing the corrosion, i.e. the electrolyte which contains the road de-icing salts in solution. This 'isolation' means that we shall be looking for a physical barrier such as a paint film. However, a paint film is permeable to both water and oxygen, and **so does not function when kept wet for long periods** (Section 2.5) and breaks down beneath mud poultices. For further details of water penetration through paint films see Section 8.11 and the mechanism of paint protection see Fig. 8.18.

There are four options open to stop corrosion.

(1) Keep the vehicle dry, thus breaking the contact in the liquid environment, and/or prevent stagnant deoxygenated water from forming in a 'water trap'.
(2) Have a highly electrically insulating film on the metal to separate the metal from the environment. This is one of the functions of a paint film (Sections 8.2 and 8.11 *et seq.;* Figs. 8.18–8.21).
(3) Keep the vehicle clean, thus eliminating the road de-icing salts which are naturally 'hygroscopic' and absorb water from the atmosphere (Section 10.5). (The salt on a car dries white on a dry day, but most of the time it is wet and cannot be seen except for a very thin film of moisture on the car body.) The presence of chlorides can accelerate the breakdown of the protective paint film and destroy the effectiveness of any corrosion inhibitors and allow corrosion to proceed.
(4) Ensure equal oxygen supply or replenishment at all surfaces. This cannot be achieved with the restrictions placed on motor vehicle manufacturing methods. One of the functions of a paint film is to provide a uniform oxygen replenishment over the whole metallic surface.

This area will be discussed again later on in the book. It is important that other interacting phenomena are also considered so that a complete picture emerges about motor vehicle corrosion. The other sections of this book will show that 'corrosion resistance' must be designed into the vehicle body. This means that there must be an awareness of good corrosion designs at the onset of vehicle design which eliminates these natural differential aeration cells. **Vehicle design and corrosion protection are inter-related properties and one cannot be achieved without the other**, e.g. see Plates 2.3, 2.8(b), 2.8(c), 4.10, 4.15(a), 4.16 and 4.17(a), see also Chapter 5.

2.7 ROAD SPRAY, MUD COMPACTION AND MOMENTUM

In order to understand how a motor vehicle corrodes, it is necessary to understand something about the spray generated by a vehicle moving at speed along a wet road. The tyre effectively 'recyles the rain' and increases any affect due to rain. The origin of the water spray is given in Fig. 2.7. It is helpful to define the size of a spray droplet by its momentum. The **momentum** M of a particle is its mass m multiplied by its velocity v. The magnitude of this mv gives a parameter which characterizes how easy it is for that particle to alter direction. The smaller $M = mv$ is the easier it is for the particle to be moved from its given direction. A leaf falling under gravity does not drop straight down but alters direction continually, whilst a heavy object drops in a straight line. The particle size, and hence momentum in the spray generated by a rotating wheel, is smaller the faster the speed of the car (Fig. 2.8). The size of droplet, often termed a micro-droplet, is given in the literature as varying inversely with the third power of the vehicle speed. Hence a car travelling at 60 mph creates a very fine spray of mud and water. For this high velocity water spray to form a micro-droplet, nature dictates, by the surface free energy theory (Section 10.8 *et seq.*), that

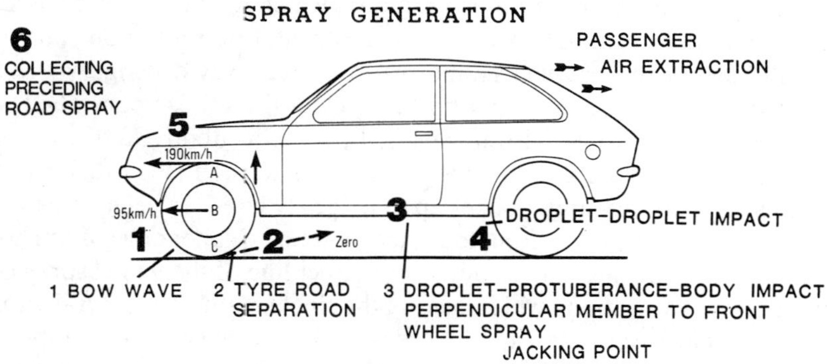

Note 60 mph=95 km/h; 120 mph=190 km/h

Fig. 2.7 — Analysis of the tyre throwing road dirt forward of the vehicle at twice the vehicle speed (see also Fig. 10.8). The figure also shows the origins of water spray which is developed: (1) The 'bow wave' produced when the tyre tries to push a finite volume of water ahead of the tyre. This is broken up into spray as the car body quickly overtakes the spray. (2) Spray is produced on the separation of the tyre from the road. Spray is also formed as the water is thrown out of the tyre tread over the whole circumference of revolution. These droplets impact the vehicle body at the vehicle speed. (3) Droplet–body impact: the water droplets which are just lifted off the road (and have zero velocity with respect to the road) produce a very fine spray on impacting any part of the car body which is perpendicular to the impacting wheel spray. The importance of this is at jacking points and other protuberances beneath the vehicle. The change of energy from kinetic energy (of one drop) to surface free energy of a number of micro-droplets of small momentum causes a fine mist and obscurity of vision which is inversely proportional to the cube power of the vehicle speed. These micro-droplets follow the pressure eddies around the vehicle body (droplet–body impact is discussed in Fig. 4.2). (4) Droplet–droplet impact with the spray from both front and rear wheels. These collisions ahead of the rear wheel produce spray. Some spray is also produced by the spray from the front wheel, colliding with the rear wheel at the point (4). (5) Water spray thrown in front of the vehicle at twice the vehicle speed, colliding with the underbody. The relative speed at collision is the vehicle speed. (6) The preceding vehicle causes a turbulent spray through which other vehicles must travel. This applies to overtaking and overtaken vehicles alike (see Figs. 5.6 and 5.7).

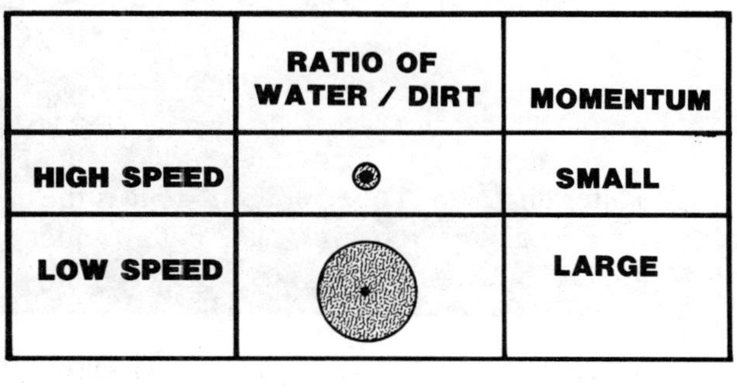

Fig. 2.8 — This figure shows diagrammatically what is meant by the size of water droplet and momentum. The figure also portrays the ratio of water to road dirt for the two speeds. Note that the water content in both droplets is exaggerated (see Fig. 10.7).

there must be a nucleus upon which the water can form its droplet. The nucleus acting for micro-droplet formation is road dirt, which may be worn road surfaces, tyre dust and de-icing salts. The small amount of water present upon each road dirt particle will have been observed on motorway or free-way driving, where there is often insufficient water for the wiper blades to clear the windscreen. The low water content of each microdroplet also means that the associated road dirt is artificially densely compacted by the phenomenon of surface tension (Fig. 10.7 and Section 10.10) into a high density mud pocket upon ledges (Plate 2.1) or in areas which are out of the way of direct washes from the wheel, a steam cleaner or a garden hose. The areas upon which they deposit are out of the direct line of the wheel spray owing to the small pressure variations which distort the linear path of their trajectory. The water-borne dirt also deposits on those areas where it is hard to give supplementary wax coatings. Another example of this is on the rear wheel arch well (see Section 4.6 and Fig. 4.6).

To summarize: the slower the car travels, the larger will be these water droplets and the higher the ratio of water:dirt which they can contain. The faster the car travels, the smaller will be these spray droplets and the less water they can contain. The wheel spray from slowly moving vehicles has a high momentum and so travels in straighter lines. The spray is not affected by pressure eddies beneath the car body.

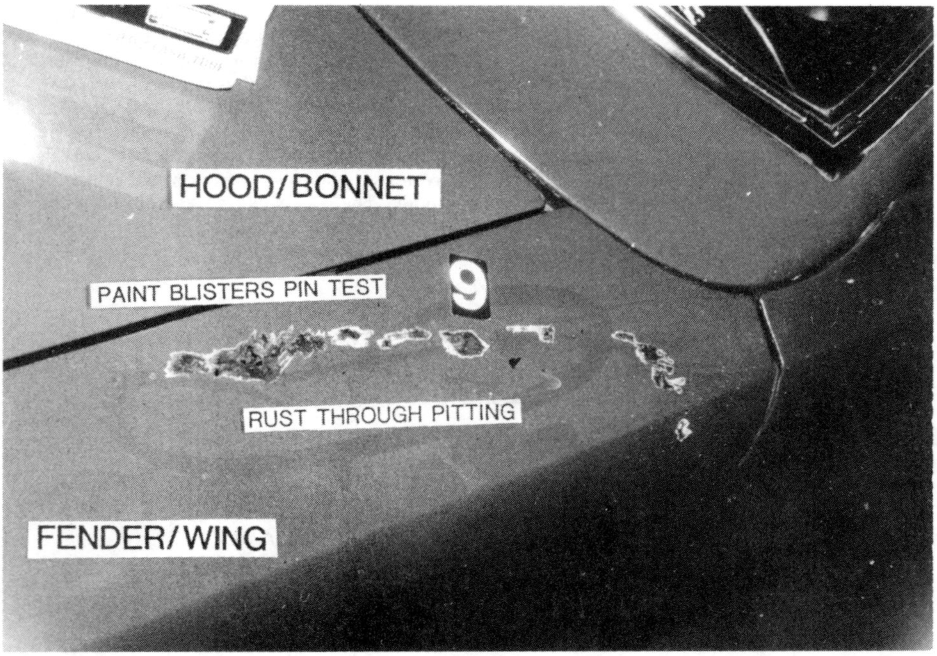

Plate 2.1(a) — A row of paint blisters on the front wing/fender as a result of metal penetration due to the underlying mud pocket. The row of pits outline the boundary between where the mud pocket is both wet and denuded of any oxygen replenishment. The anodic or dissolving area will be where there is little oxygen replenishment, whilst the cathodic area will be where oxygen is readily replenished. Both cathodic and anodic areas will have to be wet to form the electrolyte, and so the ledge ABC (Plate 2.1b) will contain all three parameters.

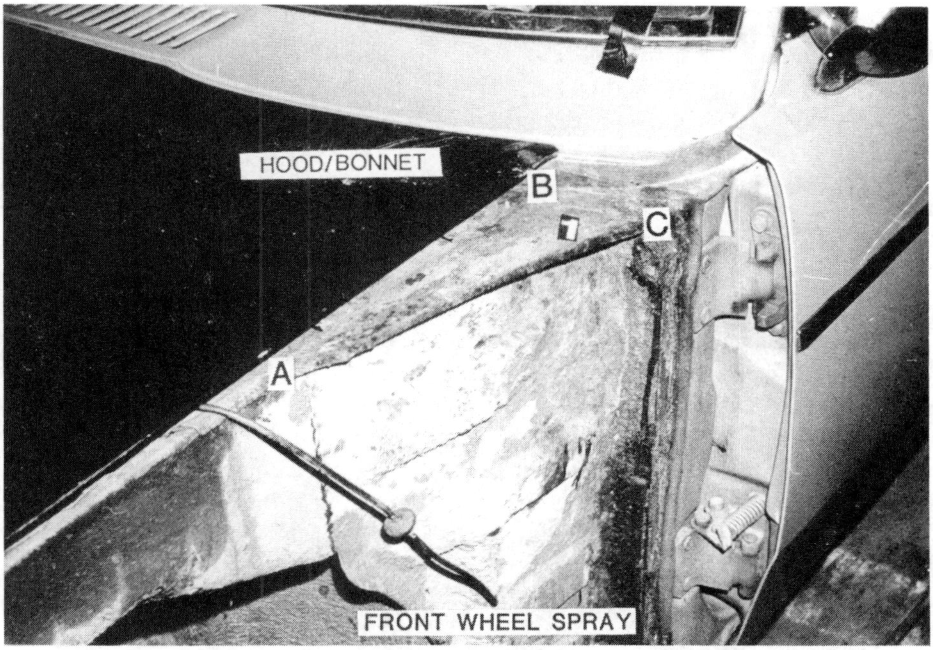

Plate 2.1(b) — Removal of the wing/fender shown in Plate 2.1(a) reveals the offending ledge which traps the mud pocket on top of it. The ledge (ABC) protects the mud pocket from the water thrown up by the rotating wheel. The ledge also acts as a shield for the mud pocket against steam cleaning.

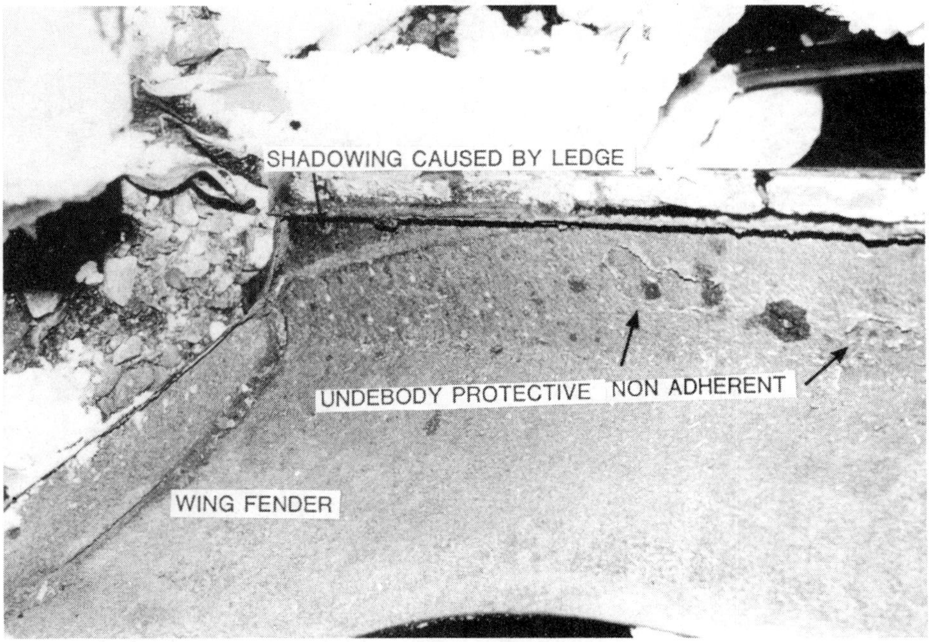

Plate 2.1(c) — This ledge also prevented the thorough application of underbody protective coatings applied during manufacture. It also presented a problem to the after-market rustproofing stations.

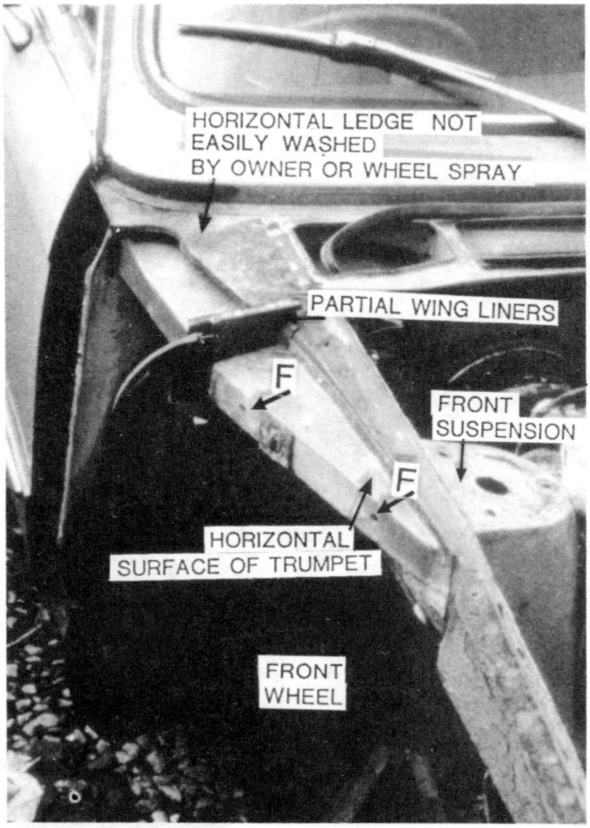

Plate 2.1(d) — In the original design of this vehicle, this horizontal ledge collected the road dirt. There was no anti-corrosion design incorporated in this area. Following customer complaints regarding the rapid corrosion, the manufacturers supplied partial lining for the wing/fender, of the type D shown in Fig. 2.9. The energy absorbing trumpet was still left to collect mud. Subsequent designs were modified with another plastic guard which was placed over the trumpet, using fixing holes shown here — F.

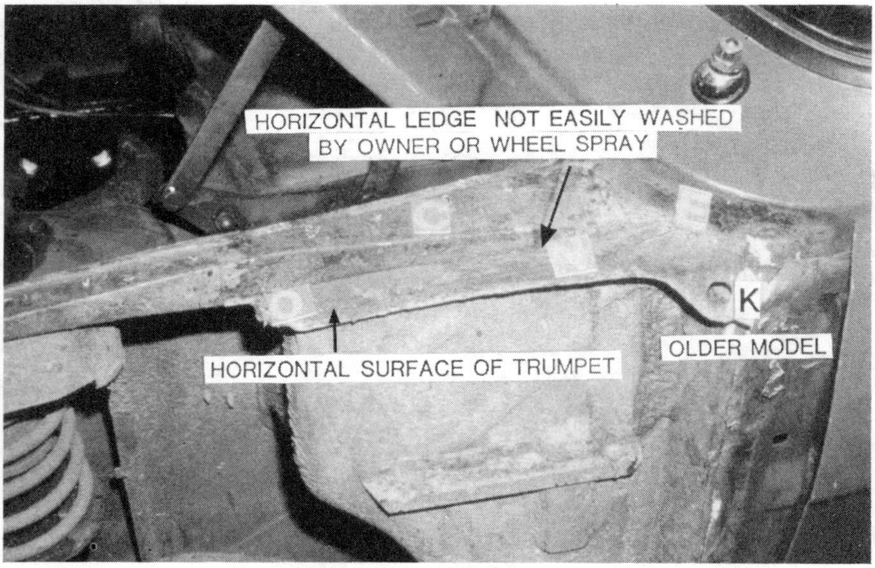

Plate 2.1(e) — Other manufacturers ledges were not so large, but they all contained these horizontal surfaces. This shows the Mark 2 design, where the ring —K— which was horizontal in the Mark 1 design is knocked down to form a 45-degree slope.

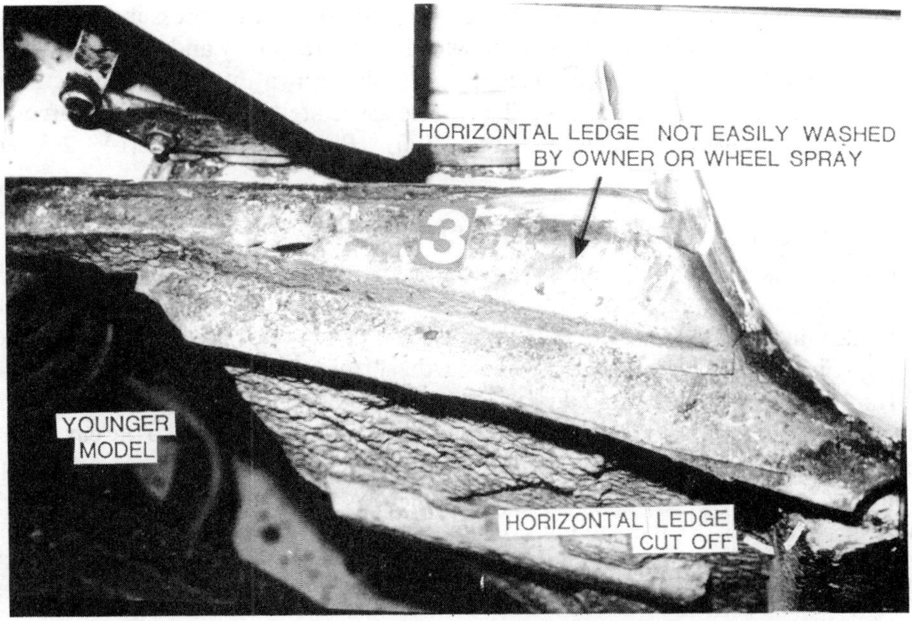

Plate 2.1(f) — Mark 3 design was to cut the ring off, but this still left a considerable horizontal area upon which mud could accumulate.

Plate 2.1(g) — Mark 4 redesigned the area to make a sloping 45-degree shelf, but road mud could still accumulate. This road spray has very little water associated with the micro-droplets and so cannot drain or run off the sloping ledge. Perhaps Mark 5 will be a complete inner wing/ fender liner which incorporates the correct scientific principles for anti-corrosion design.

Smaller momentum particles, in contrast, will be affected by pressure eddies and may drift and deposit in areas out of the wheel spray trajectory and even in cavity or box sections. These smaller particles may also be drawn into the manufacturer's process drainage holes (see Section 2.11). The associated salt accelerates corrosion rather than causing it (Sections 10.1–10.6) (Plates 2.2(a), 2.2(b), 2.3 and 2.4(a)–2.4(d)).

2.8 VELOCITY OF SPRAY FROM A ROTATING WHEEL (60 mph=95 km/h)

If a car travels at 60 mph, then the bottom of the tyre, which is in contact with the road, must have a relative speed with respect to the road of 0 mph, i.e. it must be stationary at the time of contact, otherwise the tyre would be 'skidding' (Section 10.12) (Fig. 2.7). The point of the wheel furthest from the road must be travelling at 120 mph (Fig. 10.8). Any road dirt or water thrown off at this point will be travelling forwards at 120 mph with respect to the road or 60 mph with respect to the car. Hence road dirt thrown off the top tyre will impact the car body, on areas over the head-lamps, at the vehicle speed. At the point half-way, the dirt will be thrown off at 60 mph at a tangent to the circumference. Hence the underside of the vehicle is continually bombarded from all directions with road dirt which is travelling at the vehicle speed.

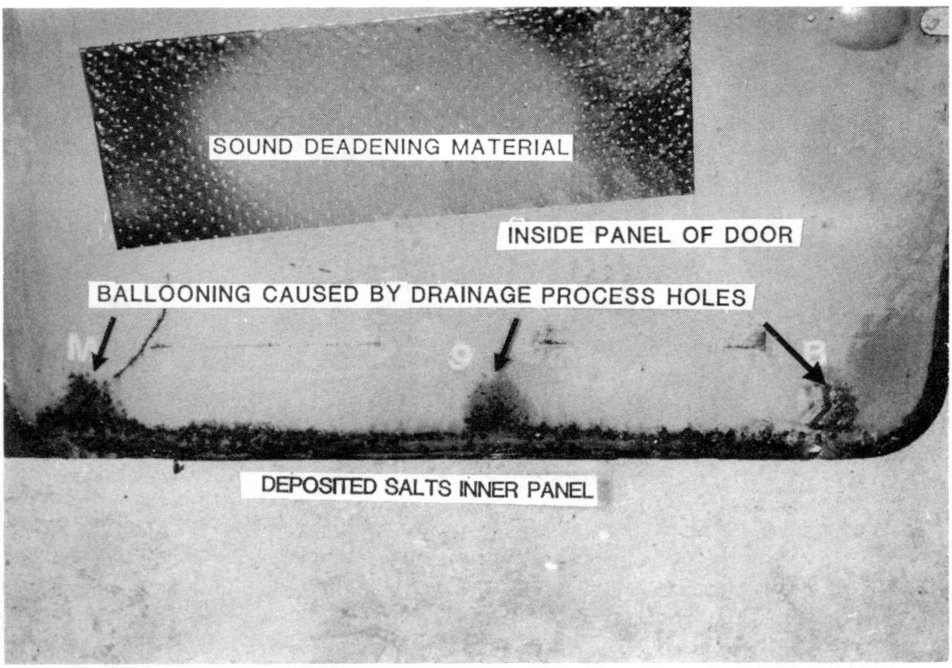

Plate 2.2(a) — The inside surface of the outer skin of the door panel shows the localised corrosion at the drainage holes. For the mechanism of deposition see Section 2.13. The ideal position for TWO of the drainage holes is at the door extremes so that water pooling is avoided when the car is garaged on an incline. The photograph also shows the corrosion within the hemming fold crevice. See Fig. 4.1(a), Fig. 4.3 and Fig. 4.4.

Plate 2.2(b) — A car door where the inner surface is zinc. This shows no brown rust staining as the zinc corrodes preferentially and the corrosion product is white and offers some protection. Section 8.2.

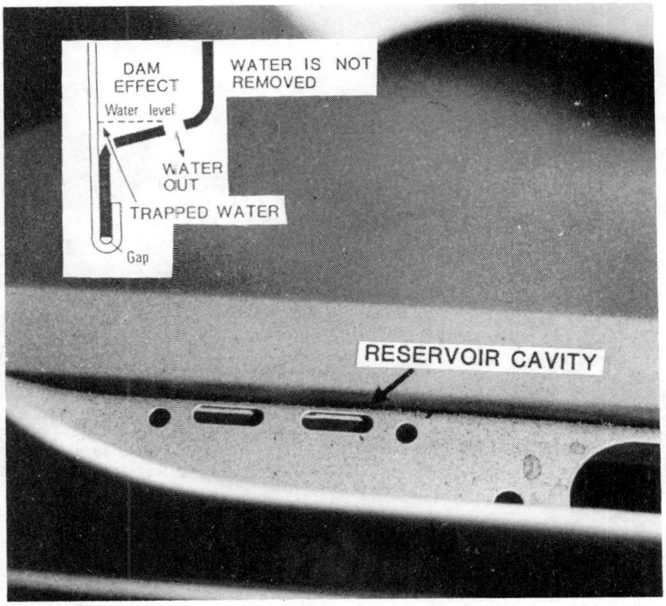

Plate 2.3 — Rain water running down the window pane (Figure 4.3(a)) can by-pass the rubber at the top of the door. This collects in a water trap at the bottom of the door because the drainage hole(s) are too high. This traps water and road de-icing salts in the crevice formed by the clinch fold. On drying these de-icing salts will be concentrated in the clinch fold, which is also the area denuded of paint (due to the poor electrocoat throw) and oxygen replenishment within the trapped liquid is restricted.

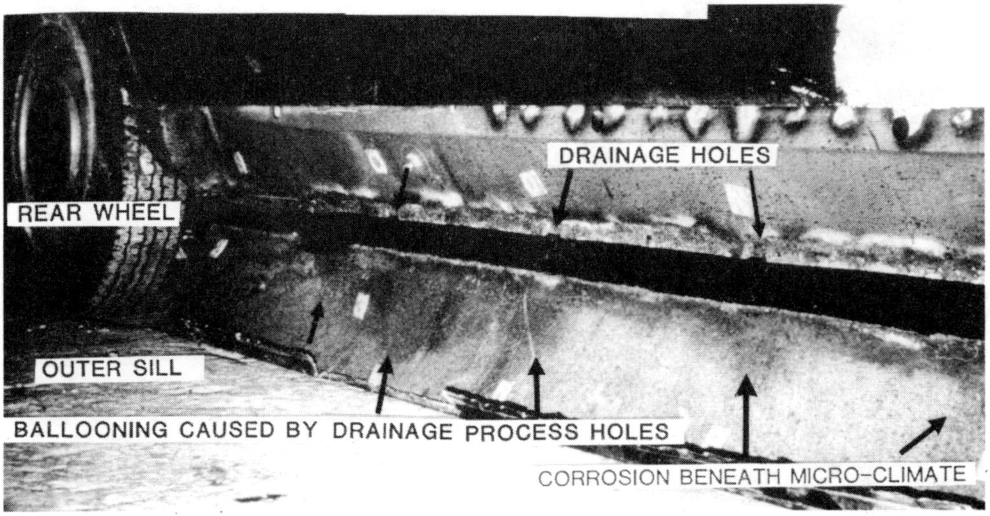

Plate 2.4(a) — The microclimate of deposited road dirt and de-icing salts can be seen localised around the process holes within the outer panel of the sill/rocker. The air motion within this box member is clearly shown by this localisation. The external appearance of the corrosion caused by this localisation of road dust is a paint blister on the outer surface when rust penetration is achieved. See Plate 2.4(b), (Fig. 2.11 for test).

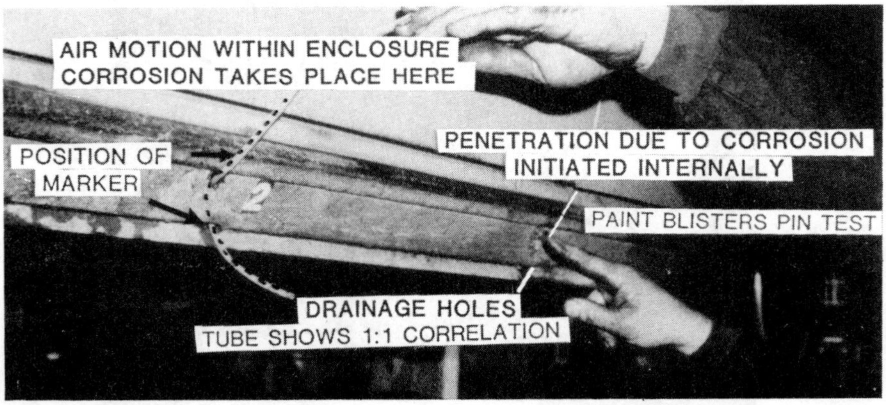

Plate 2.4(b) — These paint blisters have been punctured and enlarged with a screwdriver and a plastic marker inserted into the rust hole and associated electrocoat drainage hole. This shows the 1:1 relationship between the drainage hole and rust penetration resulting from internally initiated corrosion. This correlation indicates structural corrosion rust through corrosion rather than surface corrosion due to stone chipping. They can be differentiated by the drawing pin test (Fig. 2.11, Section 2.15). Once larger holes are produced, air and water ingress is accelerated and rapid deterioration results. Only two open drainage holes should be left open, positioned behind the front mudflap, and in front of the rear wheel. Ideally they should be protected from air ingress, (mechanism, Section 2.11 and sequence) by one-way valves. A simple DIY one-way flap valve is shown in Fig. 5.10.

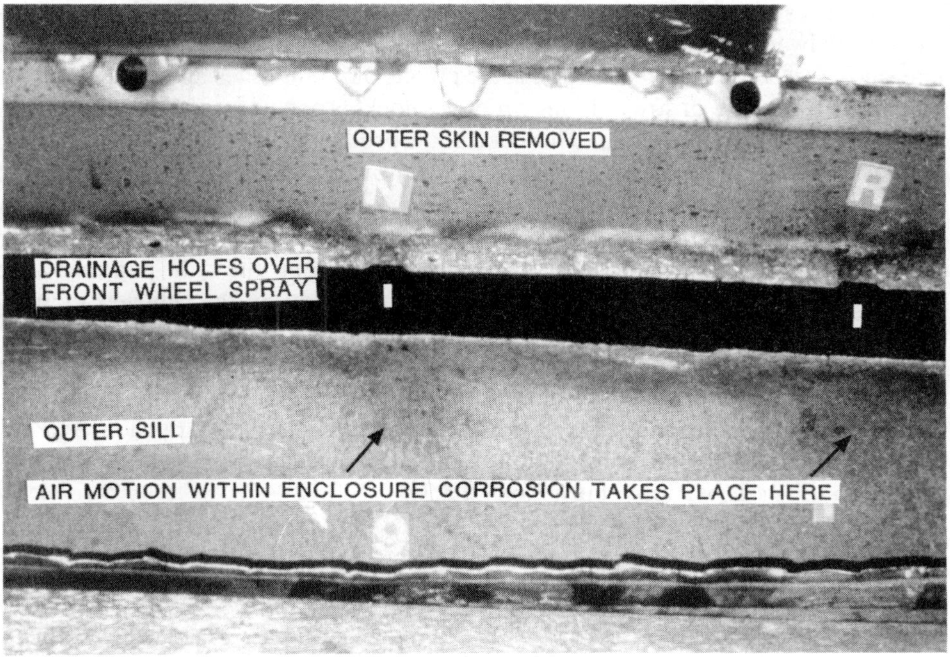

Plate 2.4(c) — A close-up photograph of the road dust depositing on the thin electrocoat (Section 3.2) and causing corrosion. The two holes are primarily to allow the electrocoat into the sill/rocker. These holes are linked into the passenger compartment (Plate 2.5) and so drag air through the sill from the air over the front wheel track.

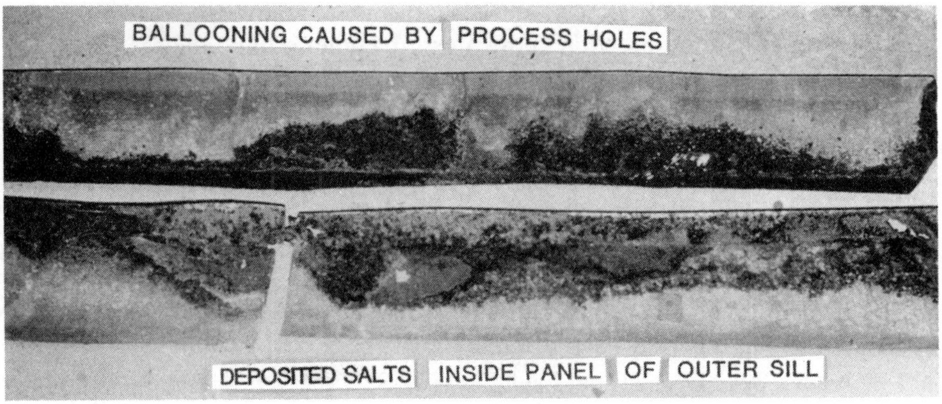

Plate 2.4(d) — A very expensive motor vehicle will have 17 coats of paint on the outer surfaces, but only one on the inner surfaces, but because these drainage holes are still a feature it will suffer from the sill/rocker corrosion brought about by the mechanism postulated in this book.

2.9 PRESSURE EDDIES

The complex structure of the motor vehicle body is far from streamlined. The glossy brochures always provide evidence of smoke tests over the central part of the vehicle. The smoke tests would not show the same effect round the sides and underneath of the car. This absence of streamlining produces pockets or currents of air which move contrary to the main flow of air. This contraflow produces pressure eddies which can deposit road dirt on surfaces which face the rear of the vehicle. A good example of this is the necessity for a rear wiper blade on tailgates to remove the road dirt deposited by this contraflow owing to pressure eddies. The pressure eddies cause turbulence which results in particulates impacting along the side panels of the vehicle. As shown with the tailgate wipers, this can be out of the vehicle's induced wind as a result of its speed and direction.

Pressure eddies are also produced along a vehicle body both by overtaking vehicles and by vehicles passing in the opposite direction on two-lane roads. The front of the vehicle produces a high pressure region, while the rear produces a low pressure region. This has considerable influence on the ingress of a salty micro-climate to the body sections (Figs. 5.6 and 5.7).

SUMMARY

(1) Road dirt thrown off the tyre from any point will impact the car body at the vehicle speed. Road dirt is thrown forward of the vehicle and will collect over lights, air vent pipes to the petrol tank and inlet hoses and associated ledges. These ledges can be out of the line of direct spray from the vehicle wheel and so are not 'naturally washed' by low speed spray. The areas where deposition takes place can be inaccessible to steam cleaning jets and associated water jets.

(2) The faster the vehicle speed, the less water is associated with the dirt and the more the dirt is compacted, i.e. the compaction can be likened to the use of a steam roller to compact moist road dirt (Section 10.10). The dried 'moist' road dirt is therefore self-compacted by the small amount of road water which is associated with each road dirt particle (Fig. 10.7).

(3) The faster the vehicle speed, the less is the momentum of the road dirt and the more likely it is to alter direction in the pressure eddies around the car.

(4) The vehicle travelling at speed produces various pressure eddies which result in turbulence beneath and alongside the vehicle. It is the turbulence which causes the micro-climate to build-up in box members.

2.10 THE MOTOR INDUSTRIES' ACCELERATED CORROSION TESTS

The motor industries' accelerated corrosion tests bear no resemblance to the conditions which the vehicle meets on the roads. Their accelerated tests (Section 10.22) involve driving the vehicle through a salt ford or dip at low speeds (20 mph) and then placing the vehicle in a humidity chamber and allowing the vehicle to 'sweat' for several hours. The large diameter spray from this 20 mph encounter does not produce the (low momentum) fine particles of spray which will be 'self compacting' on out-of-sight (or out-of-trajectory) surfaces (Fig. 4.6) or ledges (Plate 2.1). The high momentum (20 mph) droplets will not be drawn through the manufacturer's process drainage holes (as will be discussed later) by the small pressure changes around the vehicle. This procedural test therefore leads to the acceptance of bad design faults which lead to premature failure by corrosion as will be described in Chapters 4 and 5 and illustrated for example by Plate 4.15(c).

2.11 MANUFACTURER'S PROCESS HOLES

The vehicle body is spot welded together (Fig. 3.1) and painted (electro-coated) in big baths by total immersion (Plate 3.1). The paint is deposited on the metal substrate by electrolysis (Fig. 3.3 and Chapter 9). The paint must enter all box members (for the car body to sink) and, when leaving the bath, the paint must run out to reduce 'carry-out' (Plate 2.5). For this to happen, trapped paint or air must escape.

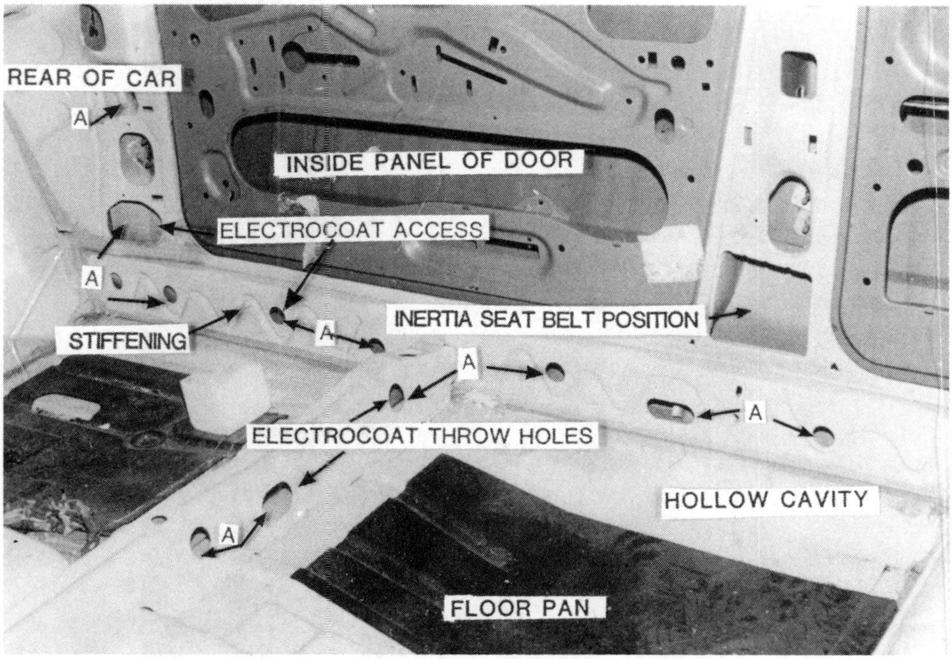

Plate 2.5 — A car body with the internal decorative panels removed, showing the holes which are required to achieve better electrocoat throw. These process holes —A— link the passenger compartment (via the sills/rockers) to the dense water spray over the front wheel track. The inner sill panel is indented to improve stiffness.

Therefore all box members have paint and air admittance and exit holes (Plate 2.6(a)). The paint has another property which is unlike the general public's understanding of a paint film. Immersion in the water-borne paint does not produce an adherent film on the surface unless it is chemically deposited by an electric current. Therefore, for the paint to be deposited on the internal surfaces of boxes or enclosed members, electro-coat 'throw holes' are required as shown in Plate 2.5.

Their function is to allow the electric current to deposit the paint upon the internal surfaces by electrolysis. Hence the passenger car compartment is linked to the external pressure eddies beneath the car by these process holes (electrocoat throw holes, paint drainage holes and air displacement holes) (Plates 2.5 and 2.6(a)). In order to investigate how these process holes allow the aggressive micro-climate to become a 'free rider' for the vehicle's life, it is best to examine the vehicle in motion during the winter environment. (Chapter 10). A further discussion of electrocating is given in Chapter 9.

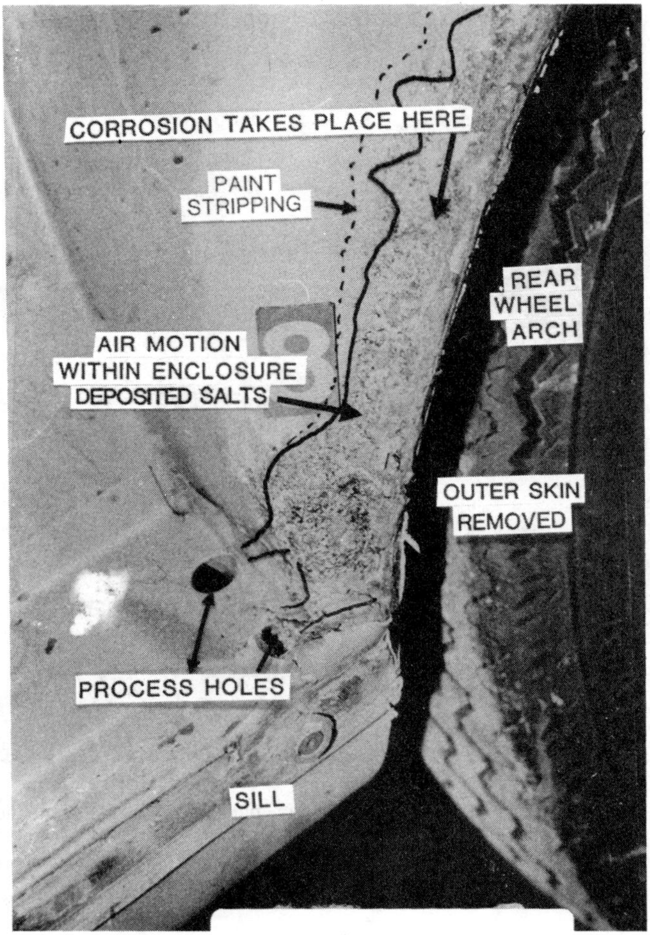

Plate 2.6(a) — The outer panel of this rear wheel arch has been removed to show the electrocoat process holes which perform the following functions: (a) an air exit (for immersion of the car body in the electrocoat bath) (b) an air admittance hole (for when the car body comes out of the bath) (c) a paint drainage hole. The air motion over the wheel arch is clearly shown by the corrosion. This air has been drawn from the front wheel track, through the sill/rocker drainage holes into the sills/rockers and then through the drainage hole shown in this plate. Throughout its passage the low momentum spray has deposited the road de-icing salts on the internal out of sight panels. See also Plate 2.6(c).

Plate 2.6(b) — The outer sill/rocker removed to show the similar extraction of the front wheel spray, through the sill rocker drainage holes, and the deposition of low momentum particles over the rear wheel arch.

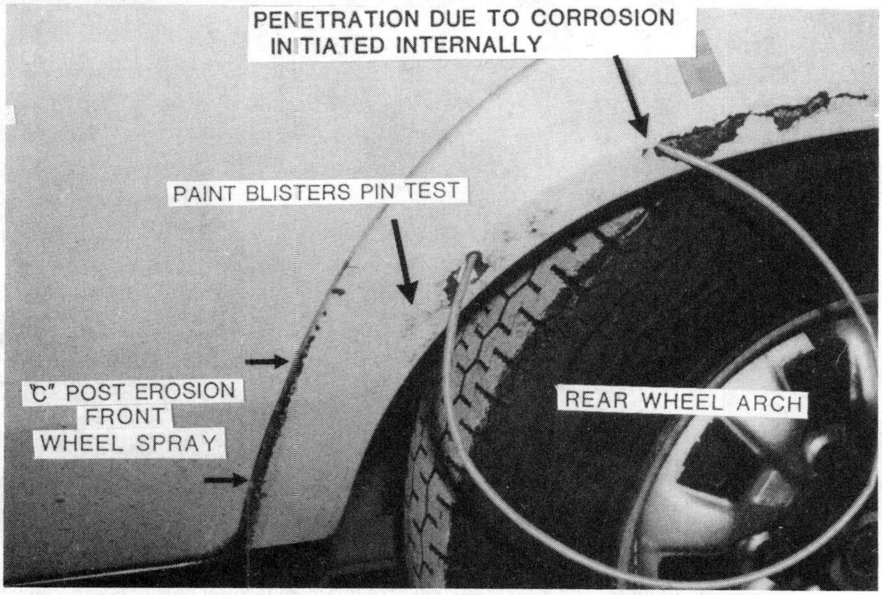

Plate 2.6(c) — The external appearance of the resulting corrosion which has been internally nucleated. Each one of these little paint blisters represents rust penetration and can be detected upon initiation by the drawing pin test (Fig. 2.11).

2.12 VEHICLE MOTION (60 mph=95 km/h; 120 mph=190 km/h)

Fig. 2.7 shows a car travelling at 60 mph (with respect to (w.r.t.) the road) and shows that water spray is thrown off from the top of the rotating wheel at twice the vehicle speed (120 mph w.r.t. the road) (see Fig. 10.8 and Section 10.12). If this velocity is made with respect to the vehicle, then the undersides of the wheel arches are constantly bombarded with road dirt and low momentum spray at the vehicle speed. This is important as design defects at low speeds do not emphasize the above point, as several manufacturers have found out (e.g. manufacturers have left open spaces over headlamp insets to the front wheel arch, others have used open (at the rear) sill–rockers which collect the forward-thrown wheel spray from the rear wheel) (Plates 2.7(a)–2.7(c)).

These former design defects are being overcome by the provision of full or partial internal wing–fender liners (stone guards) as shown in Fig. 2.9). Some of these of internal wing–fender liners are not designed correctly (B,C and D), and some corrosion-related problems will ensue owing to the accumulation of high speed spray at points X, Y and Z in Fig. 2.9 (see Plate 2.1(d)). Partial internal wing–fender liners will allow mud accumulations at RS (design C) and between N and P (design B) with the resultant corrosion which is associated with long-term mud accumulations (Sections 8.12 *et seq.*). This arises because the water thrown forward by the rotating wheel meets unstable or eddying air through which the vehicle travels. This is

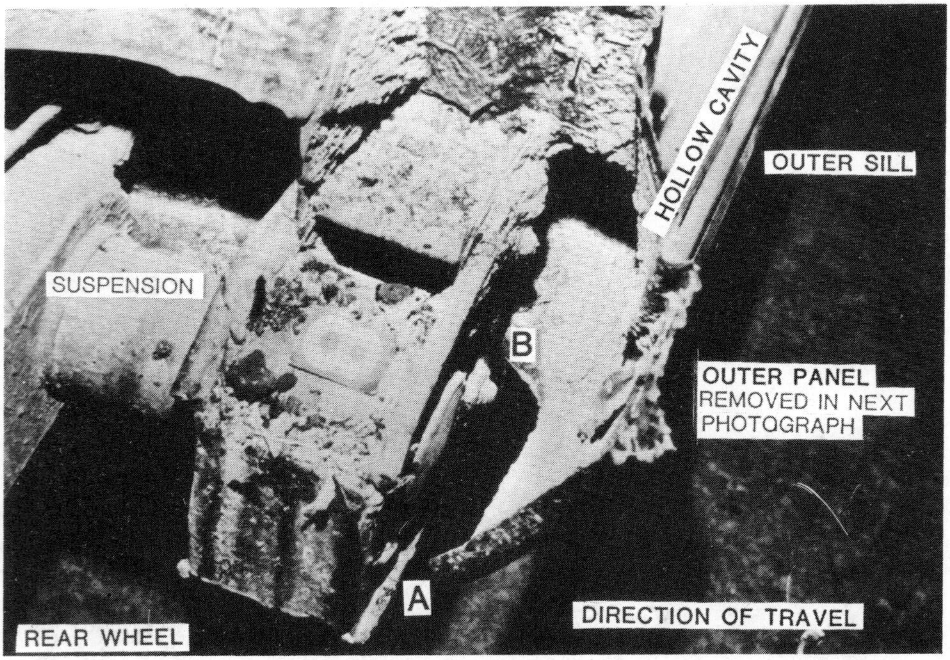

Plate 2.7(a) — The contoured (open) rear sill/rocker of this manufacturer did not take account of the low momentum particles . The higher momentum particles would be thrown out through the gap (AB) in the contoured sill, but the low momentum particles are drawn into the double panels to cause a corrosive microclimate in out-of-sight areas. See Plates 2.7(b) and (c).

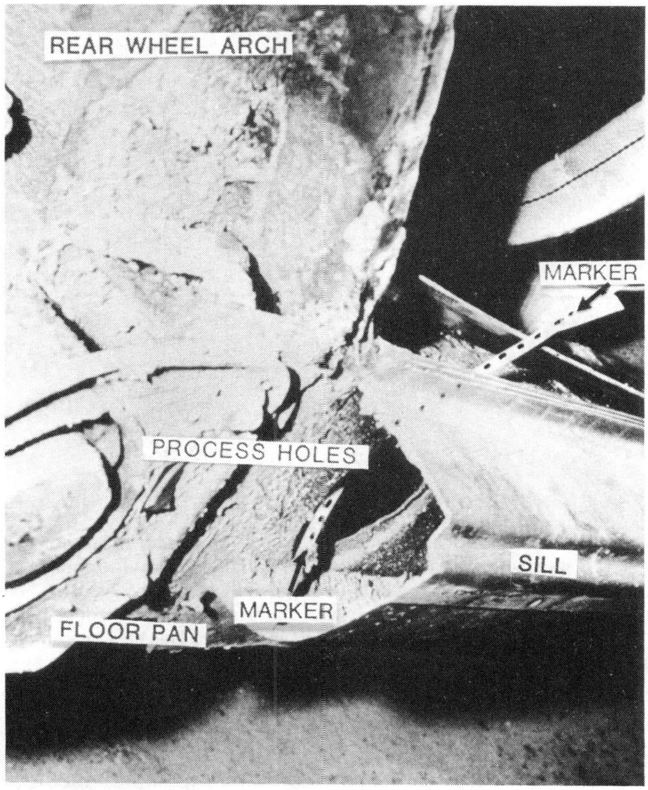

Plate 2.7(b) — 6 inches of the rear sill/rocker has been partially removed to expose the inner process holes which lead to a large cavity box member shown in Plate 2.7(c). This cavity member is joined pressure-wise to the passenger compartment by the air exit holes indicated by the plastic marker. This shows the linking of the passenger compartment to the forward thrown rear wheel spray. (Section 10.12).

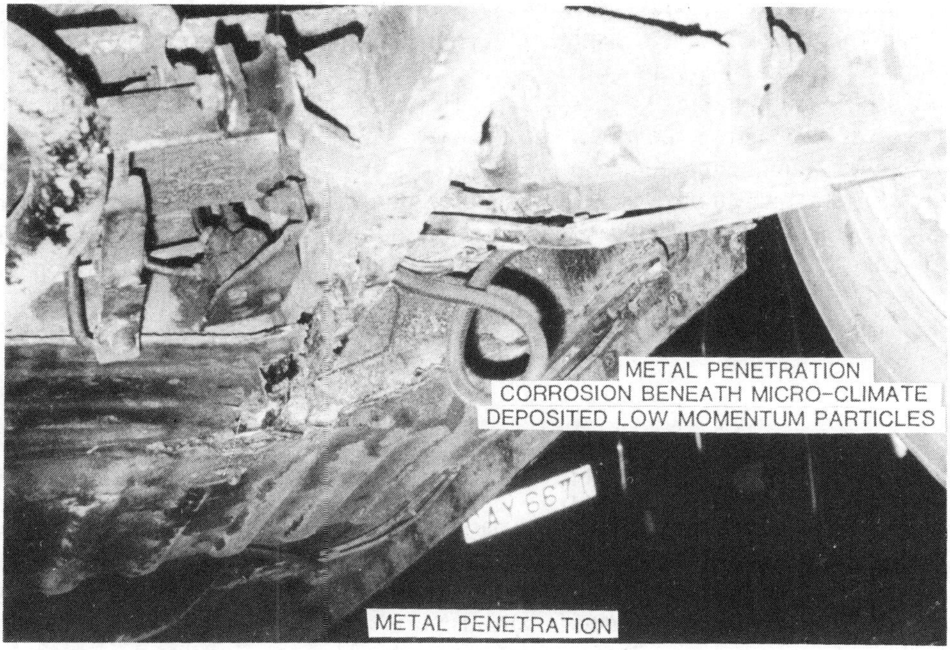

Plate 2.7(c) — The underbody of the vehicle showing the corrosion of the panel from the low momentum spray ingress.

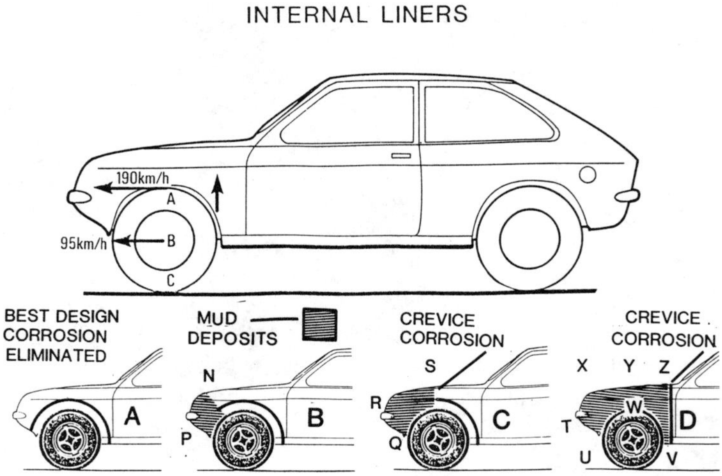

Fig. 2.9 — A series of four possible inner liners or stone guards to wings–fenders for the motor car. Only the type depicted in A is satisfactory as an anti-corrosion design owing to the water that is thrown forward of the wheel by the rotating wheel (Sections 5.8 and 10.12). Design considerations should incorporate adequate fixing points around the wing–fender in order to eliminate crevice corrosion between the metal–plastic junction or interface. This crevice can harbour road dirt and concentrate road de-icing salts.

particularly important in the cheaper cars (DL versions) which have higher specification models. Here air grills and other badly designed features may be slotted into the areas where for example the spot and fog lights of the GL version would be fitted. Plate 2.8(c) shows the air grilles which collect the preceding vehicles spray and deposit the spray within the front valance. This causes rapid corrosion of the area and is a gross oversight on the part of the designer.

Other manufacturers have left open process holes in the front of their vehicles (Plate 2.8(a)) which have picked up the preceding vehicle's road spray and caused corrosion at the front valance adjacent to the bonnet–hood. Other manufacturers have not considered the question of road spray at all and have left open drainage holes over the front wheel track (Plates 2.8(b), 4.3(a), 4.3(d), 4.8(a), 4.14(a), 4.14(b), 4.15(a), 4.24 and 4.25).

2.13 PASSENGER COMFORT AND STALE AIR EXTRACTION

Fresh heated air is introduced into the passenger car by a ram effect as a result of the car's forward velocity, or by an electric fan. The amount of air entering by this means can be controlled internally by the occupants. Stale passenger air is cotinually extracted through vents located in various parts of the vehicle (rear tailgate, rear body or doors) (Section 4.7). The amount of stale passenger air extracted cannot be controlled at all and is dependent on the vehicle speed. Hence, if the admittance of fresh air is reduced at high speeds, the extraction system will still operate as efficiently and lower pressures will be found within the passenger compartment and internal body cavity (box) members (this air continuity path is a requirement of the electrocoating processes as discussed earlier in Section 2.11). This means that replacement air (which includes low momentum road dirt and de-icing salts) is drawn through both the bottom of the doors (Plate 2.2(a)) and the sill–rockers (Plates 2.4 and 2.9) and rear quarter-panels (Plates 2.6 and 2.9(b)). The air drawn through the doors and sill–rockers originates from the regions of densest spray as these process holes to the doors and sill–rockers are directly over the front wheel track. This mechanism of building-up the aggressive microlimate on internal (out-of-sight) body panels is not shown up by the manufacturer's low speed salt dip tests (discussed in Sections 2.10 and 10.22). Another mechanism by which the micro-climate enters the vehicle is discussed in Section 2.9 and Chapters 4 and 5 (Figs. 5.6 and 5.7).

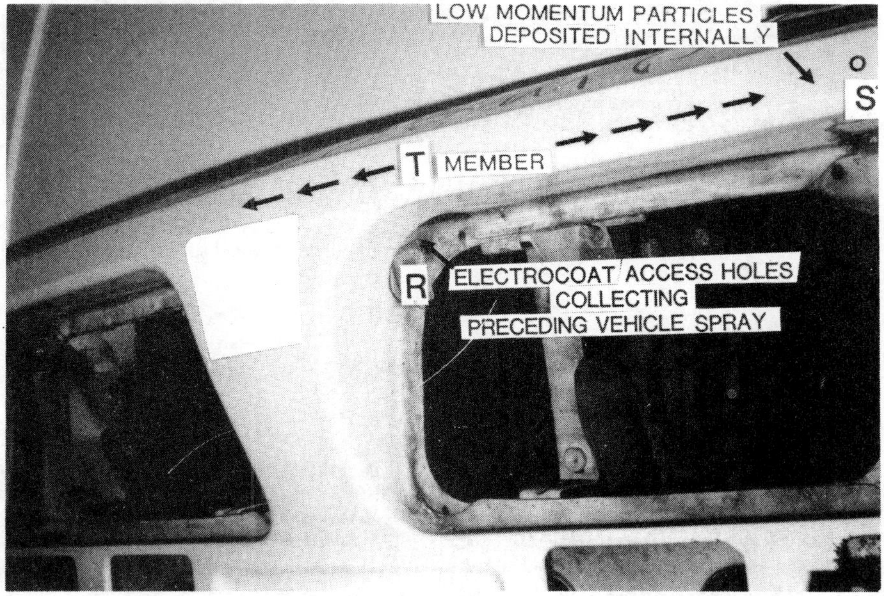

Plate 2.8(a) — An example of poor anti-corrosion design can be seen when the grill is removed from this front bonnet valance to expose the combined forming and process hole (R). Fine water spray from the vehicle ahead or mist or fog but *not* rain will be collected and deposited within the T box member. This collection causes crevice corrosion at either end of the T where the box member is constricted and large metal overlaps occur. Another electrocoat process hole, S, is located on the low pressure side of the box cavity at the two ends of the T member which supports the rear opening bonnet/hood. The original purpose of these holes is to allow the electrocoat to fill the cavity box member. However, after electrocoating, good anti-corrosion design dictates that the hole on the low pressure side should have been filled with a rubber grommet. Good anti-corrosion design also dictates that the flange R should have been cut away so that it did not produce a member at right angles to the vehicle direction which only serves to generate low momentum spray (Fig. 4.2). This member R ensures that the preceding vehicle spray is collected into the box member.

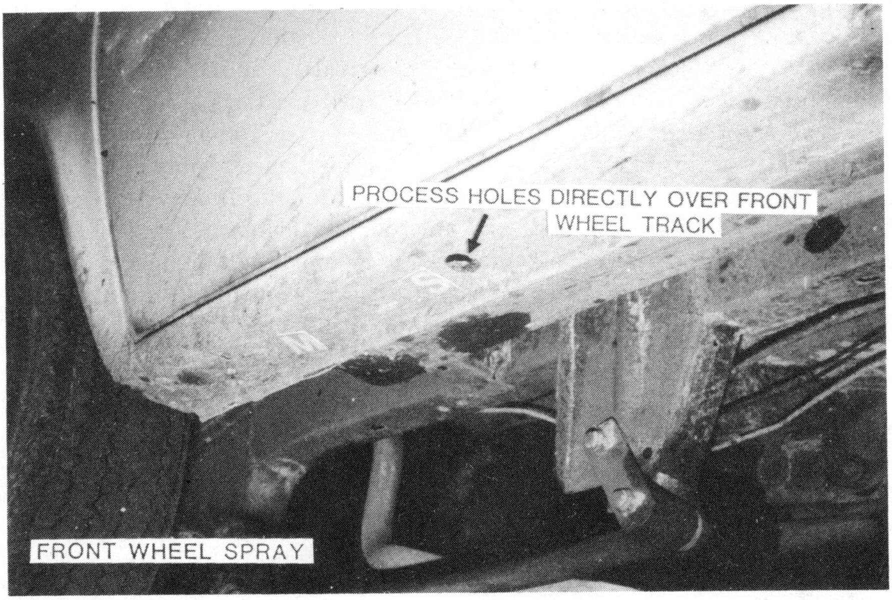

Plate 2.8(b) — A drainage hole placed directly over the front wheel track ceases to act as a drainage system as it becomes full of road dirt. This should be fitted with a one-way valve (Fig. 5.11). A cheap DIY flap valve is illustrated in Fig. 5.19, Section 5.19. Mudflaps would also prevent the ingress of the road dirt from the front wheel spray.

Plate 2.8(c) — The grills allowing the preceding vehicle spray to enter the front box members forming the structural energy absorbing members.

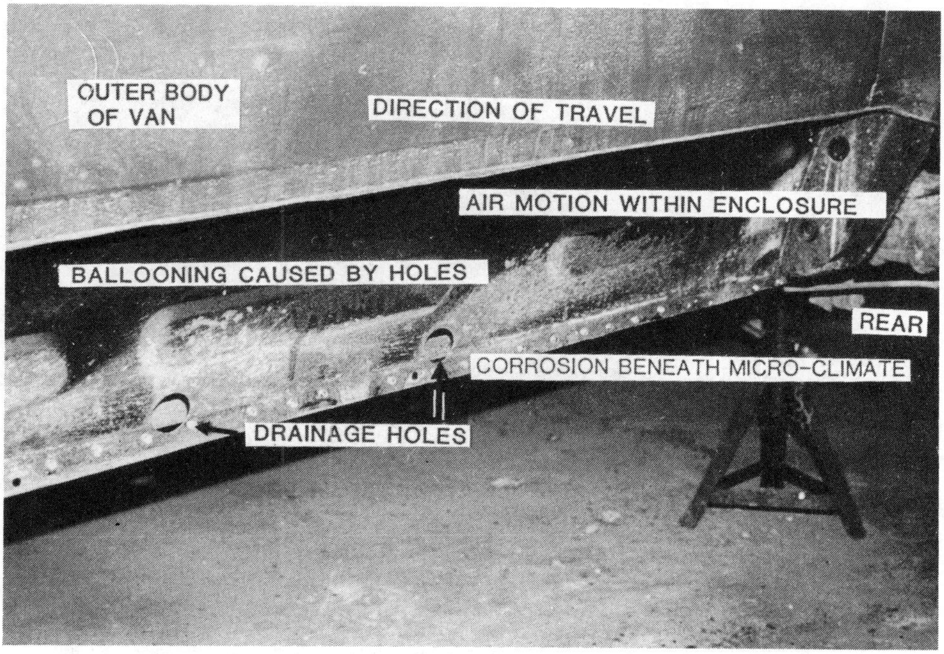

Plate 2.9(a) — The outer panel of this commercial vehicle shows the air motion within the double skin. The holes shown are the process drainage holes. These are larger to remove the greater volume of paint and to obtain the electrocoat throw into the sill/rocker. These holes are to increase the electro-coating within the enclosed panels. Good anti-corrosion design dictates that these should have been protected by a one-way flap valve.

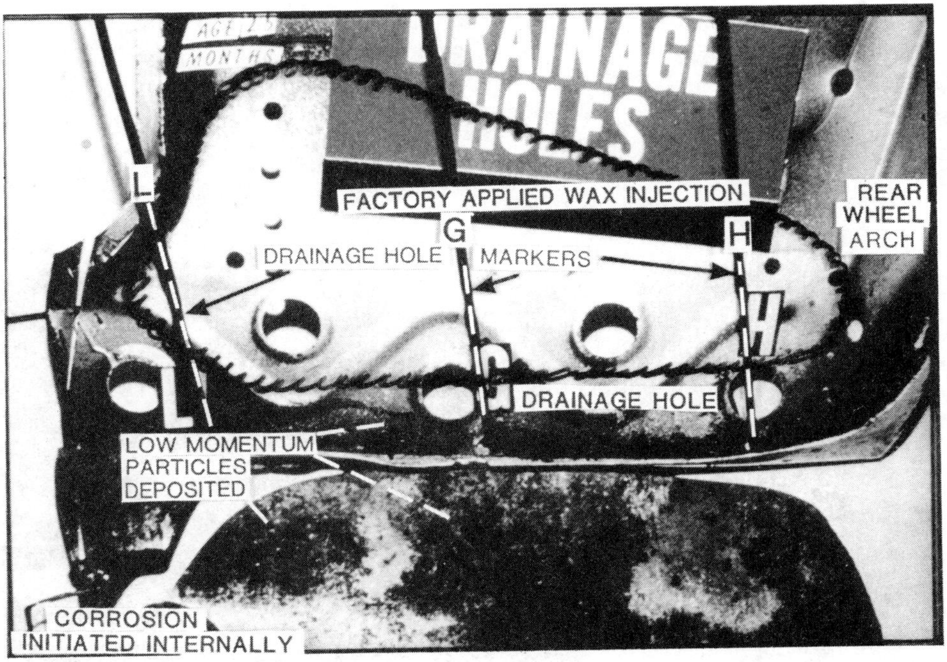

Plate 2.9(b) — The rear quarter panel completely removed from a two-year-old hatchback shows a similar effect as regards air motion within this cavity member. The rear wheel of the car is towards the right of the photograph and the electrocoat process hole, L, is at the base of the B pillar. The outer panel is laid down and occupies the lower half of the photograph. Black Tubes, L and H, go through drainage holes to the outside of the sill/rocker (and over the front wheel track). Larger holes —L,G and H— connect one side of this double cavity member to the other side and these holes have shown the air movement within this cavity by the corrosion which has taken place over the two years of service life. The brighter —silvered— area, which is outlined, is the manufacturer's on-line wax injection system which performs, in this case, a useless function as it is not applied to the interweld joint. The waxes on this vertical surface will ensure that any hygroscopic salts will run down into the interweld when they have taken up sufficient moisture from the atmosphere. See Fig. 3.7 and 3.8, If the waxes are thixotropic Section 3.5/6, Section 8.7 and 8.8) they –freeze– upon deposition and do not penetrate narrow crevices. This results in crevices which concentrate the road de-icing salts within them (Fig. 3.7). There is a tendency for motor vehicle manufacturers to use thixotropic waxes if the body is wax injected before the final body build is undertaken. The reason for this is that dripping waxes would create an unacceptable environment for the body build plant and their operatives.

2.14 THE CORROSION REACTION AND BODY BUILD

The evidence of corrosion and the associated micro-climate on internal panels produced from low mileage vehicles which have been involved in accidents has shown that the corrosion is internally initiated upon body sections as a result of this layer of road 'dust' which contains the road de-icing salts. This layer is deposited upon a very thin film of electrodeposited paint (Figs. 2.6(a) and 3.5(b) and Sections 9.6 and 9.7) which soon fails and allows corrosion to proceed, with the owner being completely unaware of the process. The internally initiated pitting corrosion proceeds until rust-through or a paint blister is the external manifestation.

The corrosion mechanism responsible is the differential aeration cell, which has already been discussed in Section 2.4. If the principles are applied to the salt particle deposited upon the internal panels of the car body, rapid pitting corrosion will ensue because, as the corrosion front gets deeper into the metal, so the oxygen availability alters. There is a bigger 'battery' effect as the difference in oxygen replenishment becomes established. The areas which are dissolving are those regions which are denuded of oxygen replenishment. These areas can be the following.

(1) Interwelds which weld together the component body parts of the vehicle to make the stress-carrying body (see Figs. 3.9 and 3.11) or two body panels forming a crevice (Figs. 3.1 and 3.9). Fig. 2.10 shows an overlap produced by a bolted section of the motor vehicle, and some of the ways of overcoming the action of the crevice as a 'hidden' battery. See for example the comparison with Plates 4.1(a), 4.1(b), 4.3(b), 4.3(c), 4.4 and compare this with your own vehicle model.
(2) Areas beneath the salt–dirt particle (Fig. 2.6).
(3) Areas at the front of a corrosion pit or rust-through area (Fig. 2.6(d) and 2.6(e)).
(4) The junction interface of screw threads and nuts, or the metal overlap where two metals are bolted together (Fig. 2.10).

These oxygen-deficient areas dissolve at the expense of those areas which are plentifully supplied with oxygen and which are usually outside the inter-weld, overlap joint, crevice or pitted area. Note: one cannot have a denuded oxygen area too far away from the oxygenated area unless there is a good aqueous conductor of electricity (Section 10.1). The de-icing salts do not cause corrosion; they catalyse the corrosion reaction for the following reasons.

(a) They are hygroscopic (Section 10.5) and attract water vapour from the atmosphere and hence maintain a moist environment around the dirt particle for a longer period of time.

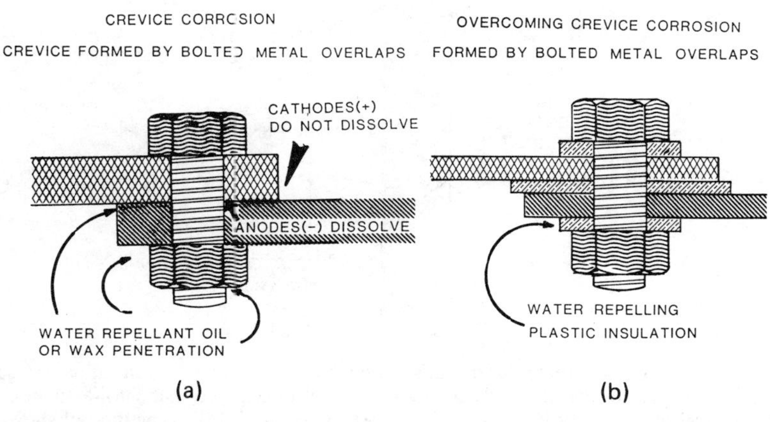

Fig. 2.10 — Crevices (see also Fig. 3.9), can be formed whenever two metals are joined. The example shown here is a bolted system (eg the bolting on of wings). The corrosion taking place at the least oxygenated regions is the reason that nuts seize on studs. One possible method of overcoming the crevice problem is to have a water-hating interface between the metals and threads. Plastic washers, oil, wax or any other hydrophobic material will suffice for the threads and junction.

(b) They provide a good conducting solution (called an electrolyte; see Sections 10.1 and 10.2) which is a basic requirement for the interaction of the two electrode reactions (Sections 6.14 and 7.2). This produces the product which is called rust (Section 6.14).
(c) The chloride ion makes the product of the dissolving area more soluble (Section 10.2).
(d) When drying out, the salt concentrates at the base of the road dirt particle or in the inter-welds (Fig. 3.7) which have restricted drying abilities and as explained above are already the dissloving areas as they suffer from an oxygen deficiency (Sections 2.4 and 6.14).

2.15 RECOGNIZING THE EARLY SIGNS OF METAL PENETRATION AND DISTINGUISHING IT FROM SURFACE-INITIATED CORROSION

Stones impacting the car body will often cause the paint to chip (Fig. 2.6(f)) or to become non-adherent to the metal Fig. 8.18). This will often form a paint blister (Sections 8.12–8.14). A similar blister is first formed by metal rust-through. The way to differentiate between the two blisters is to consider the liquid which 'weeps' from the blister. In the metal penetration case this liquid will contain a large proportion of chlorides and this will cause 'rust staining' on the paintwork below the blister. Another way to differentiate between the two, which is destructive, is to place a sharp point (e.g. of a drawing pin) in the centre of the paint blister, which will roughly coincide with the pit front, and to determine whether the point can be pushed through the brittle oxide (Section 1.17) (Fig. 2.11).

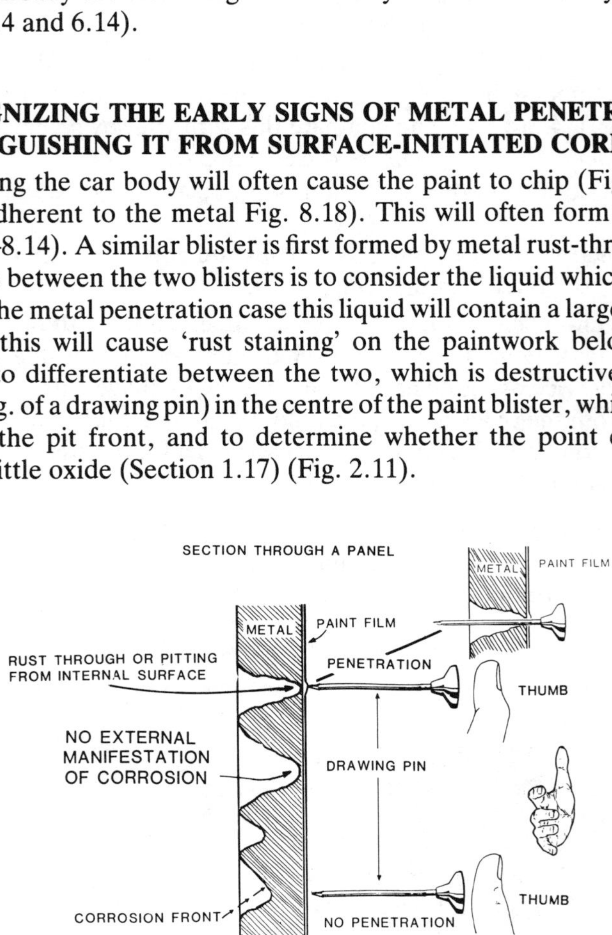

Fig. 2.11 — The problem that most motorists and garages face is whether the apparent 'surface corrosion' is in fact surface initiated and not due to 'rust-through' from the inside to the outside which is much worse. This is termed 'pitting' (see also Fig. 7.14). A method of showing up pitting corrosion is shown in a cross-section through the metal panel. A drawing pin is placed upon the suspect paint blister. If the blister is due to metal perforation by pitting, then the high stresses at the tip will puncture the rust, which is a brittle oxide, and so has little or no resistance to the high stresses which are produced by the point of the pin. Metal panels of the gauge used in motor vehicle production will not be punctured by this process, however hard the drawing pin is pressed.

The pitted metal, if corroded through, will contain the brittle oxide of rust (Section 6.14), which cannot withstand the very high point loading caused by a drawing pin pressed onto the paint excrescence (Sections 1.16 and 1.17). With moderate pressures applied to the drawing pin head, the loadings on the point tip of the pin will be much higher and will exceed the fracture strength of the oxide. 'Metal' penetration of the pin will therefore be achieved. If the pin does not penetrate the surface, after several placements of the drawing pin, then the paint blister is not due to metal penetration. This paint blister has been caused by external means, such as car-park knocks (adjacent doors), stones thrown up by contraflow (or overtaking) traffic (Fig. 5.6) or simply the pressure eddies produced around the vehicle, which cause the smaller road particles to impact the vehicle panels at right angles to the direction of motion. A metal will always be able to withstand the high pressures placed upon it by the point of the drawing pin. This test will cause no damage to the surface nucleated corrosion blister (other than puncturing the paint film). However, this is not disastrous because the paint film will let enough water through the film to allow the steel to corrode at the rate it would do without a paint film in place. See the discussion of water penetration through paint films (Section 8.13) and the method by which a paint film protects the steel surface in Sections 8.12 *et seq*.

3

Protection schemes

3.1 PRETREATMENT PROCESSES

The modern body is made with the primary thought on 'performance' in both fuel consumption and acceleration figures. This means that the body and associated power train must be as light as possible. The design requirements for this mean that the 'unitized' method of construction is preferred over the older conventional 'frame' construction. In the unitized construction there is no 'chassis' (Fig. 3.1).

The car body, as a unit itself, provides the three-dimensional strength from panels whose stiffness is increased by their shape (Sections 6.35 *et seq.*). The unitized car body is formed from preformed panels which are spot welded together to form an integral rigid unit termed a 'body in white'. In order to make an (organic) paint film mechanically adhere to the oily steel surface a pre-treatment, degreasing, and surface transformation process must be applied. The pre-treatment process consists of a spray degrease, acid etch or 'pickle' to produce a phosphate film. The most important part of this pre-treatment process is the phosphate film, whose function is to give a good 'key' for the subsequent paint film and to contain any (filiform or scab) (Fig. 8.21) corrosion due to stone chipping to a localized area and thus to prevent 'undercutting' of the paint film and the spread of corrosion. The phosphate coating (Fig. 3.2, Plate 9.1) is generally applied by an acidic phosphoric acid spray, and so only the outer cosmetic panels are treated to the phosphate film and degreasing processes. More modern pre-treatment processes are applied by a dip and spray process. For further technical details see Sections 9.1 *et seq.* and Sections 6.18 and 6.19.

3.2 ELECTROCOATING AND PROTECTION SCHEMES (See also Chapter 9)

Electrocoating (Plate 3.1) is a process very similar to electroplating, (Fig. 3.3). **Electroplating** is the process of applying a metal from a solution to a metal substrate by the passage of electricity. **Electrocoating** is the process of applying a paint film from a solution of that paint to a metal substrate by the passage of electricity. The difference between the two processes is that the film deposited by electrocoating is

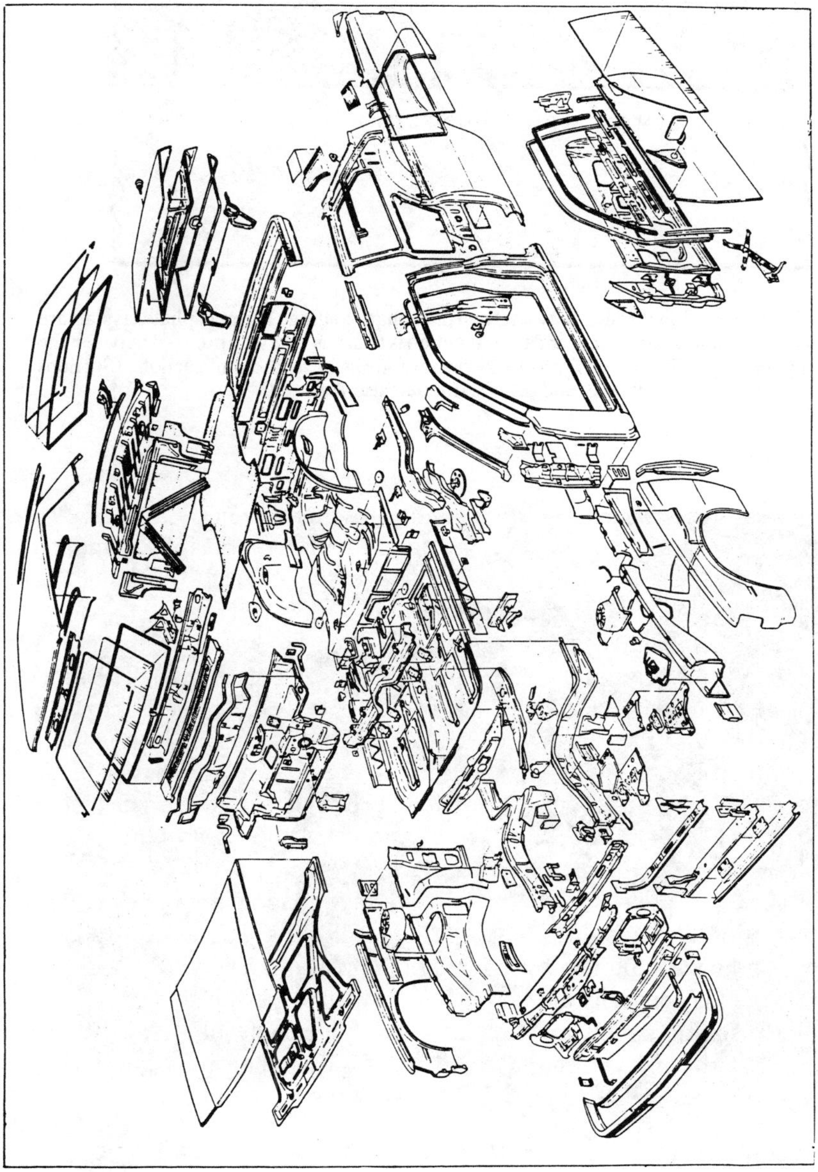

Fig. 3.1 — An exploded view of the body in white panels forming the vehicle body shell. Each panel is welded by electrical resistance spot welds (Fig. 8.7(a)) to its neighbour, thus producing a mosiac of crevices. (Figure contributed by Vauxhall Motors Luton by courtesy of General Motors.)

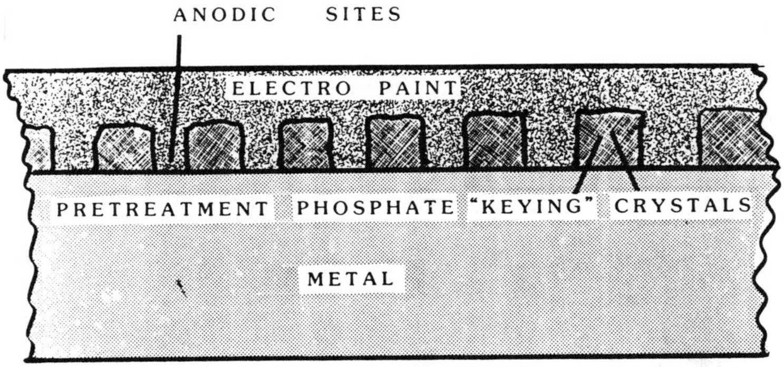

Fig. 3.2 — A cross-section through the metal and phosphate coating. The phosphate crystals are deposited from a complex acid etch, and their function is to act as a 'keying process' between the metal and paint film. The reasons for this keying are given in subsequent sections (Sections 6.18, 6.19 and 8.11–8.16). See also Plate 9.1.

Plate 3.1 — A car body entering the electrocoating bath on a conveyor belt. In order for the car body to sink, air must not be trapped in hollow box sections. Conversely on emerging air must be admitted to the hollow box sections as the body comes out front first in order for the paint to run out slowly and uniformly, without causing turbulence. Hence there are air admittance holes at the rear and front section of the horizontal box members. Notice that the car door, tailgate and bonnet–hood are slightly ajar. Plate contributed by Automobiles Citroën and acknowledged with thanks.

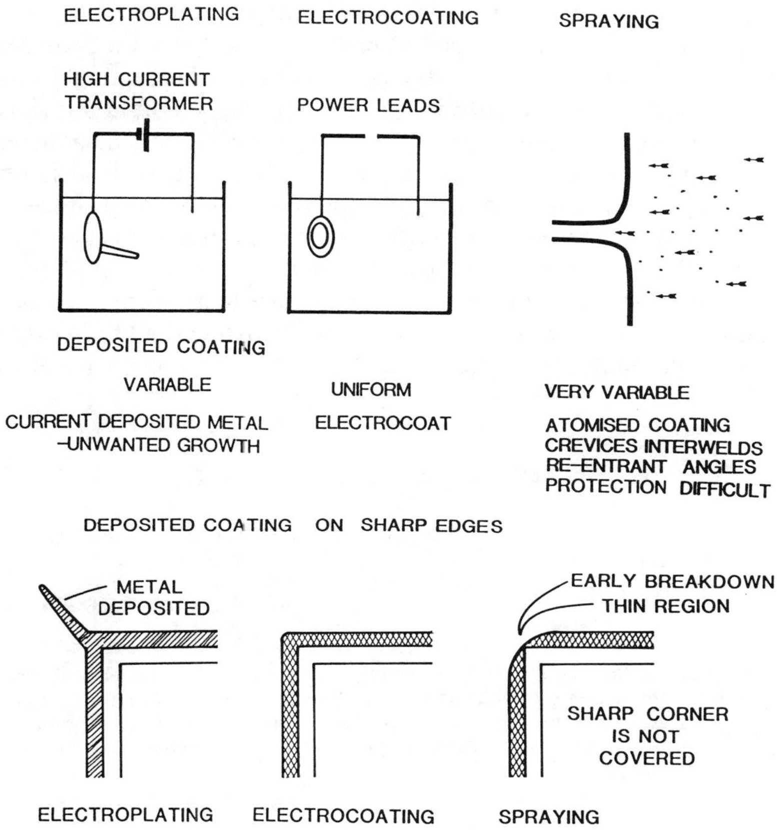

Fig. 3.3 — Comparison of (a) electroplating, (b) electrocoating and (c) conventional liquid spray application processes. (a) 'Electroplating techniques tend to produce a thicker product at the corner of the sheet metal as a result of the high current density which can be obtained. Chemical additions have to be made to the electroplating bath in order to stop this unwanted growth. (b) The electrocoating of paint films produces uniform films on the metal substrate because the deposited paints are not electrical conductors. The deposited paint film acts as its own 'thickness regulator'. (c) Conventional liquid application processes (e.g. spraying and baking), in contrast, produce thinner films on corners owing to the surface tension forces which pull the liquid back from the sharp corners. The thickness of paint film is zero on surfaces which lie in the direction of spraying. This is the reason for the mastic application given in Figs. 3.11 and 4.4. Electrocoating of paints is very much like electroplating of a metal in that there is no paint or metal deposited if no current flows. They differ in that electropainting acts as its own thickness regulator, whereas this is not true for the electroplating of metals where the metal plate is a conductor of electricity.

non-conducting and so the coating acts as its own thickness regulator. Electrocoating is the most foolproof way of painting an intricate shape (made from a metallic conductor of electricity), as there is no 'shadowing' as there is with a conventional spray gun. The electroplating of a metal, in contrast does not act as its own thickness regulator. Electropaints are formulated so that the paint is a large singly charged entity (macro-molecule), which is suspended in water. Electrocoating processes are subdivided up into cathodic and anodic electrocoats (see Chapter 9 *et seq.*). Anodic electrocoats have given way to the cathodic electrocoats in which the car body is made the cathode in a large 'electroplating bath'. Electrolysis of the paint causes the macro-molecule (Section 9.8) to lose its electrical charge whereupon the large paint unit is deposited, initially as a powder, which can be rubbed off at this stage in the process. This powder is then baked in order to form a coherent and adherent paint film. This is unlike the process of electroplating, where the metal is deposited metallurgically bonded to the substrate and does not require any heat treatment to make it adherent and coherent to the surface upon which it is deposited. The baking process melts the electrodeposited powder and the viscous liquid produces a coherent film over the whole surface. However, surface tension effects mean that the film thickness is not uniform (Fig. 3.4). This electrocoating is all the 'painting' that internal and underbody panels are given (Fig. 3.5). Underbody panels are also

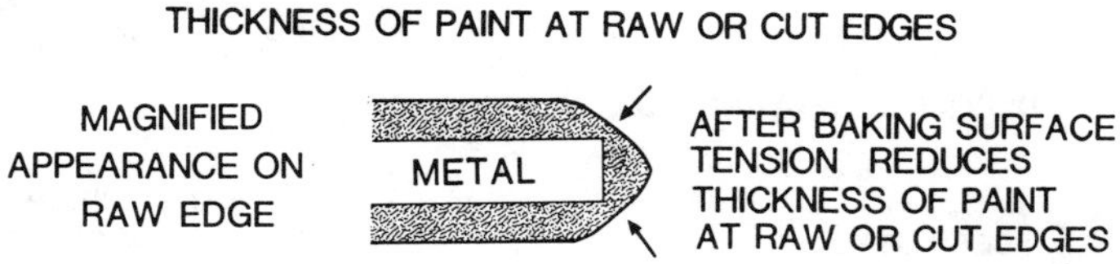

Fig. 3.4 — A section through a 'raw edge of a motor vehicle panel. Note the reduced thickness owing to surface tension effects which tend to produce curved surfaces (e.g. rain drops are perfect spheres as a result of surface tension). The electropaint is initially deposited as a powder. Electrocoats are fused to melt the powder to form a coherent adherent film.

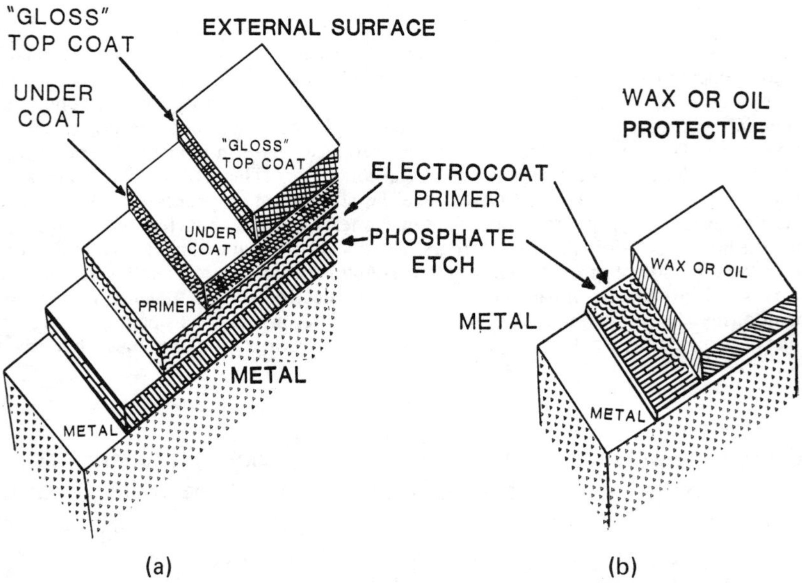

Fig. 3.5 — Comparison of the protective films on (a) the outside panels of the car body with those on (b) the inside internally 'boxed' panels. The vertical height is a representation of the film thickness.

covered with a chip resistant PVC (Fig. 11.5) plastisol coating adjacent to the front wheel tracks. The outside panels (Fig. 3.5) are given the finish coats only. When the painting of the vehicle is mentioned in advertisements and other brochure material, this point is not made and only the outside thicker coats are discussed.

3.3 THICKNESS OF ELECTROCOAT

The thickness of the paint film varies on the car body depending on how far away from the opposite (or counter) electrode are the panels in question. The thickness of deposited paint is related to the total electrical resistance between the metal panel and counter electrode, including the series resistance of the deposited electrocoat. The thickness of film is inversely proportional to the sum of (1) the resistance of the electrocoat electrolyte within the bath and the distance of the coated area from the opposite electrode plus, (2) the series resistance of any already deposited paint film (Fig. 3.3). Hence the inter-weld region is denuded of electropaint owing to the high electrical resistance that results from the narrow cross-sectional area of the electrical path within the inter-weld and the distance from the other electrode (see Plates 4.3(b) and 4.4). A similar effect would be observed in electroplating. Another similarity is the fact that there has to be a current flowing for the paint (or metal) to be deposited. Hence no current flowing means no paint. Like the metal plating baths-solutions, the paint remains in solution if no current flows. The terms used in electrocoating are similar to the terms used in electroplating. Hence the term 'throw' or 'throwing power' is used to describe how well the paint or metal is deposited at large distances from the counter electrode. The 'throwing power' of the electrocoat process is also used to describe how well the paint is deposited upon the internal surfaces of box members and inter-weld crevices.

3.3.1 Analogy of electrocoat throwing power

An analogy of the paint throwing process can be a stream of passengers entering an underground train system. As the passengers come down the steps and escalators, they must pass through a narrow door to get onto the platform. The narrow doorway is analogous to the small electrocoat throw holes. As the passengers pass through this narrow door way they can spread out along the platform facing the railway line (equivalent to the paint's being deposited). Human nature being what it is, they will stand with a gap between each passenger. The passengers will stand preferentially nearer the doorway. This is exactly how the paint is deposited; more paint is deposited nearer the electrocoat holes than further away from these holes. There will be more queuing and double and triple queues nearer the doorway when perhaps there are gaps (holidays) between the passengers further down the platform. As more passengers enter the platform, these 'gaps' will be filled. Slowly the passengers will all stand against the line. Further passengers will then start to queue behind the first arrivals. This is analogous to the paint film's thickening up. In most cases the vehicle is removed from the electrocoating bath before all the 'holidays' within the internal coat or box members have been filled.

3.4 PRESSING–SLUSHING OILS IN INTER-WELDS

The rolled sheet from the steel mills is protected by slushing oils (see the discussion of temporary protectives in Section 8.8). Their functions are to lubricate the metal during cold rolling and to prevent corrosion before the steel sheet is used in its final form. Slushing oils are also used in panel forming operations. The panels are not degreased before the panels are welded to form the car body (now called the 'body in white'). Hence these oils become trapped in the inter-weld, and being hydrophobic (which means hating water) they will not allow the water based electrocoat into the interweld crevice (see Fig. 3.6, substrate B). The body in white is degreased before electrocoating by an alkaline water-based spray (Section 3.1), but this degreasing alkaline spray does not remove the pressing oils from those inter-welds not exposed to the spray. As the alkaline degrease is water based, it is also excluded from the inter-welds and therefore does not penetrate and degrease the inter-welds. This means that on some occasions (e.g. tight inter-welds), there can be no deposition of the paint within the inter-weld owing to the fact that the water-based electrocoat solution–dispersion cannot displace the slushing oils within the 'tight' inter-welds.

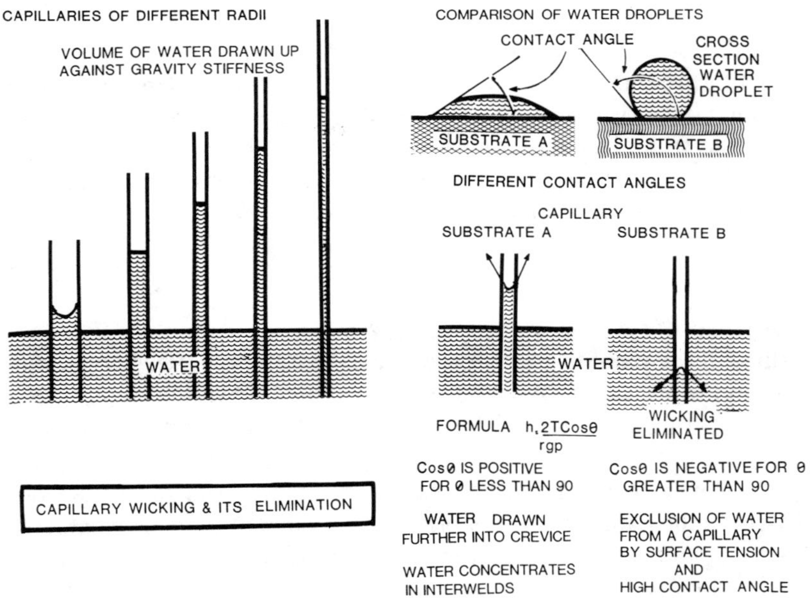

Fig. 3.6 — Capillary rise is inversely proportional to the crevice gap or capillary radius r. However, negative heights and therefore capillary exclusion can be obtained by using hydrophobic substances. (h = height; r = radius of capillary; g = acceleration due to gravity; ρ = density of liquid; T = surface tension of water or liquid in question; Cos ϕ = cosine of the contact angle of the liquid with respect to the substrate). Penetration of a liquid into a crevice depends on the angle of contact which that liquid makes with the 'coating' which is on the walls of the crevice capillary. Cos ϕ is negative for contact angles of greater than 90°.

Hence there are two factors working against this mass production painting technique.

(1) Hydrophobic slushing oils within the inter-weld prevent the water-borne electrocoat from penetrating the inter-weld crevice.

(2) If there are no slushing oils within the inter-weld and the electrocoat penetrates the inter-weld, then a very small current flows within the inter-weld crevice, (Sections 3.3 and 10.1) and little or no paint is deposited behind for example the door hinge (Plate 4.4) or within the clinch flange (Plate 2.2(a)).

3.5 WAX INJECTION (see also Sections 8.8–8.10 and 8.18)

An electrocoat paint film (Fig. 3.5) was all the protection given to the following areas: floor pan, internal panels of doors, sills–rockers, bonnet–hood etc. Because of the long electrical path in the electrocoat bath to these panels, a very thin electrocoat cover was produced. For added protection on the internal box members and areas where there is little thickness of electropaint owing to the poor throw of the electrolyte, the motor industry has undertaken a wax injection (Sections 8.8 and 8.9) programme of all the internal sections in restricted areas. The thickness of wax sprayed on is 0.0016–0.0020 in (16 thou or 0.04 mm).

The penetration of the waxes into the narrow (capillary) interwelds should be the first requirement for the wax injection system (Fig. 3.6). This arises for the following reasons.

(1) The inter-weld is an area which is denuded of any oxygen replenishment when the inter-weld contains water. This area is therefore the region where the metal dissolves and suffers preferential attack when compared with the area outside the inter-weld which is plentifully supplied with oxygen.

(2) The inter-weld is denuded of electropaint owing to the high electrical resistance caused by the narrow cross sectional area of the electrical path within the inter-weld and counter electrode. Therefore the paint deposited within the inter-weld is of negligible thickness.

(3) The corrosion mechanism produces an acidic environment at the anode site (Section 7.9) which is located within the inter-weld, which promotes the aggressive nature of the environment within the inter-weld (see Sections 6.14–6.16).

(4) Hence the most corrosive region (the crevice (see Fig. 2.10, and Plate 4.23)) has the least oxygen replenishment and the least paint thickness (if any electropaint is deposited at all, despite the pressing oils trapped within them) and becomes acidic in nature when corrosion is initiated. This acidic nature (Section 7.9) accelerates the corrosion nature of the crevice or corrosion front.

In tests (undertaken by the author, with help from British Leyland Cars (Austin Rover) and ICI Paints, Slough) on samples hung from the front door window in anodic, cathodic and slipper dip processes, it was found that an inter-weld gap of 4 thou produced the most corrosive environment owing to the possibility of the carry-over of the hygroscopic (Section 10.5) (phosphate) pre-treatment salts (Sections 3.1, 6.18 and 6.19) and Chapter 9). (A hygroscopic salt is one that attracts water from the atmosphere and becomes wet. Our table salt is sodium chloride and the manufacturers of table salts add other compounds to make it 'free running', thus overcoming its hygroscopic nature.) However, in all the specification tests for wax protectives there is no test for the capillary penetration of the temporary protectives against gravitational forces, and yet this seems to be their main advantage in combating corrosion. The science behind capillary penetration will therefore be considered in Fig. 3.6. Further details are given in Section 10.8.

3.6 CAPILLARY WICKING AND PENETRATION

The following should be noted.

(1) The term 'wax–oil protective substance' applies to those substance which are listed as 'temporary protectives' in Sections 8.8, 8.10 and 8.18. The term wax will therefore be used as an abbreviation and is also intended to refer to those hydrocarbon oils used as protectives. The difference between an oil and a wax lies in viscosity and melting point only (Section 8.9 and Fig. 8.17). The term wax–oil will therefore be used to describe these products in this non-technical section. It is not intended to refer to any particular commercial product.

(2) Corrosion protection by barrier films (such as oils and waxes) is proportional to the thickness of the barrier. High viscosity fluids will give thick films on vertical surfaces, whilst low viscosity liquids will 'drain' off vertical surfaces and produce thin films.

(3) Capillary creep is controlled by the viscosity of the bulk fluid and the surface tension forces (Section 10.8). If the fluids are subjected to heating cycles to evaporate off the solvents, the capillary creep will also be dependent on the way in which temperature influences viscosity and the rate of evaporation of the solvents.

The interaction of a liquid, containing corrosion inhibitors, with a surface is a complex reaction of several interatomic forces. It is best expressed (see Fig. 3.6), by the expression

$$h = \frac{2T \cos \phi}{r\rho g}$$

where ϕ is the contact angle which can be from 0° to 180°, T is the surface tension force, g is the acceleration due to gravity, ρ is the density, h is the height of liquid in capillary and r is the radius of the capillary.

If ϕ is less than 90°, then the liquid will rise up the capillary, and in the case of the road de-icing salts will concentrate in the crevices whose contact angle is less than 90° (Fig. 3.7(c)). If ϕ is greater than 90°, then $\cos \phi$ is negative.

The wax–oil (as the liquids on application), need to see the painted car body as a substrate where ϕ is as near to zero as possible ($\cos 0=1$) and therefore the wax–oil will enter the capillary inter-welds. When the volatiles have dried out the wax–oil is required to coat the painted surface such that the wax–oil presents a surface whose substrate gives a contact angle of greater than 90° to water and saline solutions. Hence the saline medium will be excluded from the inter-weld even though a capillary crevice is still present (Fig. 3.8). This also means that the road de-icing salts will not be concentrated within the inter-weld capillaries and the saline solution will become more mobile as the water and wax–oil substrate mutually repel each other

(Fig. 3.8). The thickness of the wax–oil coating does not play a part in this water exclusion and protection mechanism. The most important point to appreciate in this discussion is that the fluids must be thin enough (i.e. have a low viscosity) to penetrate the narrow capillaries to work subsequently to exclude water from the crevices. Water has a low viscosity, a low contact angle to surfaces that are not 'dirty' (i.e. not oily or greasy), and a high surface tension force *T* and so will gain access to any capillary not coated by the wax–oil and produce crevice corrosion with the resultant accompanying high acidity (Sections 6.17 and 7.9). Where the wax–oil is giving an extra thickness layer to the electrodeposited paint to give added corrosion resistance to the deposited salt particles, thickness will be important. To produce a thick film from a liquid on a vertical surface requires a liquid with a high viscosity. Thus there are two opposing requirements, the liquid must be of low viscosity to penetrate a crevice but be viscous enough to give a thick film on the vertical panels. Many of the after-market rustproofing stations were inadvertently misled by approval-type tests, which favoured the thickness formed by films rather than capillary penetration against gravity. In these tests capillary penetration was measured with gravity aiding penetration, with the result that in service conditions the waxes bridged the crevices. This caused accelerated failure as the saline water could often gain ingress from the other end of the capillary inter-weld–crevice–overlap.

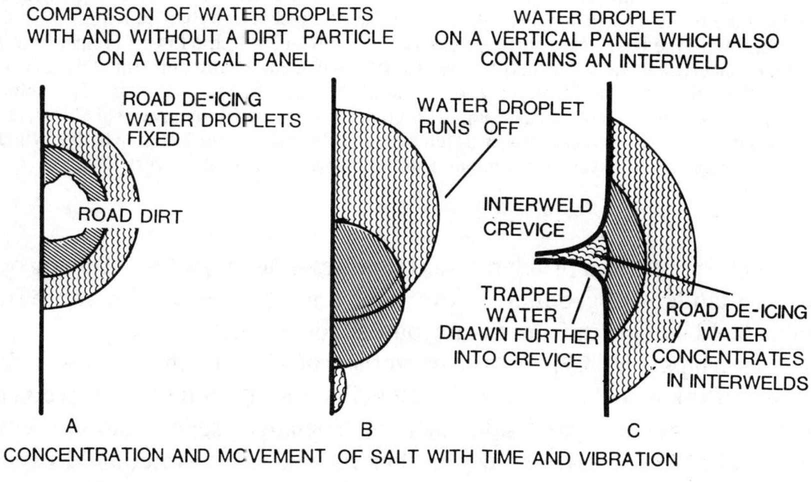

Fig. 3.7 — (a) This shows a water droplet drying with a dirt particle as a locating nucleus for the water droplet. Vibration will therefore mean that the water droplet will not run down the vertical panel as a result of gravity. Any dissolved salts such as the road de-icing salts will be concentrated at the base of the dirt particle. Compare the wax-coated surface in Fig. 3.8. (b) This shows a water droplet drying without a dirt particle to act as a locating nucleus for the water droplet. Vibration will therefore mean that the water droplet will run down the vertical panel as a result of gravity. Any dissolved salts such as the road de-icing salts will be deposited over a wide area and not concentrated in one location. Compare the wax-coated surface in Fig. 3.8. (c) This shows a water droplet drying within an inter-weld which acts as a locating nucleus for the water droplet. Vibration will therefore mean that the water droplet will not run down the vertical panel as a result of gravity. Any dissolved salts such as the road de-icing salts will be concentrated within the crevice at the last area to dry. Compare the wax-coated surface in Fig. 3.8.

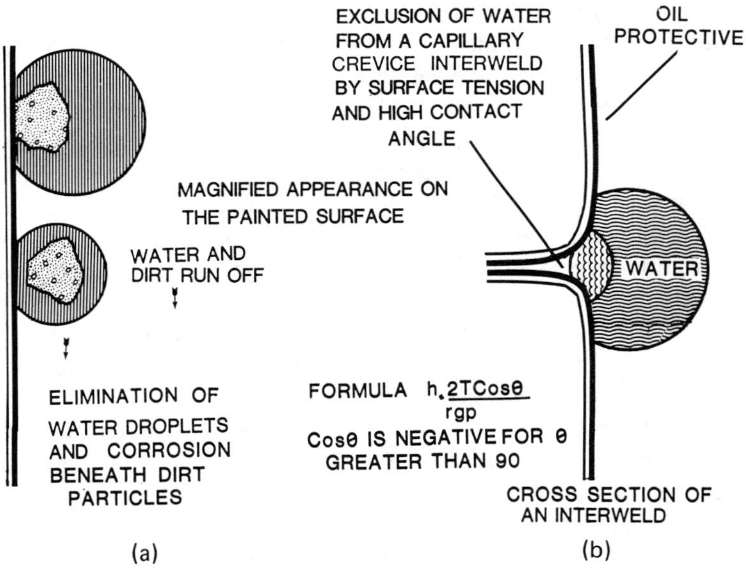

Fig. 3.8 — The same diagrams as Fig. 3.7 except that now the substrate is coated with a wax or oil which presents a hydrophobic surface to the saline droplet. (a) This shows a water droplet that will not dry with a dirt particle at the area where it was deposited. Vibration will therefore mean that the water droplet will now run down the vertical panel as a result of gravity and possibly take the dirt particle with it. Any dissolved salts such as the road de-icing salts will be concentrated at the base of the dirt particle which is now taken to the lowest point within the member concerned. This will usually be a horizontal surface, and more unfortunately, a crevice with its limited protection from electrodeposition. (b) This shows a water droplet excluded from the wax–oil-coated crevice. Assuming that it dries in this position, any dissolved salts such as the road de-icing salts will be concentrated outside the crevice at the last area to dry, which is now a well oxygenated area and so the corrosion rate will be very low.

One of the most prevalent problems was the wings–fenders, where bridging occurred beneath the fender–wing–bulkhead overlap fixing. At this location rain water and road salts could still get in from the uncoated faces of metal overlaps.

These overlaps (Fig. 3.9), are a characteristic of wing attachments etc., where the top surface of the wing beneath the bonnet (Plate 3.2) can fill the bridged capillary gap from the other uncoated side, thereby trapping a saline solution within the crevice. A differential aeration cell is thus formed in the inter-weld capillary or overlap and behind the protective coating. Drying of this trapped region further concentrated the road de-icing salts within the crevice (Sections 10.5–10.7). Thus these areas became wet when the humidity (Section 10.5) increased and they remained wetter for long periods until the restricted drying conditions could evaporate this absorbed water. The corrosion reaction must take place at the metal surface (see Section 6.14 and 7.7 and Figs 2.6, 7.16, 8.18 and 8.21) and so the corrosion process can strip the paint film (together with the thick wax film) (Plates 3.2 and 8.1) from the metal substrate surface at the metal–paint interface as shown in the laboratory test result which is shown in Fig. 3.10.

The contoured surface of say the front wing produces a complex series of capillaries where the two surfaces of separate panels have to be joined. The front

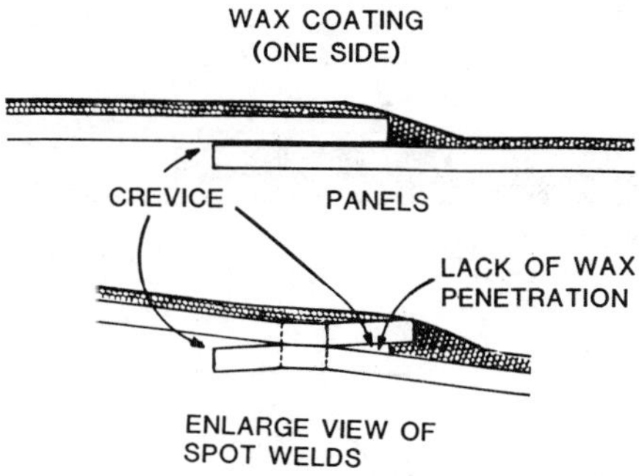

Fig. 3.9 — There are two surfaces to a metal inter-weld or overlap which can be formed either by a bolted arrangement (Fig. 2.10) or by resistance spot welding (Fig. 8.7). Thick viscous waxes will bridge the crevice on one side and, if the other side is uncoated, water and salts may get into the crevice and thereby concentrate the de-icing salts within the crevice. The fact that one side is coated and bridged means that the crevice remains wetter for longer as drying is restricted. Corrosion and paint stripping results (Sections 8.11–8.15). One way of overcoming this on bolted sections is to insert a plastic spacer (or similar hydrophobic material) into the crevice. For inter-welds a highly penetrating oil or wax must be used.

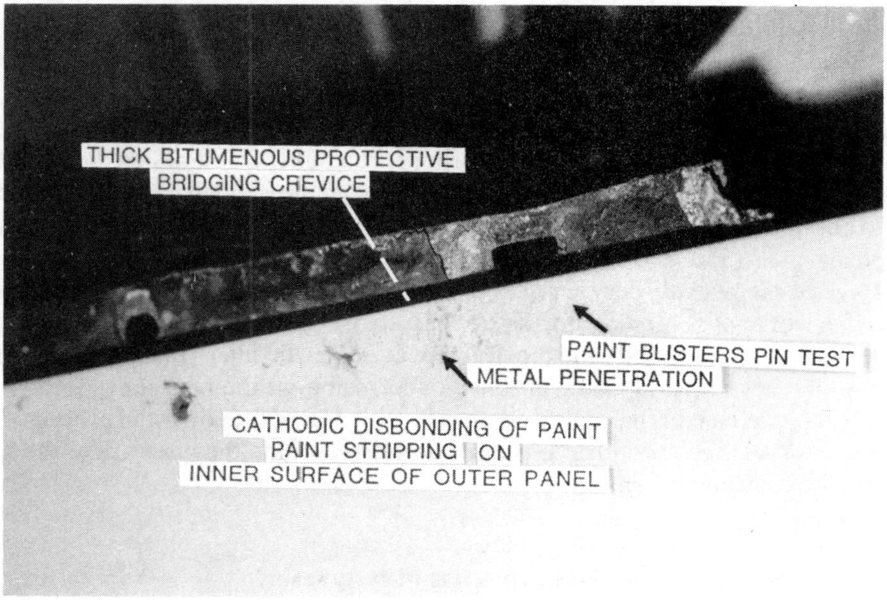

Plate 3.2 — The external surface of a wing in which corrosion was initiated between the overlap thus causing the stripping of the wax and paint films at the interface between the metal and paint. The wax coating then forms a reservoir which is constantly fed with water by osmosis (Section 8.13). This trapped saline water then causes pitting along the wing.

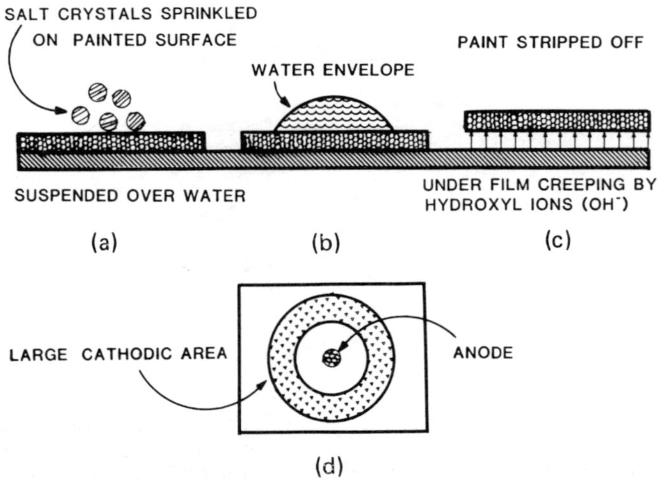

Fig. 3.10 — Road salts are hygroscopic (absorb atmospheric water vapour) (Section 10.5). This figure is a schematic view of an experiment carried out to discover how road de-icing salt would behave on a painted and undersealed surface (a). The metal sample was then placed in an enclosed environment with a bowl of water. The salt absorbed the water vapour and made a pool of water (b). After a short period of time the paint film could be stripped (c) from the metal sample from both sides (top-side and underside). Note the underside of the metal was not in contact with the sprinkled salt or the bowl of water. The cathodic reaction had made the whole paint film become non-adherent to the metallic surface (Sections 8.12 and 8.16). See also Fig. 8.18, which shows the undercutting of paint films by osmosis (Section 8.13). The corrosion cell formed is the result of a large cathode and a small anode (d) (the result after seven days).

wing to air dam seam is shown in Fig. 3.11. The crevice behind the outer surface is often of variable width and therefore harbours road dirt. Good design dictates that this crevice should be wide, with only metal contact at the (few) fixing points, in order (1) to obtain a good paint film upon the adjacent surfaces, (2) to allow rapid drying of the road water, (3) to allow for the easy removal of road salts from the crevice and (4) to allow for the two adjacent surfaces to be wax coated. In practice what happens is that the worst possible solution is applied to the problem. The motor vehicle manufacturers tend to hide the crevice beneath a mastic filler (Plate 4.23 and Fig. 3.11) which now aggravates the problem of corrosion, as the rate of evaporation is considerably reduced. This mastic also masks the onset of the corrosion process. The corrosion process can produce considerable stresses within the interface of the spot welds (Figs. 1.9 and 3.11).

3.6.1 Capillary protection DIY protection of rusty seams

Regular washing prevents this concentration of road salts in these seams. If the reader has a problem with these seams, it will be appreciated how difficult the repair can be. Often the seam breaks down periodically after painting. The ower should be advised, after steam cleaning and thoroughly drying it, to correct the external blemish as follows: Apply several applications of ‘three-in-one’ penetrating oil over a

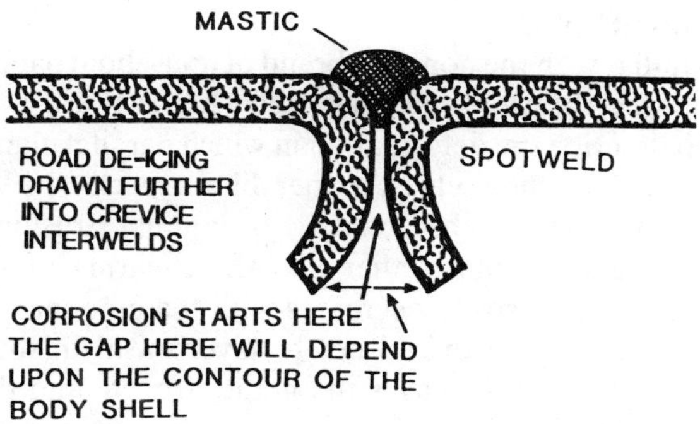

(a)

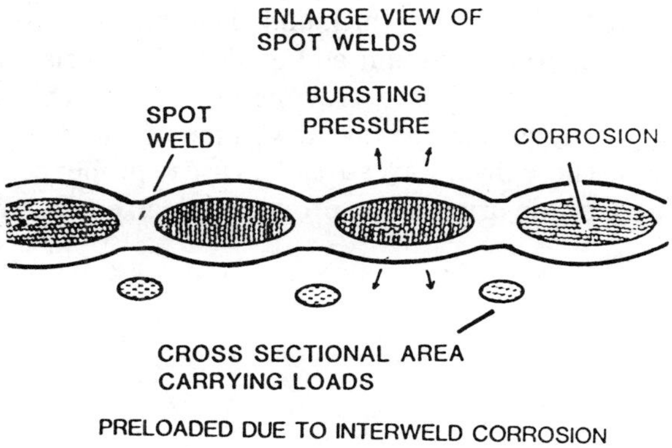

(b)

Fig. 3.11 — Inter-welds are often coated with a mastic in order to hide the onset of corrosion from the owner's sight. This is detrimental. It is far safer to advise the owner to apply a hydrophobic oil to these crevices periodically, and particularly prior to the onset of the winter. The underside can be coated with wax which will protect the variable gaps arising from the three-dimensional contour required for the shape. Once the corrosion product starts to force the inter-welds and metal panels apart, rectification is impossible. See also Section 4.3.7, Item 9.

period of a month. Note that this will only work if the vehicle is dry (i.e. garaged indoors). Remove excess oil with turpentine substitute and allow to dry. Remove rust stains with 'T-cut' or simliar and apply the touch-up paint provided at time of purchase to the seam. When this paint is hard, apply another application of three-in-one penetrating oil. This thin oil will exclude water and the de-icing salts. Subsequently three-in-one penetrating oil should be applied at six-monthly intervals and especially after steam cleaning or washing with detergents (because the three-in-one oil breaks down in the sun and is displaced by water-carrying detergents). After any application of oil, this area should be wiped clean (to prevent clothes being soiled). This will stop water getting into the crevice and eliminate the corrosion process. Wash the area behind any other external fixings (e.g. if the seam in question also contains the side or tail lights) thoroughly with a gentle spray during the winter de-icing programme.

3.7 THIOXTROPHIC WAXES

Readers will be familiar with the non-drip brand of household paints. These non-drip paints become thin when brushed and thicken up when they are not sheared. Shear is a technical term to describe the deformation in which parallel planes remain parallel whilst moving parallel to themselves, rather like a pack of playing cards being sheared. This type of flow is a characteristic of liquids. Brushing or spraying the thixotropic paint shears the film rather like the shearing of the card deck. In thixotropic mediums, the viscosity decreases with the amount of shear, becoming 'thinner' (like water). One is advised not to 'overbrush' thixotropic paints. The converse is true, — no shearing leaves the liquid 'frozen' and thick with a higher viscosity rather like treacle. They do not tend to run when applied to a vertical surface. These non-drip paints owe their properties to their *thioxotropic* nature. Certain waxes–oils can be made thixotropic. These will deposit upon the vertical surface of a car panel to give a fairly thick film. The thickness gives the car owner a feeling of confidence as to the amount and quality of material applied. However, these oils–waxes will not creep into crevices as surface tension forces will not overcome the viscosity requirements for the volume of liquid required to be pulled through the crevices. These thick waxes are of no use in proper corrosion protection. They are formulated to stop drips and to produce a clean environment in which they are applied.

4

Corrosion of specific areas

4.1 RECOGNIZING AREAS WHICH SHOW CORROSION FIRST

The areas which show corrosion on a motor vehicle are common to the manufacturer and model. Most vehicles of different manufacture show the same areas of corrosion, to a lesser or greater extent. These are shown in Fig. 4.1.

One important aspect of the momentum of a particle (Section 2.7) is the fact that the road de-icing salt can be deposited by pressure eddies (Section 2.9) out of the direct line of the wheel tracks, in areas where, at first sight, it would be inconceivable that it would be found. In these areas water from neither a hose nor steam cleaner will be able to dissolve or remove the saline environment. Examples of this are given in Plates 2.1(a), 2.4(a), and Plate 2.6. Other important aspects to recall are as follows:

(1) The inter-welds and narrow capillaries contain little or no electrocoat paint (Sections 3.2–3.4 and Chapter 9).
(2) The inter-welds can carry over the pre-treatment phosphates (Sections 3.3–3.5) and gaps of 4 thou in the inter-welds are the most prone to corrosion as a result of the carry-over of pretreatment mediums and lack of paint (Section 3.5).
(3) Any protruding member at right angles to the direction of motion will generate spray (Fig. 4.2) and may deflect the front wheel spray in all directions. Hence jacking points (Section 4.5) slung beneath the sills and adjacent to the process drainage holes will deflect the wheel spray into the sills/rockers at a much faster rate than the normal suction of the spray through the sill/rocker box members by virtue of the passenger air extraction system (Sections 2.13, 5.10 and 5.12).
(4) Ledges and vertical panels which are out of the direct wheel trajectory suffer from mud compaction owing to the small amount of water present and surface tension effects draw the particles together (Section 10.10). This compaction is out of the main wheel spray trajectory as a result of the wind/pressure eddies around the vehicle (Section 2.9).

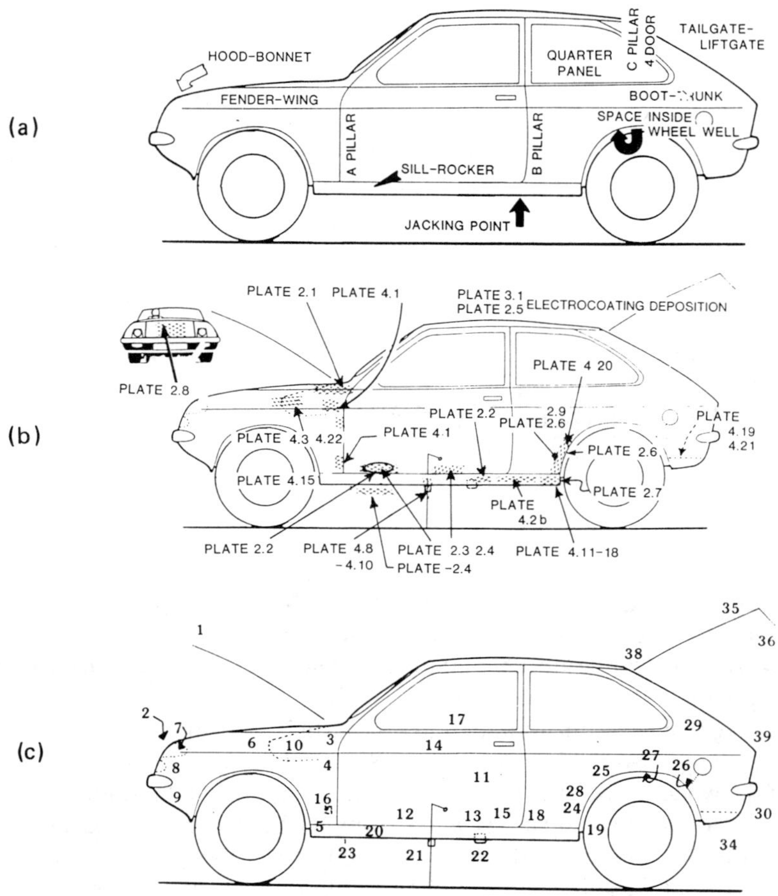

Fig. 4.1 — (a) A diagrammatic view showing the different nomenclature of sections of the vehicle body and (b)(c) the location of those areas of corrosion on a vehicle which will be discussed by number and plate number. The plate number refers to the photographs of stripped-down low mileage vehicle which have been taken from internal panels. These aspects will be discussed throughout the text of Chapter 4 by their reference number in (c).

(5) Narrow or closely spaced panels will not have a thick electrocoat primary coat on them (e.g. Fig. 4.3(b)) and will also be hard to wax coat because of the angle which they make with the sprayed wax (Fig. 3.3(c)). Plate 4.4 shows this denudation of paint film within the hinge crevice.

(6) Raw edges of the formed steel panel will have a thin layer of paint covering the edges (Fig. 3.4).

These aspects of the vehicle and the likely corrosion areas are given in Fig. 4.1 and discussed in detail in the following sections. It is important to appreciate the design aspects of metallic corrosion and the effect that design will have on the longevity of a motor vehicle or any other product. Design affects corrosion and

	RATIO OF WATER / DIRT	MOMENTUM
HIGH SPEED		SMALL
LOW SPEED		LARGE

(a)

HIGH VELOCITY IMPACT OF WATER DROPLETS

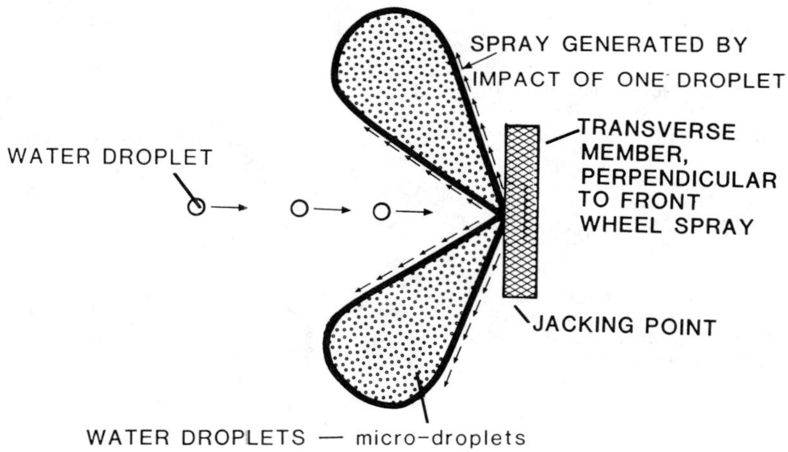

(b)

Fig. 4.2 — (a) A diagrammtic view of the meaning of low momentum particles and ratio of water to dirt. (b) A body member (e.g. a jacking point) at right-angles to the vehicle's direction of motion will 'collide' with the front wheel road spray (Fig. 2.7). The diagram shows the relative motion of the member with respect to the wheel spray. The relative kinetic energy of the spray is transfered into surface energy of the droplet on impact. This is a schematic diagram showing the spray generated when one droplet is turned into a cloud of micro-droplets. This mechanism of spray generation is important for the water ingress into sills–rockers through the holes associated with the jacking points. This is important and is not acknowledged by the motor vehicle manufacturers at the design stage. This spray generation is discussed again in Sections 4.5 and 10.12. Similar arguments can be deduced for other members which are adjacent to process drainage holes.

so corrosion should be considered at the design stage. Repair methods can lead to 'bad corrosion design' and the formation of crevices. This means that all engineers associated with the motor vehicle from purchaser (sales) to repairer must have a basic understanding of the true mechanism and the myths surrounding the subject of corrosion.

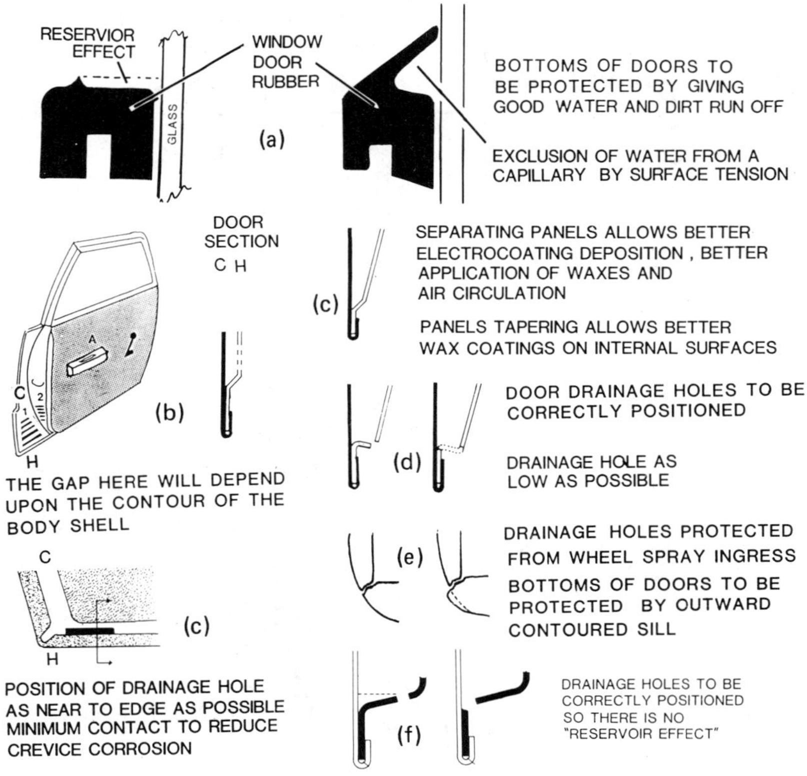

Fig. 4.3 — The fine details of the structure concerning the bottoms of the doors are important because a large amount of water by-passes the window rubbers carrying dissolved road salts with it. (a) Most window rubbers are badly designed as they trap water against the window glass as shown here. Other designs (not shown) channel water to either side of the window glass. Shutting the door causes this water to run along the rubber to the outer door shuts and to enter the crevices by capillary action as there is no wax along the door seams (CH in (b)). (b) The door section at CH is usually very narrow as it usually has to incorporate the rear wheel arch or B pillar. This means that the electrocoat deposition is poor, and thin protective paint coats are produced. Better designs would be produced, with greater deposition of the electrocoat, if the air extraction system is incorporated within this part of the door at positions (1) or (2). Better wax injection coatings (using air-assisted spraying) can also be achieved with this 'opening' shown in position (1). Door drainage holes should be at the corners of the door as a vehicle is never garaged on horizontal ground. (c) Separation of the panels allows better wax injection onto tapering panels as opposed to parallel panels. (d) Drainage holes should be as low as possible and fairly large and placed at the nearest place to the corners of the doors. This allows maximum drainage if the vehicle is garaged on a slope. Drainage holes should also be placed at locations beneath the extremes of the door window rubbers to allow the water to run straight out. This is particularly important on the outer surfaces of the doors owing to the centrifugal forces produced when the doors are shut. (e) The bottom of the door ought to be protected by outwardly curving sill–rocker members to avoid wheel spray ingress through the door bottoms. See Plate 2.2(a) for the evidence for this ingress. (f) Many manufacturers do not have the door drainage holes at the lowest point in the innermost panel of the car door. The hole is not deep enough with respect to the bend or contour and so traps water by a reservoir effect (Plates 2.3 and 4.6).

The examination of specific areas shows how the corrosion theory developed earlier may be applied to the corrosion of all motor vehicles. In order to understand this section, an elementary knowledge of some of the techniques used in construction or design is required. Corrosion and design are intimately linked. American and English terminology is used for the parts of the motor vehicle.

4.2 BONNETS–HOODS

See Fig. 4.1(c): item (1) perimeter of bonnet–hood, especially trailing and leading edges; item (2), front valance radiator grill.

4.2.1 General

The method of constructing a bonnet is to use one sheet of metal, the outer, which is pressed to shape and strengthened by an internal web which is held to the outer by a clinch fold or 'hemming fold' lap joint, as shown in Fig. 4.4. In some models of manufacture this overlap may appear to be 'filled' with mastic, but this is not so. This mastic, in most cases, is applied after the pressing operation to make the clinch fold. The function of this mastic is to stop the paint on the raw edges of the metal overlap from breaking down and exhibiting the first signs of corrosion, which the manufacturers must rectify under the terms of the Canadian Voluntary Anti-corrosion Code (Section 1.12).

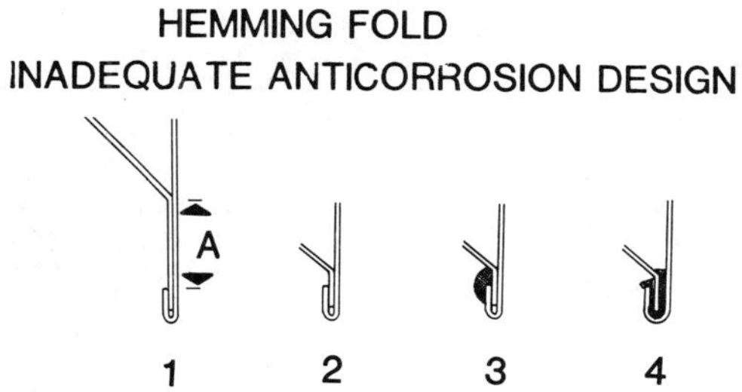

Fig. 4.4 — A schematic view of the formation of the hemming fold. Some of the problems are as follows. (1) If the distance A of the metal overlap is too large then there is a large area which can 'wick' the water. The greater is the distance A, the more water can be held and therefore the longer this area will remain wet. The water will be denuded in oxygen and will form the anodic area (dissoving area). (Remember: the oxygenated area is the cathode or (positive) terminal of the 'corrosion battery'. The deoxygenated area and the area which dissolves are the anode or (negative) terminal of the 'corrosion battery'.) (2) Better design, but now the raw edge will be the Achilles heel of the production method. This is due to the area of thin paint at the raw edges as a result of spraying (Figs. 3.3 and 3.4). (3) The 'raw edge' can be protected by a bead of sealant mastic which now increases the thickness of the protective at the raw edge. (4) The whole clinch fold should be made with a bead of mastic (or sealant) (Section 11.27) within the crevices formed by the fold. The sealant–mastic has to be heated and so it should not form 'blow holes' by the loss of the lower melting point volatiles. This completely seals the crevices from the external aggressive environment. It should be noted that the mastic should not absorb the pretreatment acids (Sections 6.18 and 6.19).

4.2.2 Suppression of vibration

In order to prevent the outer panel from vibrating and causing 'drumming', the outer panel is separated from the inner web with 'pads' or 'blobs' of sealant mastic (Section 11.27), which sets very hard when the whole body is stoved to cure the electrocoat. This mastic often shrinks and comes away from either outer or inner web and may subsequently reveal bare unpainted steel.(For application to doors see Plates 4.5(a) 4.5(d).)

4.2.3 Electrocoat drainage

The vehicle is often electrocoated with the bonnet–hood open and so there are 'paint–drainage' holes. For example, on front-opening bonnets–hoods, there are drainage holes on the hemming fold adjacent to the windscreen. This may mean that the hemming fold to the front of the vehicle contains very badly designed drainage holes. The opposite will be true for the rear-opening bonnets–hoods. In front opening bonnet–hoods, this causes the whole of the front lip to corrode by the following mechanism. The radiator offers a partial resistance to the flow of air through the radiator cooling fins, and so some of the preceding vehicles wheel spray is deflected into the bonnet–hood (Fig. 4.5). Hence the leading-edge hemming fold,

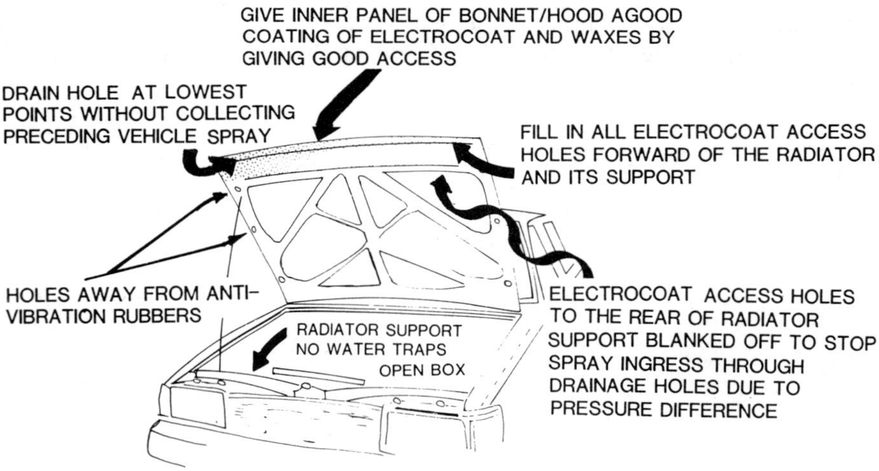

Fig. 4.5 — With the bonnets–hoods of vehicles becoming lower at the leading edge to achieve 'streamlining', the gap between the two panels (inner and outer) becomes squeezed. This requires that particular attention be paid to design with regard to the application of the electropaint within these members. The radiator (head and side lights) offers some resistance to the passage of air, so that the preceding vehicle's rear wheel spray can be forced up into the bonnet through the manufacturer's electrocoat throw holes (necessary to allow electric current into the double skin member to achieve paint deposition). In order to stop this ingress these holes ought to be fitted with grommets and blanked off after electrocoating and prior to service use. Holes in the bonnet to the rear of the radiator also ought to be filled so that a low pressure here does not allow salt-laden air to be drawn through the drainage holes. All drainage holes must be at the lowest point. Painted steel to the front of the vehicle on the surface of the bonnet or hood should be kept to a minimum in order to reduce stone chipping to the leading edge of the bonnet and other valances. This region of the car is always in the preceding vehicle's wheel spray. Hence extensive 'stone' or small particle abrasive chipping will result. The radiator support member should be an open structure to ensure adequate thickness of electropainting and to overcome humid, hot (proximity to the radiator) box sections.

being over the radiator, collects a lot of the de-icing salts and cannot drain properly, owing to the badly designed drainage holes (see for example Plate 2.3). Corrosion proceeds along the interface of the crevice–fold. With bonnets–hoods which open and are hinged along the front (rear-opening bonnets–hoods), a similar mechanism holds, except that often there are more corrosion design faults to the front valance to which the hinge is fixed. This valance also forms the surround to the front radiator grill of the vehicle. Plate 2.8(a) shows such a valance with an incorrectly designed process hole which 'collects' the preceding vehicle's wheel spray and corrodes the whole of this area from internally nucleated corrosion. There are many examples where the vehicle collects the preceding vehicle's wheel spray (Fig. 5.6), and these are more noticeable where the cheaper model vehicle is not fitted with the accessory fog lights as an example. The space for these lights is filled with an 'air vent grille' or similar, which then collects the preceding vehicle's spray (Plate 2.8(c)).

4.3 WINGS–FENDERS

See Fig. 4.1(c): item (3), beneath the wing–fender and adjacent to the 'A' pillar; item (4), crevice on wing–A pillar; item (5), mud trap between foot well and wing; item (6), suspension struts; item, (7), headlamps; item (8), side lamps; item (9), juncture of wing–fender to front valance; item (10), energy-absorbing structure (trumpets).

4.3.1 General

The wing–fender is considered a 'cosmetic' panel in that it does not form part of the energy-absorbing members during an accident (see Section 1.16). Wings–fenders do not therefore form part of any imposed structural survey for roadworthiness. External jagged edges produced as a result of corrosion do form a safety feature. Pedestrians may be injured by these areas when pedestrian and vehicle traffic come into contact.

The method of constructing and fixing a wing differs from manufacturer to manufacturer. The manufacturers, in designing wings, are moving away from the spot-welded method of attachment towards bolt-on fixing methods, which give rise to cheaper (insurance) repair costs. These bolt-on wings–fenders tend to be pressed to shape from one sheet and the stiffening of the wing–fender is obtained mainly by their shape. Other methods involve reinforcing brackets or the use of pieces of steel brought together in a clinch fold along the A pillar (Plates 4.1(a) and 4.1(b) (see Fig. 4.1(c), item (4)). These bolt-on wings–fenders are attached to the 'body-in-white' prior to electrocoating. Most manufacturers now protect the wings–fenders by a plastic 'inner wing–fender liner' stone guard or under-wings which are some form of partial internal plastic guards (Fig. 2.9). Some of these partial internal plastic guards do not take account of the road dirt being thrown forward at the vehicle speed and being deposited on the front valance panel. Sections 2.12 and 10.12).

4.3.2 Mud pocket beneath wing–fender (Fig. 4.1(c), item (3))

The mud pocket beneath the wing–fender and adjacent to the A pillar has been discussed in Section 2.7 and Plate 2.1. It should be emphasized here that even those vehicles washed frequently have trouble with this area owing to the protection

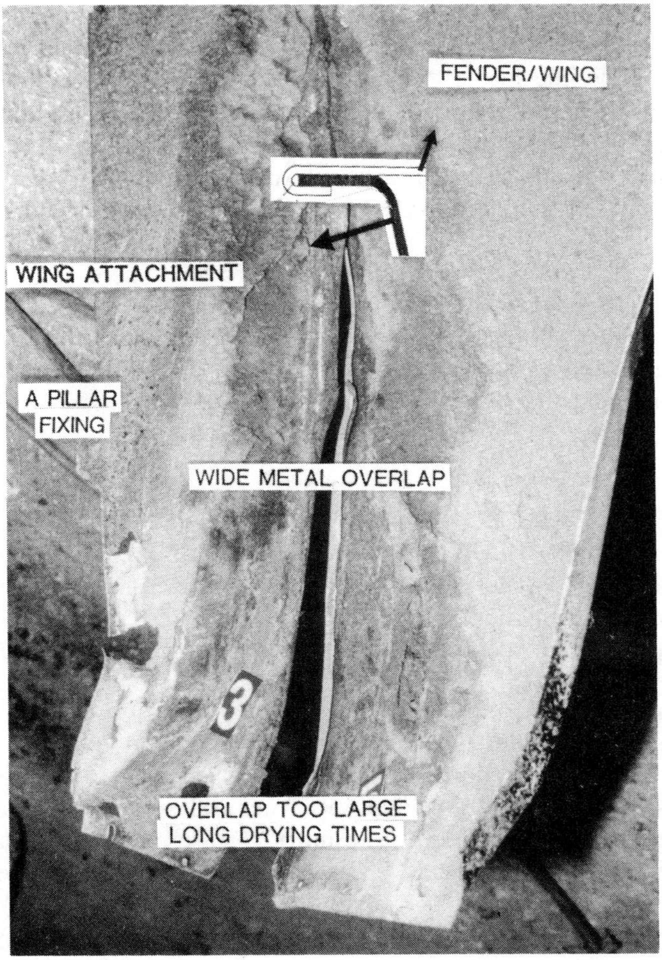

Plate 4.1(a) — To accommodate the open car door, the front wing is contoured. This often produces a large metal overlap, which as shown in Fig. 4.4 is bad anti-corrosion design. This photograph shows a considerable area of corrosion in this position.

offered to mud accumulation by the ledge (Plates 2.1(b) and 2.1(d)). This is therefore one of the corrosion design faults which have been classified as 'bad design'. Few owners or garage hands would appreciate that the hose or steam cleaner would have to go so far up the inside of the wing to remove this mud pocket.

In some manufacturers' models, this cavity has been blocked off with a metal shield (Plate 4.4). However, the cavity is host to the preceding vehicle's wheel spray owing to the radiused 'hole' required to form the wing/fender. These forward-facing holes should be blocked with mastic and, where the drainage is into the wheel arch, frequent washing undertaken to ensure that the hole is free of mud accumulations (see Plates 4.3(a) and 4.3(d) for examples of bad design where the drainage hole to other box members is over the front wheel and therefore filled with road dirt.

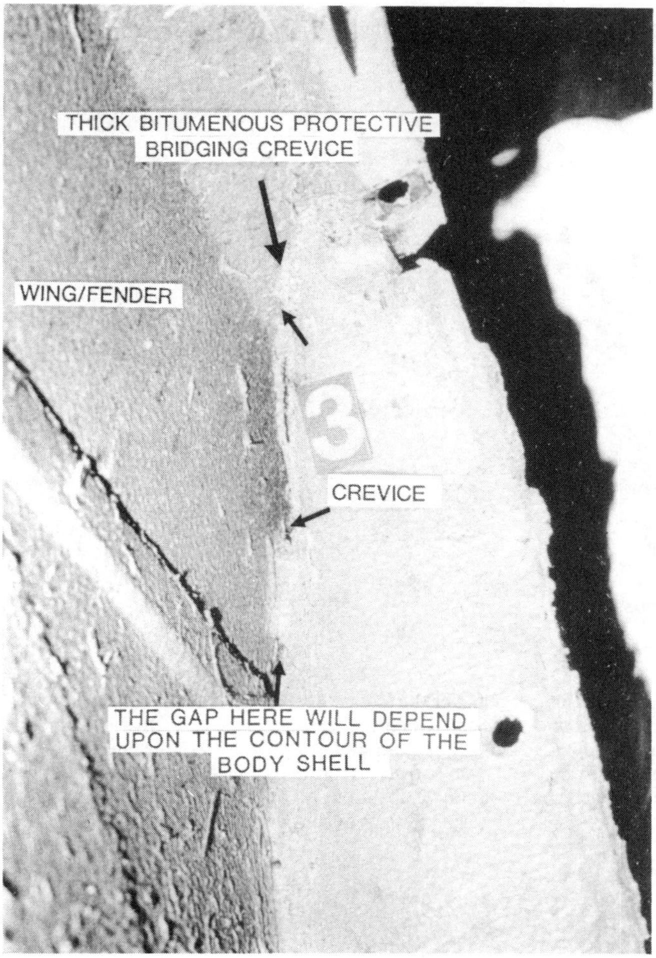

Plate 4.1(b) — The crevice produced along the A pillar is often covered by the liberal application of a manufacturers under-coating. Because this is usually very viscous it does not penetrate into the crevice. Shown here is a large void or 'holiday' which will allow the ingress of road de-icing salts into the crevice.

4.3.3 Crevice on wing–A pillar (Fig. 4.1(c), item (4))

The wings–fenders in the more dated body designs are attached by spot welds. These are often made from two panels. In this case, the two panels are joined near to the A pillar (windscreen pillar) by a clinch fold or 'hemming fold lap' joint, as shown in Fig. 4.4 and photographically in Plates 4.1(a) and 4.1(b). In some models, this overlap may appear to be 'covered' with under-body 'underseal' compound or mastic. However, this mastic, in most cases, is applied from a liquid which is too viscous to enter the capillary (Plate 4.1(b)) formed by the clinch fold. Water and road de-icing salts from either the road spray or the windscreen soon penetrate the crevice from the 'other side' and corrosion is initiated along the crevice line. For efectiveness both sides of the crevice must be coated and penetration of the crevice by the waxes must

Plate 4.2(a) — A manufacturer's energy absorbing front trumpet with a very large process hole and heater scuttle drainage area over the front wheel spray. The road dirt is not washed out of the ledge formed by this poor design with the result that the whole of this area corrodes away. In this photograph, a low speed collision has caused the fracturing of the energy absorber which is now a brittle iron oxide rust. Also shown are crevices which have been filled with a sponge rubber material which has the ability to absorb water. The overlap in question is formed by the wing fitment to the car body.

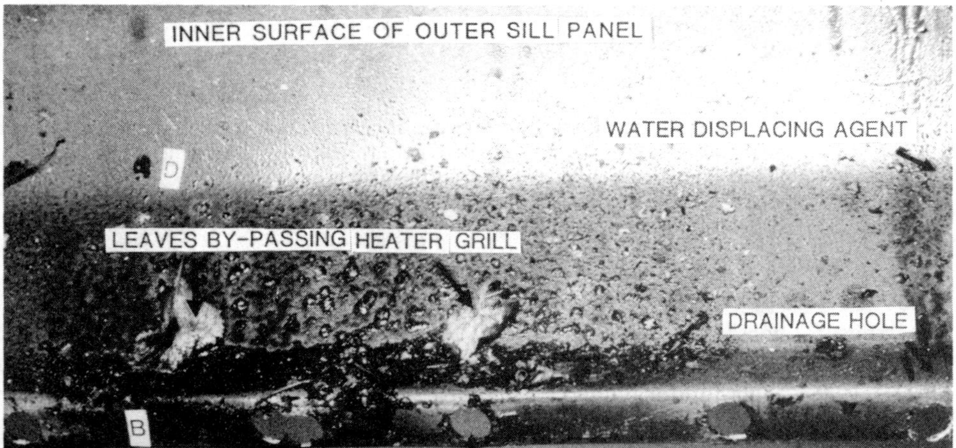

Plate 4.2(b) — Some vehicles have the air from the heater scuttle vented down the 'a' pillar and then into the sills–rockers. The heater scuttle grill allows small leaves and pine needles to pass through into the sill. The amount of this foreign debris will depend upon the area where the vehicle is parked. This photograph shows a new vehicle rocker with a leaf set into the after market wax injection.

be achieved. See also Sections 3.2–3.4 for a discussion of the electropaint within the inter-weld crevice.

4.3.4 Mud trap between foot well and wing (Fig. 4.1(c), item (5))

The interface between the curvature of the outer wing and the floor pan makes an ideal trap for the road dirt. The areas in items (4) and (5) are at present being protected by plastic guards (internal wing–fender liners–underwings), but extra care should be exercised at item (5) as some heater scuttles discharge rain water and other debris, e.g. leaves (Plates 4.2(a) and 4.2(b)) into this area. The leaves tend to block up the natural drainage of this area as the internal wing–fender liners (Fig. 2.9) have not yet been provided with an exit point for leaves, pine needles etc. which can get into the heater scuttle through the heater scuttle screening grill.

4.3.5 Suspension struts (Fig. 4.1(c), item(6))

The suspension struts are a very vulnerable area for corrosion as a result of the method of reinforcement which produces metal overlaps and crevices. High load factors are imposed upon the complicated unitary structure for the support of the suspension loads. Crevice corrosion, caused by adjacent metal overlaps produced to obtain the required strength, together with mud accumulations out of the main wheel spray are the biggest problem with which to deal. Most manufacturers should be protecting their suspension struts by internal wing–fender liners, but care should be exercised as some inner wing–fender liners are only partial liners where the designer has not taken account of the mechanics of the rotating wheel (Section 10.12) and the associated pressure eddies (Section 2.9) produced by the underbody attachments. These pressure eddies can cause fine spray to be deposited behind the wing–fender liner and where it will reside throughout the vehicle life in out-of-sight areas.

Coil suspension mountings are another area for corrosion owing to poor drainage from the lower spring platforms. These platforms collect the low momentum mud, which does not drain away. The small drainage holes which are incorporated in the welded structure soon become blocked and then the high momentum spray keeps the area very wet. If the drainage area was correctly designed, this high momentum spray would help to wash the low momentum spray off and away from the platform. Several manufacturers will be involved in heavy costs if the proposed regulations on product liability are enacted.

4.3.6 Headlamps and side lamps (Fig. 4.1(c), items (7) and (8))

Cavities over headlamps and the body of side lights present areas where road dirt may reside as the road dirt being thrown forward by the rotating wheel (see Section 10.12). If the wing–fender liner stops short of this area, mud accumulations will be produced out of the wheel spray trajectory. These will cause corrosion if they are not removed regularly by hosing or steam cleaning (Section 4.13). In some new vehicles some of the cavities over the headlights have been blanked off with a piece of metal which then is covered with mastic to prevent the road dirt from causing a problem. However, the designer has overlooked the crevice on the other side of the blanking metal, which collects the road de-icing salts from the preceding vehicle's road spray. Therefore areas behind the side lights and headlights should be checked on 'new

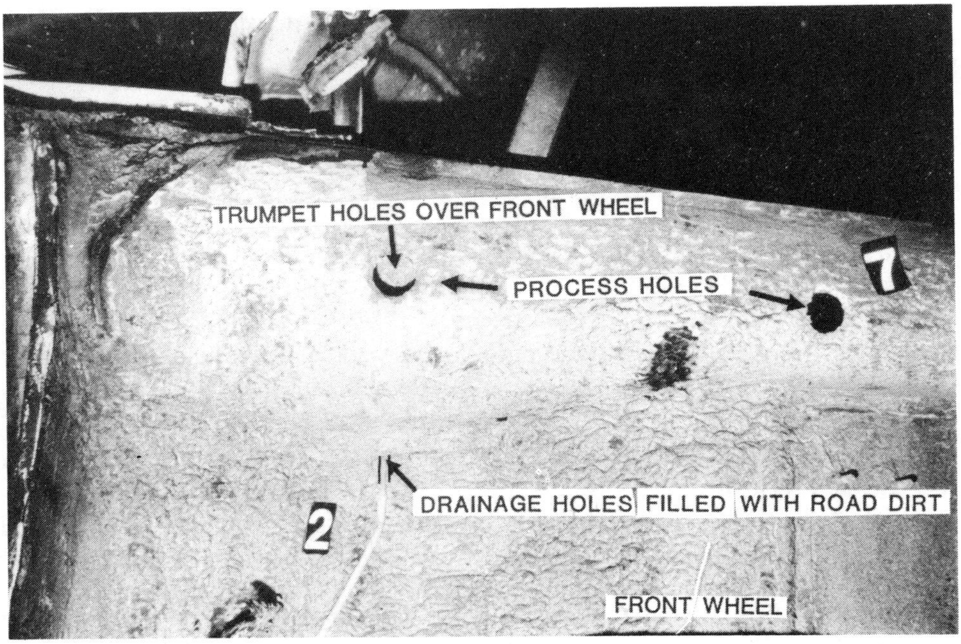

Plate 4.3(a) — The outer wing–fender has been removed to show the structure of the energy absorbing 'trumpets'. These trumpets contain two holes which are directly over the front wheel and thus collect a lot of road dirt and road de-icing salts. Electrocoat drainage of this section is provided within the interweld seam which is also over the front wheel and so becomes blocked very early in the vehicle's life.

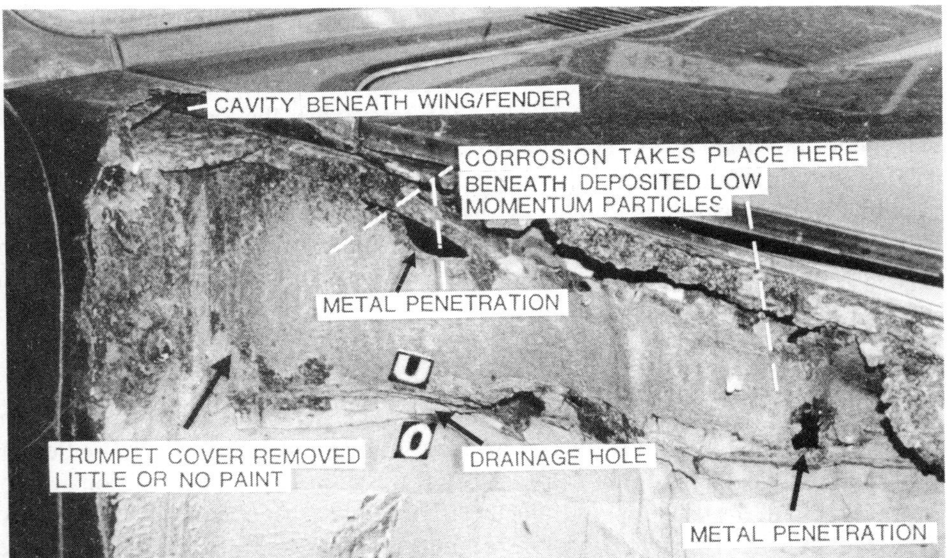

Plate 4.3(b) — The trumpet is removed to show the corrosion taking place as a result of the thin electrocoat in the cavity. Point U is in the trumpet cavity and O outside. Between O and U is the very small drainage hole whose orientation is such that it collects and traps the wheel spray. Note the cavity at the wing top which forms a mud trap for the low momentum spray.

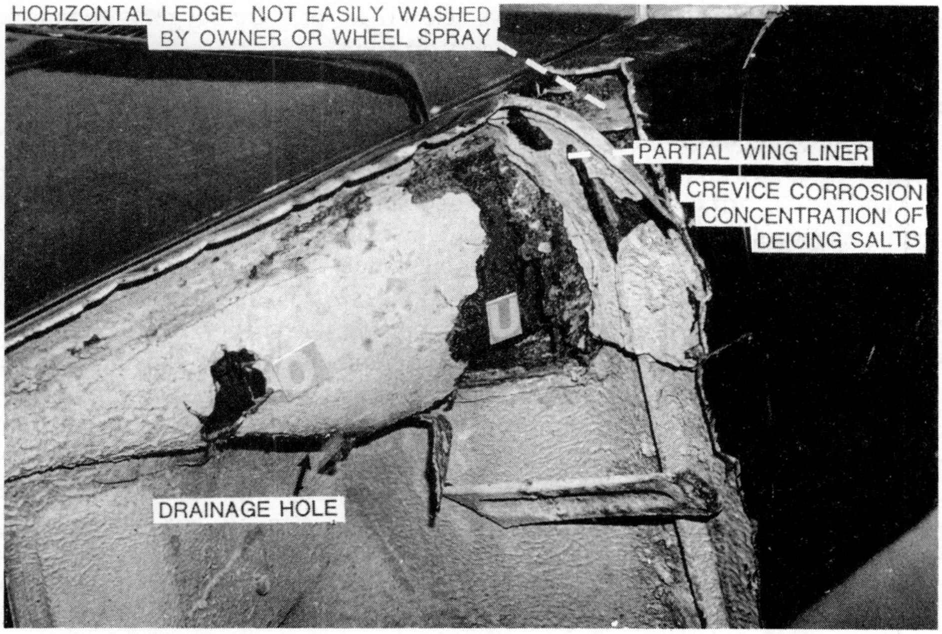

Plate 4.3(c) — Note the cavity at the wing top has now been separated from the dense front wheel spray by a partial cavity closure. This rubber seal to the outer wing–fender causes crevice corrosion along the line of contact.

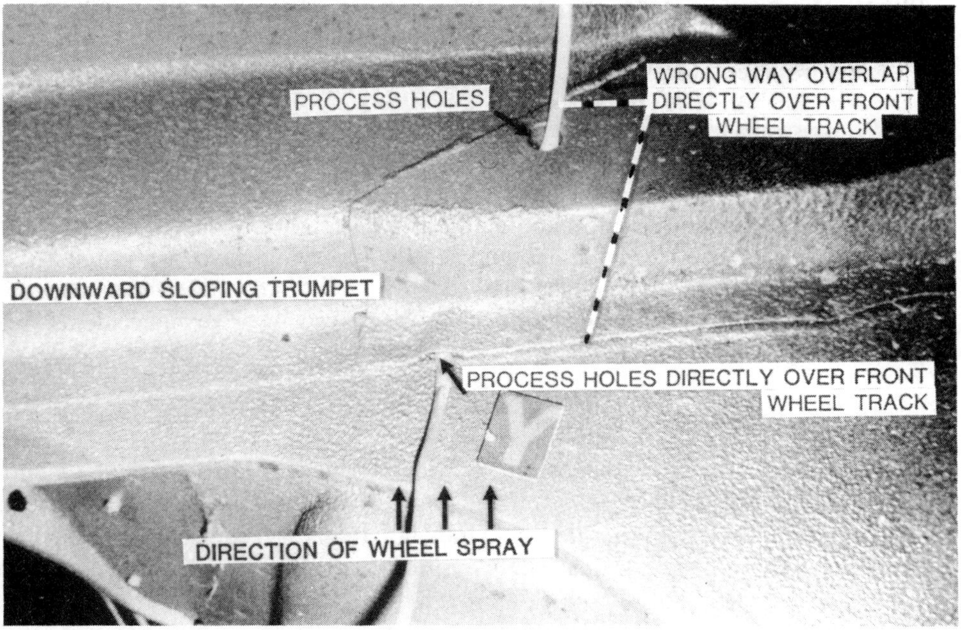

Plate 4.3(d) — A more recent model contains similar design faults with the drainage hole to the energy absorbing trumpet still directly over the front wheel. The plate also shows a wrong way overlap beneath the wheel arch. This is because there are opposing requirements for gravity drainage and spray direction. Rectification can only be achieved by a continuous weld all the way along the crevice.

designs'. In some cases the preceding vehicle's road spray can lodge on top of the plastic wing–fender liner and cause crevice corrosion.

4.3.7 Juncture of wing–fender to front valance (Fig. 4.1(c), item (9))

In older cars this was formed by spot welding, with only the outer seam, which is seen, being filled by mastic (Fig. 3.11). This prevented the owner from realizing that his vehicle was corroding until the complete seam broke down and then there was nothing that could be done as the rust forced the seams apart. This resulted from the internal seam (on the wheel side of the wing–fender), which is not coated with a mastic, and is open to the ingress of road spray. This open seam–crevice is very poor anti-corrosion detailing. More modern methods of attachment are by bolting, but there is still the same problem, that the internal crevice is denuded of paint if the gap is small because these wings–fenders, in the majority of cases, are bolted on before electrocoating. These seams can also carry over some of the pre-treatment salt, which are hygroscopic (Sections 3.5 and 5.5, Chapter 9 and Section 10.5). These pre-treatment salts can accelerate corrosion by adsorbing atmospheric water and providing an electrolyte within the crevice. Better designs to look out for which would help to eliminate this form of crevice corrosion would be to have a fairly large gap with a (plastic or non-conducting) washer between the two juxtaposed metal surfaces at the position of bolting. The presence of this washer would indicate good design (as the plastic washer will reduce the crevice length and therefore the amount of water present) and the fact that the wing was bolted on after electrocoating. Thus there would be a paint film between the interfaces (compare Figures 2.10(a) and 2.10(b)). This would prevent the 'differential oxygen' corrosion cell from working between the two adjacent steel surfaces; as the plastic is 'hydrophobic', it will keep out the environmental water.

4.3.8 Energy-absorbing structure (trumpets) (Fig. 4.1(c), item (10))

These energy-absorbing boxes have, in the past, caused considerable corrosion problems as the manufacturers' process holes (both drainage holes and holes for lifting the vehicle around during manufacture) have been over the front wheels. Mud accumulations, blocked drain holes and restricted drying have prevented these structural areas from drying out (see Plates 4.3(a)–4.3(d) and 4.2(a)). The boxes are a structural member of the vehicle and act as an energy absorber in accidents (see Figs. 1.7–1.9 and Section 1.16).

4.4 DOORS (Plates 4.4–4.7 and Fig. 4.3)

See Fig. 4.1(c): (11), corrosion along crevices formed by disbondment of anti-drumming materials or internal strengthening struts; item (12), corrosion around clinch fold; item (13), door drainage hole too high; item (14), corrosion around door brightwork; item (15) door drainage hole positions; item (16), corrosion around door hinges; (17) corrosion around door rubbers; item (18), erosion of door edges; item (19), rear door sections adjacent to rear wheel arch.

4.4.1 General

The method of constructing a door is to use two sheets of metal which are pressed to shape and held together by a 'hemming fold' lap joint, as shown in Figs. 4.3 and 4.4.

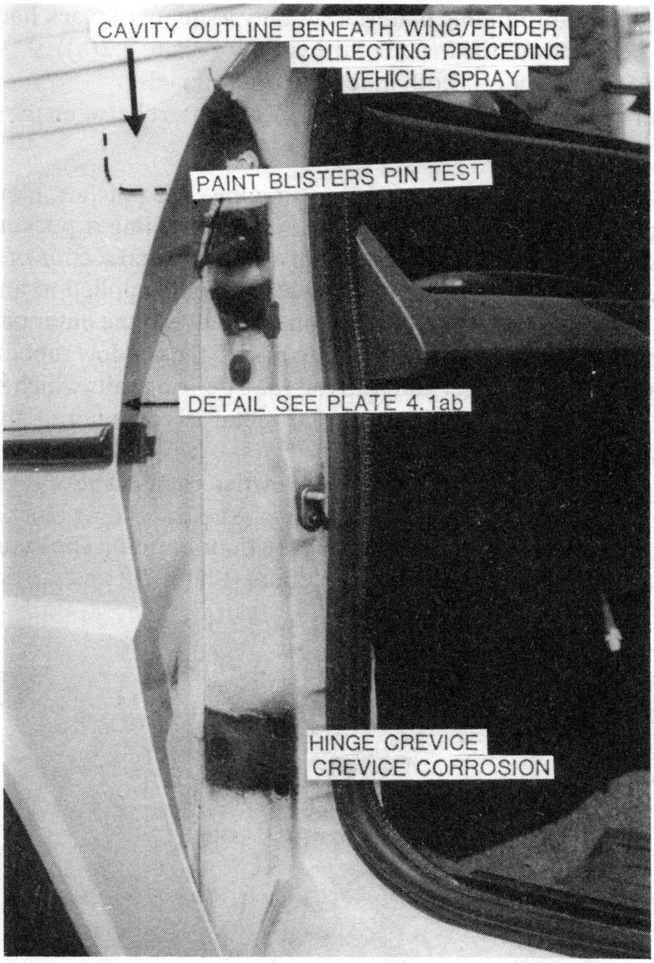

Plate 4.4 — Hinges must be bolted up hard to support the door. Consequently no electric current can flow in this area and so no electro-paint is deposited. Methods of overcoming the crevice corrosion vary with manufacturer. Most use a thin bead of mastic around the hinged areas and attempt to keep water away from the region. Also shown on this photograph is a small cavity member which is outlined. Unfortunately water can enter this member by the motion of the vehicle through the preceding vehicle spray, and also by the rain when the vehicle stands facing up a small incline.

This overlap may appear to be 'filled' with mastic, but this mastic is applied after pressing and in most cases to one side. The function of the mastic is to stop the electropaint and top coats on the raw edges of the metal overlap from breaking down and showing the first signs of corrosion. In order to prevent the panels from vibrating and causing 'drumming', the panels can be stiffened by any of the following mechanisms:

(1) By an internal metallic section applied internally within the door. This stiffener is attached to the outer door panel by either continuous or patches of mastic (Plates 4.5(a)–4.5(d)).

(2) By a piece of anti-drumming material (usually a square of black hard bitumastic compound) stuck onto the inside of the car door (Pate 2.2(a)).
(3) By contouring of either or both inner and outer panels.
(4) Much older designs of cars have a sponge rubber insert between the strut and the outer door panel.

Method (3), above, is the best mode of stiffening. The patches, method (2), and sealant mastic blobs, method (1), often tear away, forming a pocket, during the electrocoat baking operation, leaving no paint on the internal section of the car door. This non-adhesion is primarily due to the sealants being applied to an oily surface (Sections 3.4 and 11.26.6). The continuous mastic between the outer panel and door can lead to the trapping of water which has by-passed the window rubber. The mastic patches, method (2), can collect and trap the road de-icing salts which run down the window, and rapid corrosion ensues as the metal is corroded at the water line produced by the new pocket.

It should be noted that the manufacturer's owner's manual always gives advice to the owner which implies that the car door drainage holes become blocked with mud. This is not the main cause of corrosion along the door bottom. The main cause is the combination of high drainage holes to the doors (Plates 2.3 and 4.6) and incorrect

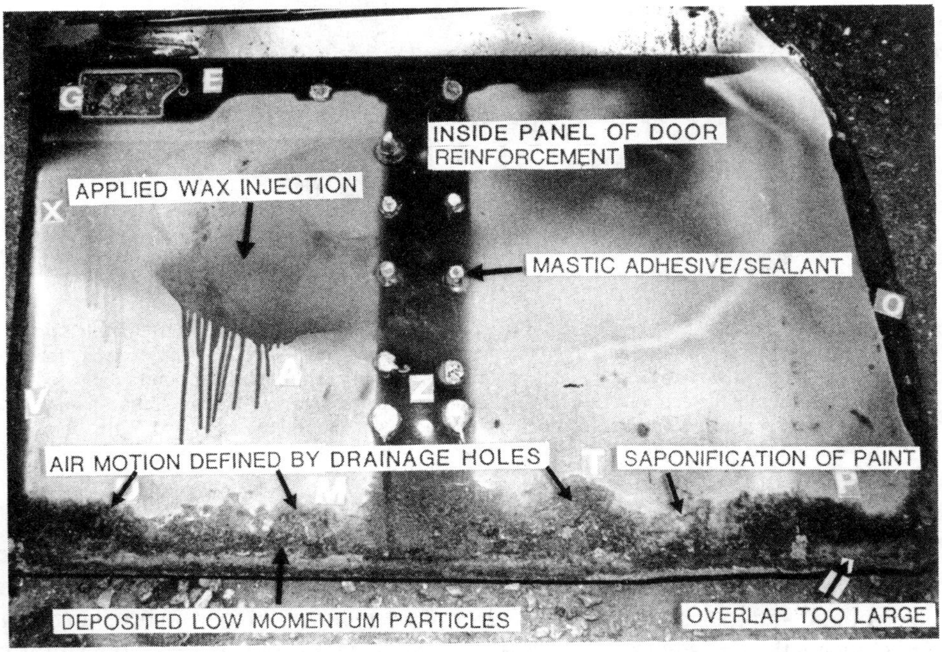

Plate 4.5(a) — The plate shows several areas of corrosion due to the ingress of the road de-icing salts which are drawn up through the door drainage holes. In this case the corrosion is more wide-spread and has masked the initial corrosion associated with the deposits at the drainage holes themselves. This lateral spread of corrosion is produced by the mechanism put forward in Figs. 8.18-8.21. The panel also shows evidence of inadequate wax injection. The wax really wants to be concentrated where the paint film is the thinnest which is the lower clinch fold. The row of mastic blobs show where an internal strengthening member was situated.

location (Plate 4.7(b) and Fig. 4.3; see Section 4.4.6, item (15)). Furthermore, if mud does get into the drainage hole, then it is inadequate design against corrosion (Sections 5.2, 10.15 and Fig. 5.11).

The corrosion areas of the doors are as follows.

4.4.2 Internal disbondment crevices Fig. 4.1(c), item (11))

A water trap is formed by the internal disbondment of the sound-deadening material (Plate 2.2(a)) or the mastic between the internal door strut and door (Plates 4.5(b)–4.5(d)). The internal sections of the door structure can cause the trapping of

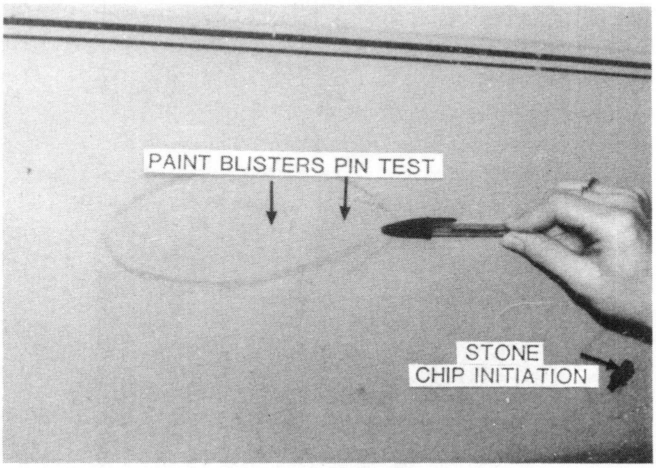

Plate 4.5(b) — The outer surface has been photographed to show the very faint paint blisters which are in line with the biro top.

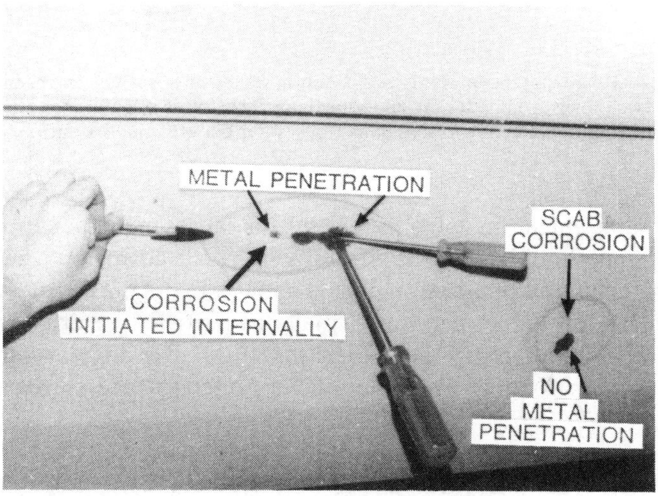

Plate 4.5(c) — These paint blisters were tested for rust penetration using the pin test (Fig. 2.11). The pin hole has been enlarged and a screw driver used as a marker.

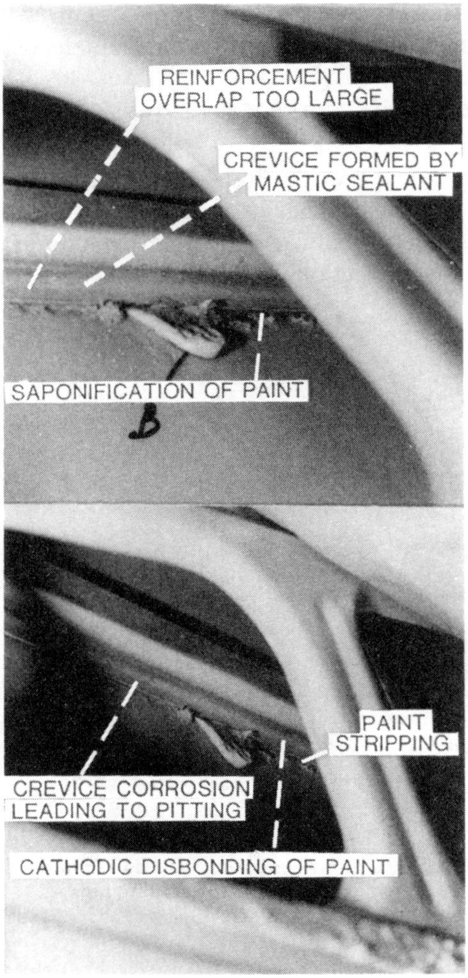

Plate 4.5(d) — Removing the internal door panel shows the mastic which has been applied to prevent the two panels from vibrating and squeaking. This forms a continuous fillet along the lower reinforcement and so readily traps water between the two surfaces.

road de-icing salts which run down the glass. This causes metal penetration half-way up the door, and should not be confused with 'scab corrosion' due to car park damage, stone chips and/or abrasion or side impacts in the pressure eddies down the vehicle side. This turbulence can cause small dirt particles (or dried salt) to impact the car body at right angles to the vehicle direction. The angular particle size can cause very high point loading stresses on the impacted surface(s) (Sections 1.17 and 2.15).

4.4.3 Corrosion around clinch fold (Fig. 4.1(c), item (12))

Corrosion takes place at the exterior of the door clinch fold as a result of a lack of paint in the crevice formed by the hemming fold owing to the poor electrocoat throw

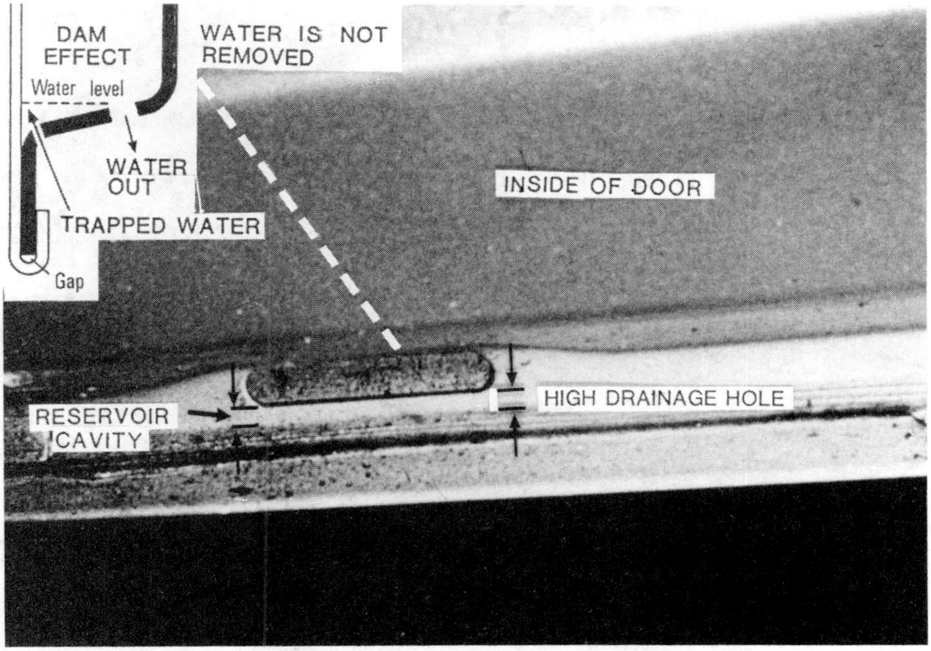

Plate 4.6 — Process holes to doors have a high internal lip which causes water to remain in the internal door crevice for longer periods than is necessary. A cross section of this is shown for clarity. This form of water trap causes crevice corrosion in between the clinch fold (see Plates 2.2(a) and 2.2(b)).

into these areas. This area usually exhibits the first signs of corrosion and, once started, it is difficult to stop. It is accelerated by the concentration of de-icing salts in the crevice of the clinch fold (Section 10.15). (See Section 3.4 concerning the lack of electropaint in the crevice, Sections 3.3, 6.14 and 7.9 for the corrosion mechanism, and Section 10.15 for the reasons for the concentration of de-icing salts in crevice areas). The best procedure is to apply a highly penetrating hydrophobic cavity material both internally and externally and to wash regularly during the winter months (Section 5.20).

4.4.4 Door drainage hole too high (Fig. 4.1(c)), item (13))

Because the window does not make a good fit on the inside ledge as it does on the outer rubbers, road de-icing salts may be taken into the inside of the doors by the vehicle motion (Plate 2.2; Section 2.12) as well as by salty water running down the windows. Some window rubbers are actually designed to produce a well of water against the window glass and to discharge it to the inside of the door. In most model makes, the door drainage holes are too high (see Fig. 4.3(f) and Plates 2.3 and 4.6) and not correctly sited (see Section 4.4.4 and 4.4.6). This internal deposition of salt on the inside panel of the door causes the road de-icing salts to concentrate within the crevice hemming fold and internally initiates rust through to the external door panel (Plates 2.2(a), 2.2(b) and 2.3). In order to differentiate quickly between externally

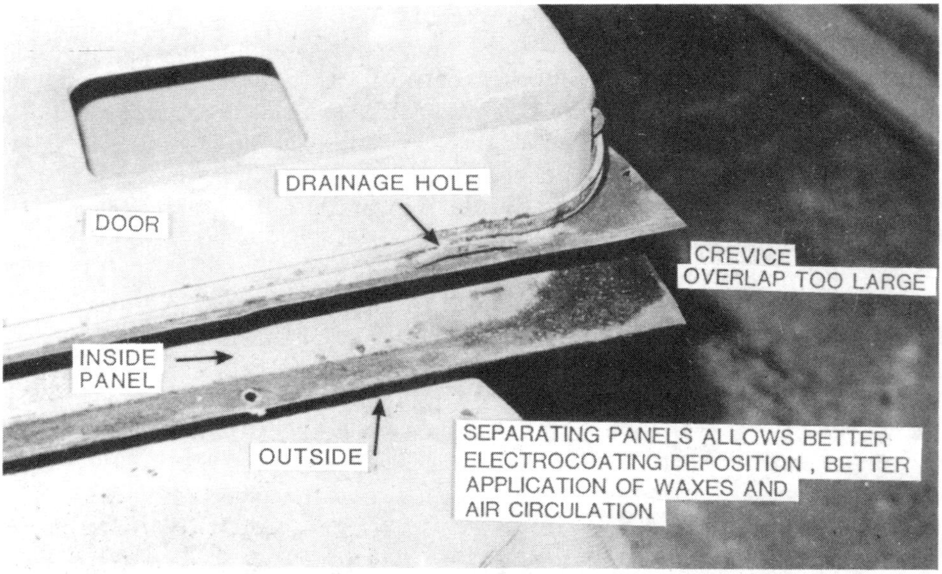

Plate 4.7(a) — The plate shows bad corrosion design because the drainage holes are not at the extreme corners of the doors and the metal overlap is too large. This large overlap creates a reservoir of water for the corrosion process.

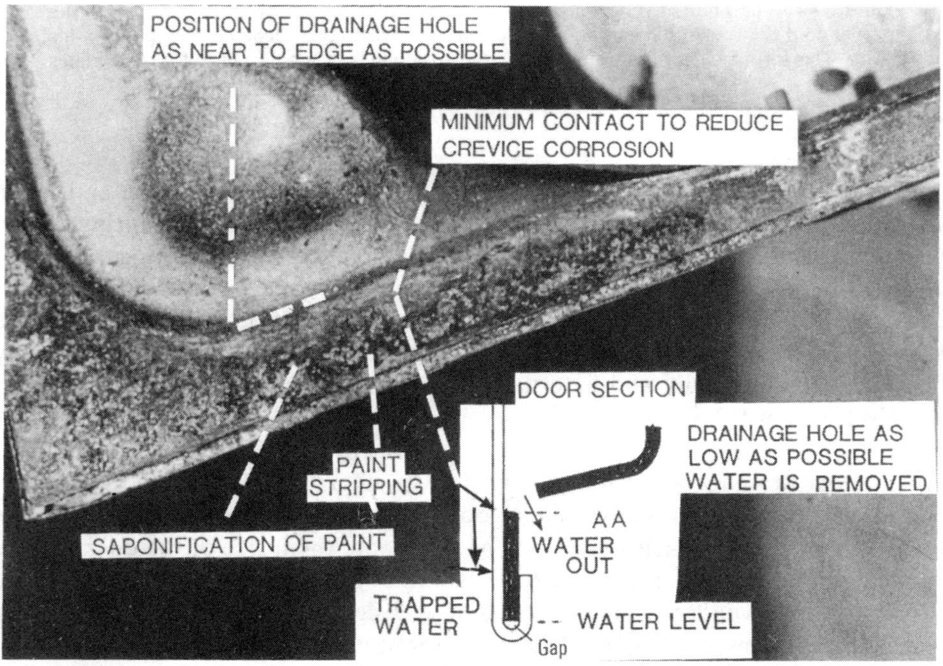

Plate 4.7(b) — Lack of paint within the door hemming fold means that crevice corrosion may take place with the cathodic stripping of the paint film (Section 8.12). The photograph shows the hemming fold clinch flange with part of the return fold ground away to show the reservoir effect formed at the bottom of the clinch fold. When this fault is combined with either a high drainage hole or large overlaps, severe corrosion and paint stripping results. The inset diagram shows a well designed drainage hole in the correct position.

nucleated scab corrosion by road stones thrown up in the turbulent airstream (see Section 2.15) or car park knocks etc. and internally nucleated rust through corrosion, it is best to examine for a 1:1 relationship between the door drainage holes and the external corrosion ('scab' corrosion) on the outside panel. A 1:1 relationship should be taken as the first sign of penetration of metal by internally initiated rusting. Before concluding this, the internal door panel should be removed in order to substantiate the finding. The mechanism given in Section 2.15 may be quicker than removing the internal door trim which may require a special tool to remove the window winder from its stud.

4.4.5 **Corrosion around door brightwork** (Fig. 4.1(c), item (14))

The decorative brightwork on motor vehicles can cause galvanic corrosion (see Section 7.15). The cheapest way to overcome the resultant corrosion on existing manufactured marques is to insert a hydrophobic oil or wax from above the decorative brightwork and to wipe off the excess (a wax is preferred as an oil will in time be displaced). This will not be as good as removing the decorative brightwork and inserting a polythene 'washer' (or electrical insulating tape) between all the fixing points. In the newer models, plastic clips or adhesive strips are used, which overcomes the problem of galvanic corrosion. It should be noted that poor contact of the adhesive strips can cause the local 'pooling' of water behind the strips and the concentration of de-icing salts within the region.

4.4.6 **Door drainage hole positions** (Fig. 4.1(c), item (15))

If the vehicle is 'garaged' outside or washed on a slope, the rain water can collect at the bottom of the car door, and in one corner of the door if the drainage holes are not at both corners or extremes of the door (Fig. 4.3(c)). This is bad anti-corrosion design detailing. This design defect cannot be remedied easily by wax coatings. External drilling is the only sure means of rectification, but the adverse results may be reduced by opening the vehicle door after each 'wetting'.

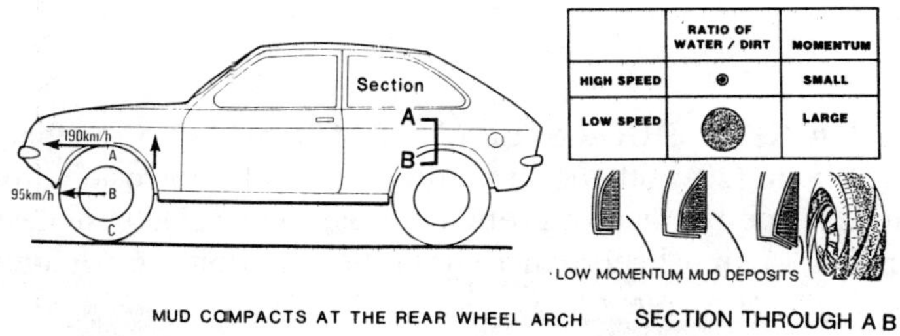

Fig. 4.6 — The section of the rear wheel arch is illustrated in the three diagrams. This area is well out of the trajectory of the wheel spray, but it becomes laden with dirt as a result of the pressure eddies beneath the wheel. These pressure eddies take dirt with a very low water content and deposit it upon the vertical members. The small amount of water carried means that the dirt particles are 'self-compacted' (Section 10.10) rather than draining down to the lower edges of the wheel arch. Attempts are made by the manufacturers to slope this area, but because of the low water content of the road dirt, run-off is not achieved. This area is the most common area to overlook in vehicle maintenance as it difficult to steam clean and hose away the road dirt.

4.4.7 Corrosion around door hinges (Fig. 4.1(c), item (16))
Lack of electrocoat behind the door hinges (Plate 4.4), can give rise to corrosion with the added disadvantage of 'paint stripping' (Sections 8.11 and 8.13; Figs. 8.18 and 8.21) around door hinges. This arises as a result of the very narrow crevices which cause a lack of throw of the electrocoat behind the hinge and is one good reason for the application of a penetrating hydrophobic wax–oil to the crevice. It may be necessary to loosen the top screws of the hinge.

4.4.8 Corrosion around door rubbers (Fig. 4.1(c), item (17))
Corrosion can be nucleated around the door and glass rubbers by the concentration of de-icing salts beneath rubbers and can be eliminated by the application of a penetrating hydrophobic waxes–oils beneath the rubbers (see Sections 8.8–8.10).

4.4.9 Erosion of door edges (Fig. 4.1(c), item (18))
Erosion of door edges is due to the front wheel spray's catching the leading edge of the next door. It is also usually due to the poor alignment of the two doors along the vehicle direction.

4.4.10 Rear door sections adjacent to rear wheel arch (Fig. 4.1(c), item (19))
Erosion of the rear door wheel arches is due to the front wheel spray's catching the leading edge of the wheel arch. It is usually due to a combination of poor anti-corrosion design and/or the poor alignment of the trailing edge of the rear door with the wheel arch along the vehicle direction.

4.4.11 Summary
The best advice for combating the corrosion of doors is to inject with a very thin highly penetrating hydrophobic wax–oil (Sections 8.8–8.10) and to wash regularly on both 'sides of the door', paying particular attention around the door returns and folds. Open the doors to allow water to drain from the door to allow for both badly designed and badly positioned drainage holes. This is most important if the vehicle is garaged on uneven or sloping ground.

4.5 SILLS, ROCKERS, DOOR STEP AND JACKING POINTS (Plates 4.8–4.14)
See Fig. 4.1(c): item (20), sill–rocker section; item (21), corrosion around jacking points; item (22), perpendicular protuberances; item (23), best design for sill–rocker; item (24), rear wheel arch returns. See Section 4.6 for outward flared sills–rockers (Fig. 4.1(c), item (28)).

4.5.1 Sill–rocker section (Fig. 4.1(c), item (20))
In Section 2.13 dealing with passenger comfort, it was suggested that an air continuity path between the sills–rockers and passenger compartment was a requirement of the electrocoating processes as discussed in Section 2.11 and Chapter 9. This air path is shown in the photograph of a vehicle body without any internal fixings (Plate 2.5). This means that replacement air (which includes low momentum road dirt and de-icing salts) can be drawn through the sills–rockers (Plate 2.4) and over the

rear quarter panels or rear wheel arches (Plates 2.6(a) and 2.6(b)). The air drawn through the sills–rockers originates from the regions of densest spray as these electrocoat process holes to the sills–rockers are directly over the front wheel track (Plates 4.15(c), 4.16, 4.17, 4.18). This mechanism of building-up the aggressive microclimate on internal (out-of-sight) body panels is not appreciated by the automotive designers as it is not shown up by the manufacturer's low speed salt dip tests which have been their characteristic means of accelerating their corrosion tests (discussed in Sections 2.10 and 10.22). The ingress of the aggressive environment becomes more acute as the vehicle passenger compartment or body becomes bigger (simply because the volume of paint contained in the body is greater and, for the same manufacturing vehicle line speed, the faster will have to be the emptying of the electropaint). Hence, taking as an example a commercial vehicle (Plate 2.9(a)), these drainage holes have to be very much larger than in large passenger cars simply to empty the electrocoat from the van body in a given time. Therefore, when driven, a greater volume of air, and therefore more wheel spray, can be drawn into the sill/rocker sections. This is shown in Plate 2.9(a) for a commercial vehicle where the outer panel has been removed to show the internal microclimate and associated corrosion.

4.5.2 Corrosion around jacking points (Fig. 4.1(c), item (21))

The jacking points are usually incorporated into the sills–rockers and the adjacent area often requires reinforcement. Therefore, usually associated with the jacking point, or in very close proximity to the jacking point is an electrocoat drainage hole (Plates 4.8(a)–4.8(c)). Most manufacturer's jacking points are usually slung beneath the sill–rocker and so present an extra perpendicular protuberance in the high speed wheel spray, which atomizes on impact with the protuberance (Plates 4.9(a), and 4.9(b), 4.10 and 4.11; Fig. 4.2). This atomized spray is sucked into the drainage hole adjacent to the jacking points and causes corrosion of the internal jacking point reinforcement, so that with time the corrosion process allows the whole jacking area to collapse as a result of the loss of strength by corrosion (Plates 4.8(c) and 4.9(a)). For best results against corrosion, any drainage holes adjacent to a jacking point (and forward facing) should be filled with mastic (ordinary builders' mastic is sufficient) (Plates 4.8(b), 4.9(a), 4.9(b), 4.10 and 4.11). This will prevent the front wheel spray from being 'jettisoned'into the sill–rocker member by the protruding jacking point reinforcement–member.

4.5.3 Perpendicular protuberances (Fig. 4.1(c), item (22))

Inter-weld flanges within the sill (Plates 4.17(a) and 4.17(b)) present a perpendicular protuberance in the high speed wheel spray, which atomizes on impact with the protuberance. This atomized spray is sucked into the drainage hole which is adjacent to the inter-weld and causes corrosion of the sill–rocker member from the aggressive microclimate contained within the box section. For best results against corrosion, any drainage holes adjacent to a protruding inter-weld should be filled with mastic, a rubber grommet, or a one-way valve (Section 5.19; Fig. 5.10). This will prevent the front wheel spray from being 'jettisoned' into the sill–rocker member.

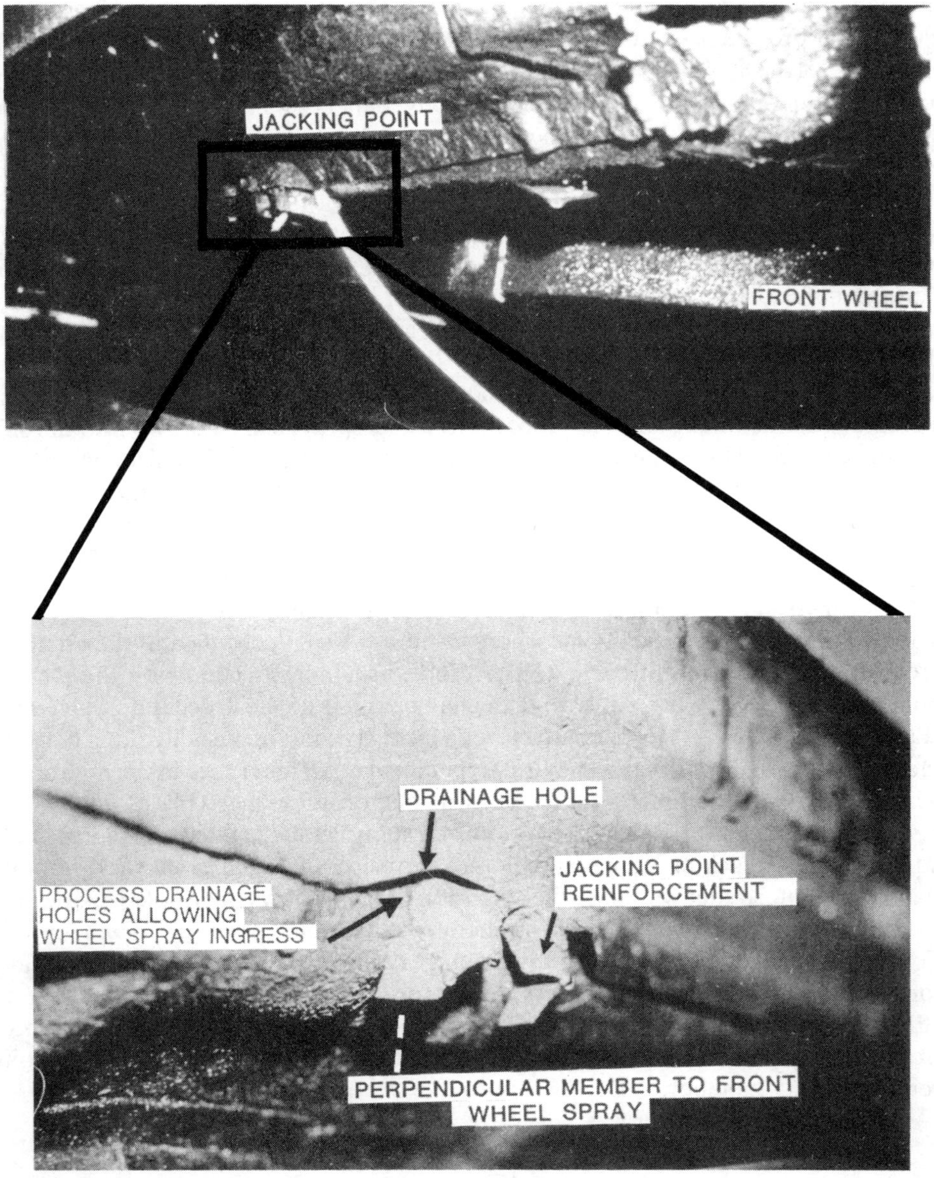

Plate 4.8(a) — The juxtaposition of drainage holes with underbody protuberances create a cloud of micro-droplets when impacted by the vehicle at high speed (Fig. 4.2). In some cases these protuberances form part of the internal body suspension, but in most cases they are under-slung jacking points as depicted in these plates. Whatever their function, their presence is contrary to good anti-corrosion design. Their adverse effect is not detected by the motor manufacturers low speed accelerated tests which are described in Chapter 10. This plate shows a view of the jacking point from the road–wheel interface.

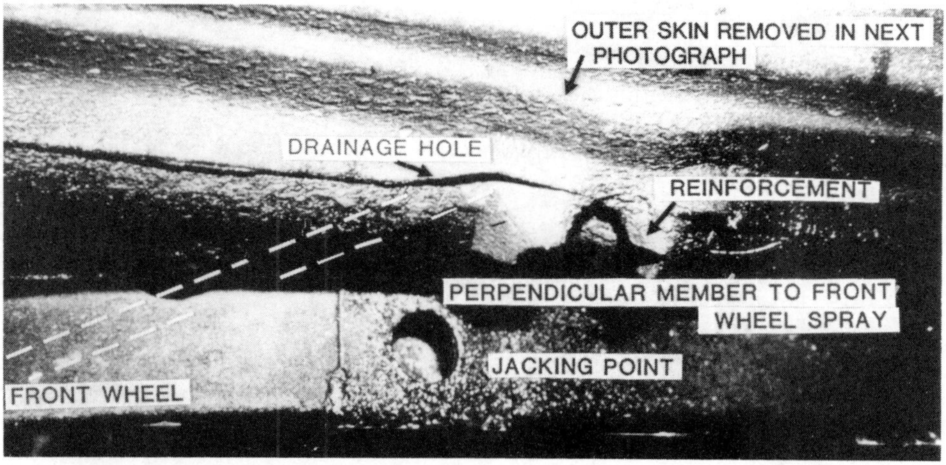

Plate 4.8(b) — View of the same jacking point from the side showing the electro-coat drainage hole to the sill/rocker associated with it. In this case the reinforcement to the jacking point forms a protuberance beneath the vehicle body which scoops and jettisons the front wheel spray into the box members as shown in Plate 4.8(c). The process is aided by the extraction of stale passenger air (Fig. 2.7) from the passenger compartment and the pressure difference when vehicles overtake at high speed (Figs. 5.5 and 5.6). These forward facing drainage holes should be filled by the manufacturer as detailed in Fig. 5.11 and Table 5.4

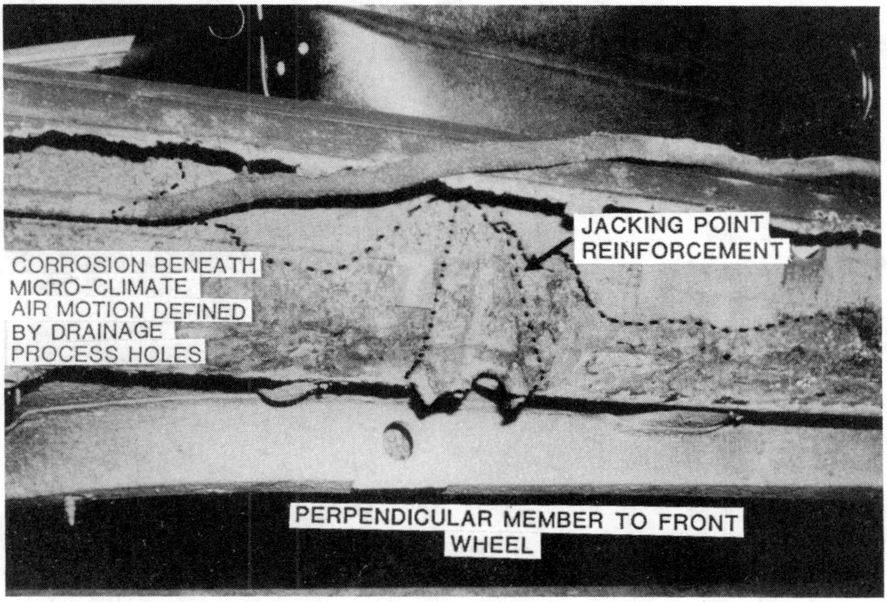

Plate 4.8(c) — The external panel to the sill–rocker to Plates 4.8(a) and 4.8(b) has been removed to show the de-icing salts and corrosion around the jacking point reinforcement. Internal, out of sight corrosion will cause the jacking point to collapse when loaded.

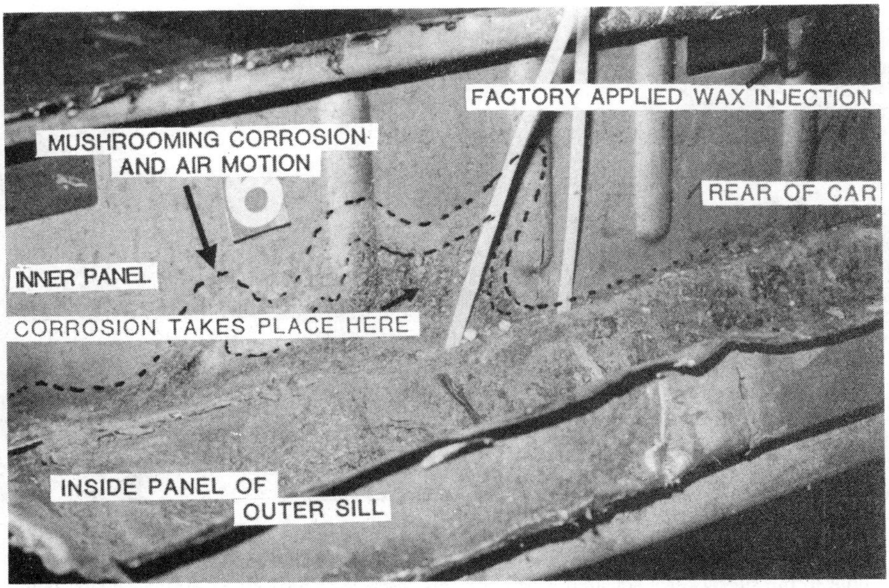

Plate 4.9(a) — Jacking point drainage holes seen from inside the sill–rocker. The outer sill has been cut away at the top and pulled down so that both internal surfaces can be seen. Tubes have been placed in the drainage holes on either side of the jacking point, and the evidence for salt and water ingress causing corrosion is clearly shown. There is also evidence of wax injection but only a small amount of wax has been sprayed due to the thixotropic nature of the wax.

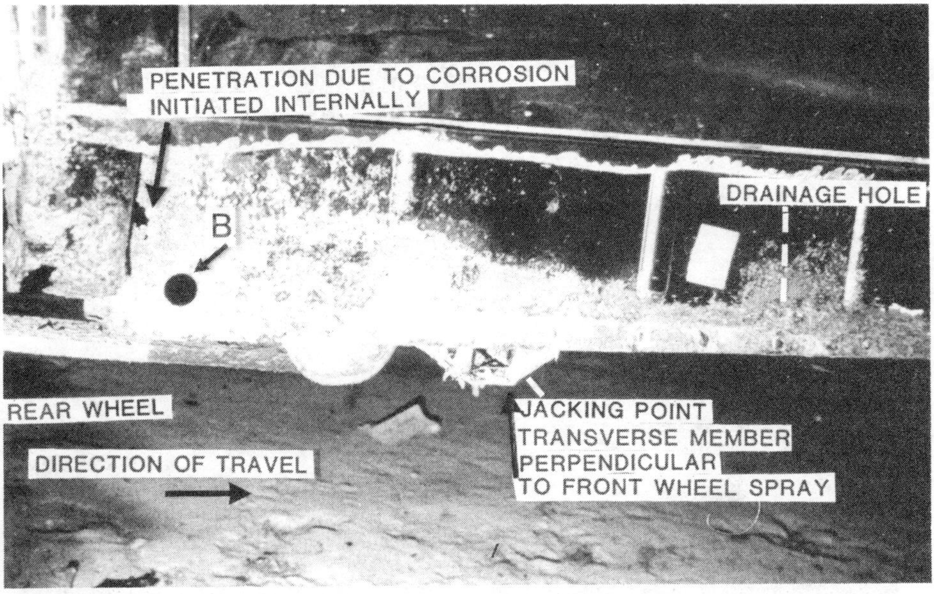

Plate 4.9(b) — The outer sill has now been totally removed to show the rear jacking point, (which is designed the same as the one illustrated in Fig. 4.9(a)). However this plate shows a hole in the inner sill, which leads to the low pressure area ahead of the rear wheel. The deposited low momentum micro-droplets show the air motion which is from the drainage holes to exit at the hole in the inner sill. This causes severe corrosion around the inner rear wheel panel and spring hanger. Compare the design fault with that illustrated in Plate 4.14(a).

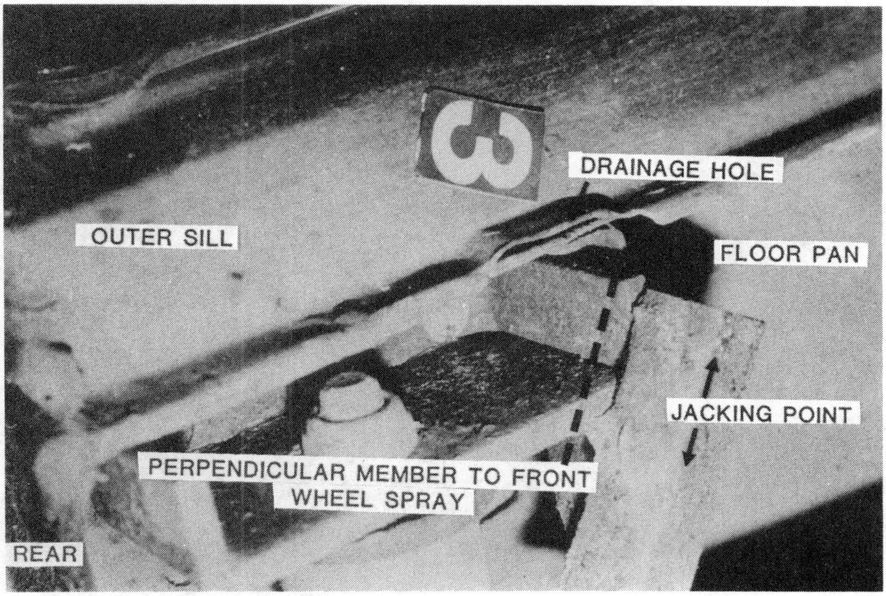

Plate 4.10 — This plate shows the jacking points of a different manufacturer. Similarly the juxtaposition of drainage hole and jacking point reinforcement jettison the wheel spray into the box members.

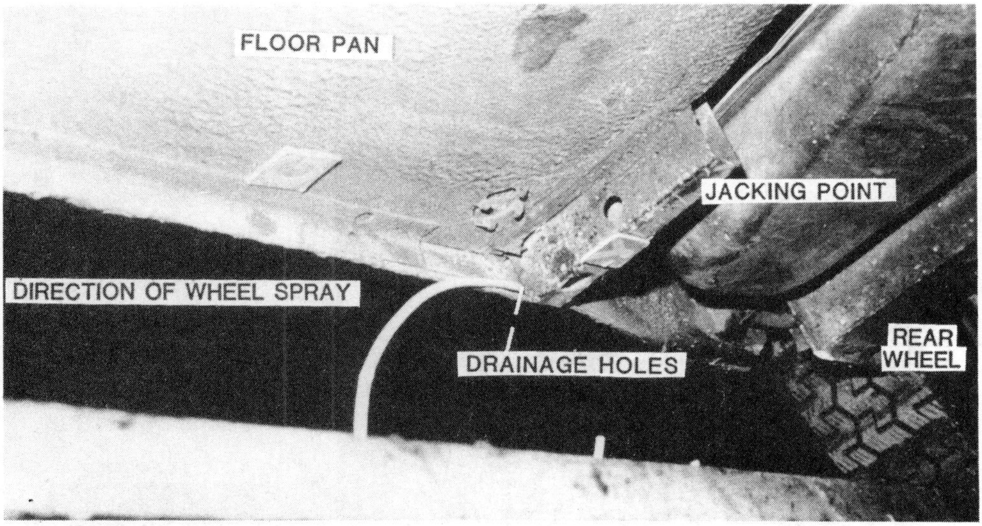

Plate 4.11 — External view of the drainage hole associated with a new model vehicle showing the proximity of the sill–rocker drainage hole to the externally slung but protruding jacking point.

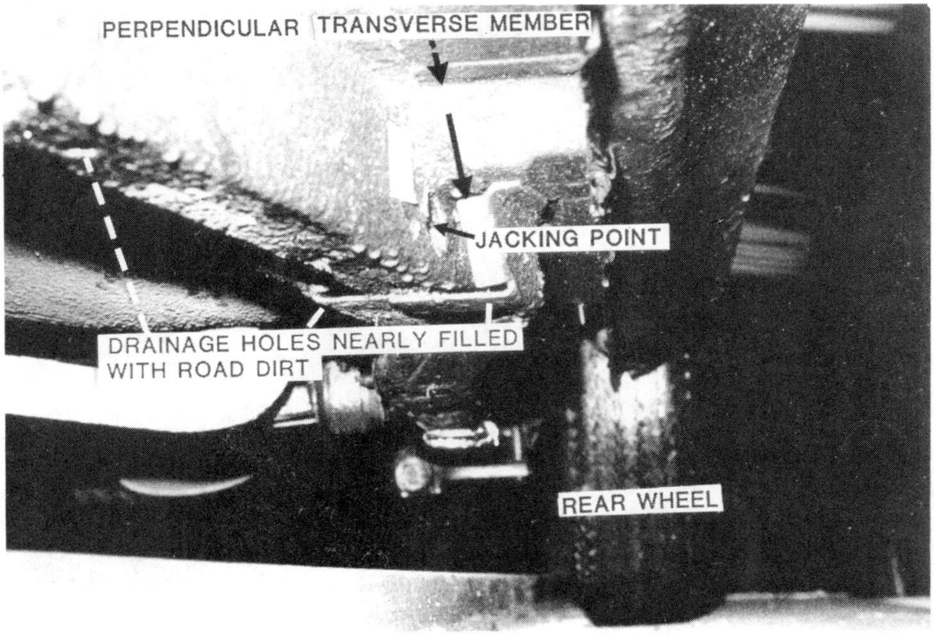

Plate 4.12(a) — This plate highlights a bad design fault which was not shown up by the motor manufacturer's 20 mph accelerated tests procedures. This is a view looking from the front wheel towards the rear wheel and shows the box member which runs across the front wheel spray. The member stops half way along the car body (adjacent to the B pillar) to form a jacking point (Plate 4.12(b)). Behind the jacking point hole is a perpendicular member which acts as a spray generator. The circular hole for the jacking arm presents a certain cross sectional area to the front wheel spray (Fig. 5.5).

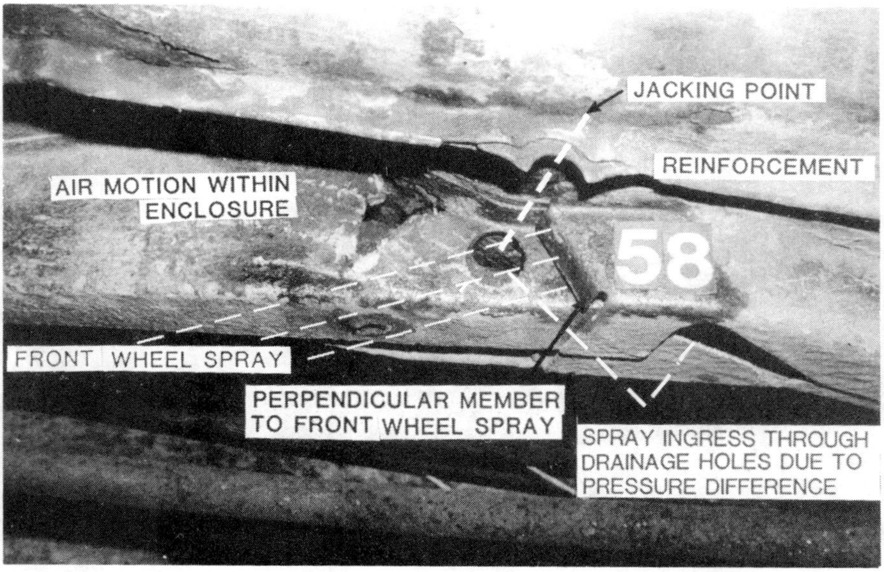

Plate 4.12(b) — A view of the jacking point taken at right angles to the vehicle showing the protruding member which squashes the road spray into the jacking point hole. The mud deposits can be seen within the holes.

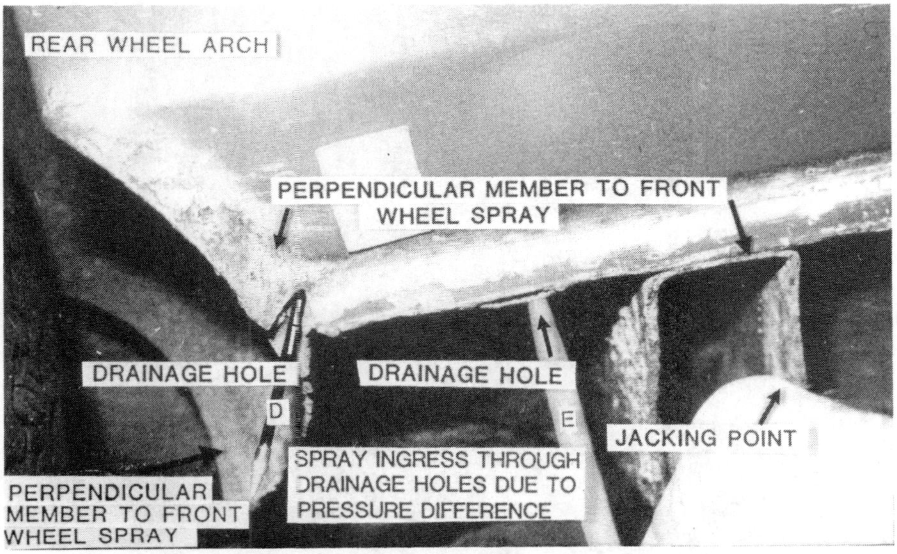

Plate 4.13(a) — It is a requirement that the outer sill–rocker is welded onto the inner rear wheel arch. Here the welding has been undertaken on an outward contoured member. Electrocoat drainage and various pressing requirements have produced a hole (shown up with the biro 'D', within the outward flared member. The second pointer 'E' shows a drainage hole which is in the low pressure area of the protruding jacking point. Thus any air will be sucked out of this drainage hole only to be replaced by spray generated at the high pressure area 'D'. A secondary bad design fault is the protruding rear wheel arch which is perpendicular to the direction of motion.

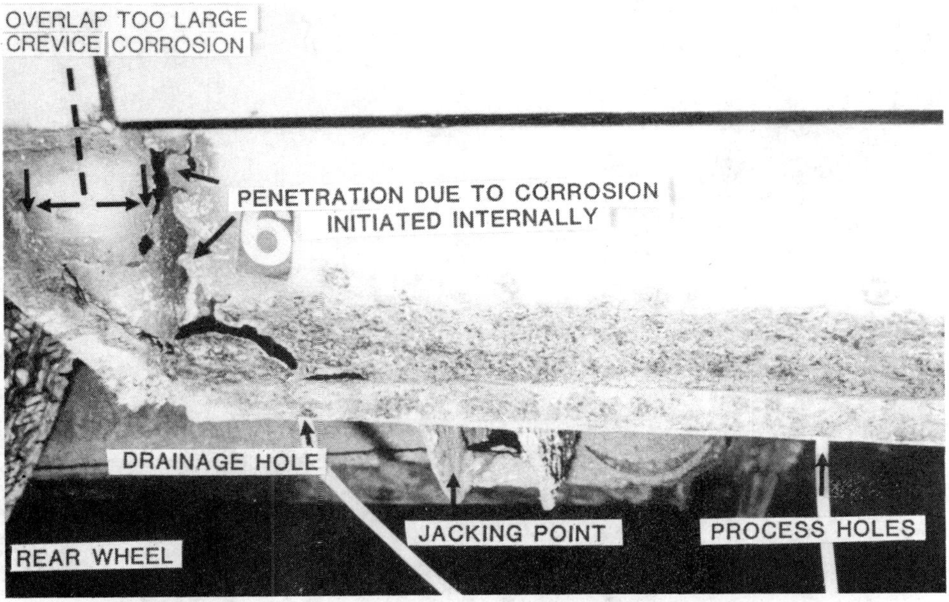

Plate 4.13(b) — Although these sill–rockers are pressure vented from the heater scuttle, a lot of leaves and other debris can be mixed in the aggressive road environment which comes through the holes shown in Plate 4.2(b). Hence the crevice at the rear of the sill rocker has become blocked causing complete rust penetration.

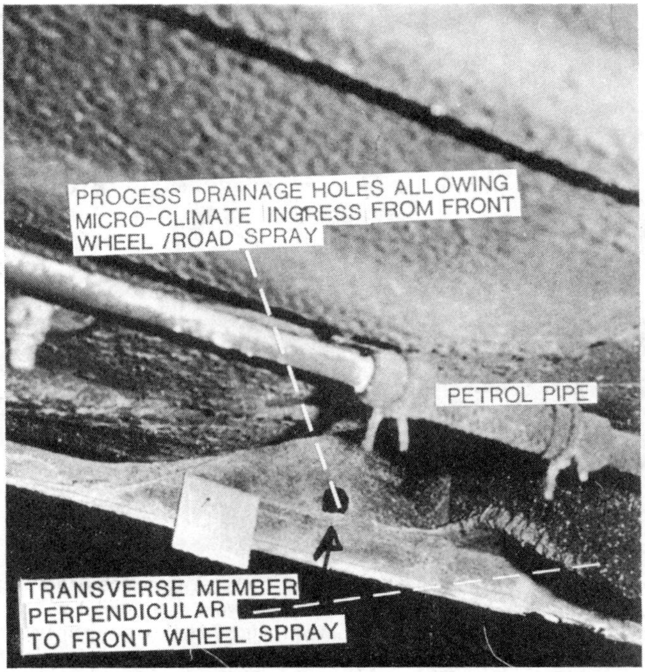

Plate 4.14(a) — A badly positioned process hole which is to the rear of a cross member. The spray and water run off is shown by the dirt residue which partially covers the circular hole.

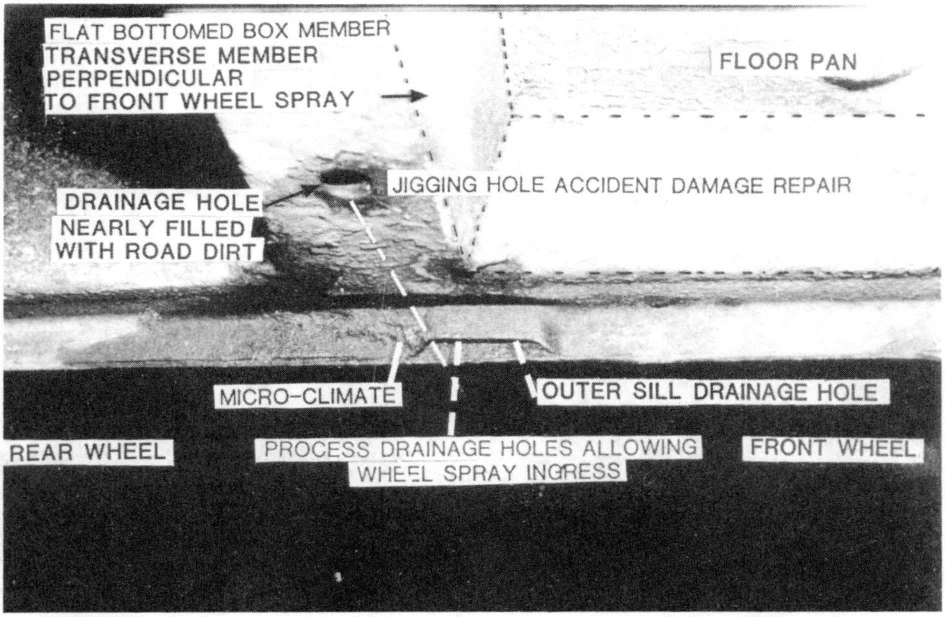

Plate 4.14(b) — Box members which are slung below the floor pan act as sources for the generation of micro-droplets (Fig. 4.2). The transverse box member is completely filled with mud because this drainage hole presents a very small angle to the front wheel spray (Fig. 5.5).

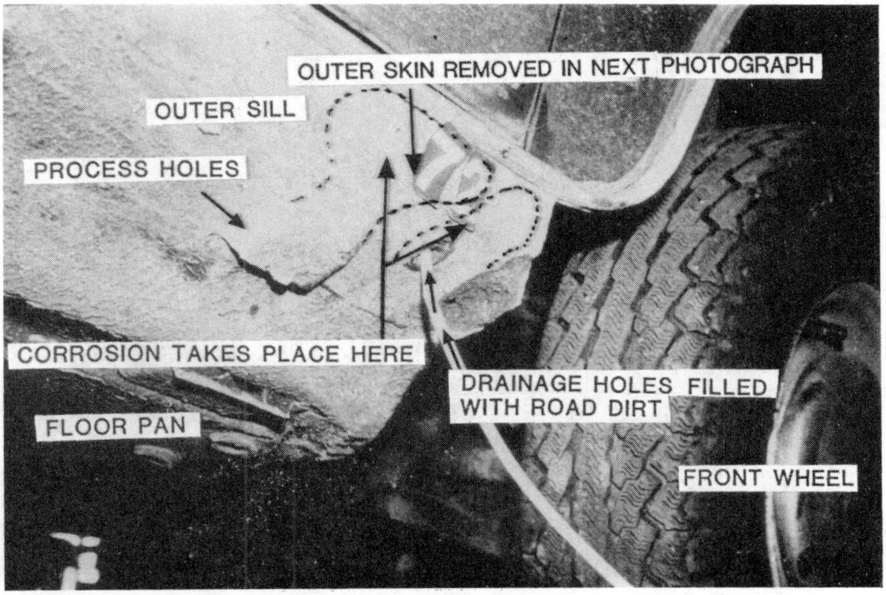

Plate 4.15(a) — Perhaps the worst example of the complete disregard for anti-corrosion design is the drainage hole which is directly over the front wheel spray. The position of this drainage hole caused a number of problems with wax injection warranties as this drainage leads to a cavity which is nearly self contained (Plate 4.15(b).

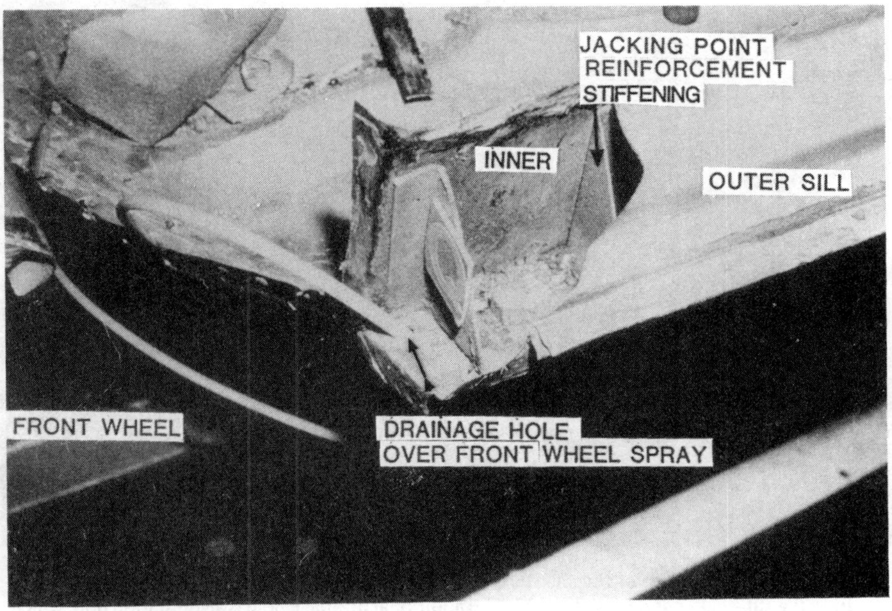

Plate 4.15(b) — This plate shows the outer sill–rocker panel of Plate 4.15(a) removed to expose the inner cavity box member which forms the reinforcement for the jacking point.

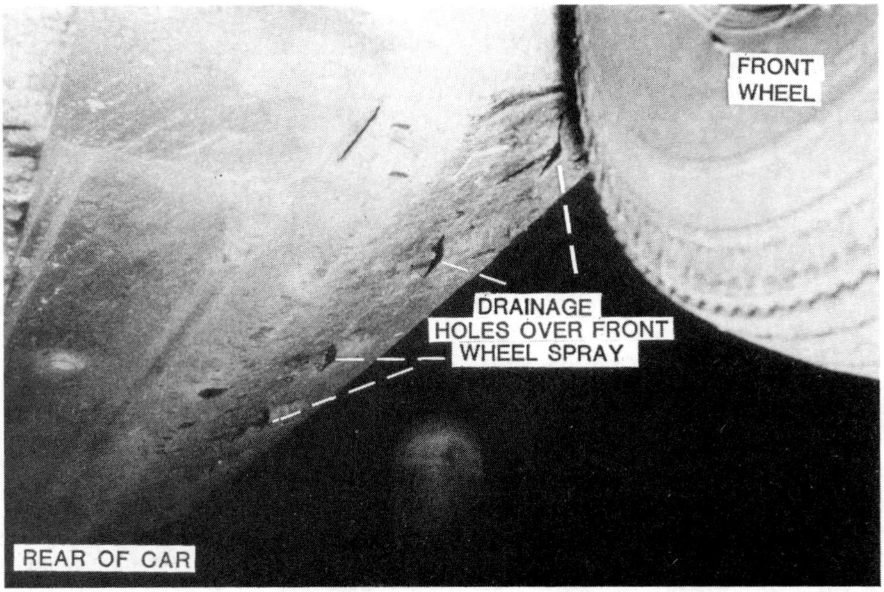

Plate 4.15(c) — If the road de-icing salts entered the sill–rocker by means of large momentum particles, these protected sill–rocker drainage holes would be the ideal solution. However the ingress is not by large momentum particles, but by small momentum particles which can easily change direction in the pressure eddies beneath the vehicle. These small momentum particles can be pushed into the sill–rocker by the pressure variations as vehicles overtake one another (Figs. 5.6 and 5.7).

Plate 4.16 — A close up of the sill–rocker showing the original manufacturer's drainage hole within the sill–rocker as denoted by 'A'. The additional drainage hole is a ventilation hole which it was thought, would reduce corrosion by making the sill–rocker box member less of a sweat box. This additional drainage hole increased the flow of road de-icing salts into the cavity member during the winter de-icing programme.

4.5.4 Sill–rocker and rear wheel arch returns (Fig. 4.1(c), items (23) and (24))
Other drainage holes (Plate 4.16) within the sill–rocker suck up the road de-icing salts and therefore accelerate the ingress of the aggressive microclimate within the sill–rocker structure. Only two drainage holes should be left open within this area, one at the front (and protected from the front wheel spray by a mudflap) and the other at the rear (protected by a cheap one-way (air) valve which can be made from ordinary electrical insulating tape (Section 5.19). Mudflaps should be fitted as these protect the sill–rocker from both road and wheel spray. It should, however, be remembered that the 'go-faster' type of mudflap which is very short and which is being used at present does not perform in the manner which is given above because it is too short. If mudflaps are fitted to the wheel arch, check for road de-icing salts in the crevice formed by the mudflap and wheel arch return (Item (24)), as these will be concentrated within this crevice (see Section 10.15). Once again, a high creep wax and regular washing are advised. It should be noted that long mudflaps also cause greater pressure eddies around the sides and doors and this gives rise to a higher incidence of external paint damage when the vehicle is driven at high speeds.

One should expect imminent sill–rocker corrosion and penetration and an examination of the sill–rocker for a 1:1 relationship with drainage hole and external 'scab' corrosion which can arise from the turbulence around the vehicle body. This turbulence causes road debris to impact the surface of the car body at right angles to its motion. Methods of differentiation are given in Section 2.15 (see Fig. 4.1(c), item (20)). However, Plates 4.14–4.18 give other causes of corrosion within the sill–rocker member.

4.6 REAR WHEEL ARCHES AND WHEEL WELLS (TWO- AND FOUR-DOOR MODELS) (Plates 2.6(a)–2.6(c), 2.9(b))

See Fig. 4.1(c): item, (24), open area at the base of the C pillar; item (25), rear wheel arch pitting corrosion; item (26), corrosion due to petrol pipes–breathers; item (27), rear wheel arch lip return; item (28), outward flared wheel arches.

4.6.1 Rear wheel arches (four-door models) (Fig. 4.1(c), item (24))
The outer panel to the rear wheel arch of four-door vehicles can have an open seam at the bottom of the C pillar adjacent to the sill–rocker. This hole arises from the double cavity formed by the inner and outer rear wheel arches. The larger external panel meets the rear 'dog leg' at the base of the sill–rocker where the manufacturer applies mastic to 'fill' this small crevice. As the mastic is very viscous, it does not flow into the crevice and open spaces are often left. These open areas will allow road spray to be sucked up into this area and cause either crevice corrosion at the steel–mastic–steel interfaces or metal penetration along the rear dog leg.

In some cases the manufacturers have pushed foam into the rear dog leg in order to reduce noise. This foam, however, filters out the low momentum spray containing the de-icing salts which would normally deposit within the rear wheel and this is concentrated at the foam–metal interface and thus causes metal penetration at the interface.

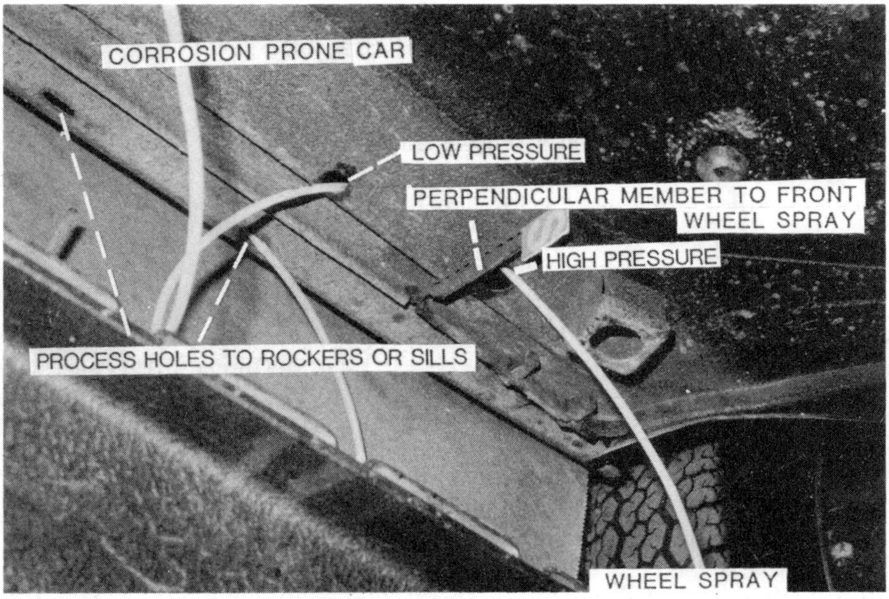

Plate 4.17(a) — The design faults illustrated are (1) a large number of holes in the sill–rocker which were to ventilate the box members to reduce corrosion. (Applying the mechanism of micro-climate build up given in this book (Section 2.13, Figs. 5.6 and 5.7) would predict a greater ingress of contaminants.) (2) A process hole between the front wheel and an adjacent transverse interweld. This acted as a scoop to jettison the front wheel spray into the sill–rocker. Severe corrosion resulted.

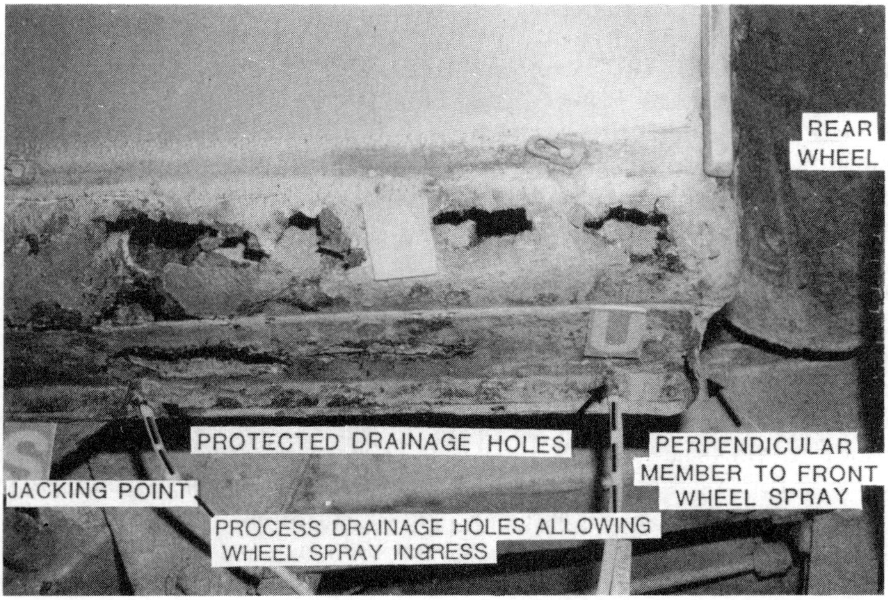

Plate 4.17(b) — Markers show the position of the 'protected' drainage holes to the sill-rocker. The perpendicular member to the front wheel spray caused the low momentum spray to be pressurised as the vehicle moves along at speed. The 'pressurised spray' enters the drainage holes due to the lower pressure within the sill-rocker as a result of the passenger air extraction system depressurising the whole cavity structure.

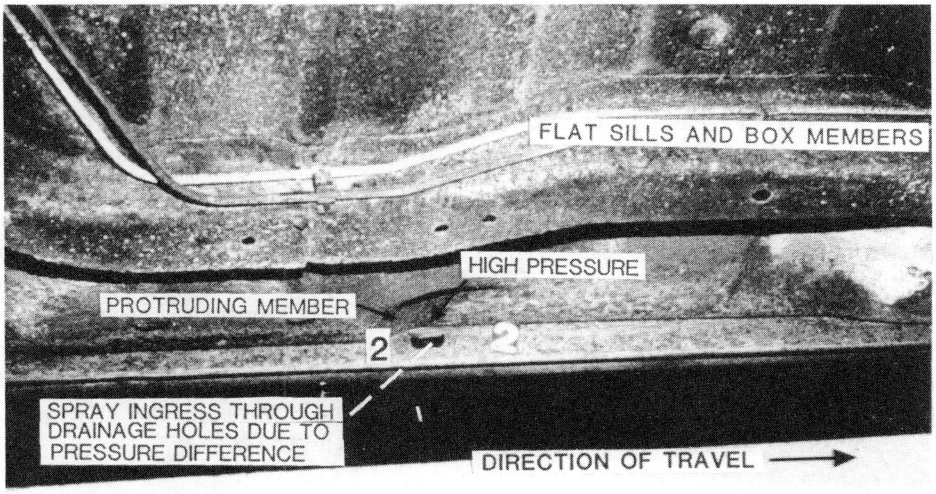

Plate 4.18 — A view of a different manufacturer's sill–rocker member over the front wheel track where the spray developed is very dense. There are many sill–rocker drainage holes in this area and some are located in the worst possible positions, e.g. those in front of a panel curvature denoted by '2'. These holes are ventilation holes to keep the box member dry. In fact these ventilation holes allowed a greater ingress of road salts into this rust prone area. These ventilation holes show how it is possible to grasp at ill-founded concepts, namely ventilation and drying. The solution resides in the opposite to increased ventilation, that is exclusion of the micro-climate. This can only be achieved by correct design.

4.6.2 Rear wheel arches (two- and four-door models) (Fig. 4.1(c), item (25))

In Section 2.13 dealing with passenger comfort it was suggested that an air continuity path between the sills–rockers and passenger compartment was a requirement of the electrocoating processes. During manufacture, the vehicle enters the electrocoating bath at an angle of 15° (Plate 3.1) and, for the electrocoat to enter the sill–rocker box member, an air exit hole must be placed on the top of the sill–rocker into the cavity made by the rear wheel arch lower panel (closest to the rear wheel) and the rear wheel arch outer panel (closest to the rear door). This hole, shown in Plate 2.6(a), allows the air drawn through the sills/rockers, originating from the regions of densest spray, to deposit over the rear wheel arch where the electrocoat is very thin owing to the poor throw (Chapter 9). This mechanism of build-up of the aggressive microclimate on internal (out-of-sight) body panels is not appreciated by the automotive designers, garage repair trade and the general public. This deposition of road salts causes internally nucleated corrosion around the outer panel of the rear wheel arch (Plate 2.6(c)). This corrosion is very difficult to rectify because it is internally nucleated in an inaccessible region (Plates 2.6(a), 2.6(b) and 2.9(b)). No amount of surface preparation from the outside (Plate 2.6(c)) will arrest the corrosion process. As in all cases of corrosion it is easier to stop it by good design and adequate protection than it is to stop corrosion when it has started.

4.6.3 Corrosion due to petrol pipes–breathers (Fig. 4.1(c), item (26))

Road dirt collections as a result of design defects in petrol tank breathers and/or petrol pipe fillers. These areas should be washed periodically and it is imperative to ensure that a good continuous anti-corrosion coating is present (unaffected by shadowing if the application is by after body build and assembly techniques).

4.6.4 Rear wheel arch lip return (four-door models) (Fig. 4.1(c), item (27))

Some vehicles have a small vertical line at the top of the rear wheel arch which marks the juncture of the outer panel and the panel which forms the rear dog leg. This vertical line would normally contain a spot weld if it were not for the fact that the spot weld would leave a 'dimple' (Fig. 8.7). The overlap is not brazed (as it should be to satisfy the principles for good design (Fig. 5.4) and is covered with mastic in order to prolong the onset of the visible corrosion staining. The brown corrosion stains arise from this overlap because it is effectively joined to the low momentum mud poultice beneath the rear wheel arch (Fig. 4.6). Furthermore, this overlap can be flexed and the brittle mastic can be broken by the slamming of the rear doors. This aspect and the repair of the area are discussed in Sections 5.11.6 and 5.20.

4.6.5 Rear wheel arch lip return (Fig. 4.1(c), item (27))

In Section 2.7, the size of a spray droplet was defined by its momentum = mass multiplied by velocity. The mass was the combined mass of the water and road dirt droplet. The magnitude of this momentum $M = mv$ gives a parameter which characterizes how easy it is for that particle to alter direction and the smaller the quantity $M = mv$ the easier it is for the particle to follow the turbulent air around the moving vehicle. Nature also dictates, by the surface free energy theory, that there must be a nucleus upon which the water can form (Section 10.10). Thus the small amount of water present upon each road dirt particle means that there is insufficient to wash away and clear the road dirt which is deposited out of the wheel spray. This low water content also means that this road dirt is artificially more densely compacted by the phenomenon of surface tension into a high density mud pocket upon, and around, the rear wheel arch lip return. This is an area which is out of direct water washes from the wheel, a steam cleaning gun (Section 4.13; Fig. 4.6) or a garden hose. Unless very special precautions are taken, this wheel arch return is also out of the direction of spray from the hydrophobic wax–oil spray head and the manufacturers' PVC plastisol underbody protection.

The slower the car travels the larger will be these water droplets and the higher the ratio of water:dirt which they can contain, and so the rear wheel arch is only cleaned in the direct line (its trajectory) of wheel spray. The wheel spray from slowly moving vehicles has a high momentum and so travels in straighter lines and these high momentum water droplets are not affected greatly by the pressure eddies beneath the car body.

4.6.6 Outward flared wheel arches (Fig. 4.1(c), item (28))

Corrosion of this area can be caused by a badly fitting rear door so that, when closed, it does not protect the rear wheel arch from the erosion effects of the spray from the front wheels. The remedy is to alter the closure of the rear door so that the outside panel remains flush with the rear wheel arch.

A more common cause of corrosion is an outward flared wheel arch, which can catch directly the front wheel spray and give rise to erosion–corrosion. This usually happens very quickly, and to meet the terms of the Canadian Voluntary Anti-corrosion Code (Section 1.12), the manufacturer's have placed black plastic guards on these areas of erosion (Plate 4.20). The fitting of front mud flaps tends to increase the turbulence of the spray from the front wheel. If the vehicle is not designed with an outward flared wheel arch, and the corrosion around the wheel arch is not due to paint damage by 'turbulent grit particles or scab corrosion', then the corrosion problems may be severe as they are caused by road de-icing salts' being dragged through sills/rockers and deposited, internally, along this crevice area (see also Section 4.6.2, and Plate 2.6(a)). The first signs of metal penetration by corrosion are usually beneath the 'extractor vent' where atmospheric moisture can keep the de-icing salts damp for long periods of time and during outside overnight 'garaging'. One should expect imminent sill–rocker corrosion and penetration and an examination of the sill–rocker for a 1:1 relationship with drainage hole and external 'scab' corrosion which can arise from the turbulence around the vehicle body. This turbulence causes road debris to impact the surface of the car body at right angles to its motion. Methods of differentiation are given in Section 2.15.

4.7 PASSENGER AIR EXTRACTION GRILLS (Fig. 4.1(c), item (29))

Passenger air extraction grills can be placed in the following locations:

(1) On the outside of the tailgate and open to the driving rain and wind. The vehicle will have a high pressure region and a low pressure region which can cause the moist air to be blown through the passenger air extraction system into the tailgate or boot–trunk area. The internal trim in these areas need not offer a great deal of 'wind resistance' to the passage of wet air.
(2) Underneath the tailgate and not open to the driving rain.
(3) In the side panels of the vehicle and open to the driving rain. Once again, pressure differences can develop across the car body.
(4) Behind the roof line and above the rear window. This will cause least corrosion as the pressure difference is not developed when the vehicle is garaged outside.
(5) Adjacent to the boot–trunk.
(6) At the trailing edge of the car doors (Fig. 4.3(b)).
(7) Behind the rear light clusters.

Air extraction systems work more efficiently at high speed. This is perhaps more noticeable in the summer, when a pleasant environment can be achieved within the car with all the windows shut, providing that the vehicle is travelling fast. At lower speeds it is necessary to either open the windows or put on the electric fan to keep the same environment. Extraction grills within any of the above positions will cause the deposition of salts in the sills–rockers by this method (Section 2.13) but another mechanism of road spray ingress to these areas of the rear wheel arch is discussed in Section 5.12 (Fig. 5.6).

Corrosion can be caused by a nearby passenger air extraction grill. If corrosion is most advanced beneath the grill, suspect water penetration, via the grill, into the area together with a lack of hydrophobic wax penetration and/or application (Plate 4.19(a) and 4.19(b)). Water penetration can be a result of pressure eddies due to over-taking vehicles Fig. 5.6. Cheap one-way air valves can be made (Section 5.19,

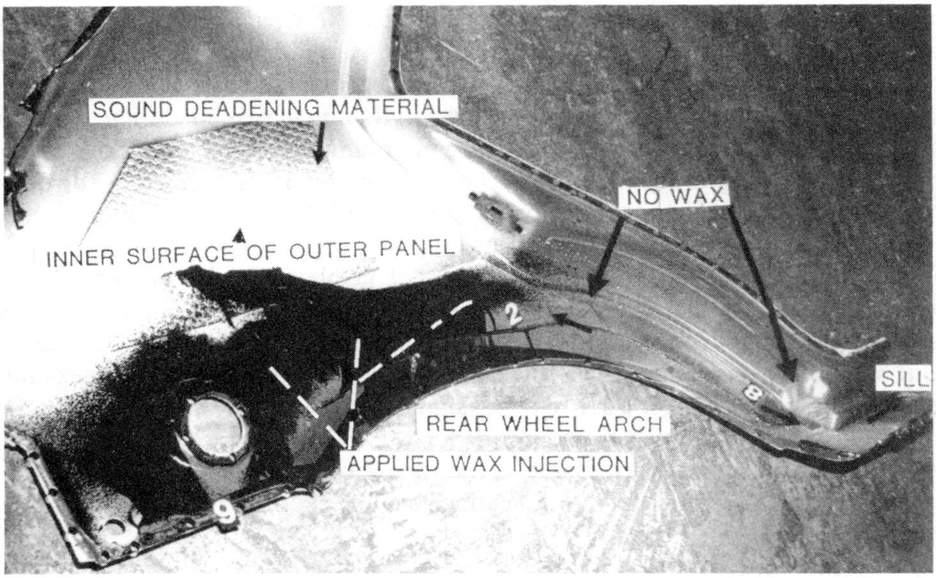

Plate 4.19(a) — A manufacturer offering a six year warranty with this type of application will want a good lawyer. The photograph shows a poorly covered internal surface of the outer panel taken from over the rear wheel arch where all the salty air stream passes during driving in the winter. The manufacturer's demand that the waxes should not run out and drip whilst the rest of the assembly is in progress. This requires a very thick material, which as a result is difficult to spray evenly. Note that these thixotropic waxes will not penetrate the crevice interwelds and will produce bridged areas (Fig. 3.9).

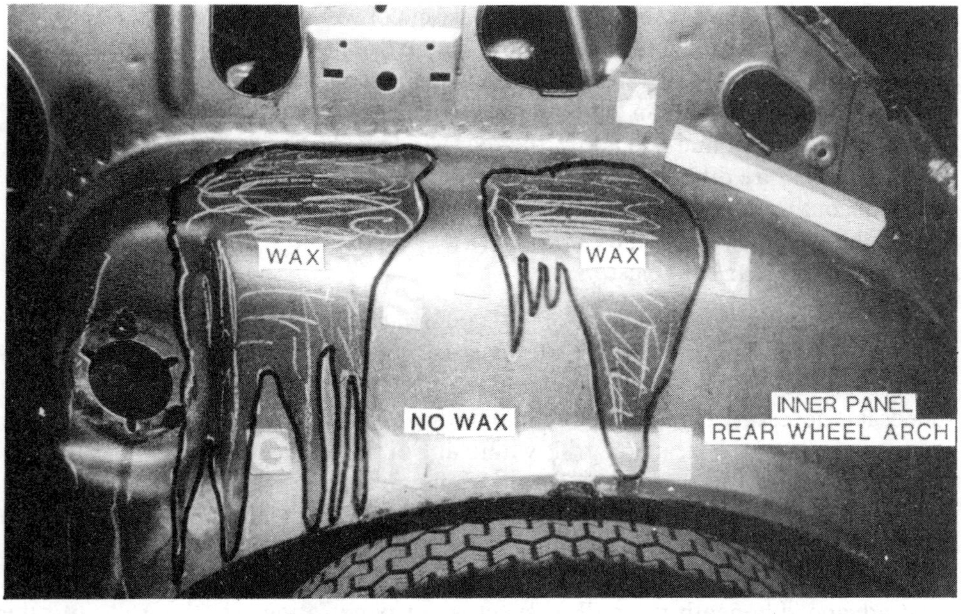

Plate 4.19(b) — This plate shows the result of the incorrect application of waxes. Internal cavity wax injection using thixotropic waxes is outlined to show the limited cover within the area where the electrocoat is thinnest. The waxes are not present in the interweld crevice where the road de-icing salts are deposited. Plate 2.6(c) showed the external view of how the corrosion appears as a result of the salts deposited in this interface Plates 2.6(a) and 2.6(b). This corrosion could be prevented by the correct application of waxes.

Plate 4.20 — Outward flared wheel arches are bad corrosion design features. Here the flared area is protected by a plastic guard to reduce the claims under the 12–18 months paint warranty period. The plastic tube highlights the crevice produced between the plastic and car body. This plastic attachment is an external manifestation of bad design by the manufacturer. It should be noted that crevices are still produced if the other surface is plastic. (It is not such a bad corrosion feature as the all steel crevice, but it still allows the differential aeration cell to operate). The crevice effect is reduced provided that the plastic guard does not vibrate and chatter at high speeds and cause paint damage when the trapped dirt is squashed between the metal and plastic interface.

Figs. 5.6 and 5.10) and stuck onto the air extractor in order to arrest the ingress of the micro-climate. Otherwise, ensure that the area is coated well with waxes on all model makes to reduce the corrosion. Examine the air extraction system and determine whether it is sufficiently well designed and sealed to the metal body to keep out rain water on a windy night and when the car is parked overnight on sloping, or uneven ground.

4.8 REAR VALENCE (Plates 4.21(a) and 4.21(b))

See Fig. 4.1(c): item (30), rear valance; item (31), around the rear spring hangers; item (32), rear box section; item (33), petrol tank and vertical recesses; item (34), spare wheel well–tool kit well.

4.8.1 General

The rear valance, on most model makes, acts as a scoop to the front and rear wheel road spray (Plate 4.21(b)). Only the smaller water droplets are lifted and deposited out of the wheel tracks by the pressure eddies around the vehicle. In some cases, on

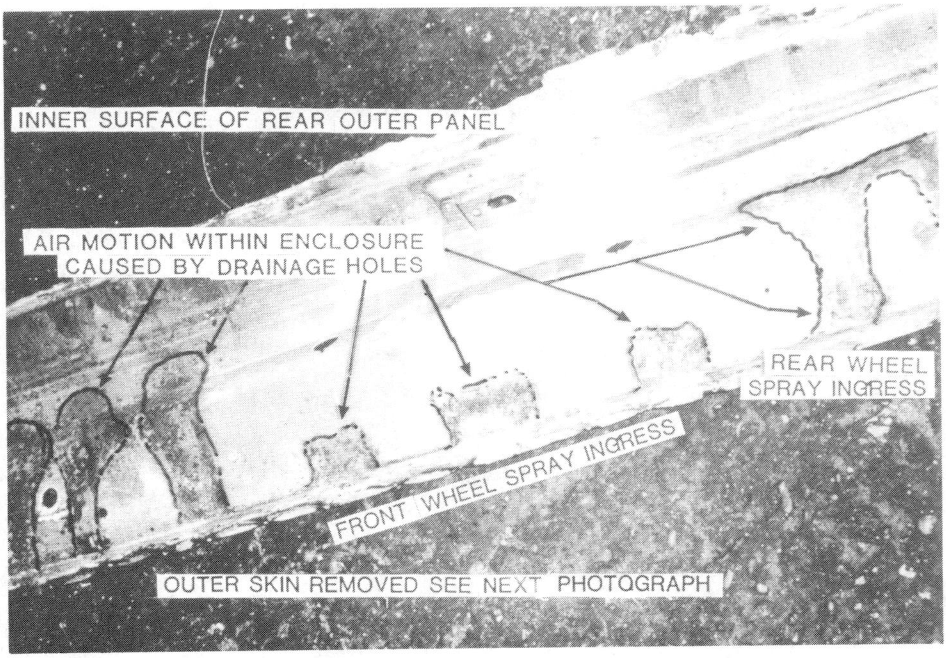

Plate 4.21(a) — The plate shows the air ingress through the drainage holes which are incorporated in the box member of the rear valance. As shown in the next plate, these areas are all linked to the passenger compartment by the principles associated with the sinking of the vehicle in the electrocoating bath.

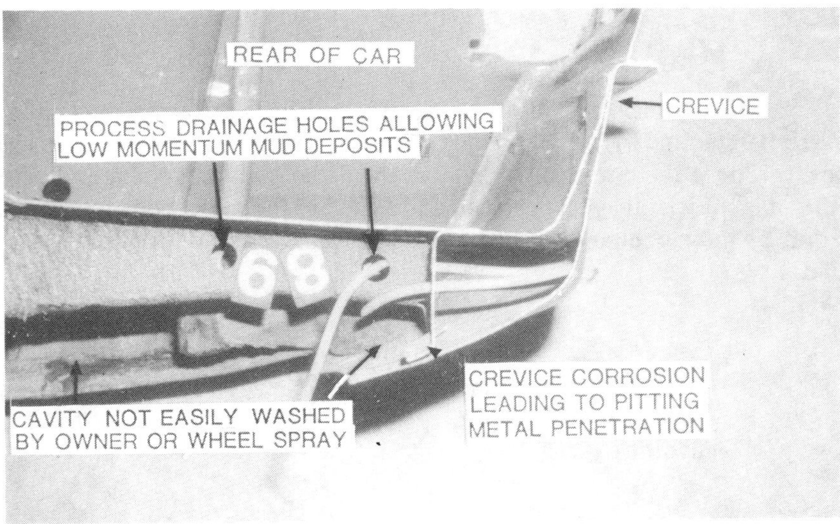

Plate 4.21(b) — A section through the rear valance of the car. The plastic tubes show that the holes are all linked together. This is a requirement for the electrocoating process (Sections 9.0–9.10, Plate 3.1).

small cars, the manufacturers' provision of rear wheel flaps attached to the rear valance skirt forces the larger droplets from the rear wheel spray to be scooped up by the rear valance drainage holes (item (30)) (Plate 4.21(a)). For commercial vehicles, see Plates 4.24 and 4.25.

The smaller amount of water present upon each road dirt particle is insufficient to wash away and clear the road dirt which is deposited out of the wheel track. This deposition is aided by the turbulent air around the rear of the car. Thus mud poultices are retained around the rear spring hangers (item (31)) and rear box section (item (32)). All these areas are out of the direct water line spray from the larger water droplets from the trajectory of the rear wheel spray.

This low water content also means that this road dirt is artificially more densely compacted by the phenomenon of surface tension into high density mud pockets (Sections 10.8–10.10). These rear wheel arch mud traps are formed by the following: rear box section–petrol tank (item (33)) and spare wheel well (item (34)) and abrupt changes in profile, e.g. where petrol tanks are positioned in low slung vehicles (item (33)). Any member slung beneath the vehicle causes pressure eddies around it with the deposition of road de-icing salts on the vertical surface which faces the rear of the car. (This is the same phenomena as that of road dirt being deposited on the rear tailgate which does not face the direction of motion.) These are areas, therefore, which are out of direct water washes from the wheel, a steam cleaning gun (see Section 4.13) or a garden hose, and unless very special precautions are taken this area is also out of the direction of spray from a hydrophobic wax spray head.

The slower the car travels the larger will be these water spray droplets and the higher the ratio of water:dirt which they can contain, and so the rear wheel arch is only cleaned in the direct line of wheel spray. These cavity sections (item (34)) must have electrocoat holes which are filled with rubber grommets, but the grommets are fairly easily displaced by the weights usually carried in the rear wheel arch wells. Missing grommets in this area cause the well behind the rear wheels to collect a thin film of road de-icing salts, which can be kept wet with water, road dirt and poor ventilation as this area often houses the spare wheel.

4.9 THE BOOT/TRUNK LID AND TAILGATE–LIFTGATES

See Fig. 4.1(c): item (35), internal disbondment of the mastic; item (36), clinch fold; item (37), ingress of salts due to pressure eddies; item (38), corrosion around hinges; item (39), horizontal ledges.

4.9.1 General

The boot–trunk lid and tailgates are constructed in a similar manner to a door. The two sheets of metal which are pressed to shape and held together by a 'hemming fold' lap joint, as shown in Fig. 4.4. Once again this overlap may appear to be 'filled' with mastic, but this mastic is applied after pressing, in most cases. The function of the mastic is to stop the electropaint and top coats on the raw edges of the metal overlap from breaking down and showing the first signs of corrosion. In order to prevent the panels from vibrating and causing 'drumming', the panels are treated as follows:

(1) They are stiffened by allowing the inner section to be attached to the outer boot panel by patches of adhesive–sealant mastic. These patches often tear away

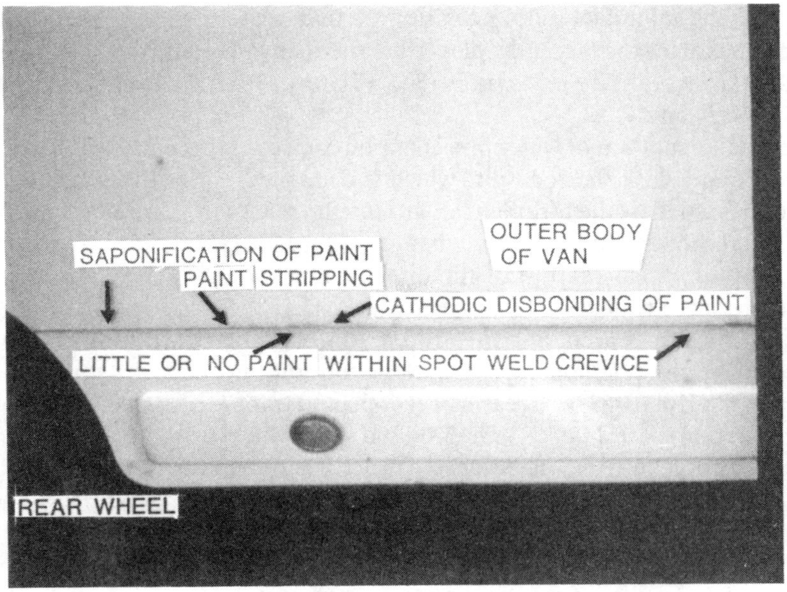

Plate 4.22 — Interwelds and metal overlaps are bad corrosion design. These panels should be either made from a full steel sheet or continuously welded along the interface in order to overcome the corrosion problem.

during the electrocoat baking operation leaving no paint on the internal section of the car boot. Because the boot lid 'frosts' over earlier than any other part of the car, it can often 'collect water' by condensing the passengers' moisture vapour on the cold metallic surface as the stale passenger air is drawn out of the car body. This is particularly noticeable on the raked tailgates and boots which will allow snow to accumulate. This snow does not fall off when the vehicle is in motion.

(2) They are stiffened by contouring.

The corrosion areas are the following:

4.9.2 Mastic (Fig. 4.1(c), item (35))

Corrosion takes place after the internal disbondment of the mastic between the two panels forming the boot.

4.9.3 Clinch fold (Fig. 4.1(c), item (36))

Corrosion takes place in the exterior of the tailgate–boot clinch fold as a result of a lack of paint in the crevice formed by the hemming fold. This arises from the poor electrocoat throw into these crevice areas and the concentration of de-icing salts in the crevice areas.

4.9.4 Panel junctions

Crevice corrosion occurs in panel junctions where the boot floor joins to the boot valance. Side panels on commercial vehicles often show this lack of electrocoat and

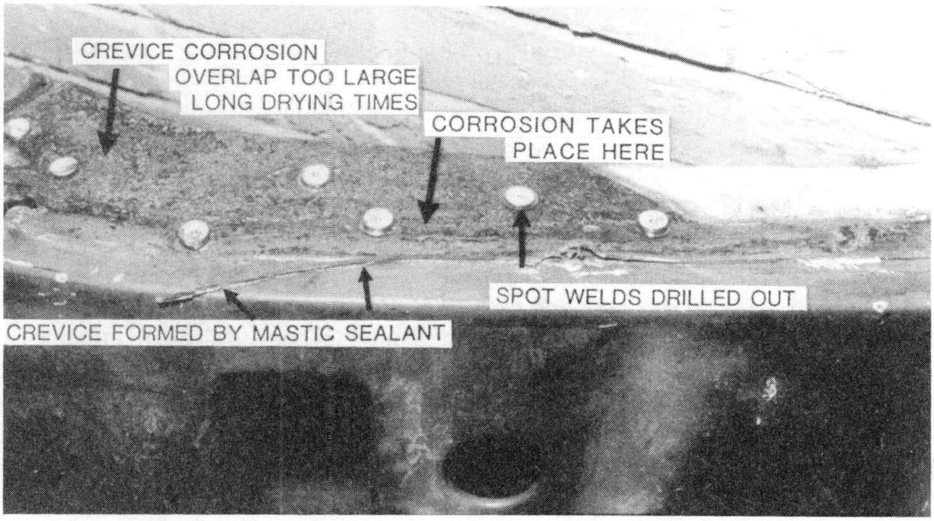

Plate 4.23 — Wide metal overlaps cause accelerated corrosion and they are bad design. The adverse effect of these can be reduced by allowing a thin oil to penetrate the crevice.

spraying (Fig. 3.3(c)) within crevices (Plate 4.22). This area usually exhibits the first signs of corrosion and, once started, is difficult to stop (see the discussion of the corrosion mechanism in Sections 6.14, 7.1–7.16, 2.2, 10.15 and 10.16 for the reasons for the concentration of de-icing salts in crevice areas). The best procedure is to apply a highly penetrating hydrophobic cavity wax or oil material (Sections 8.8–8.10) both internally and externally and to wash regularly during the winter months (Section 4.13). Large metal overlaps should be avoided as these form ready made corrosion sites as shown in Plate 4.23.

4.9.5 **Ingress of salts** (Fig. 4.1(c), item (37))

Road de-icing salts may be taken into the inside of the tailgate–boot through the drainage holes by pressure eddies caused by the vehicle's speed. In most model makes, the tailgate–boot drainage holes are too high (Plates 2.3 and 4.6). This internal deposition of salt on the inside panel of the tailgate–boot causes the road de-icing salts to concentrate within the crevice hemming fold and internally initiates metal penetration by rust-through of the external tailgate–boot panel. The crevice corrosion is also accompanied by paint stripping (Figs. 8.18-8.21; Sections 8.12 and 8.13).

4.9.6 **Corrosion around hinges** (Fig. 4.1(c), item (38))

Lack of electrocoat behind the tailgate hinges can give rise to corrosion with the added disadvantage of 'paint stripping' (Sections 8.12 and 8.13) around tailgate hinges. This is due to the very narrow crevices which cause a lack of throw of the electrocoat behind the hinge and is one good reason for the application of a penetrating hydrophobic wax, after loosening one of the screws to the hinge, if necessary.

4.9.7 Horizontal ledges (Fig. 4.1(c), item (39))

Horizontal ledges which are in the same horizontal plane as crevice inter-welds and juxtaposed to the boot lid closure can cause the accumulation of water between the boot lid and the closure shuts. Surface tension phenomena prevent these areas from losing the accumulated water. These are areas which dry slowly, and hence the road de-icing salts are concentrated in these areas. Corrosion is usually manifested very early in the vehicle life by the break-down of the paint film on the 'raw edges' of the boot–trunk lid. Rust stains also bleed from the crevice inter-weld.

4.9.8 Summary — boot/trunk and tailgate area

The best advice for combating the corrosion of tailgate–boot–trunk is to inject with a very thin, highly penetrating hydrophobic wax–oil and to allow the wax–oil or oil to run immediately to the lower clinch fold by closing the tailgate–boot after wax–oil injection. It is advisable to wash regularly on both 'sides of the tailgate–boot juncture', paying particular attention around the lower clinch folds. The lower clinch folds remain the wettest, as all the water vapour condensed on the inside panel will run down to these areas. In running down the internal panels, any de-icing salts will be washed to the clinch folds where they will be concentrated (Sections 10.14–10.16).

On horizontal surfaces, provided that the 'dirty' colour of the waxes–oils, and the fact that they collect dirt, can be tolerated, a coating of wax–oil will make the water 'more mobile' and this will help to shed any unwanted water, thereby encouraging the drying of these areas.

4.10 FRONT CHASSIS AND UNDERNEATH

See Fig. 4.1(c): item (40), sub-frame; item (41), integral box sections; item (42), crumple zones; item (43), wish-bones–box sections; item (44), transverse box sections — front and rear.

4.10.1 Subframe (Fig. 4.1(c), item (40))

The method of constructing a chassis differs from manufacturer to manufacturer, e.g. sub-frame (item (40)) or integral box sections (item (41)). The sub-frames are not attached to the body at the electro-painting stage. The manufacturers, in designing the front chassis, are moving towards incorporating 'crumple zones', which 'tell' the structure where to collapse during an accident (see Section 1.16). Most manufacturers should be placing the electrocoat holes away from the front wheel spray and are not producing flanges which are adjacent to process holes, as these scoop up the preceding vehicle's wheel spray (Plate 2.8) or its own wheel spray (Plates 4.17 and 4.21(a)). Subframes are dealt with in Section 5.17).

4.10.2 Wish-bones–box sections (Fig. 4.1(c), item (43))

Wish-bones and other parts made from steel sheet to give an open box will exhibit pressure differences between the open ends. This is the organ pipe effect which will give rise to a micro-climate within the box member. There will be a lower pressure on the out-board end of the open member and so the inner open end will collect the wheel spray of the preceding vehicle and deposit it within the member (see Figure 5.9). Some later structures of this type are filled with polystyrene foam (or similar) to prevent this from happening.

4.11 TRANSVERSE BOX SECTIONS (Fig. 4.1(c), item (44))

Transverse box sections to the vehicle direction (Plates 4.14(a) and 4.14(b) are abraded by air–salt–grit particles as a result of the vehicle motion and collect dense mud poultices owing to impaction as a result of high speed driving. Each of the transverse sections must have two process holes, one to let the air in and out and the other to let the paint in and out. Some air holes vent into the passenger compartment, whereas in other model makes they vent to the top of the transverse box member. Either way there will be pressure eddies and differences between these holes (Section 2.13 described how the road spray was drawn through the electrocoat drainage holes by the extraction of the stale passenger air from the vehicle). Corrosion, internally initiated, is therefore seen earliest around the electrocoat drainage holes within these transverse members. In newer designs most transverse box sections are placed within the passenger compartment (e.g. as shown in Plate 2.5 beneath the driver's seat) so that it does not form a protruding member beneath the car. The electropaint within these internally housed transverse members is drained into the passenger compartment and so eliminates the need for any additional protection.

4.11.1 Rear box sections

Rear box sections often carry unequal loads, e.g. where anti-roll bars are incorporated with the solid rear axle. These cavities will require stiffening in order to carry the extra loads imposed on this side by the anti–roll bar attached to the axle. This stiffening is carried out by extra diaphragms, which reinforce the rectangular configuration of the box section. This means that most box sections cannot be wax injected in one operation, and extra holes must be drilled into each section. Electrocoat drainage is achieved by contouring the diaphragms in the box section but these stiffening sections mean that wax injection is difficult without extra access holes.

The juxtaposition of protruding structures such as jacking points, reinforcements for front and rear trailing arm suspension units, rear coil spring suspension units and inter-weld flanges which are perpendicular to the vehicle motion, with their adjacent electrocoat drainage holes, causes the road spray to be 'jettisoned' into the box members. Most sill–rocker box members consist of two sections (Fig. 5.7) and it is not possible to wax inject both cavities from the one insertion.

4.12 CORROSION BENEATH THE MANUFACTURER'S ANTI-CHIP PROTECTION SCHEMES

This is perhaps the most serious of the corrosion mechanisms because it is proceeding unchecked beneath the non-adherent coating. The coherent coating coating keeps the underlying metal wet and provides the necessary differential aeration which is required for the corrosion mechanism (Fig. 8.18). This arises as follows. Stones or loose road chippings can be thrown up against the manufacturer's anti-chip, anti-abrasion coating which was applied at the factory and on top of the electrocoat (note: there is no final gloss or intermediate coats of paint on the underside body, inner wheel arches, wheel wells and inside door sections). If the impact is sufficiently high, taking into account point loadings, to break the adhesion between the substrate and paint film at the point of impact, then the tough anti–abrasive coating becomes a liability. This tough coating keeps the paint film intact so that there appears to be no problem at the point of impact. All paint films are permeable to both water and oxygen, and their method of protection is to limit the corrosion current which flows in the liquid environment. The paint film fails in this respect once it becomes non adherent to the substrate. This subject is dealt with in greater detail in Section 8.12.

4.13 STEAM CLEANING AND ACCIDENT REPAIRS

Steam cleaning is considered to be a very efficient way of cleaning away the road accumulations beneath the car, but there are certain disadvantages. These arise from the very high jet pressures (which also applies to high pressure water wash systems)

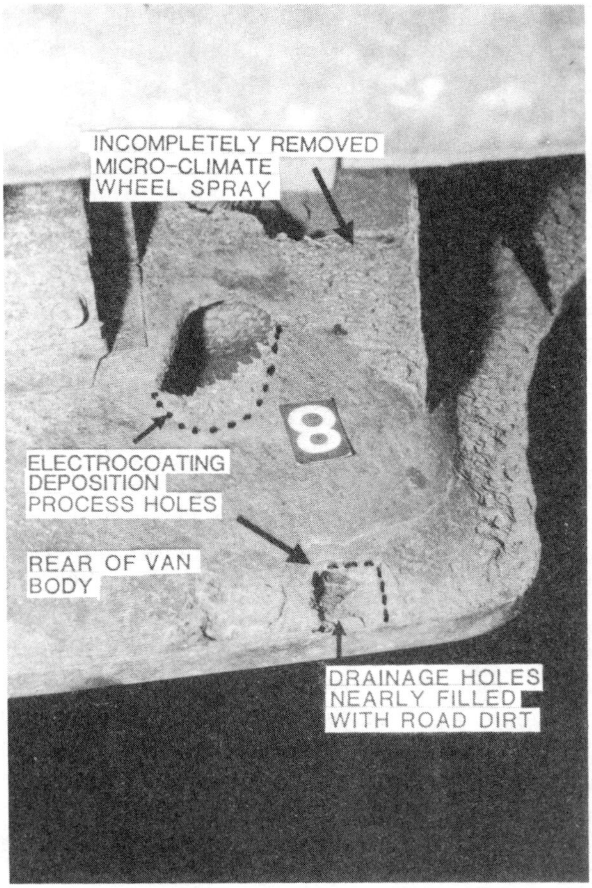

Plate 4.24 — For a commercial vehicle to be painted in the electrocoat baths, a larger volume of liquid must flow through the process holes. This means that the holes must be larger. Where these holes are badly placed from a corrosion point of view, they let in more road de-icing salts, and therefore the corrosion is often much worse. This is the process hole behind the rear wheel. The original hole is outlined. The lower drainage hole is partially blocked by road dirt ingress through the upper drainage hole.

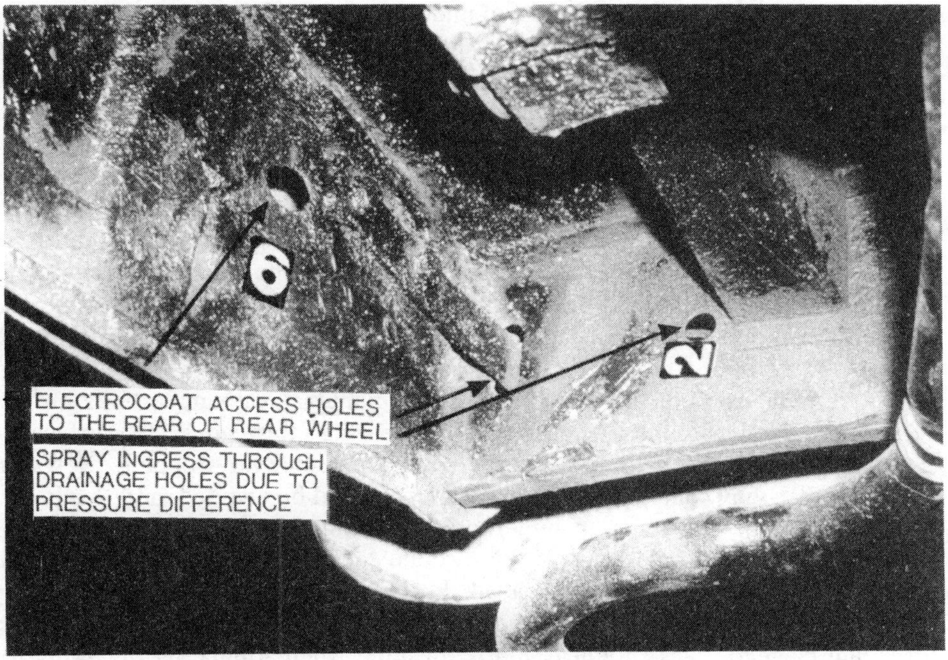

Plate 4.25 — Another commercial vehicle showing the large process holes under the rear wheel arch. The hole '6' is in an angled panel such that the hole collects the rear wheel spray. The smaller hole '2' is directly behind the rear wheel and also collects the spray. This results in rapid corrosion of these panels.

in combination with the relatively high temperatures (with respect to the softening points of the underbody waxes). These high temperatures, together with the fact the 'underbody waxes' or manufacturer's protective schemes have been in contact with the moist road dirt for a long period of time, cause the waxes to become 'soft' and they are easily removed by the fierce water–steam jet. The high temperatures also cause the waxes to soften and the steam cleaning pressure can cause the the road dirt to be "impacted" into the soft wax surface. Steam cleaning is inferior to the periodic washing from a 'soft' fine spray obtained with a garden water hose, which the owner should periodically undertake during the winter. To minimize damage of the underbody protection scheme, it is suggested that the jet temperature should not be higher than 90°C (190°F) and the jet should not be too near the body surface (say 10 inches). This automatically causes problems in certain locations, e.g. under the wheel arch, for the following reasons.

(1) The lance is rather cumbersome and the gap between the wheel and its arch means that the above guide cannot be achieved.
(2) The lance is rather cumbersome and the jet head cannot be oriented to remove mud around wheel arch returns and re-entrant angles (Fig. 4.6).

The philosophy of a good corrosion protection scheme should be that the owner–customer removes these road de-icing salts on a regular basis, and the owner should be made aware that his failure to undertake this task will mean higher paint repair–wax retreatment costs owing to the aggressive nature of stream cleaning together with the fact the 'underbody waxes' do become 'soft' when in contact with the moist road dirt for a long period of time. The owner also ought to be aware that the paint system become permeable to the road de-icing salts when they are in moist contact with mud poultices. The vehicle owner ought to be informed or aware that these waxes in their 'soft' condition are easily removed by the fierce water–steam jet. However, the manufacturer also should be aware that designs which harbour road salts are unacceptable and are bad designs. Several plates have been given for these bad designs. Plates 4.24 and 4.25 show examples of bad designs in commercial vehicles where regular steam cleaning or owner care will make no difference to the longevity of the product.

Where large areas of 'underbody waxes' or manufacturer's underbody protectives are removed, care must be exercised in preparing a throughly clean surface for the subsequent respraying of the underbody. Any possibility of trapping dirt within the new underbody coating and the old coating should be avoided, and this is best achieved by applying touch-up underbody by means of a brush as opposed to spraying. This is particularly important around the rear wheel arches (Fig. 4.6). Large areas, of course, must be sprayed. In the matter of vehicle retreatment by injecting a cavity (firm) wax material onto existing waxes, the internal surfaces are not steam cleaned. This should be borne in mind when stipulating or asking for periodic retreatments as the retreatments can be trapping pockets of road de-icing salts and therefore keeping in area wet for longer periods of time. The waxes stop the trapped pockets of road dirt from drying out.

5

Elements of good design

5.1 ASPECTS OF GOOD DESIGN

Many attempts have been made to assess the true costs of corrosion. It is easy to calculate the costs to industry of a plant shut-down due to a malfunction of a plant, but it is not so easy to calculate the costs of a malfunctioning domestic boiler owned by a member of the public, as it is expected that the public just 'grin and bear' the inconvenience and any added costs. Degradation of products by corrosion has become an accepted and unavoidable part of life, especially for those products made from steel. Rust is the natural form for the metal iron to take unless prevented by the correct application of scientific principles by the manufacturers of the products made from iron and steel. This image appears to have its origins in the corrosion of the motor vehicle. This acceptance has been generated largely through an ignorance of design against corrosion and the costs of corrosion, which are not borne by the manufacturer. It can safely be assumed that the costs of corrosion of the motor vehicle are 1.43 times (Section 1.7) the depreciation of the motor vehicle. The conversion of rust back to steel requires a large amount of energy. The transport of the large amount of steel as the vehicle body requires energy. As fuel costs have risen, so have the manufacturers' attempts to make lighter and more efficient engines and total systems. The U.S. Government stipulated in the Energy Protection Act (EPA) the expected fuel consumption of various vehicles, and an example of this is shown in Fig. 5.1 for the years 1975 and 1985. There were heavy penalties for not complying with this directive. Accompanying this goal of achieving better fuel consumptions has been the effort to make lighter cars, so that performance figures can be improved without losing out on the fuel consumption figures. (At present a saving of 0.6 mpg may be achieved per 100 lb reduction in weight. See Table 1.14). This reduction in weight (Fig. 5.2) has been achieved by using thinner gauge steels and by combining mild steel with low alloy high strength steels which are discussed in Section 6.20 *et seq*.

The price of fuel also affects the purchaser's choice of vehicle and engine size. The public demand for different engine mixes has been provided by General Motors over the same period, as shown in Fig. 5.3. The use of thinner steels, however will necessitate a more active investigation into corrosion of the vehicle body.

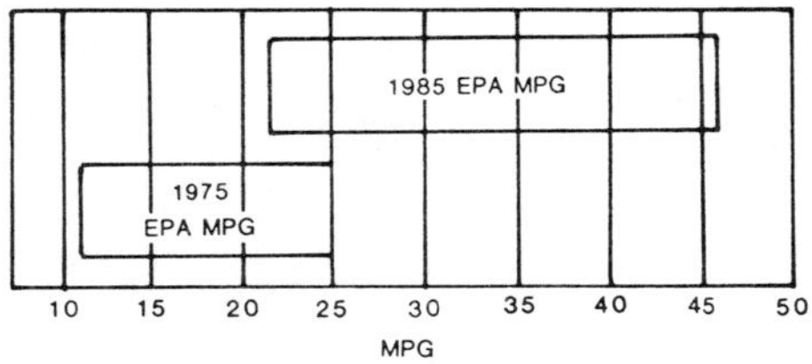

Fig. 5.1 — Fuel economy improvements. From 1975 to 1985, General Motors has improved the range of fuel economy of its cars dramatically, and future cars will be even more fuel efficient. (Reproduced courtesy of General Motors, 1986 General Motors Public Interest Report, May 15, 1986.)

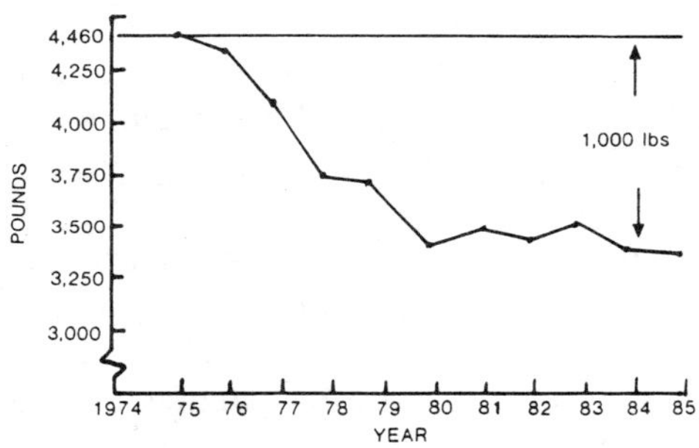

Fig. 5.2 — General Motors production fleet average test weights. An average vehicle weight reduction of about 1000 lb, or 22%, since 1974 has significantly improved fuel economy. (Reproduced courtesy of General Motors, 1986 General Motors Public Interest Report, May 15, 1986.)

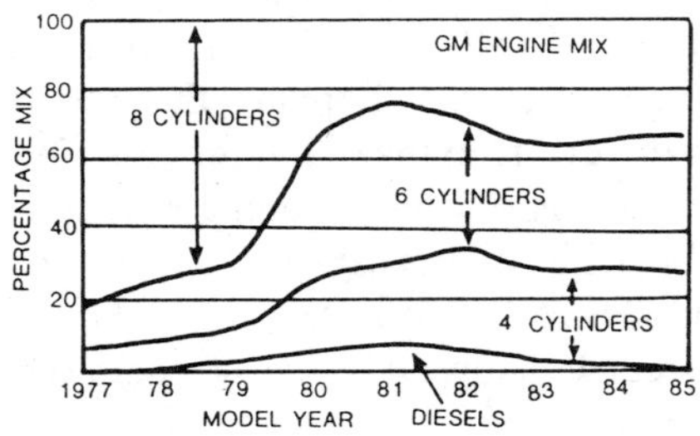

Fig. 5.3 — Genaral Motors engine mix. Consumer choice of engines reflects the price of gasoline as well as General Motors' efforts to encourage fuel economy. (Reproduced courtesy of General Motors, 1986 General Motors Public Interest Report, May 15, 1986.)

The Department of Trade and Industry set up a committee to report on corrosion and protection in the industries of the United Kingdom. This report was published in 1971, and it found that only a limited number of firms were corrosion conscious. The majority had no idea of the costs of corrosion as these were hidden under the heading of a general maintenance cost. The possible savings which could accrue if the existing knowledge of corrosion, assuming no further research, were applied, is reproduced in Table 5.1. The factors and cost of corrosion in the automotive industry are given in Tables 5.2 and 5.3.

Some further research into corrosion would produce significantly higher savings than those given above. As a result of these findings, the Department of Trade and Industry have produced guidance notes on designs to combat corrosion. (Department of Trade and Industry, 1981). A small section of their guidance notes, which is more relevant to motor vehicle corrosion, is reproduced in Fig. 5.4. These values refer to static structures but the principles are applicable to moving structures such as the motor vehicle, if we assume the average speed of a vehicle travelling is 60 mph, (90 km/h or 88 ft/ s). In travelling at this speed, a drainage hole in the front wheel spray (e.g. Plates 2.4(a), 2.4(b), 2.9(a), 4.8(a), 4.11, 4.12(a), 4.15(a), 4.16, 4.17(a), 4.25) will scoop up a large volume of water in one second from the lightest shower of rain (Fig. 5.5). This arises because the 'rainfall' is 'recycled' by the spray generated as the vehicle travels at speed on the wet roads. The detailed design of the front of the vehicle which travels in the spray generated by the preceding vehicle has been omitted by most designers, e.g. Plates 2.8(a) and 2.8(c).

Table 5.1 — Expenditure and estimated potential savings in various industries at 1971 prices

Industry or agency	*Estimated cost* (£M)	*Estimated potential saving*	
		(£M)	*Expressed as % of estimated cost*
Transport	350	100	29
Building and construction	250	50	20
Food	40	4	10
General engineering	110	35	32
Government depts. and agencies	55	20	36
Marine	280	55	20
Metal refining and semi-frabrication	15	2	13
Oil and chemical	180	15	8
Power	60	25	42
Water supply	25	4	16
Total	1365	310	23

The Hoar Committee (Hoar, 1971) looked at various industries listed in this table and estimated their potential savings if known solutions were applied to corrosion.
Table reproduced from L. L. Schrier, Corrosion, Vol. 2, 1977, Table 10.1, by courtesy of Newnes Butterworth.

Table 5.2 — Elements of the costs of corrosion

Capital costs
Replacement of equipment and building
Excess capacity
Redundant equipment
Control costs
Maintenance and repair
Corrosion control
Design costs
Material of construction
Corrosion allowance
Special processing
Associated costs
Loss of product
Technical support
Insurance
Parts and equipment inventory

It is often hard to put a cost on the corrosion of the motor vehicle body owing to the structure of the industry and its tax subsidy (Section 1.7 and 1.8).
Table reproduced from R. Baboian economic and environmental factors, Automotive Corrosion by De-icing Salts, 1981, p. 18 Table 1, by courtesy of the National Association of Corrosion Engineers.

Table 5.3 — Design costs for automobile corrosion

	Cost $ millions
Paints and allied products	182.3
Mild steel–carbon steel	212.0
Alloy steel	182.3
Stainless steel	75.9
Coke–Coke oven products	30.4
Primary copper	136.7
Primary aluminium	212.6
Coating and plating	364.5
Total	972.7

If, hypothetically, corrosion did not exist, then the motor manufacturing industry in 1975 could spend much less ($973 million) in design costs of corrosion. Hence the 'hidden' costs of corrosion must be the added costs to combat the corrosion. The actual plant required for applying the corrosion protection schemes is shown in Figs. 9.1, 9.2, 9.3(b) and 9.4(d).
Values from NBS GCR78–122, and Battelle Columbus Laboratories Input/Output Tables, Economic Effects of Metallic Corrosion in the United States, Appendix C, PB–279–430, 1978.
Table reproduced from R. Baboian, Economic and environmental factors, Automotive Corrosion by De-icing Salts, 1981, p. 18, Table 2, by courtesy of the National Association of Corrosion Engineers.

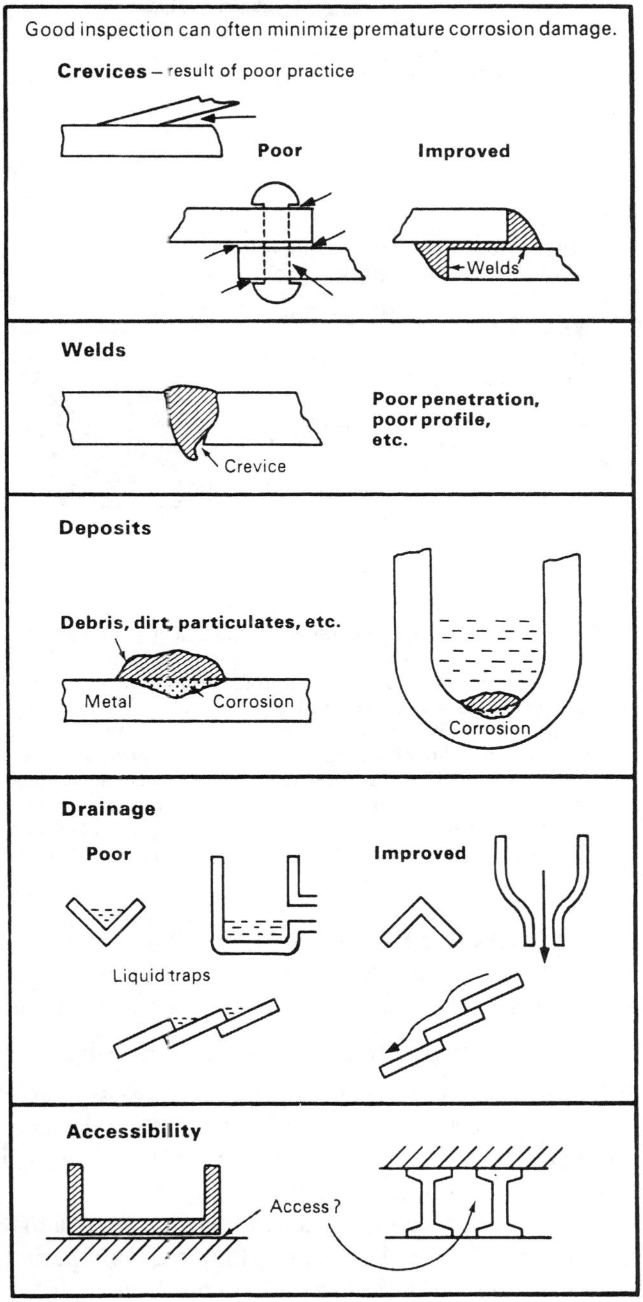

Fig. 5.4 — Some elementary design parameters put out by the Department of Industry showing the principles of design against corrosion. These parameters should be modified to take account of the fact that the motor vehicle is a moving box and subject to areas of differing pressures and pressure eddies.

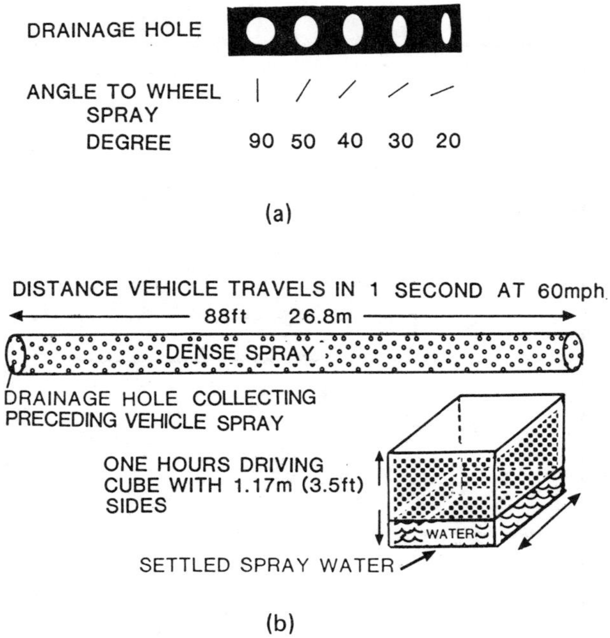

Fig. 5.5 — (a) A manufacturer's process hole viewed at right angles to the sheet panel is a circle. If this makes an angle to the front wheel spray as shown in Plate 2.8(b), the apparent area of cross section is reduced as shown here. Assume a cross-sectional area of 1 cm^2. (b) The cross-sectional area presented to the wheel is reduced by a factor of ten but over an hour's driving a considerable amount of spray is collected. At a steady speed of 60 mph (88 ft/s or 26.8 m/s) the drainage hole will collect in an hour's driving 1.61 m^3 of wheel spray, which is equivalent to a cube with 1.17 m (3.5 ft) sides. The total amount of 'settled' water can be some hundred-fold greater than the rainfall, simply because the rain is recycled by the vehicle motion. A similar argument holds for wrong way overlaps which will collect water spray either from the preceding vehicle (Plates 2.8(a) and 2.8(c)) or by drainage due to gravity (Plate 4.3(d)).

5.2 DESIGN AND CORROSION CONTROL

Corrosion is best reduced by Design. Good design at the drawing board in the first instance is no more costly than bad design. In fact the maxim should be as follows:

Good design is free, bad design is expensive

A severely corroded vehicle on the road negates all the glossy advertisements.

Prototype design is necessary. No coating will have a lifetime in a salt spray cabinet of over 1000 h unless it is very thick (in which case it will not penetrate crevices) or very good, in which case it will be expensive.

According to the Canadian voluntary code for vehicle corrosion (Section 1.12), 1 year (Cosmetic corrosion) is 8760 h, 6 years (Body warranty period) is 52 560 h, 8 years (Structural corrosion) is 70 080 h and 12 years (Median life) is 105 120 hours.

Therefore, when purchasing, repairing or designing the vehicle, one fundamental question ought to be, does the design (or repair) take account of the factors which are all scientifically derived and discussed and listed in this book? Examples are (1) temperature cycles, (2) drainage hole positions, (3) passenger air extraction mechanisms and efficiency and their positions with regard to overtaking vehicles, (4) crevices, and their ability to 'switch on' a battery which is powered by differential aeration, (5) out-of-sight road de-icing salts which are deposited as a result of pressure eddies, (6) mud compactions due to surface tension effects and (7) wrong-way overlaps and the lack of electrocoat within crevices.

These design parameters will now be discussed in further detail.

5.3 PROTECTION AVAILABLE

5.3.1 Material selection

This will be dealt with in Chapter 8. Section 8.1 and Fig. 8.1 show the parameters which need to be considered.

5.3.2 Coatings

Coatings only have a limited life in adverse conditions. These will be discussed technically in Chapter 8.

5.3.3 Environment

This may be made less aggressive by designing out water traps, crevices, inter-welds, metal overlaps or joins, and large areas of metal contacts which reduce the rate of drying. Inter-welds in horizontal panels, or horizontal ledges in areas of the car body, should be avoided as these do not drain naturally by gravity. Sharp changes in contours, e.g. right-angles, also restrict drainage under gravity.

5.3.4 Absorbent materials

Absorbent materials, such as foams, have been used by all manufacturers. They have been observed in sills and rear dog legs (Section 4.6, item (24)) and as 'seals' between the front wings and 'A' pillar of vehicles (Plate 4.2(a)). Their presence is detrimental to corrosion and should be avoided. The choice of materials should be limited to those which provide a hydrophobic surface or have a large contact angle with water.

5.3.5 Glass fibre

Glass fibre without resin has been observed in the tops of 'A' pillars to try and eliminate the mud pocket caused beneath the wing (Plates 2.1(b) and 2.1(d)).

5.3.6 Dissimilar metals

Metallic contacts between dissimilar metals should be avoided, but where dissimilar metals are a necessity plastic insulating washers or adhesive sealants should be inserted between the metallic surfaces. Large cathodes in contact with small anodes should be avoided. (see Fig. 2.10). Chapter 3 developed the concept that the car body was built up with doors, bonnet and boot fixed onto the passenger compartment as the 'body-white'. This was then painted (Plate 3.1) by the flow of an electrical current within the water-borne paint and termed electrocoating. Electrocoating is a process similar to the electroplating of metals. The fundamental mechanisms behind the deposition of the paint by electrocoating are developed in Chapter 9. For the electropaint to enter the car body, and to be electrically deposited on the metal of the car body, process holes must be punched into the car body and cavity members. These process holes were of two kinds.

(1) Holes to allow the paint to enter (Plate 2.7(b)) and any trapped air to escape from the member (Plate 2.6(a)).

(2) Holes 'A' in Plate 2.5 for the electrical current to enter the box member. The current flowing in the ionic form needs to be presented with the least electrical resistance in order to deposit the paint as a powder. The holes 'A' in Plate 2.5 are called electrocoat 'throw' holes as they increase the 'throw' or covering power of the electrocoat bath. This is analogous to the electrolysis and deposition of a metal by electroplating techniques. (See Section 3.3.1.)

These process holes are essentially circles cut out of the members which can be covered by the internal trim as was shown by 'A' in Plate 2.5. The thickness of the paint coat depends on the amount of current per unit area from the car body to the counter electrode. This is discussed further in Chapter 9.

Pressure eddies beneath the car body, and the continuous extraction of stale passenger air, causes water (in the form of micro-droplets and small particles of road de-icing salts) to enter the process drainage holes as a result of their small momentum. These are usually deposited on areas which were also denuded or had little electropaint deposition owing to the high electrical resistance between counter-electrode and car body. This high resistance arises as a result of the path that the current has to follow in the aqueous solution.

Pressure eddies due to vehicles overtaking at speed or in contra-flow situations cause rapid fluctuations between high and low pressures as the vehicles pass (Fig. 5.6). This causes the road spray to enter the drainage holes which are over the front

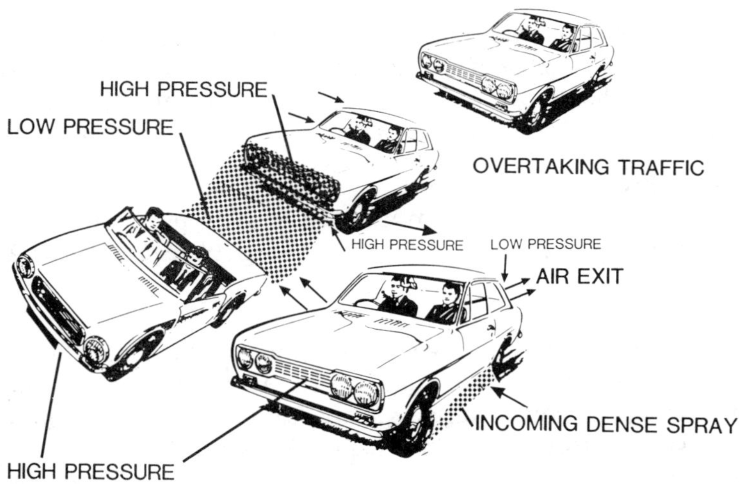

Fig. 5.6 — A vehicle travelling at speed will produce a high pressure at the front and a low pressure at the rear. The high and low pressures around and along the vehicle travelling at speed can be complex. These pressure eddies are more complex when overtaking or contra-flow conditions are considered. Here an overtaking stream of vehicles causes high pressures on the overtaking side, while on the nearside there will be a low pressure. Passenger air extraction systems and drainage holes over the front wheel track will allow a lot of road spray to enter the drainage holes on the high pressure side. This situation is reversed when the rear of the vehicle comes alongside. The figure also shows that a car in front causes a lot of wheel spray. A following vehicle has to drive through this dense spray which becomes a high pressure area for the following vehicle by virtue of its own speed through the turbulent air which the preceding vehicle causes.

wheel track and therefore collect the densest spray. This is particularly damaging if the vehicle floor pan has a continuous perimeter box member as shown in Fig. 5.7. The resulting corrosion is by a battery mechanism which was charged by there being an electrolyte and a difference in oxygen availability. This battery (Sections 2.5 and 2.6) was of the 'everlasting' type, providing that there was a metal to be consumed to provide the negative terminal, oxygen and an electrolyte to provide for the positive terminal.

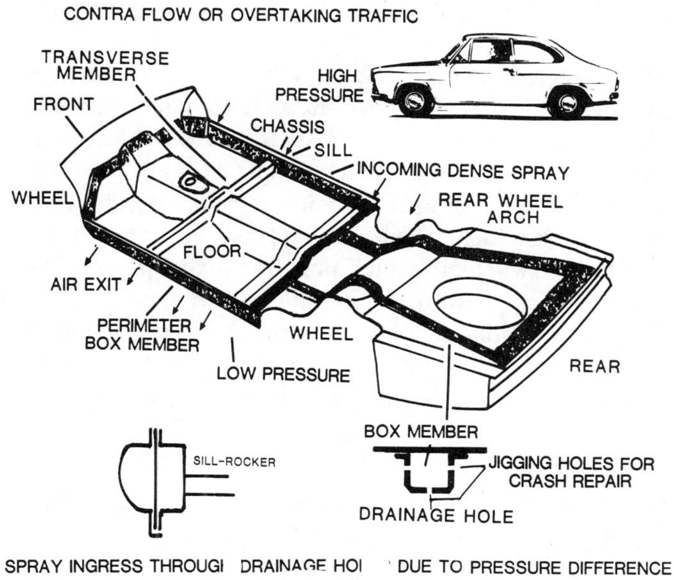

Fig. 5.7 — The floor pan is the main structure of the vehicle. Upon this prime structure are hung various cosmetic exterior panels in order to produce a 'new' design. This basic floor pan, illustrated here, can allow air motion from the right-hand side to the left-hand side and vice versa. The diagrams indicate the location of the ring floor pan and the pressure build-up when vehicles overtake each other and in contra-flowing traffic. Detail of the sill–rocker construction is also shown as a double cavity.

5.4 GOOD DESIGN FOR PROTECTIVE COATINGS

Hence if the above concepts are applied to the design of the vehicle the following parameters arise. One elementary good design parameter would be to increase the thickness of protection coatings on those areas which only received the first protective coating (one and only paint film—the electrocoat (Fig. 3.5)). This is achieved by giving a thicker electrocoat by providing more current access holes to narrow compartments (Plate 2.5). These holes should also both provide air movement from regions of the least dense spray and also promote more rapid drying in the event of them becoming wet. These requirements can all be achieved in one swoop by providing stale passenger air extraction grilles in doors in position 1 rather than position 2 (Fig. 4.3(b)). Position 1 allows the narrow crevice along the edge to be opened up to the electrocoating current and gives the manufacturer the ability to apply wax through the resulting grills. Any water entering such areas should be able to drain out through drainage holes at the bottom of the doors which should be large and at the corners of the doors so as to provide complete drainage on whatever slope the vehicle is parked. This would appear to be an elementary good design consideration, but a casual look at most manufacturers' cars will show that this is not always considered. The presence of the passenger air extraction system in the doors increases the path for extraction air through the sill–rocker drainage holes and does not allow overtaking vehicle spray to enter the rear wheel arch seam (Plates 2.6(a) and 2.6(b)).

Drainage holes, in certain areas of the car body, should be incompletely stamped out metal ellipses. These are termed 'protected' drain holes and can be used where there is little danger that the wax, pine needles and road dust will block off the drainage aperture. These holes could also be 'tucked' out of the direct spray of the front wheel or the preceding vehicle by being recessed behind panels which are contoured to achieve this ideal. There should not be holes at the rear of any transverse member which will tend to suck air from a higher pressure as a result of the vehicle motion to a lower pressure as a result of a transverse member, e.g. the transverse support for the radiator cooling system. As immersion in the electrocoating bath requires these holes, they must be filled up with plastic grommets before the vehicle is offered for service life. Alternatively, the box member can be replaced by an open '⎍' form.

5.5 ASPECTS OF GOOD DESIGN—WRONG-WAY OVERLAPS

The vehicle body has to made from a number of panels joined together to form the stress-carrying unit. Some of these joins are formed by a butt joint, whilst others are formed from metal overlaps. The butt-welded structure can provide a surface at right-angles to the front wheel spray (e.g. Plate 4.17(a)), and the overlap joints can be of the wrong-way overlap type (e.g. Plates 2.8(a) and 4.3(d)). These 'wrong-way overlaps' will collect the spray from either the front wheel or the preceding vehicle (Plate 2.8(a); Figs. 3.9 and 4.5). Rectification can only be made at the factory and if a one way overlap is imperative, as they often are (e.g. (a) on 'A' pillars where the roof is attached to the lower frame or (b) simply because there are two conflicting conditions which are found beneath the wheel arches, i.e. wheel spray and drainage due to gravity), then welding or brazing is the only successful means of rectification (Fig. 5.4).

5.6 GOOD DESIGNS IN BONNETS

The method of constructing a bonnet–hood is discussed in Section 4.2. A well-designed bonnet from the point of view of corrosion will have the following characteristics.

(1) The external panel of sheet of metal will not present a perpendicular surface or a large surface area to the preceding vehicle's wheel spray.

(2) The internal strengthening web should be fairly large with 'cut-aways' to increase the thickness of the electrocoat. The adhesive mastic supports should make

the minimum number of contacts with the outer panel to minimize the formation of crevice corrosion.

(3) The perimeter member will make the minimum contact in the clinch fold area to the outer panel by a clinch fold or 'hemming fold' lap joint, as shown in Fig. 4.4(4).

(4) The clinch fold will be filled with non-blowing mastic, which will not adsorb the pretreatment salts.

(5) The raw edges of the clinch fold should be protected by mastic applied after the pressing operation to make the clinch fold. This added mastic will stop the paint on the raw edges of the metal overlap from breaking down and exhibiting the first signs of corrosion. If the new vehicle does not have this form of corrosion protection it is worthwhile touching this up with touch up paint before it breaks down. The time spent in the early days is amply repaid in the time spent in rectifying or advertising the vehicle for resale.

(6) The mastic added to the internal member to prevent the outer panel from vibrating and causing 'drumming' should not shrink when the electrocoat is stoved so that it comes away from either outer or inner web as this will reveal bare unpainted steel. A wax coating after the paint baking system is a good requirement to prevent breakdown as a result of this shrinkage.

(7) There should be no possibility that rain water, or water from a high pressure wash, will enter the bonnet reinforcements through the fresh air grills for the passenger compartment should these be inserted into the bonnet. The vehicle should also be able to stand on an incline without rain water entering these areas through the fresh air grills.

5.7 ELECTROCOAT DRAINAGE

The vehicle is often electrocoated with the bonnet–hood in the partially open position and so the 'paint drainage' holes will operate in this position but they will not drain in the closed position, when the car is on level ground. This arises because the bonnet drains when the vehicle body comes out of the bath at 30°–45° to the horizontal. In some bonnet designs some of the drainage holes will collect the preceding vehicle's wheel spray as this spray is deflected by added protuberances over the front headlamps, bonnet–hood catches or radiator support mechanism. A well designed bonnet–hood will overcome these limitations and defects.

The radiator offers a partial resistance to the flow of air through the radiator cooling fins, and so some of the preceding vehicle's wheel spray is deflected into the bonnet–hood (Fig. 4.5). Hence the open holes in the underside of the leading edge and hemming fold, a requirement for a thicker electrocoat, must be modified for service conditions by being fitted with plastic grommets.

Corrosion proceeds along the crevice or fold formed between the outer panel and the perimeter box member and so these members should be capable of being wax injected to reinforce the coating applied to this area with adequate barrier films. Rubber stops, used to stop the bonnet resonating and fixed midway in the trumpet –upper wing–fender, must not coincide with an electrocoat hole in the length of the bonnet. These rubber stops would deflect air up into these large electrocoat holes.

The front valance should not have electrocoat drainage holes, or unfilled wax injection holes, to the rear of the transverse member (low pressure side) and/or the radiator, which would cause the preceding vehicle's spray and air to enter the crevices at the front at speed.

5.8 GOOD WING DESIGN

The wing–fender is considered a 'cosmetic' panel in that it does not form part of the energy-absorbing members during an accident and therefore does not form part of any imposed structural survey for roadworthiness. The road spray generated is of very small momentum and has little associated water and so this associated road dirt, which is deposited out of the direct wheel spray, will not run down under gravity where it is deposited out of the front wheel spray. Therefore the wings–fenders should be protected by a full plastic 'inner wing–fender liner' (Fig. 2.9). Partial underwings or flaps, which have been added in some model updates, do not take account of road dirt being thrown forward at the vehicle speed and being deposited on the side of the wheel arch and out of the direct wheel spray.

The method of constructing and fixing a wing–fender differs from manufacturer to manufacturer. Well-designed fixings will use bolt-on techniques such that the contact area between the adjacent metals is at a minimum. The resultant crevices, which will only exist at the bolt–on fixing, will (if correctly designed) contain a large gap into which a wax can penetrate and therefore present a hydrophobic surface to the road de-icing salts. These well-designed bolt-on wings–fenders will be pressed to shape from one sheet such that the stiffening of the wing–fender is obtained mainly by their contoured shape. There is no need to provide closure down the 'A' pillar if the full plastic wing liner is fitted. By this means there will be no major crevices to harbour the road de-icing salts and those crevices formed beneath the bolt-on fixings will dry much more quickly if the area of contact is kept small. The corrosion rate of the wing would be several orders of magnitude less than the degradation rate on present designs.

5.9 GOOD DESIGNS FOR DOORS AND TAILGATES

When a bucket full of water is rotated in a vertical plane at high speed the water does not fall out. This principle is also applicable to doors and tailgates which when slammed develop a considerable angular momentum which then has to be dissipated when the door closes. The rate of change in momentum is the force exerted, and this force can cause considerable damage to nearby seams (e.g. the seam on the 'C' pillar at the rear wheel arch or the seam produced at the juncture of the rear valance and quarter panels). However, the centrifugal force also causes the water lying in door handles and in the window rubbers to be centrifuged out and to collect in the outer perimeter of the vehicle door. Most manufacturers do not wax inject along this outer crevice. (Most manufacturers wax inject about 4 inches along the bottom of the doors.)

The internal structure of the door is wetted by three processes (Fig. 4.3). These are as follows.

(1) Water runs down the windows. Rain by-passes the rubbers around the lower window section and enters the bottom of the door. Therefore road de-icing salts which are deposited on the glass by contra-flow or overtaking traffic may be washed down to concentrate in the lower crevices.

(2) The design of the rubber around the window is such that water running down the window is made to 'pool' on the rubber and therefore by-passes the rubber (Fig. 4.3). The by-passing takes place very easily at the two ends of the rubber window strip.

(3) The air flow to replace the extracted passenger air is into the doors through the bottom of the drainage holes in the doors, which are over the front wheel track. The front wheel track is a region of the densest spray. This ingress is shown in Plate 2.2(a). Most manufacturers are using one-side zinc-coated steels for these panels (Sections 7.5, 7.15, 8.3 and 8.4; Fig. 7.13; Plate 2.2(b)).

The method of constructing a door therefore ought to take into account this wetting from these three points. The method of construction is to use two sheets of metal which are pressed to shape and held together by a 'hemming fold' lap joint, as shown in Fig. 4.4. This overlap may appear to be 'filled' with mastic, but this mastic is applied after pressing in most cases. The function of the mastic is to stop the electropaint and top coats on the raw edges of the metal overlap from breaking down and showing the first signs of corrosion (Fig. 3.4). This internal crevice needs adequate protection and ideally should be made with an internal mastic (Fig. 4.4(4)) which does not aerate on heating or absorb aqueous solutions. In order to prevent the panels from vibrating and causing 'drumming', the panels can be stiffened by any of the mechanisms discussed in Section 4.4.1.

Method (3) in Section 4.4.1 is the best mode of stiffening. The patches, method (2), and mastic blobs, method (1), often tear away, forming a pocket, during the electrocoat baking operation, leaving no paint on the internal section of the car door. The continuous mastic between the outer panel and door can lead to the trapping of water (Plates 4.5(b)–4.5(d)) which has by-passed the window rubber. The mastic patches, method (2), can collect and trap the road de-icing salts which run down the window and are scattered on the internal surfaces every time the door is slammed. Rapid corrosion ensues as the metal is corroded at the water line produced by the new pocket.

Note: the manufacturer's owner's manual always gives advice to the owner which implies that the car door drainage holes become blocked with mud. This is not the main cause of corrosion along the door bottom. The main cause is the high drainage holes to the doors (Plates 2.2(a), 2.3 and 4.6). Furthermore, if mud does get into the drainage hole, then the design against corrosion is inadequate.

5.10 GOOD DESIGN FOR SILLS–ROCKERS

Road de-icing salts enter the sills or rockers by several mechanisms.

(1) Water drains from opening rear quarter lights into the sill–rocker box member between the 'B' and 'C' pillars. Note: some sliding roof suppliers and fitters also drain the sliding roofs down the 'A' pillar into the front sill box member.

(2) Road spray enters the 'B' pillar by a wrong-way overlap at the 'B' pillar. The front wheel road spray is forced into the wrong-way overlap by the vehicle speed and direction.

(3) The sill–rocker member, in some vehicles, is a hollow box which runs round the perimeter of the passenger floor. For electrocoating this member has to be a continuous hollow ring box member. This means that large volumes of air may be forced into the sill–rocker member when the vehicle overtakes another at speed or

when two vehicles pass from opposite directions. This is diagrammatically shown in Fig. 5.7. This pressure variation is often noticed more when driving past high sided vehicles. Smaller vehicles will also produce a sufficiently large pressure difference to force wheel spray through these low members.

(4) The suction of the passenger air extraction vents causes air from the front wheel track to be drawn into the sill–rocker drainage holes which are all over the front wheel track, e.g. Plates 4.8–4.12.

(5) Spray from the front wheel hits the protruding jacking points and entering through the electrocoat drainage holes.

(6) Driven rain can enter on one high pressure side to exit on the leeward side, which is at a lower pressure, if passenger air vents are not correctly designed.

(7) De-icing salts can enter the sill–rocker as a result of pressure eddies between the off–side and near–side due to overtaking or contraflow traffic (Fig. 5.6).

In Section 2.13, dealing with passenger comfort, it was suggested that an air continuity path between the sills–rockers and passenger compartment was a requirement of the electrocoating processes as discussed in Section 2.11 and Chapter 9. This air path is shown in the photograph of a vehicle body without any internal fixings (Plate 2.5). This means that replacement air (which includes low momentum road dirt and de-icing salts) can be drawn through the sills–rockers (Plates 2.4(a)–2.4(d)) and over the rear quarter panels or rear wheel arches (Plate 2.6). The air drawn through the sills–rockers originates from the regions of densest spray as these electrocoat process holes to the sills–rockers are directly over the front wheel track. The ingress of the aggressive environment becomes more acute as the vehicle passenger compartment or body becomes bigger (simply because the volume of paint contained in the body is greater and, for the same manufacturing vehicle line speed, the faster will have to be the emptying of the electropaint). Hence, taking as an example a commercial vehicle (Plate 2.9(a)), these drainage holes have to be very much larger than in large passenger cars simply to empty the electrocoat from the van body in a given time. Thus, when driven, a greater volume of air and therefore more wheel spray can be drawn into the sill–rocker sections. This is shown in Plate 2.9 for a commercial vehicle where the outer panel has been removed to show the internal micro-climate and associated corrosion.

For a well-designed vehicle body, these holes should be filled with a rubber grommet or taped* over beneath the carpet, to reduce the volume of air which can be drawn through the members. Underneath the vehicle all drainage holes (except two of the sill–rocker holes) should be blocked off. (This cannot be done if either the rear quarter lights or the sliding roof drain into the box member between the 'B' and 'C' posts. This is bad design, but unfortunately it does happen and must be altered if the sill drainage holes are to be blocked off.)

---

*Some manufacturers do specify that these holes are taped over in the factory after electrocoating. However, the location of the tape is not always over the holes 'A' in Plate 2.5. Rubber grommets are the only positive way of removing this pressure connection. The blocking up of these holes also will lead to some beneficial effects such as a reduction in road noise within the vehicle. The average sound reduction index SRI(av) is given by the following formula (for readers new to to acoustics, please refer to a textbook on noise):

$$\mathrm{SRI}_{av} = 10 \log_{10} 1/\tau_{(av)};\ \tau_{av} = \frac{A_1\tau_1 + A_0\tau_0}{A_1 + A_0};$$

For a hole $\tau_0 = 1$; τ_1 is very small therefore neglect $A_1\tau_1$

$$\therefore\ \mathrm{SRI} = 10 \log_{10} \frac{A_1 + A_0}{A_0}.$$

A_1 = area of panel; A_0 = area of hole.

5.11 GOOD DESIGN—REAR WHEEL ARCHES AND WHEEL WELLS (TWO- AND FOUR-DOOR MODELS)

The corrosion of the rear wheel arches occurs through the following mechanisms.

(1) The air continuity path between the sills–rockers and passenger compartment promotes corrosion. This air continuity path is a requirement of the electrocoating processes (Section 2.11; Plate 3.1). These paths should be blocked off by rubber grommets or with tape. This mechanism of build-up of the aggressive microclimate on internal (out-of-sight) body panels is discussed in Section 2.11. The deposition of road salts causes internally nucleated corrosion around the outer panel of the rear wheel arch (Plate 2.6(c)). This corrosion is very difficult to rectify because it is internally nucleated in an inaccessible region (Plates 2.6(a) and 2.6(b)). No amount of surface preparation from the outside will arrest the corrosion process. As in all cases of corrosion it is easier to prevent it by good design and adequate protection than it is to stop it when it has started. See also Section 5.12.

(2) The rear wheel arch may be slightly proud of the rear doors or quarter panel, which means that the high velocity wheel spray from the front wheel abrades the paint at this proud area. This is more noticeable if the car is a dark colour as the modern cathodic electrocoats are of a very pale colour. Hence these stone chips can be seen as white specks against the darker background. Rectification should be undertaken before this erosion process leads to corrosion and becomes worse. The area can be protected with plastic guards, but it would be better design if these guards were not there and the areas were not proud to start with. These plastic guards collect the front wheel spray by their motion through the front wheel spray (Plate 4.20) and cause the guard–paint–metal interface to remain wet for long periods as the plastic guards restrict evaporation. Note: Ill-fitting plastic items will cause crevice corrosion, the severity of which is less than that of the similar crevice formed by two adjacent metal surfaces.

(3) Only two drainage holes should be left open within the sill–rocker to reduce the amount of low momentum micro-droplets being carried over the rear wheel arch. The two left open should be located as follows: one at the front of the sill–rocker (and protected from the front wheel spray by a mudflap) and the other at the rear (protected by a cheap one-way (air) valve which can be made from ordinary electrical insulating tape (Section 5.19; Fig. 5.10).

(4) Mud flaps should be fitted as these protect the sill–rocker from both road and wheel spray. Note: it should be pointed out that the 'go faster' type of mud flap which is being used at present by some manufacturers does not perform in the required manner because it is too short.

(5) Some rear wheel arch quarter panels are not sealed or welded at the bottom where this joins the sill–rocker (see Section 4.6.1, item (24)). The open crevice will allow the road de-icing salts to gain entrance to the area of the inner wheel arch.

(6) Some vehicle's rear wheel arches are made up of small panels welded together to form a whole unit. Unfortunately, spot welding produces distortion and leaves a dimple where this has taken place. The alternative usually undertaken by the manufacturer is to leave this one spot weld out and to apply 'mastic' to the area. This mastic sealer ages and becomes brittle. This is bad design as the area breaks down as a result of the different thermal expansions which take place because some of the panels are shaded from the sun by the door overlap whilst adjacent panels are not

shaded at all. Also, a swinging door has considerable momentum and this change of angular momentum when the door shuts and stops swinging is considerable. This can break the brittle mastic and open up the overlap. The only correct design involves continuous welding or brazing of this overlap area (Fig. 5.4). In some cases the area offers a wrong-way overlap to the outer panel, whilst inside the wheel arch the wrong-way overlap is to the rear wheel. Better designs would incorporate the whole of this rear quarter panel as one unit. However, this becomes expensive in accident repairs as it means the whole side panel must be replaced for sometimes minor damage.

(7) Corrosion can arise through misplaced passenger air extraction grilles because the designer has not considered the mechanism of corrosion together with the build-up of the microclimate: this is discussed in Section 5.12.

5.12 PASSENGER AIR EXTRACTION SYSTEMS

Passenger air extraction systems ensure that the passenger compartment air is exchanged frequently. This is necessary in order to keep the driver alert. The designer can place the extraction system in any low pressure area. However, there are some areas which are better from a corrosion design point of view than others. The positions open to the designer for these extractor vents are as follows:

(1) on the outside of the tailgate and open to the driving rain
(2) underneath the tailgate and not open to the driving rain
(3) side panels of vehicle and open to the driving rain
(4) behind and underneath the roof and above the rear window
(5) adjacent to the boot–trunk
(6) at the trailing edge of the car doors (Fig. 4.3(a)).

The last of these is the best design as it allows for more wax injection in the internal crevice.

Extraction grilles within any of the above positions will also cause the deposition of salts in the sills–rockers and these areas (Plate 2.6) of the rear wheel arch. Some manufacturers have attempted to prevent this mechanism by applying small wads of foam within the sill–rocker–rear dog leg area. However this has effectively filtered out the road de-icing salts to concentrate them at the interface of the foam and internal metal surface. The air extraction system should be examined to determine whether it is sufficiently well designed to keep out rain water when the car is parked overnight on sloping or uneven ground for the worst possible wind direction. Each extractor should incorporate a one-way valve so that moist air cannot be blown through the extractors. One-way valves would also reduce the discharge of road salts within the cavity members from the pressure eddies due to overtaking vehicles. A cheap DIY one-way valve is described in Section 5.19.

The worst position for the passenger air extraction system is on opposite sides of the vehicle and linked into the rear wheel arch. In this location there is ready access through the boot–trunk–luggage compartment to the other side of the vehicle (i.e. the offside of the car is linked to the nearside wheel arch through the passenger compartment or boot). The reason why this is the worst possible position can be explained by considering Fig. 5.6 which shows an overtaking line of vehicles, each of which produces its own pressure eddy depending on whether we are considering the front (locally a high pressure region) or the rear (locally a low pressure region). As these high pressure regions pass the offside vehicle, then the spray is passed through the vehicle body to exit on the low pressure nearside of the overtaken vehicle. The air motion is reversed as the low pressure area (rear) comes alongside the overtaken vehicle's passenger extraction system.

5.13 REAR VALANCE AREA

The rear valance, on most model makes, acts as a scoop to the front and rear wheel road spray (Plate 4.21(b)). A good design will overcome this problem and either keep all butt welds within the vehicle or certainly not arrange the panel join to form a perpendicular member at right-angles to the vehicle direction. Only the smaller water droplets are lifted by the wheel to the height to be scooped up by the rear valance (Plates 4.21(a) and 4.21(b)). This means that the small amount of water present upon each road dirt particle is insufficient to wash away and clear the road dirt which is deposited. This deposition is aided by the turbulent air around the rear of the car. Thus mud poultices are retained around the rear spring hangers and rear box section. All these areas are out of the direct water line spray from the larger water droplets from the trajectory of the rear wheel spray and can be further protected by mudflaps.

The low water content of the high speed spray also means that this road dirt is artificially more densely compacted by the phenomenon of surface tension into a high density mud pocket. These high density mud pockets are formed upon the rear wheel arch mud traps produced by suspension housings, rear box section and spring hangers, petrol tank and spare wheel and by abrupt changes in profile, e.g. where petrol tanks are positioned in low-slung vehicles. Any member slung beneath the vehicle causes pressure eddies and the deposition of road de-icing salts on the vertical surface which faces the rear of the car. (This is the same phenomenon as the deposition of road dirt on the rear tailgate which does not face the direction of motion.)

5.14 THE BOOT–TRUNK LID—TAILGATES–LIFTGATES

The boot–trunk lid and tailgates are constructed in a similar manner to the door and therefore the same aspects will apply.

5.15 THE REAR WINDOW RUBBERS AND TAILGATE WINDOWS

The fixing of windows in by rubbers and the resultant associated corrosion are different in the cases of fixed vehicle panels and of those which are hinged and capable of being slammed. The corrosion is due to the following.

5.15.1 Stationary panels

Where the fixed window is stationary there is very little corrosion. Water seeps in between the metal and rubber interface and remains in this interface. Corrosion is often limited to the interface between the available oxygen and perimeter of the rubber.

5.15.2 Hinged panels

Where the panel is moved and can be slammed, the capillary wicked water is thrown out of the crevice interface between the metal and rubber and lodges as a ball of water in the lowest point as a result of the centrifugal force of the slamming. This causes corrosion well below the perimeter of the rubber. A well-designed vehicle will have 'bleed holes' to allow this water to be 'centrifuged' out by the slamming. A badly designed glass attachment will not drain and so corrosion will be the result. This is particularly important in the lower raked hatch-back tailgates (Fig. 5.8).

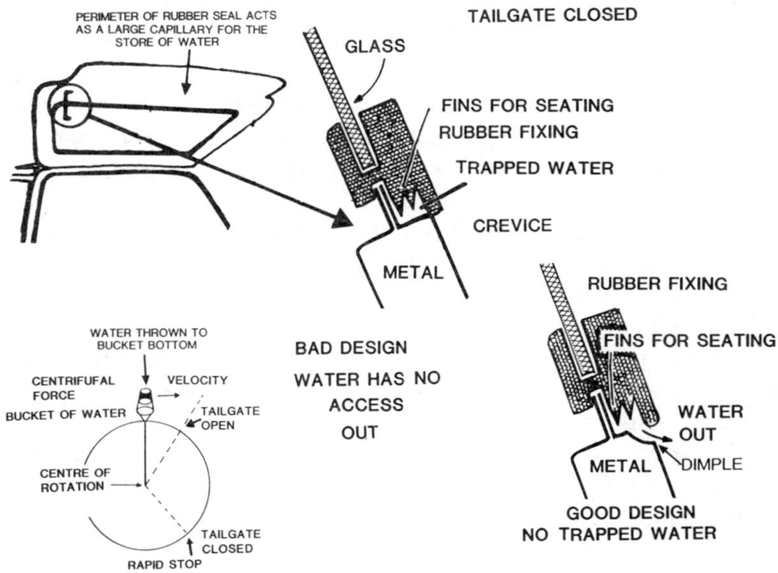

Fig. 5.8 — (a) The complex crevices produced around the sealing of a window into the frame can trap small amounts of water around the perimeter of the seal. (b) this small amount of water can be centrifuged out of these crevices into the larger pools in areas where it will collect and initiate corrosion. (c) Small 'dimples' (illustrated) (or 'holes') should be introduced to drain these areas at the bottom of the glass–rubber seal. In less steeply inclined tailgates this form of drainage cannot be used as these 'dimples' will act as local areas of water ingress. In these cases drainage must be into the well-designed tailgate and through the lowest point. This 'lower' point should also take into account where the water is forced by centrifugal forces. Bi-metallic coupling or the production of crevices to drain the water from this area should not be attempted as it is 'bad design' and will cause paint blistering around bimetallic couple which will spread to beneath the rubbers. Similar arguments hold for vehicle doors which are also subjected to centrifugal forces which can 'fling' water into unexpected crevice areas.

5.15.3 Condensation

Both boots and tailgates must be designed with condensation in mind. This condensation arises from two aspects: passengers and wet clothing or garden refuse carried in the boot. Condensed water, from any source, must be able to drain out.

5.16 THE SUSPENSION UNITS

The design of the suspension unit must take into account the drainage of water from the unit when the spring is in position. The whole suspension unit is out of the direct wheel spray but is subjected to the high speed low momentum spray which contains a small amount of water upon each road dirt particle. These particles contain insufficient water to wash away the road dirt which is deposited. This deposition is aided by the turbulent air around and beneath the car. This low water content also means that this road dirt is artificially more densely compacted by the phenomenon of surface tension into a high density mud pocket (Fig. 10.7). As the low momentum spray has little water associated with the dirt, it also follows that the drainage holes in the lower spring platform become quickly blocked. This means that the spring platform becomes the site for poultice corrosion and the collapse of the whole suspension system.

5.17 SUBFRAMES AND OTHER BOX MEMBERS

Most subframes, and components made from pressed steel panels to replace forged components (e.g. track control arms) beneath the car, give rise to corrosion. This arises because the component undergoes pressure changes along its length owing to the following.

(1) The open ends nearer the wheel often have two openings which surround the location of the lower ball joint assembly (Fig. 5.9(c)). These two openings, although in the same plane, have a pressure difference between them. This pressure difference is produced when the vehicle motion is combined with the air which is dragged round by the rotation of the wheel. Hence one opening will be at a higher pressure than the other.

(2) The open ends at the two ends of the box member are not in the same horizontal plane and are therefore subjected to different air speeds. The higher, more protected, opening will have a different pressure from the lower, more exposed, opening. This pressure difference will allow the preceding vehicle's spray to enter the members and to deposit road salts within the box member and out of the owner's sight Fig. 5.9).

Item (1) above can be rectified by blocking off the forward hole and leaving the rearward hole open, while item (2) above can be rectified by blocking off the uppermost end and leaving the lowermost hole open.

5.18 THE (DO-IT-YOURSELF) RECTIFICATION OF IMPROPER DESIGNS TO RESIST CORROSION

If the bad (corrosion) designs are left without rectification, the vehicle will prematurely age by the accumulation of road de-icing salts. The road de-icing salts will concentrate in crevices and keep the crevices moist for a much longer time. Therefore the rectification processes recommended are as follows.

(1) Inject wax or oil. This can be done at the after-market stage, by the manufacturer or by DIY.

(2) Fill wrong-way laps with weld*, braze† or mastic sealant material.

*This must be undertaken by the manufacturer at the body-build stage. Accident repair garages should consider these principles as corrosion sets in very quickly after accident repairs.

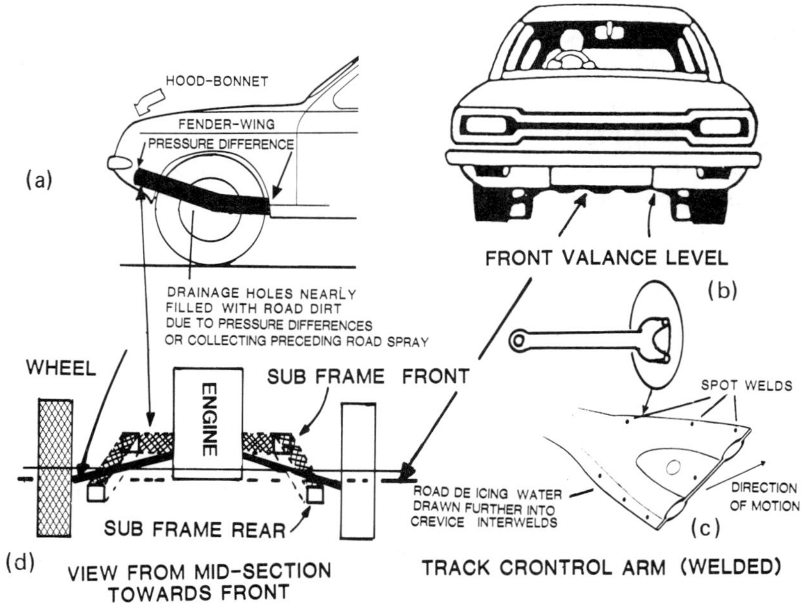

Fig. 5.9 — (a) Engine mountings (or subframes) can be underslung, parallel to the direction of travel, but in positions where there is a difference in air speed. This difference in air speed causes pressure differences. Outriggers which lie across the direction of motion are subjected to differences in pressure caused by overtaking or contra-flow traffic. (b) Track rod ends may be made from forged metal or from pressed steel plate, which is then spot welded together. The diagram shows schematically the lower end of the track rod end with the pressure differences due to their unequal lengths and proximity to the rotating wheel. The fact that one end (engine side) is shielded by the front valance while the outboard end is within the turbulent air at a point below the wheel's rotation will cause spray ingress. (c) Pressure differences due to the track control arm open ends' being in two different planes and therefore subjected to different wind speeds owing to the vehicle motion. (d) Subframes are also made this way. Care must be taken not to collect the preceding vehicle's wheel spray by drainage or attachment holes.

(3) Fill crevices and paint drainage holes with mastic (builder's mastic is good enough). Pay particular attention where these drainage holes are adjacent to jacking points.

(4) Fit mud-flaps. This can be done at the after-market stage, by the manufacturer or by DIY.

(5) It is preferable that the manufacturers block off electrocoat holes with rubber grommets at the point of assembly and after electrocoating. On existing vehicles, use PVC tape (PVC tape is polyvinyl chloride tape or electrical insulating tape (Section 5.19)) to cover over holes which collect the preceding vehicle's spray.

(6) Make one-way valves for circular drainage holes from PVC tape.

5.19 ONE WAY VALVES

Previous chapters developed the concept that the car body was built up with doors, bonnet and boot fixed onto the passenger compartment as the 'body-white'. This was then painted by electrocoating. For the electropaint to enter the car body, process holes or 'drainage holes' must be punched into box members to allow the paint to enter and any trapped air to escape. The process holes are essentially circles cut off the box member. Their function has been wrongly promulgated by agents of the motor trade as holes to let out water from the road spray. This false representation has been accepted by the general public. Pressure eddies beneath the car body and the continuous extraction of stale passenger air produces water in the form of low momentum micro-droplets. These micro-droplets are associated with small particles of road de-icing salts and enter the process drainage holes as a result of this passenger extraction system.

This ingress of road spray can be stopped, without stopping the drainage aspect of the 'drainage holes', by cheap one-way drainage valves which can be made from electrical insulation tape. These one-way valves will prevent the ingress of the aggressive micro-climate into the box members and so reduce internally nucleated corrosion. This is shown in Fig. 5.10. The electrical insulation tape is bent partially back on itself so that both adhesive sides are in contact. This forms the 'one way flap' when the non-overlapping adhesive section is stuck to the box member. Pressures greater outside the sill will automatically close the flap. When the vehicle is stationary, the weight of the double electrical tape will cause the valve to open and allow any road spray or water to drain out. Hence this device acts as a one-way valve,

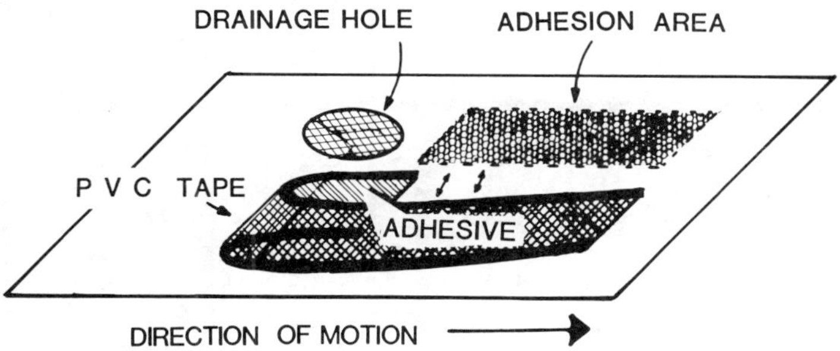

Fig. 5.10 — The DIY manufacture of one-way drainage valves for badly positioned drainage holes. These may be made from ordinary PVC electrical insulating tape which is partially doubled back on itself to form a 'flap' valve. Trapped water can operate the flap valve and drain out but, if the pressure on the lower side exceeds the pressure within the box member, the valve will close up.

allowing water to drain out, while at the same time stopping the ingress of the fine spray. It is important to ensure at least two outlets in case of unsuspected water ingress into these areas by pressure eddies which can be developed by the high speed of the vehicle. Other holes within the same box section can be completely covered by tape or filled by other more suitable means (Table 5.4). Fig. 5.11 summarizes the items of good design which should be found in any motor vehicle.

Table 5.4 — Owner rectification of design faults

DESIGN OVERSIGHTS — DRAINAGE HOLE LOCATION] = VERTICAL	RECTIFICATION
Engine — Drainage hole facing engine compartment	Leave open
Wheel — Drainage hole to wheel spray, either front or back, where the drainage hole is vertical	Method A, B or C
To inside facing holes out of wheel spray	Method B
To drainage hole in a vertical plane where flow of water due to gravity and wheel direction are in the OPPOSITE direction . . .	
Drainage hole in the horizontal plane in spray	Method B
Wrong way Laps & Crevices to wheel spray	Method D
Shielded drain hole to sill; Open drain hole to sill; Shielded or open interweld joint to sill	Leave front and rear open. Fit mud flaps. Fill the rest with mastic after wax injection
Open Drainage hole to sill in the horizontal plane in spray	Method B
Jacking point. Fill in drainage holes to the front of the jacking point; Jacking point causing pressure changes adjacent to the drainage holes within the sill	Method D
Jacking point protuberance adjacent to D.H.	Method E Method D

Adhesive — Adhesive turned back on itself

← Direction of motion of car for all diagrams.

Methods

A Completely covered and stuck all round

B Only half stuck on, with the other half turned back to stick to itself. OBJECT: to provide one way ventilation

C As method B but only to be used where gravity and wheel spray, ie direction of rotation, are acting together

D — Fill with mastic

Method E — drain hole, tube, mastic, Jacking point

Notes
- Ensure mud poultices are removed
- Do not make a sweat box out of the enclosed member
- Fill in all crevices with mastic *after* wax injection
- Fit mud flaps, but ensure that the vertical surface does not come behind an electrocoat drainage hole
- Check operation of one way valves (B, C) periodically
- Clean area to be stuck, give a good adhesive area to the front of the car

This table shows the areas (e.g. jacking points on most vehicles; see Plates 4.8–4.18) and rectification points to which the owner could attend should any of the point made in Fig. 5.11 be missing.

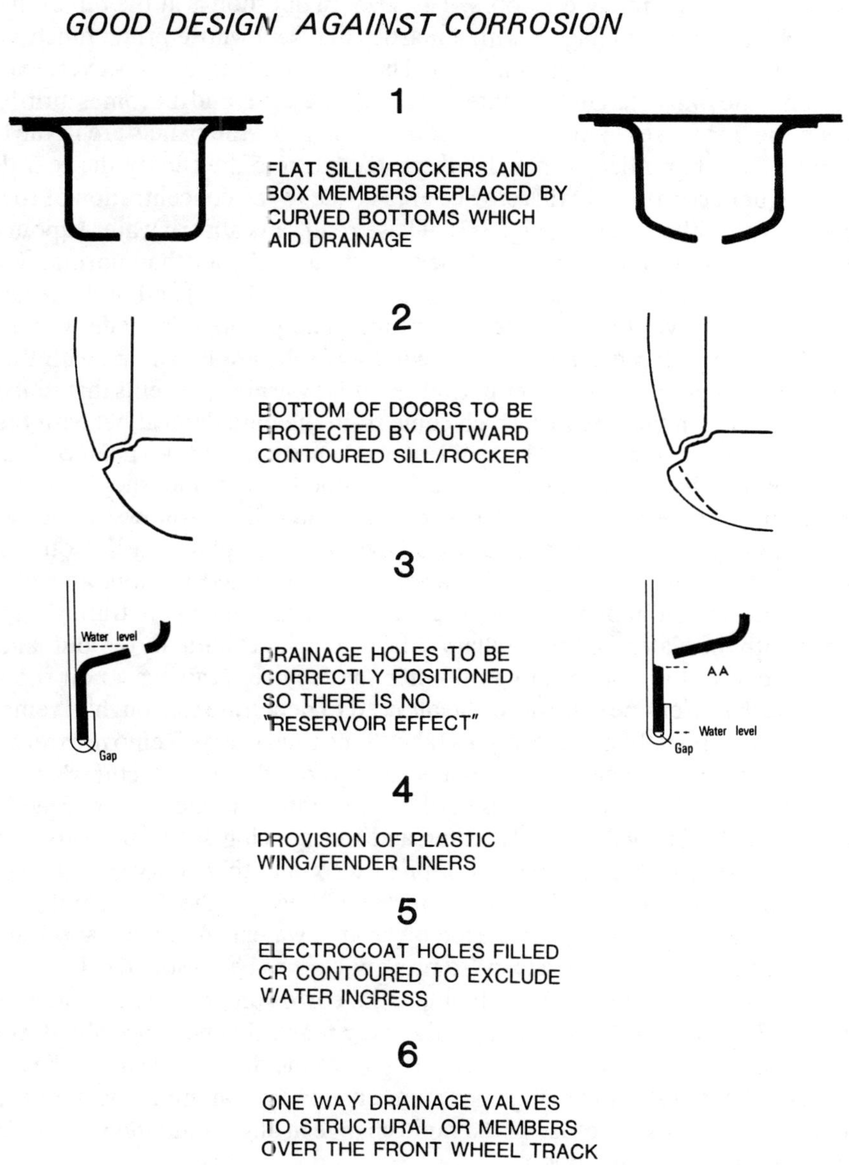

Fig. 5.11 — A figure summarizing and showing the basic 'good design points' which should be incorporated in all vehicle designs.

5.20 CREVICES FORMED BY METAL OVERLAPS

This 'C' post in many manufacturers' vehicles is very susceptible to damage as a result of the occupants' slamming the doors. The seam in question is at the top of the rear wheel arch and is 'camouflaged' with a mastic (Fig. 4.4) whose prime function is to suppress any crevice corrosion from view. The inter-weld mastic, however, excludes water (and the waxes). Unfortunately, the mastic ages and becomes brittle and breaks if the doors are slammed on a hot day as some of the panels are in the direct sunlight while other adjacent panels are protected from the sun by the rear doors. The subsequent corrosion in this crevice area is due to the concentration of road de-icing salts within the crevices (Fig. 3.7). These road salts absorb water vapour from the atmosphere and thus promote condensation much earlier than normal. Evaporation is also slower from these salty crevices (see Table 10.12) and so the result is to keep the crevices wet for long periods of time. The problem is made worse if the vehicle is not regularly cleaned as these corrosive salts are in contact with the bare unpainted metal for a much longer period. Regular washing prevents this concentration of road salts in these seams which come from the mud deposit beneath the rear wheel arch returns (Fig. 4.6). This will lead to the stripping of the paint over a wide area adjacent to the crevice unless the fault is rectified. The reader should be advised, after steam cleaning and drying, to correct the blemish as follows: apply several applications of three-in-one penetrating oil over a period of a month. Note this will only work if the vehicle and crevice are dry (i.e. garaged indoors and the mud deposits removed from beneath the wheel arch returns). The penetrating oil has to compete with the water for the capillary. The water will eventually dry out, and only when it is dry will the penetrating oil enter the crevice. Any excess oil should be removed with turpentine substitute and allowed to penetrate thoroughly: remember that capillaries and tight metal overlaps take a long time to dry. Remove external rust stains with 'T'-cut or a similar compound and apply the manufacturer's touch-up paint to the seam. When this paint is hard, apply another application of three-in-one penetrating oil. This will exclude water and the de-icing salts, and subsequently three-in-one penetrating oil should be applied at six-monthly intervals and especially after steam cleaning or washing with detergents (because the oil is emulsified and breaks down in the sun and is displaced by 'soapy' water). After any application of oil, this area should be wiped clean (to prevent clothes being soiled). This will stop water getting into the crevice and eliminate the corrosion process. It is advisable to wash the wheel arch returns with a gentle spray regularly and especially during the winter de-icing programme. It is advisable to have the doors adjusted if they do not shut with a slight push from being open some six inches. Slamming the door causes a large force on this area because of the rapid change in angular momentum which can be obtained with swinging doors.

5.21 SUMMARY OF SOME COMMON RUST TRAPS AND THEIR CAUSE, DESIGN, RECTIFICATION AND SUBSEQUENT CARE BY THE OWNER–GARAGE

See the items in Fig. 4.1. In each case the reference point is given, followed by a description of the corrosion and its cause and advice concerning its rectification.

5.21.1 Front valance and wings

(1) Clinch fold around bonnet (Fig 4.4) Crevice corrosion; drainage holes too high; spray and immediately drop bonnet.

(2) Front valence — radiator grill. Corrosion by mud accumulation either from preceding vehicle spray or vehicle spray thrown ahead of wheel (Fig. 10.8).

(3) Cavity formed by styling (Plate 2.1): ledge. Area will be difficult to spray by waxes initially; therefore, pay particular care when these vehicles are considered for second-hand purchase or in for retreatment–repair. There is a possibility of a blockage of heater scuttle drainage by leaves—remove leaves.

(4) Junction of wing (Plate 4.1): crevice corrosion and mud poultice where internal wing–fender liners are not fitted; wash regularly. For good design total inner wing–fender liners are essential. Crevice separated by wax or plastic washer.

(5) Blockage by leaves at lowest position where the internal wing–fender liner and mud-flaps meet the outer wing. Remove leaves and cut internal wheel liner to allow adequate drainage and the natural removal of leaves.

(6) Attachment points for front suspension: crevice corrosion. Wash regularly. If an internal wheel liner is fitted check for road de-icing salts in the crevice from drainage from the bonnet area (this is important during winter work).

(7) Top of headlamps: deposits arising either from the preceding vehicle's spray or because the wheel in question gives rise to pitting beneath deposits. Wash regularly through the front aperture over the headlamp eyebrows if internal wing–fender liners are fitted or from beneath if inner wing–fender liners are not fitted. For good design total inner wing–fender liners are essential. There should be no crevices formed by (a) metal-to-metal contacts coated with mastic as shown in Fig. 3.11 or (b) metal-to-rubber contacts which result from partial internal wing liners (Fig. 2.9 and Plate 4.3(c)).

(8) Behind front panel: mud poultice where incomplete internal wheel liner is fitted (Fig. 2.9). Wash regularly. For good design total inner wing–fender liners are essential. Crevice separated by wax or plastic washer.

(8) Forward-facing direction: stone chips from preceding vehicle. Repair as soon as possible, otherwise the non-adherent paint will aid differential oxygenation (Figs. 2.6 and 8.18).

(9) Front valance seam: crevice corrosion owing to lack of electrocoat. Penetration of waxes is essential—wash regularly. For good design on bolted structures a wide crevice is required except at bolted positions. There will be no paint unless the wing is electrocoated off the bolted position. Crevice separated by wax or plastic washer (Figs 2.10 and 3.11). Does an air grill (Plate 2.8(c)) collect the preceding vehicle's spray or is the wheel in question giving rise to pitting beneath deposits which enter?-Block up the grill.

(9) Front radiator grill: the radiator deflects the preceding vehicle's wheel spray into box members (Figs. 4.5 and 5.6). Steam clean if observed, then wax inject and tape over the offending holes.

(10) Front wheel arch returns: de-icing salts owing to wind turbulence around the internal wing liner if fitted. Otherwise as rear wheel arch (item (25)). Wash regularly.

5.21.2 Doors

(11) Internal disbondment of sound deadening material (Plate 2.2(a)) or mastic–door strut within door structure (Plate 4.5). The adhesive mastic (Plate 4.5) may also be trapping the manufacturer's pretreatment salts—advise manufacturer.

(12) Exterior of door clinch fold (Figs. 4.3 and 4.4): lack of paint in inter-weld. First signs of corrosion. Apply penetrating oil or thin wax material internally and externally; wash regularly.

(12) Raw edges to door. If a new vehicle, paint–wax coat before corrosion is initiated.

(13) Inside doors: door drainage holes too high. Examine for corrosion and a 1:1 correlation with drainage hole at the bottom of the door for the first signs of metal penetration. External 'scab' corrosion on the outside panel is a high possibility if this 1:1 correlation is not observed. Check by the pin test (Fig. 2.11) or removal of the door trim. The first signs of penetration of a metal by internally initiated rust through will always occur around the drainage holes in all members (Plates 2.2 and 2.4). Claim from the manufacturer if within the Canadian voluntary code conditions (Section 1.12).

(14) Decorative brightwork: galvanic corrosion. Wax from above and wipe off excess; wash regularly. If repainting an old vehicle consider replacing the brightwork with PVC tape between the brightwork and newly painted surface. If the vehicle is new, claim from the manufacturer.

(15) Door drainage positions (Section 4.4.6).

(16) Paint stripping around door hinges (Plate 4.4): lack of electrocoat behind hinge. Apply penetrating oil or wax after loosening the top screws if necessary to get the oil or wax into the crevice.

(17) Around door–glass rubbers: concentration of de-icing salts beneath rubbers. Wash regularly.

5.21.3 Sills–rockers

(21) Jacking points (Plates 4.8–4.13): drainage holes adjacent to jacking points so that front wheel spray is scooped up. Fill forward holes with mastic (Table 5.4).

(21) Drainage holes: road de-icing salt ingress (Plates 4.8–4.13). Are mud flaps fitted? If not, fit them. If mudflaps are fitted check for road de-icing salts in the crevice formed by the mudflap and wheel arch return. Wash regularly.

5.21.4 Rear wheel arch

(18) Abrasion due to road spray. Mudflaps should either be fitted or renewed or be replaced if they are too short.

(24) Wrong-way overlap or an area which is not welded. Braze area. Permanent (maintenance-free) rectification is impossible by any other means. Temporary rectification can be achieved with penetrating oil (Section 5.20).

(28) Outward-flared wheel arch? Erosion from the front wheel spray: claim from manufacturer if within the Canadian voluntary code conditions. Fit mud flaps. If not an outward-flared wheel arch, and the cause is not 'scab corrosion' (pin test; Fig. 2.11), then the problem may be more severe as this may be corrosion of the dog leg and is caused by salts being dragged through sills (Plate 2.6). Expect imminent sill

penetration: examine sill for 1:1 relationship of corrosion with drainage hole and external 'scab' corrosion (Plate 2.4(b)). See Section 4.6, items (25) and (28).

(25) Passenger air extraction grille near? Suspect overtaking vehicle's spray or water penetration into area together with lack of wax penetration (Plate 4.19). Ensure that area is covered well on all models when accident repair work is carried out. Could passenger air extraction system be modified with a one-way valve (Section 5.19)?

(27) Rear wheel arch returns: road de-icing salts around ledge (Fig. 4.6).

5.21.5 Seams

Wrong-way overlap or an area which is not welded. Braze area. Spot welds make crevice areas and capillary action keeps the seam wet for long periods. Permanent (maintenance free) rectification is impossible by any other means. Temporary rectification can be achieved with hydrophobic penetrating oil (Sections 3.6, 3.52 and 5.20).

5.21.6 Wheels etc.

(29) Chrome discs (most are plated plastic): these will suffer erosion if set outside the wheel tracks. Metal discs suffer internally initiated corrosion if the drainage used for electroplating is high (Plate 2.3) or blocked.

(30) Chrome looking rusty or mud poultice behind bumper bar? Defect in the chrome plating (see Figs. 8.3 and 8.4). Wash regularly behind the bumper, which may act as a scoop.

5.21.7 Rear

(39) Around light cluster: galvanic corrosion or salts concentrating in the crevice overlap on drying. Apply penetrating waxes–oils.

REFERENCES

Department of Trade and Industry. (1981), Corrosion Control Checklist for Design Offices. Corrosion: Stop it now.

Hoar, T. P. (1971). Report of the Committee on Corrosion and Protection, HMSO, London.

6

Background chemistry of materials

6.1 CHEMICAL SHORTHAND NOTATION BONDS IN MATERIALS — INTRODUCTION

Science is a difficult subject with a shorthand and jargon of its own. It takes some time to assimilate both the jargon and scientific concepts as a subject. Therefore it will take time for the non scientist to understand some of the points made in these following chapters. The science has been taken out of the text, in order that a non–scientist can still understand the corrosion of the motor vehicle. However, some readers may like to read and understand the scientific modelling which is so characteristic of science. This may take some time as, like walking, you can be told, as a two year old, that it is possible to walk on two legs, but it takes a lot of practice to be able to do it. These sections, like walking to the non–scientist, will only be assimilated with constant reading around the subject. It is the author's intention to make the knowledge easier to understand by subdividing the chapters into sections which can be cross referenced in the text. By this means it is hoped to make the chapters more readable to all levels of scientist. A bibliography of further reading is given.

Science is man's way of understanding the environment around himself. The sciences are man's way of creating a pattern of thought, which is a hypothesis, about our environment. This hypothesis is then 'tested' by experiment and gradually the modelling of our environment becomes 'law' — the laws of nature. If these 'tests' show that the modelling is correct, the hypothesis becomes a law. A knowledge of science helps man to plan other scientific events such as the moon landing and moon walk. All the scientific principles are tested by such events. Science is built on models: these models are tested and, if proved right, are brought into the realms of fact and become the basis of our commonsense. Not all the scientific models have proved the test of time and become 'commonsense' and, indeed, not everyone wants this basis for their commonsense. If, however, we want to be forewarned about a possible interaction based on science, then a knowledge of the scientific modelling is imperative.

Some of the scientific modelling is well known, but other models are less well known. For example, the model of the earth as a flat plate held man's imagination for centuries. The world is now modelled as a large number of satellites (of which the Earth is one) revolving in orbits around the sun.

The structure of matter is a model which has been refined over the ages to give man a basis on which atomic bombs and nuclear energy have been developed. It is worth following this model through its development.

6.2 THE STRUCTURE OF MATTER

When man first looked at a material, he decided that the fundamental building block of that material was the smallest unit to which it could be broken or cleaved. Hence a crystal which grew from solution and had the external form of a cube as its faces was thought to be composed of a fundamental unit which itself was cubic in shape (Fig. 6.1(a)). Likewise, all the geological crystals found in nature were composed of smaller crystals all packed together. Historically, before the advent of microscopes, scientists could grow very large crystals from solutions, and their external faces

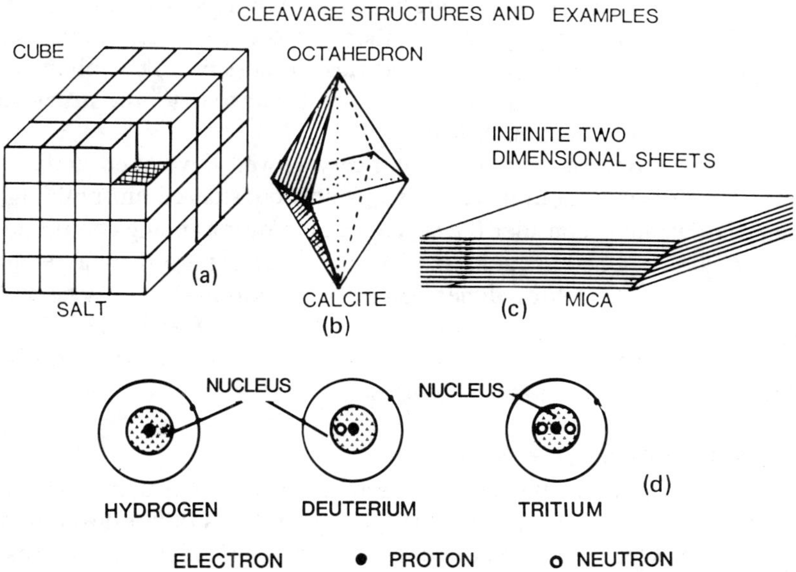

Fig. 6.1 — The first model postulated by scientists resulted from the 'fracturing' of the crystal into its smallest fundamental units. Geologists know this process as cleaving. The cleavage planes upon which fracture took place always bore a constant angle to other cleavage planes however big or small the fracture entities were. Therefore, if this process could be continued 'indefinitely', the fundamental unit could be obtained. (a) A larger crystal was made up of these smaller units all stacked together, without leaving empty (void) spaces. (b) Calcite would cleave as octahedra, which unfortunately do not stack together without leaving voids. (c) Mica would have an infite two-dimensional 'fundamental unit'. (d) The modelling of the hydrogen atom, showing the nucleus which contains the proton and neutrons which form the isotopes of hydrogen, (deuterium and tritium are often called heavy hydrogen). In modelling the atom, the scientist often represents the spherical atom as a dot to represent the centre of the sphere. If chemical bonds are being modelled, the centre of the sphere is omitted and only the outer orbital electrons shown (Section 6.6).

always bore a constant angle with other crystals of the same material. For a time this seemed to be a very good basis on which to explain structure, as it was very easy to consider all materials as being made up of smaller, miniature crystals with plane flat surfaces.

It was soon realised that this method of modelling a material had severe disadvantages:

(1) Some materials, which crystallized as octahedrons (Fig. 6.1(b)), could not be stacked together to form a larger octahedron without leaving voids in the bigger crystal. This meant, according to this model, that the density of small crystals should be different to the density of larger crystals. This is obviously a nonsense result of the modelling and is not supported in nature. Hence this form of modelling was discarded as it did not correlate with nature.

(2) The fundamental unit of something lamellar, such as mica (Fig. 6.1(c)), using this modelling has a fundamental unit which is an **infinite** two dimensional sheet. This was clearly absurd for a 'building block' which is a fundamental unit.

In order to overcome these disadvantages, a radical new approach was postulated which has become the basis for our modelling processes.

In the 'modern' theory, matter is considered to be composed of spheres, which are packed together to form the solid material. We now call these spheres **atoms**. This postulation had the advantage that there is a constant ratio of free space (26%) when equally sized spheres are packed as close as possible. The volume of this free space is very small and will not allow a water molecule to enter the structure (Section 2.1, Fig.2.1. (See also Sections 6.8 and 6.9).

By 1848 the concept of a crystal lattice was well developed and, in order to produce the fundamental **unit cell**, which is now considered our building block, of our material, we must consider the structure of matter as being composed of atoms (Section 6.20) or ions (Sections 6.5(7) and 6.11) arranged in a regular three–dimensional array. This three-dimensional array of atoms–ions may be outlined by a **parallelepiped** which forms the **unit cell** (see Sections 6.8 and 6.11). Fig. 6.9 gives representation of the unit cell concept for graphite and diamond and Fig. 6.14 the unit cells for the common ionic structures.

6.3 THE ATOMS AS SPHERES

These spheres were thought to be composed of 'fundamental matter' which could not be broken down. Our concept of these spheres to-day is rather different as we now know that by nuclear reactions these 'fundamental particles' can be broken down. The scientist considers them to be made up of three separate particles, which 'pigeon hole' the atom into a periodic table (Tables 6.1–6.5). Before dealing with the periodic table, it is best to consider the structure of these atoms.

The modelling of the atom is rather similar to the modelling of the world. The central entity of the atom is the nucleus (equivalent to the Sun). The nucleus of the atom contains particles of mass number 1 which are positively charged. These are called the protons, (abbreviated p). The atomic nucleus contains another particle of mass number 1 which has no charge. This is called the neutron, (abbreviated n). The atom also contains particles of mass much smaller than the mass of the proton and neutron and of a negative charge. This we call the electron, (abbreviated e). The mass of the electron is 1/1840th of that of the neutron or proton. The electron revolves around the nucleus in defined energy orbits (called subshells) rather like the satellites Jupiter, Mars and the Moon revolve around the sun in orbits. The electron is considered as the means by which electric currents are conducted in metallic conductors.

The model of the atom must describe how each atom is different from all other atoms. Our modelling suggests that the lightest atom known (hydrogen; abbreviated as H) consists of 1 proton and 1 electron only (Fig. 6.1(d)). If we want to model the atom of deuterium (D) which is essentially (chemically) the same as hydrogen, then we have 1 proton, 1 **neutron**, and 1 electron. The model describes the addition of a neutron as forming an **isotope** of the atom and this has the same chemical nature and activity as the hydrogen atom with fewer neutrons. Tritium is another isotope of hydrogen and has the same chemical properties as hydrogen. This is the modelling to explain nature's similar elements which are of different weights.

The next heaviest atom is helium (abbreviated He), which has an atomic number of 2 and an atomic weight of 4. In our modelling this is related to 2 protons and 2 neutrons (making up a mass of 4) and 2 electrons to balance the charge of the 2 protons. This element, which is a gas, is **chemically inert** and so our model reflects this by stating that this is a stable electron configuration, which cannot take part in any further chemical reaction. This 'stability' is ascribed to the electron state, which will be considered later (Section 6.6).

The next heaviest atom is lithium (abbreviated Li), which has an atomic number of 3 and an atomic weight of 6. In our modelling this is related to 3 protons and 3 neutrons and 3 electrons to balance the 3 protons. Chemically this element, which is a solid and a metal, is **very reactive** and so our model reflects this by stating that this is an **unstable** electron configuration, which can take a further part in some chemical reaction. This element has distinctive alkaline reactions (Secion 6.16.)

Subsequent elements are also built up in a similar way and these are summarized as follows.

6.4 THE ATOM'S CONSTITUTION

Each atom differs from its neighbour by the number of **protons** as shown in Table 6.1.

This table could be continued throughout the range of elements up to an atomic number of 92 and from which our engineering and geological materials are derived. Each element owes its 'individuality' to the number of protons in the atom. The chemical behaviour of the element is dependent on its electron configuration.

The important points are as follows.

(1) The **atomic number** gives the number of protons or electrons within the atom.

(2) A change in the number of **protons** gives a new chemically different element. Note this is not so with the number of either neutrons or electrons.

(3) The neutron is responsible for the number of **isotopes** and the fractional atomic weights (see chlorine).

(4) There is a gradual change from **alkalinity to acidity** as the atoms get more protons.

(5) There is an **abrupt** change from acidity to inertness to strong alkalinity at certain atomic numbers, e.g. inertness occurs at atomic numbers 2, 10, 18, etc.

This is now discussed in the periodic table.

Table 6.1 — Part of the periodic table. The structure of the atom in relationship to the number of protons, neutrons and electrons. Atomic numbers 1 (hydrogen) and 17 (chlorine) are chosen to show the existence of isotopes and how fractional atomic weights arise

Atomic number	Atomic weight	Numbers of Protons	Neutrons	Electrons	Chemical reactivity of oxide or element	Chemical name	Abbreviation
1	1	1	0	1	Acid	Hydrogen[a]	H
1	1	1	1	1	Acid	Deuterium[a]	D
1	1	1	2	1	Acid	Tritium[a]	T
2	4	2	2	2	Inert	Heliuim	He
3	6	3	3	3	Alkaline	Lithium	Li
4	7	4	3	4	Less alkaline	Beryllium	Be
5	11	5	6	5	Very slightly acid	Boron	B
6	12	6	6	6	Slightly acidic	Carbon	C
7	14	7	7	7	More acidic	Nitrogen	N
8	16	8	8	8	—	Oxygen	O
9	19	9	10	9	Very acidic	Fluorine	F
10	20	10	10	10	Inert	Neon	Ne
11	23	11	12	11	Very alkaline	Sodium	Na
12	24	12	12	12	Less alkaline	Magnesium	Mg
13	27	13	14	13	Alkaline and acidic	Aluminium	Al
14	28	14	14	14	Alkaline and acidic	Silicon	Si
15	31	15	16	15	Acidic	Phosphorous	P
16	32	16	16	16	Very acidic	Sulphur	S
17	35	17	18	17	Very acidic	Chlorine[b]	Cl
17	37	17	20	17	Very acidic	Chlorine[b]	Cl
18	40	20	20	18	Inert	Argon	Ar
19	39	19	20	19	Very alkaline	Potassium	K
20	40	20	20	20	Less alkaline	Calcium	Ca

a These are isotopes of hydrogen. They all have the same chemistry but are of different mass.
b These are two isotopes of chlorine, of which chlorine gas is composed. Chlorine gas contains 75% of the isotope with 18 neutrons and 25% of the isotope with 20 neutrons: its average atomic weight is 35.5.

6.5 PERIODIC TABLE

The periodic table (Table 6.2) was the result of many arguments which arose as our forefathers tried to arrange the elements in some sort of logical order from apparent disorder. This is one of the fundamentals of science, getting some sort of order into our way of thinking about our environment as we 'discover' the complexity of it. This complexity gives us a number of apparently unrelated facts. **Science is the correlation of these facts into a theory**. Originally the 'order' within the elements was on a weight basis, and similar chemical elements were placed beneath other similar elements in groups or vertical columns. The horizontal lines are called periods. As some elements did not fit into this order gaps were left for elements yet to be discovered, i.e. this ordering gave scientists elements to look for. One of these new elements was 'eka silicon' (which means similar to silicon), which we now call germanium (Ge).

There are more elements in the universe than these to consider in the Periodic Table, but the essential points to note are as follows.

(1) The strongly **alkaline** elements are to the **left** of the periodic table (in groups 1 and 2) and the acidic elements to the right (in groups 5, 6 and 7).

Table 6.2 — The structure of the periodic table. The atoms are arranged in ascending weights and chemical activity. Like chemical activity tends to be grouped in the vertical columns. This leaves a group of engineering elements as a separate group on their own. These are referred to as the transition elements

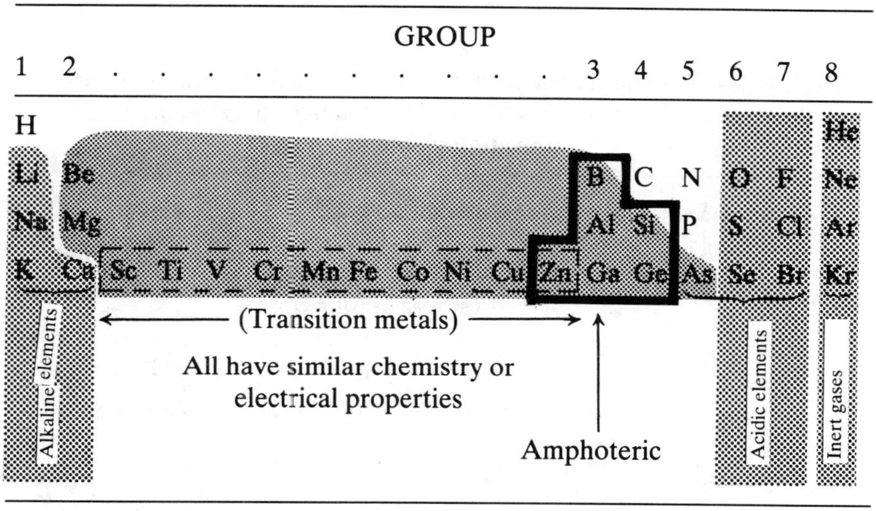

(2) The **inert** gases are all beneath each other in group 8 of the table.

(3) The elements which are all metallic in nature are grouped in the periodic table as shown by the hatched area.

(4) The non-metals are also grouped together as the acidic elements.

(5) Between the metals and the non-metals are a group of elements whose chemistry is both acidic and alkaline in nature, i.e. they are **amphoteric**. For example aluminium forms aluminates and aluminium compounds. The oxide is amphoteric. Silicon also forms silicates and silicon compounds. The ending '−ate' is used to denote that the compound formed has oxygen in the unit (see for example Figs. 6.10 and 6.13).

(6) The chemical combining power or valency of the element is some function of the group in which it is placed. For groups 1–5 the valency is the group number. For groups 5–8, the valency is the group number subtracted from 8. Hence sodium (Na; group 1) is monovalent, magnesium (Mg; group 2) is divalent, fluorine (F; group 7) is monovalent (8−7=1) and oxygen (O; group 6) is divalent 8−6=2. The number of (external) electrons in each group is the group number. Hence fluorine (F) contains 7 external electrons as does bromine (Br) and chlorine (Cl) (Table 6.3).

(7) The valency is the ease at which an element gains or loses an electron to become an **ion** in the ionic bond (Section 6.11), or, in the case of the covalent bond (Section 6.10), it is the number of atoms required as **near neighbours** to achieve this stable octet of the inert gas electron configuration.

(8) The valency of the inert gases are by definition zero.

The above can be summarized as follows.

(1) The number of protons determines the type of atom and its nomenclature.

(2) The number of electrons increases with atomic number (i.e. with the number of protons) (Table 6.3).

(3) The electrons govern the **reactivity** of the element.

Table 6.3 — The similar chemical activity is modelled on the similarity of the external electron orbits. This is continued in Table 6.8 for the transition elements

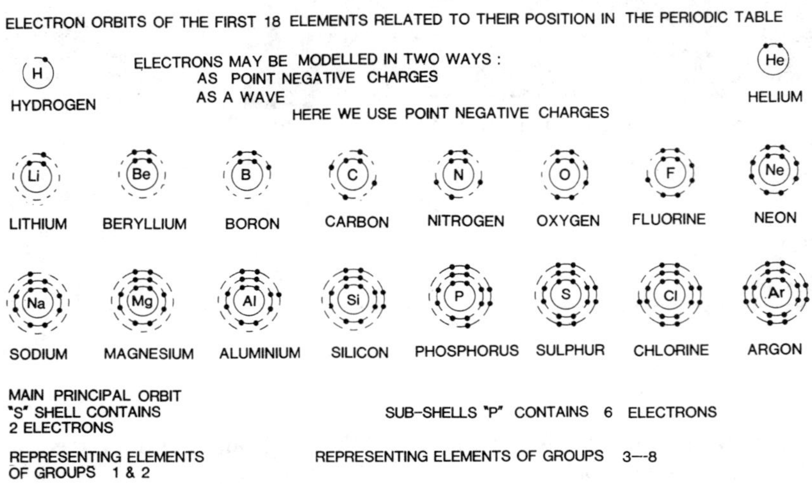

6.6 ELECTRON SHELLS

Sodium is a highly reactive metal and the modelling of this reactivity is to postulate that sodium metal wants to get rid of the eleventh electron in order that it can have the inert gas electron configuration of neon; 10 protons, 10 electrons. If the sodium does get rid of this electron it will have 11 protons and 10 electrons and so will have a resultant positive charge of +1. This is termed an **ion**. It is also modelled as a sphere. The radius of the sphere becomes smaller as shown in Table 6.4. Magnesium, the next element along, is less reactive and the model is that magnesium would like to lose two electrons to get to the electron configuration of the previous inert gas (neon). This requires more energy and as a result magnesium (Mg) is less reactive than sodium. When magnesium does lose the two electrons it has a resultant charge of +2 (the difference between 12 protons and 10 electrons), and so these remaining 10 electrons are attracted more strongly by the 12 protons. The radius of the charged sphere, now also called an ion, becomes smaller than the overall neutral atom. The **readiness** of the atom to **ionize** is one of the parameters required for corrosion (Section 7.1 *et seq.*)

Table 6.4 — Ionic and atomic radii (in nanometers). The atomic radii are one-half the distance of closest approach in the element. The ionic radii are those for six-fold coordination (See Fig. 6.14(b)).

	1		2		3		4		5		6		7		8						0	
1													H^-	0.154							He	
													H	0.046								
2	Li	0.152	Be	0.112	B	0.097	C	0.077	N	0.071	O^{2-}	0.132	F^-	0.133							Ne	0.160
	Li^+	0.078	Be^{2+}	0.034			C^{4+}	0.02	N^{5+}	.01–.02	O	0.060										
							Si^{4-}	0.198			S^{2-}	0.174										
3	Na	0.186	Mg	0.160	Al	0.143	Si	0.117	P		S	0.104	Cl^-	0.181							Ar	0.191
	Na^+	0.098	Mg^{2+}	0.078	Al^{3+}	0.057	Si^{4+}	0.039	P^{6+}	.03–.04	S^{6+}	0.034	Cl	0.107								
									V	0.130			Mn	0.118								
	K	0.231	Ca	0.196	Sc	0.151	Ti	0.146	V^{3+}	0.065	Cr	0.125	Mn^{2+}	0.091	Fe	0.124	Co	0.125	Ni	0.124		
	K^+	0.133	Ca^{2+}	0.106	Sc^{3+}	0.069	Ti^{3+}	0.083	V^{4+}	0.061	Cr^{3+}	0.064	Mn^{3+}	0.070	Fe^{2+}	0.083	Co^{2+}	0.082	Ni^{2+}	0.078		
4	$(NH_4^+$	0.143)					Ti^{4+}	0.064	V^{5+}	ca.0.04	Cr^{6+}	.03–.04	Mn^{4+}	0.052	Fe^{3+}	0.067						
	Cu	0.128	Zn	0.133	Ga	0.122	Ge	0.122	As	0.125	Se^{2-}	0.191	Br^-	0.196								
	Cu^+	0.096	Zn^{2+}	0.083	Ga^{3+}	0.062	Ge^{4+}	0.044	As^{3+}	0.069	Se	0.116	Br	0.119							Kr	0.201
									As^{5+}	ca.0.040	Se^{6+}	.03–.04										
	Rb	0.243	Sr	0.215	Y	0.181	Zr	0.156	Nb	0.143	Mo	0.136			Ru	0.133	Rh	0.134	Pd	0.137		
	Rb^+	0.149	Sr^{2+}	0.127	Y^{3+}	0.106	Zr^{4+}	0.087	Nb^{4+}	0.069	Mo^{4+}	0.068	Tc		Ru^{4+}	0.065	Rh^{3+}	0.068				
5									Nb^{5+}	0.069												
	Ag	0.144	Cd	0.149	In	0.162	Sn^{4-}	0.215	Sb	0.145	Te^{2-}	0.211	I^-	0.220								
	Ag^+	0.133	Cd^{2+}	0.103	In^{3+}	0.092	Sn	0.140	Sb^{3+}	0.090	Te	0.143	I	0.136							Xe	0.220
							Sn^{4+}	0.074			Te^{4+}	0.089	I^{6+}	0.094								
	Cs	0.262	Ba	0.217	La	0.186	Ce	0.182	Pr	0.181	Nd	0.180										
	Cs^+	0.165	Ba^{2+}	0.143	La^{3+}	0.122	Ce^{3+}	0.118	Pr^{3+}	0.116	Nd^{3+}	0.115										
							Ce^{4+}	0.102	Pr^{4+}	0.100												
6	Sm^{3+}	0.113	Eu^{3+}	0.113	Gd^{3+}	0.111	Tb^{3+}	0.109	Dy^{3+}	0.107	Ho^{3+}	0.105	Er	0.186								
							Tb^{4+}	0.089					Er^{3+}	0.104								
	Tm^{3+}	0.104	Yb^{3+}	0.100	Lu^{3+}	0.99	Hf	0.158	Ta	0.143	W	0.136	Re		Os	0.135	Ir	0.135	Pt	0.138		
							Hf^{4+}	0.084	Ta^{5+}	0.068	W^{4+}	0.068			Os^{4+}	0.067	Ir^{4+}	0.066				
							Pb^{4-}	0.215														
	Au	0.144	Hg	0.150	Tl	0.170	Pb	0.175	Bi	0.155	Po		At								Rn	
	Au^+	0.137	Hg^{3+}	0.112	Tl^+	0.149	Pb^{2+}	0.132														
					Tl^{3+}	0.105	Pb^{4+}	0.084														
7	Fr		Ra		Ac		Th	0.180	Pa		U	0.138										
							Th^{4+}	0.110			U^{4+}	0.105										

Reproduced by permission from R. C. Evans, *Introduction to Crystal Chemistry*, Cambridge University Press, London, 1952, p. 171, Table 36.

Similar arguments apply to aluminium and silicon. Silicon (group 4) could lose 4 electrons (and have the neon electron configuration) and therefore have a resultant charge of +4. However it could gain four electrons (and have the argon (Ar) electron configuration) thus having a resultant of −4 charges. The size of the sphere with a charge of −4 would be larger than the neutral sphere as the excess electrons repel each other. Conversely, the size of the silicon ion with a charge of +4 is smaller than the neutral atom as the excess protons over electrons pull the electrons into the centre of the ion (see Table 6.4).

With phosphorous (P; group 4), it is much easier to gain 3 electrons (to go to the following inert gas electron configuration), than to lose 5 electrons (to go to the preceding inert gas electron configuration). Similar arguments pertain to sulphur (S; group 6). These elements therefore form negative ions in preference to positive ions (Section 6.11). This is the basis of valency and bonding (Section 6.7). The valency of the element is the same as the number of the group to which it is ascribed up to group 4. After group 4 the valency is the group number subtracted from 8.

The modelling of the electrons in the orbits must explain or model the following.

(1) The reactivity of the various elements

(2) The differences between metals and non-metals.

(3) The presence of the **transition elements**, (Table 6.2) and some of the more complicated chemical elements referred to as the **rare earth** elements (Table 6.5).

(4) The difference that some elements are of an alkali nature and others are of an acidic nature as well as the 'chemistry' of the elements (amphoteric elements) between the two extremes of acidity and alkalinity.

Table 6.5 — The periodic table

GROUP IA	IIA	IIIB	IVB	VB	VIB	VIIB	GROUP VIII			IB	IIB	IIIA	IVA	VA	VIA	VIIA	VIIIA
1 H 1.0																	2 He 4.0
3 Li 6.9	4 Be 9.0											5 B 10.8	6 C 12.0	7 N 14.0	8 O 16.0	9 F 19.0	10 Ne 20.0
11 Na 22.9	12 Mg 24.3											13 Al 26.9	14 Si 28.0	15 P 30.9	16 S 32.0	17 Cl 35.4	18 Ar 39.9
19 K 39.1	20 Ca 40.0	21 Sc 44.9	22 Ti 47.9	23 V 50.9	24 Cr 52.0	25 Mn 54.9	26 Fe 55.8	27 Co 58.9	28 Ni 58.7	29 Cu 63.5	30 Zn 65.3	31 Ga 69.7	32 Ge 72.6	33 As 74.9	34 Se 78.9	35 Br 79.9	36 Kr 83.8
37 Rb 85.4	38 Sr 87.6	39 Y 88.9	40 Zr 91.2	41 Nb 92.9	42 Mo 95.9	43 Tc 97.0	44 Ru 101.1	45 Rh 102.9	46 Pd 106.4	47 Ag 107.8	48 Cd 112.4	49 In 114.8	50 Sn 118.7	51 Sb 121.7	52 Te 127.6	53 I 126.9	54 Xe 131.3
55 Cs 132.9	56 Ba 137.3	57 La 138.9	72 Hf 178.5	73 Ta 180.9	74 W 183.8	75 Re 186.2	76 Os 190.2	77 Ir 192.2	78 Pt 195.0	79 Au 197.0	80 Hg 200.6	81 Tl 204.3	82 Pb 207.2	83 Bi 208.9	84 Po 209.0	85 At 210.0	86 Rn 222.0
87 Fr 223.0	88 Ra 226.0	89 Ac 227.0															
				58 Ce 140.1	59 Pr 140.9	60 Nd 144.2	61 Pm 145.0	62 Sm 150.3	63 Eu 152.0	64 Gd 157.2	65 Tb 158.9	66 Dy 162.5	67 Ho 164.9	68 Er 167.2	69 Tm 168.9	70 Yb 173.0	71 Lu 174.9
				90 Th 232.0	91 Pa 231.0	92 U 238.0	93–103 Trans Uranic elements which are unstable										

The atoms are arranged according to atomic activity. Similar chemical activities tend to be grouped in the vertical columns. This leaves groups of elements as a separate group on their own. These are referred to as the first transition elements (Sc–Zn), the second transition elements (Y–Pd) and the rare earth elements (Ce–Lu). Table 6.6 attempts to show how these are modelled.

The electron modelling is as follows. There are several spherical energy levels or subshells around the nucleus. These energy levels are given a further shorthand: the derivation of the codes is not important.

The 'codes' are 1s, 2s, 2p, 3s, 3p, 3d, 4s etc. The number refers to the period, i.e. 1 is for hydrogen and helium and so the 's' level can only have two electrons before it is full. The number 2 refers to the elements Li–Ne. The 2s refers to the elements lithium and beryllium and the 2p for the six elements B–Ne. The model therefore is that the 'p' level can only have 6 electrons in it. There are other subshells which are denoted by 'd' and 'f' subshells. The number of electrons contained in these subshells is given in Table 6.6. Each subshell fills up in a particular order which is shown in Table 6.6 and illustrated in Table 6.7.

Table 6.6 — The electron orbits depicted in Table 6.3 are further modified to contain subshells which many are modelled to contain the number of electrons shown and their order of occupancy. Note that internal subshells are filled later in the periodic table (e.g. compare subshell 4f and 5s)

	Number of electrons in the following subshells														
Subshell (radially)	1s	2s	2p	3s	3p	3d	4s	4p	4d	4f	5s	5p	5d	6s	6p
Electron capacity	2	2	6	2	6	10	2	6	10	14	2	6	10	2	6
Filling order	1	2	3	4	5	7	6	8	10	13	9	11	14	12	15

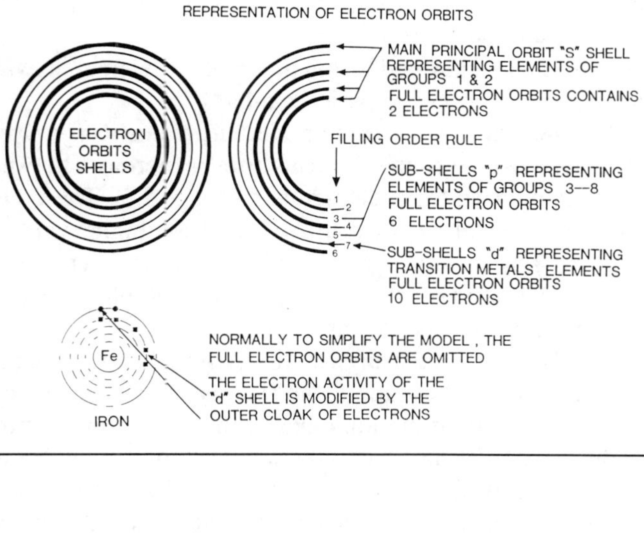

When electrons fill the 4f electron subshell the elements produced will have very similiar chemical properties because there are several other subshells **outside** this inner 4f level. The outer levels which produce this 'camouflage' are the 6s, 5p and 5s. The model of the atom with its electrons, shown as dots, is shown in Table 6.3. The filling up of the electrons in the innermost shell is shown in Table 6.8

Table 6.7 — The filling of the subshells is shown for a few elements of the periodic table

Subshell		1s	2s	2p	3s	3p	3d	4s	4p	4d	4f	5s	5p	5d	6s	6p
Electron capacity		2	2	6	2	6	10	2	6	10	14	2	6	10	2	6
Filling order		1	2	3	4	5	7	6	8	10	13	9	11	14	12	15
Element	Atomic	Number of electrons in the subshells for each element														
	Number	1s	2s	2p	3s	3p	3d	4s	4p	4d	4f	5s	5p	5d	6s	6p
H	1	1														
He	2	2														
Li	3	2	1													
C	6	2	2	2												
Cl	17	2	2	6	2	5										
K	19	2	2	6	2	6	0[a]	1								
Ti	22	2	2	6	2	6	2[b]	2	0							

a Note that the 3d shell is not being filled here.
b Note that the 3d shell is now being filled beneath the outer 4s shell, thus explaining the similarity of the transition metals.

6.7 VALENCY AND CHEMICAL COMBINATION

There are four types of bonds which bind all these different elements together to form the wide range of engineering and geological materials.

(1) The **metallic bond** is formed only with the metallic elements, i.e. elements from groups 1–3 and the transition elements in the 'pure unalloyed' state (e.g. copper), or in the alloyed state as in brass. Brass is an alloy formed between copper (Cu) and zinc (Zn) (Section 6.20).

(2) The **covalent bond** is formed by the non metallic elements, which are usually those elements in groups 4–7 of the periodic table.

(3) The **ionic bond** is formed when non-metallic elements (groups 5–7) combine with the metallic elements of groups 1–3. This bond is therefore between elements of group 1 or 2 combining with those elements in groups 5–7. These elements can be classed as electropositive elements (groups 1 and 2) combining with electronegative elements (groups 5–7) by the attraction of unlike charges (Section 6.11).

(4) **A van der Waals** bond is a residual bond and is therefore very weak and easily broken by mechanical or thermal means.

The strength of the bond is measured by the melting point of the bonded material and this gives, to a first approximation, the following order of bond strengths: the covalent bond is stronger than the ionic bond; the ionic bond is stronger than the metallic bond; the van der Waals bond is the weakest bond of all. Nature does have a system whereby there can be mixed bonding within the same molecular structure. This is shown in Fig. 6.13 which depicts both sulphuric acid and phosphoric acid as covalently bonded. However, when these acids dissolve in water, the hydrogen becomes 'ionically bonded' and forms hydrogen ions. It is difficult for man to make use of materials with a 'mixed' bond system, such as a covalently bonded paint system adjacent to a metallic bonded metal. This combination presents a problem of **adherence** or 'keying' of the paint film onto the metal surface. This is considered in detail in Section 6.18, 6.19 and 8.11 and Chapter 9.

Table 6.8 — Similar chemical activity is modelled on the similarity of the external electron orbits. Here the transition elements are modelled to show the two external electrons in the 4s subshell while the internal 3d subshell is filled up with 10 electrons, corresponding to the 10 transition elements

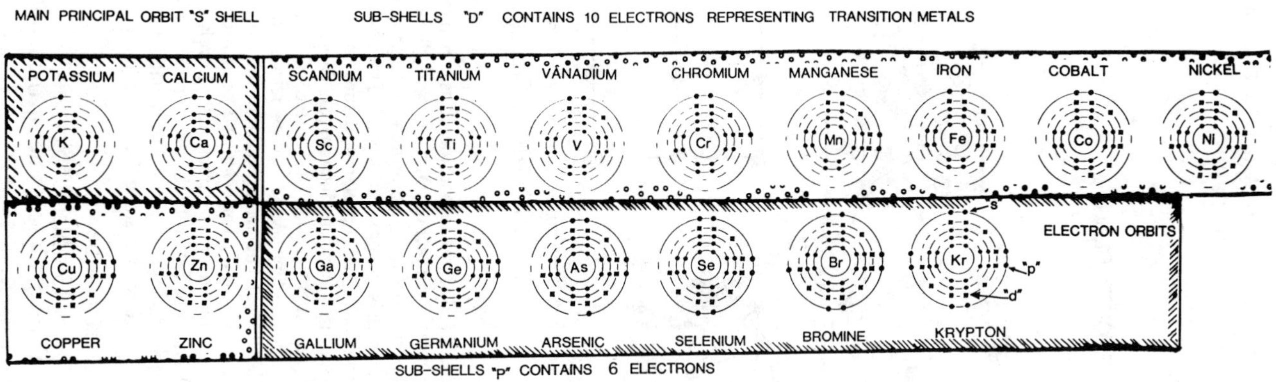

6.8 METALLIC BONDING

The atoms of the metal are closely packed with 12 nearest neighbours in a three-dimensional array as shown in Fig. 6.2. This three-dimensional array is bonded together by the metallic bond which is a bond between the same elements or elements of the same chemical nature.

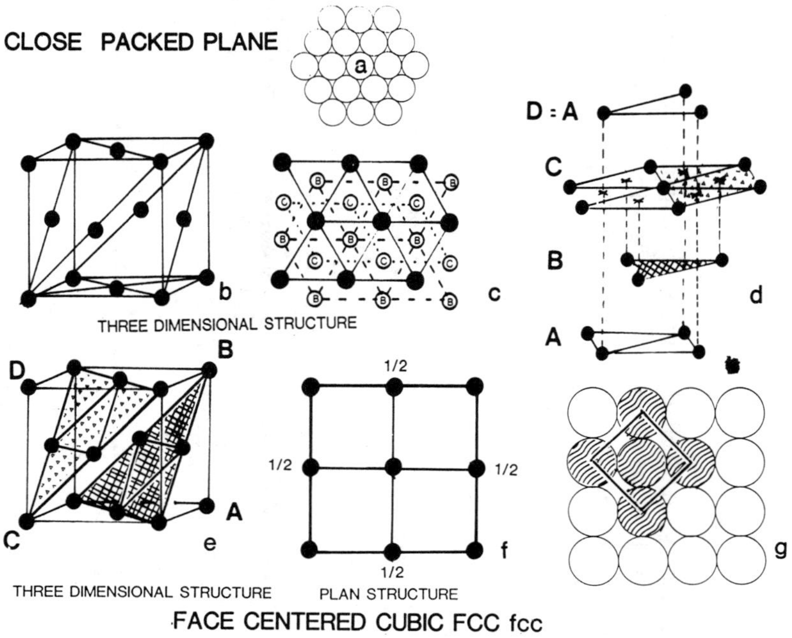

Fig. 6.2 — (a) The metal atoms are considered as hard non-deformable spheres packed together so that each sphere is in contact with the adjacent spheres. This is called a close-packed plane. (b) The centres of the spheres are shown in the diagram to represent the fundamental unit for the f.c.c. lattice, which is composed of several of the close-packed planes (shown in (a)) stacked together as shown in (c) as the plan structure and (d) as an exploded three-dimensional structure viewed at right angles to diagram (c). The orientation of this fundamental unit to the close-packed planes A, B (hatched), C (dotted), D is shown in (d) and (e). (d) The plane D is shown directly above plane A and so is re-named plane A. The packing sequence is ABCABCABCABC. (e) The arrangement of the close-packed planes A, B, C and D in the f.c.c. lattice shown in (e). (f) The scientist's plan structure of the f.c.c. lattice, where the Z coordinate is represented by the fraction $\frac{1}{2}$. Levels 0 and 1 are omitted for clarity and convention. (g) The cube face with the 'free' space available in the structure. An observation which will be required later (e.g. Fig. 6.24), is that not all directions, are close-packed, e.g. the direction along the cube edge is not close packed.

The metallic elements lose some of their outer electrons. These electrons are donated to form an **electron cloud** (Figs. 6.3 and 6.4). The number of electrons lost, or donated to this electron cloud depends on the group which the element occupies. For example, in brass, copper will donate one electron whilst zinc will donate two electrons. The metallic bond is a non-directional bond. This electron cloud is free to move through the three dimensional array of positive charges under a small potential gradient, and hence metals are conductors of electricity. The movement of the electrons is the flow of an electric current in the **opposite** direction to the flow of the **electrons**. This is one of the historical hiccups in electrical theory which arises because the flow of an electrical current was discovered before it was known what

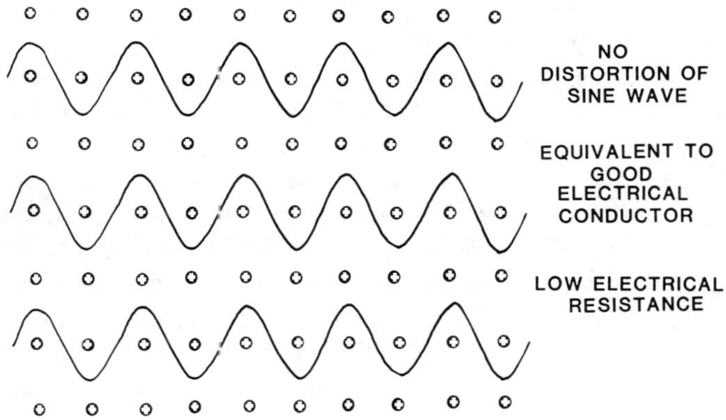

Fig. 6.3 — The structure of a metal is diagrammatically modelled as an array of postiviely charged centres representing the atoms which have given up their loosely bound electrons to an electron cloud. This electron cloud, modelled here as a sine wave travelling through the positive array, is the bonding agent which cements all the positive charges into a three-dimensional array and gives a metal its characteristic good electrical properties. The positive charge on the array of positive centres is modelled on a first approximation as the group in which the elements appear (Table 6.5).

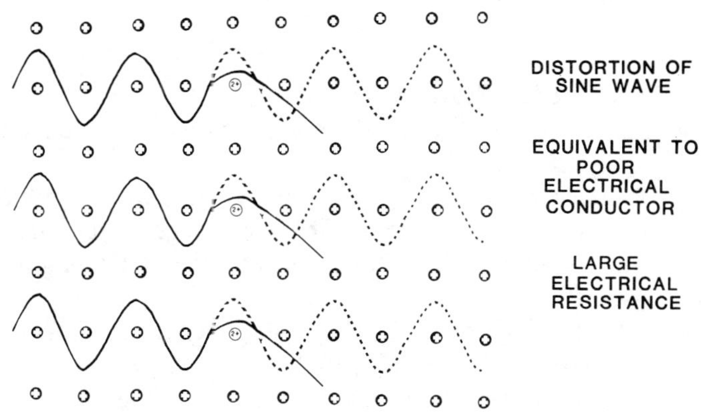

Fig. 6.4 — When a metal A from group 1 is alloyed (Section 6.20 *et seq.*) with another metal B which is often from a different periodic group, say from group 2, then the number of electrons donated to this electron cloud is different from that of 'host' or base metal. The electron cloud now suffers a resistance to its flow through the different positive atoms such that the cloud will stay longer around the centres with a 2+ charge. If the flow of the electron cloud is modelled as a sinusoidal wave passing through the metal, then this sinusoidal wave is distorted as it passes the 'foreign' higher charged centres as shown in the model. The alloy is said to be a soild solution where metal B replaces metal A in the lattice points of metal A's structure (this is an example of substitutional solid solution (Section 6.20)).

'entity' carried the current. The flow of current was therefore given an arbitrary direction from a positive terminal in a battery to the negative terminal. In actual fact the flow of charge carriers (electrons) is from the negative to the positive terminal (Sections 6.14, 7.3 and 2.2; Figs. 2.4–2.6). Electrons can **only** travel easily in a metallic lattice. This fact has important consequences for corrosion and corrosion protection (Sections 8.1 and 8.2).

Metals which are **alloyed** with another metal out of their group will possess an electrical resistance. Consider brass, an alloy of copper (group 1) and zinc (group 2) (Table 6.5). The electrical resistance arises because the alloy metal zinc (Zn) is in the higher group (group 2). Zinc will contribute two electrons to this cloud and so leave a lattice site of charge +2, whereas the metal from group 1 will only contribute one electron and leave a lattice site charge of +1 (Fig. 6.4). The electron cloud is further modelled to be a sinusoidal wave passing through the array of positive sites. Hence the sinusoidal waveform of the electron cloud will tend to be distorted around the resultant +2 charges where the zinc atoms are situated. Thus the periodic charged array will offer some distortion to the easy movement of the electron cloud under small potential differences. This model is diagrammatically shown in two dimensions in Figs. 6.3 and 6.4. One important consequence of this model is that the number of electrons given to the electron cloud is dependent on the group in which the element is located. Hence as a first approximation, those elements in group 1 (copper (Cu), silver (Ag) and gold (Au)), will not suffer in resistivity terms by alloying with each other as they are all in group 1. It is only detrimental in terms of resistivity when alloying takes place across groups, e.g. group 1 element with group 2 element. See Table 6.5 for the complete periodic table of all the elements and their assigned groups.

The degree of ionization is dependent also on the group in which the element is ascribed by its atomic number.

6.9 SCIENTIFIC STRUCTURES - METALS

There are essentially three structures which concern us at this elementary stage. These are:

(1) The face-centred cubic (f.c.c.) structure (Fig. 6.2) is the closest-packed structure of equally sized spheres or atoms. The percentage of free space is 26%. It is formed by **close-packed planes** (Fig. 6.2(a)) stacked in the following order:

ABC(D=A)BCABCABC

(D=A) means that the fourth plane D is spatially over the first plane A and so is referred to as the A plane as shown in Fig. 6.2(d).

(2) The hexagonal close-packed (h.c.p.) structure (Fig. 6.5) like the f.c.c. structure, is also a close-packed structure of equally sized spheres or atoms. The percentage of free space is **26%**. It is formed by close-packed planes (Fig. 6.5(c)) being stacked in the following order:

AB(C=A)ABABABAB

(C=A) means that the third plane C is spatially over the first plane A and so is referred to as the A plane as shown in Fig. 6.5.

(3) The body-centred cubic (b.c.c.) structure (Fig. 6.6) **is not close packed**. The percentage of free space is **32%**. This structural unit **cannot** be described as a series of

close packed planes. The structures are shown in Fig. 6.6 and, because the atoms are all metallic, they are shown in the same way (as filled in dots).

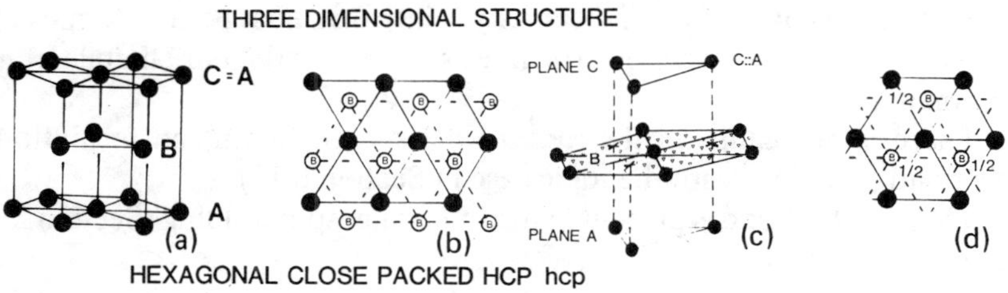

Fig. 6.5 — (a)–(c) The h.c.p. lattice is formed by the same close-packed planes shown in Fig. 6.2(a), but their stacking becomes coincident every alternate layer and so plane C is re-named plane A. The packing sequence is ABABAB as shown in (c) and (d). (d) The name 'hexagonal-close packed' is derived from the unit cell as viewed along the Z coordinate. Levels 0 and 1 are omitted for clarity.

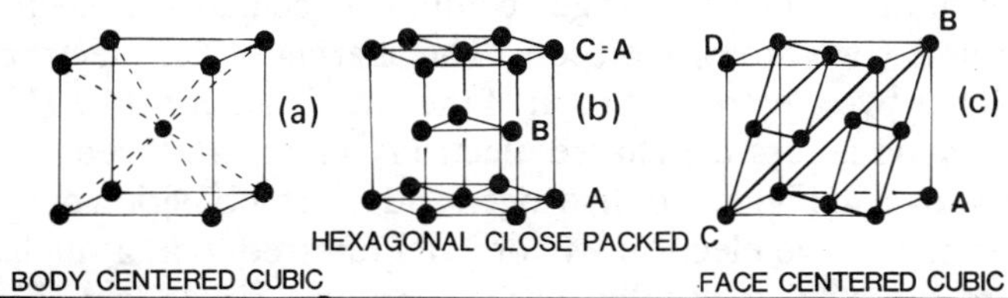

	BCC bcc	HCP hcp	FCC fcc
VACANT FREE SPACE	32%	26%	26%

Fig. 6.6 — The b.c.c. lattice is not formed by the stacking of close packed planes (Fig. 6.2(a)) in three dimensions. It is a more open structure and consequently an alloy formed by metals which have different inherent primary structures will 'fight' over which structure to make in solid solution. this leads to very strong metallic structures and an alteration in the chemical properties of the alloy. Therefore the corrosion properties of an alloy are different from the two base metals. Note that the percentage of free space quoted here for each structure does not lead to water ingress (Section 2.1). The percentage free space is calculated from a consideration of the packing of hard (non-deformable) spheres.

6.9.1 Alloys

An alloy structure such as brass is a **random replacement** of the copper atoms (Cu) by zinc atoms (Zn) and there is **no differentiation of sites** at which one would find copper or zinc. The difference in size factor of the atoms as hard non-deformable spheres tolerated by this substitutional replacement is **15%**. This size difference (Table 6.4) is important in all the structures to be examined when alloys are considered (see Section 6.20).

6.9.2 Summary of properties of the metallic bond

(1) The bond is non-directional.

(2) Metals are mechanically deformable, and so atoms must 'slide' over each other.

(3) It is a relatively weak bond and therefore metals have relatively low melting points.

(4) Metals are conductors of electricity (electrons must be free to move).

(5) Alloys possess a higher electrical resistance than the pure metals from which they are made.

(6) There is **no change in the radii** of the sphere (atom) on formation of the metallic bond (compare with the ionic case in Section 6.11).

(7) The metallic bond is present in the following: spot welding (Section 2.11; Fig. 8.7), brazing and soldering.

6.10 COVALENT BONDS

The covalent bond is formed between the **non-metals**, i.e. the elements of groups 4–7. Each case will be examined separately.

6.10.1 Non-metals: Groups 6 and 7 and hydrogen

Each atom will want to get the inert gas electron configuration for chemical stability. Thus two atoms of the element hydrogen combine to share their electrons and thus produce a diatomic (two-atom) molecule. The sharing of the electrons produce a molecule which has two electrons in each of the atom's outer orbit. This produces a **very directional bond** across the shared electrons which are 'fixed' at the point of sharing. This is shown in Fig. 6.7 for hydrogen and the combination of hydrogen with carbon. The sharing of two electrons is normally referred to as a single bond and is fairly difficult to break. It is usually represented by a single line. Fig. 6.8 shows oxygen as oxygen O_2 where there is a double bond (see later) and ozone O_3.

Two atoms of the element fluorine or chlorine (or any other element in group 7) combine to share only one of their outer 7 electrons and thus produce a diatomic molecule. This, therefore, produces a very directional bond between the two chlorine atoms Cl_2 or fluorine atoms F_2.

The **number of nearest neighbours** within the structure is given by the number of electrons in the inert gas following the element minus the group in which the element occurs. Hence oxygen (O; group 6) should have two nearest neighbours ($8-6=2$) on this model rule. Oxygen exists as two different molecular units (molecules) called oxygen (present in air) and ozone (present in the upper atmosphere). Ozone is the example which follows the above rule, but the oxygen atom can also share more than one electron. Oxygen will have to share two electrons from each atom (this is shown in Fig. 6.8) in order to make up the stable 8 which is required for the inert gas configuration. This means that there are 4 shared electrons and this is called a double bond. The double bond is not easily broken, and is represented by two lines (Figure 11.1). Oxygen is represented as O_2 or O=O, where the = sign represents a double bond. These diatomic gases (two atoms combined) or tri-atomic (three atoms combined) are only bonded in a **very directional bond** across the shared electrons. The atoms are **not** bonded in a three-dimensional network because there are no spare electron spaces or electrons in the orbit as all the atoms have the stable 8 electrons. These 'materials' are therefore **gases**. The bonds between the molecules of these gases are weak van der Waals bonds which will be considered in Section 6.12.

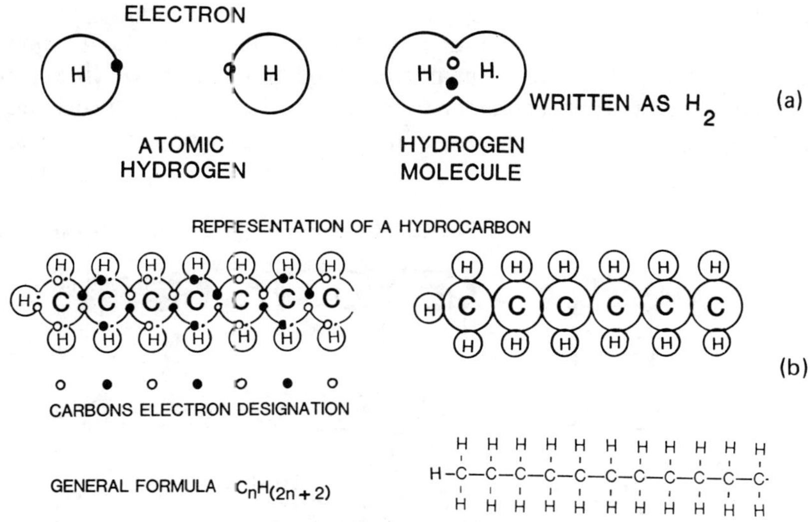

Fig. 6.7 — (a) Hydrogen cannot exist as atomic hydrogen H. It is far too reactive. Hydrogen gas exists as a di-atomic molecule as shown where each hydrogen atom has obtained the inert gas electron configuration of the next inert element helium. Note that the hydrogen molecule is not charged even though there are now two electrons in each shell. This is explained by the fact that the electrons are only half associated with each atom and so the di-atomic molecule is overall electrically neutral. (b) The sharing of electrons both between carbon and carbon atoms and between carbon and hydrogen atoms making up the hydrocarbon. Each carbon electron is represented differently to facilitate identification. These electrons are usually further shortened as lines, one line denoting the sharing of two electrons.

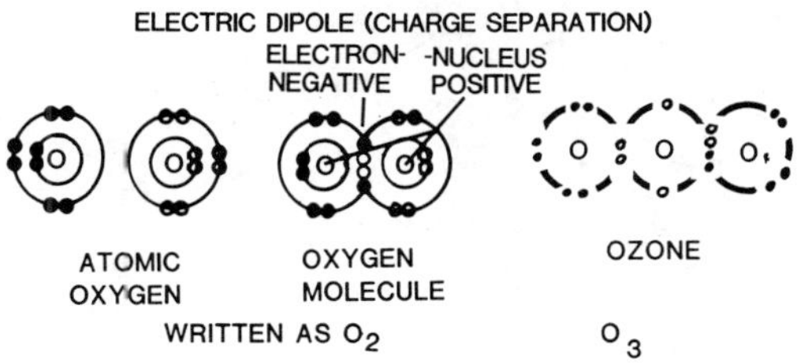

Fig. 6.8 — Oxygen requires two electrons to make up the electron configuration of eight electrons for the next inert gas. It does this by the donation by each atom in the di-atomic molecule of two electrons to the bond, thus making a directional bond with four shared electrons. Note that the oxygen molecule is not charged even though there are now eight electrons in the outer shell. This is explained by the fact that the electrons are only half associated with each atom and so the di-atomic molecule is overall electrically neutral. Note that there is no bond to any other oxygen atom apart from the covalently joined adjacent atom which exists near to the di-atomic molecule. If the electrons are considered as discrete entities, and their positiion as shown, then there is an electrical dipole (alternate positive and negative charges). The existence of this dipole is the 'bond model' required for the residual van der Waals bond. By convention, it is permissible to leave the filled electron levels (corresponding to the previous inert gas) as these inner electrons do not take part in any chemical reaction. This omission is shown in the ozone molecule.

6.10.2 Group 4 elements

These elements produce a three-dimensional structure by covalently bonding in a directional manner by having four nearest neighbours as shown in Fig. 6.9. Four electrons are shared from each of the four atoms which are spatially oriented around each other atom. It is a directional bond. A covalently bonded material is a **bad**

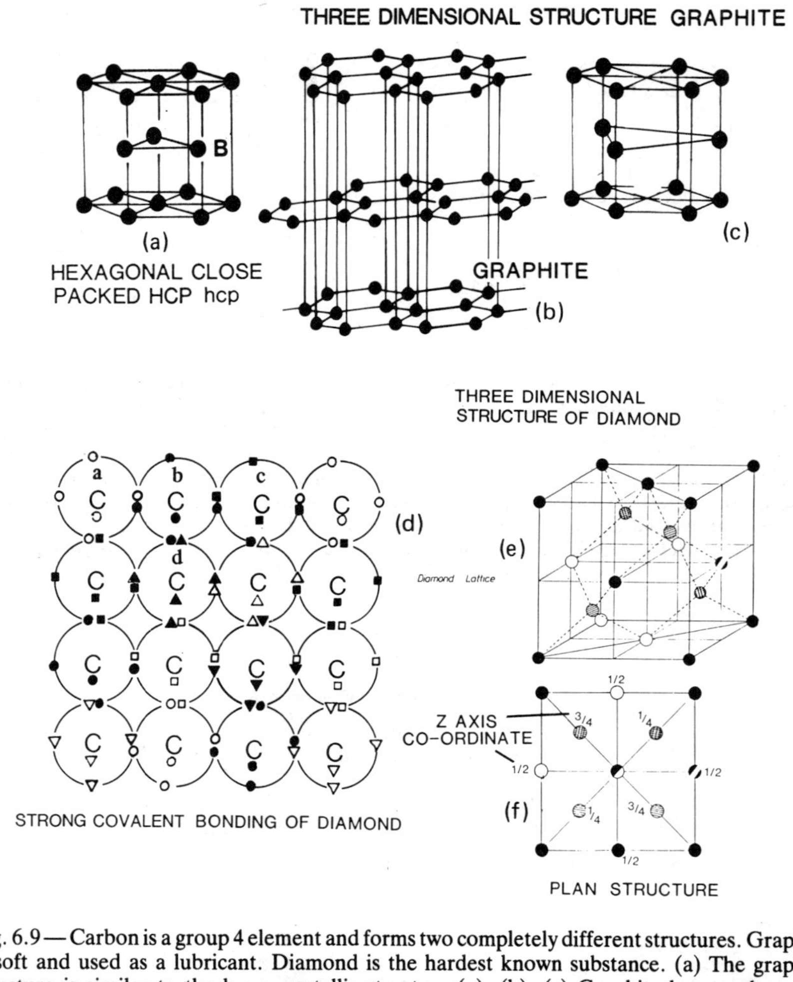

Fig. 6.9 — Carbon is a group 4 element and forms two completely different structures. Graphite is soft and used as a lubricant. Diamond is the hardest known substance. (a) The graphite structure is similar to the h.c.p. metallic structure (a). (b), (c) Graphite has an elongated structure in one direction which gives rise to this softness. (d) Carbon forms the three-dimensional bond by sharing two electrons between the atoms which can be bonded in a three-dimensional array with four nearest neighbours modelled here as a two-dimensional planar structure. This would model it as a weak, easily cleaved molcule. (e), (f) The three-dimensional structure of diamond, which is a very strong material having a melting point of over 3000°C. Note that once again the carbon atoms are not charged even though there are now eight electrons in the outer shell. By convention, it is permissible to leave out the filled electron levels corresponding to the previous inert gas as these inner electrons do not take part in any chelmical reaction. Note that diagram (d) can be transferred to (e) and (f) by considering (d) as being rotated through 45° and then superimposed on (e) and (f), i.e. atom **a** is at levels 1 and 0; atom **b** at level $\frac{3}{4}$; atom **c** at levels 0 and 1 and atom **d** at level $\frac{1}{2}$. By this means (d) is equivalent to diamond's being bonded in three dimensions rather than in two dimensions.

conductor of electricity because these electrons are '**fixed**' between each atom and cannot move away (under low potential differences). Covalently bonded materials are therefore insulators. Under high potentials this electrical insulation can be broken down as the electrons are stripped off the bond. Carbon can have two crystal structures. One is called diamond and the other graphite. The bond strength is related to the melting point and mechanical properties. Diamond is one of the hardest materials known, and is an insulator, yet graphite is one of the softest materials used commercially (used in pencils) and is an electrical conductor and is used for carbon commutator brushes.

6.10.3 Covalent structures

There are several simple structures for the covalently bonded materials as each atom has to be surrounded by 8−N atoms in order to obtain the electron configuration of the inert gas which follows the atom in the periodic table (Table 6.5). Therefore there are essentially different three-dimensional structures for each group and combination of groups.

The most prevalent structures are those between silicon (Si; group 4) and oxygen (O; group 6). These are the silicates (Fig. 6.10). Hence, applying the 8−N) rule, each

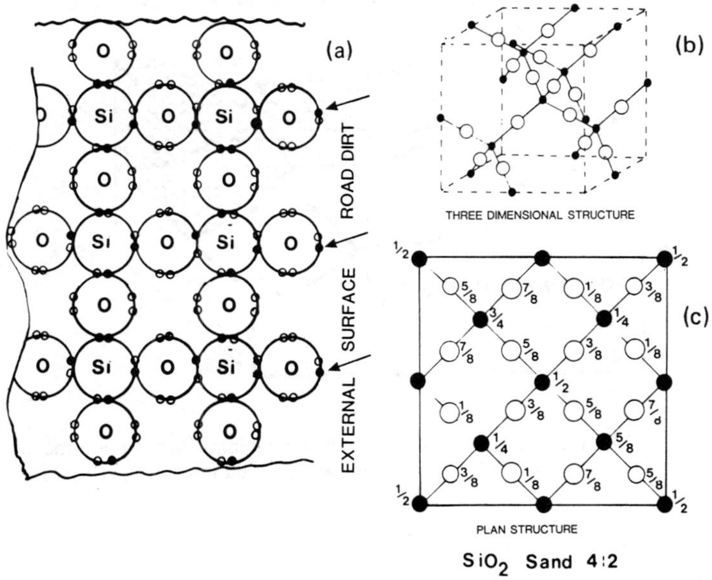

Fig. 6.10—The structure of the most abundant 'unit' on this earth, i.e. the silicate unit. Silicates from the basis of the road aggregate and therefore the 'road dirt' as the road surface is worn away by both the alternating freeze–thaw cycles and the traffic: (a) the scientist's model and (b), (c) the three-dimensional representation. At the surface of the silicate unit there is an electron deficiency which is made up by an ionic bond. this is a donated electron and so the surface of the silicate becomes negatively charged. In rivers this negative charge prevents the silt from settling until the river reaches the sea where this negative charge can be neutralized by the Na^+ ion in the sea water. This silt deposits to form our continental shelf (Fig. 6.11). In the same way, the road de-icing salts allow the road dirt to compact by neutralizing the surface charge on the silicate unit. (There are other scientific reasons for the self-compacting of mud pockets, and these given in Figs. 10.6 and 10.7 (Section 10.8 and 10.10)).

atom of silicon must have 8−4=4 nearest neighbours, and each oxygen atom must have 8−6=2 nearest neighbours. This is shown in Fig. 6.10. Problems occur at the free surface of the material and here there must be electron **donation** from another atom and so this is an 'ionic bond'. Thus each grain of silicate or silt has a surface (negative) charge. Geologically, silt particles are suspended by this negative charge in fresh river water, but this negative charge is neutralized in saline sea water (by the absorption of Na^+ ions) and deposition of the silt occurs, giving the continental shelf around our shores (Fig. 6.11). Similarly, each grain of fractured road aggregate (or road dirt — a silicate) has a surface charge and the ability to neutralize this negative charge is important for the 'settlement' or compaction of the particle in saline water (dissolved de-icing salts). Hence the de-icing salts are responsible for the compaction of road mud accumulations and particles when deposited in the presence of road de-icing salts. Thus ionic and covalent bonds can exist in the same structural unit and have some important part to play in the '**openness**' of some deposited particulate materials.

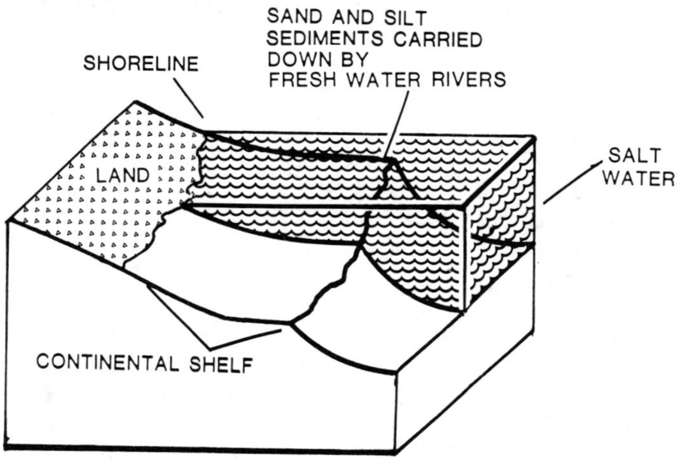

Fig. 6.11 — A map of England would show the continental shelf around the coast. Here a section of the land mass around the coast shows the distribution of silt particles in the river and sea water.

If the inert gas configuration can be obtained by the two atoms collectively obtaining their 8 electrons, this nearest-neighbour rule does not apply. For example aluminium nitride (AlN) (aluminium is in group 3, nitrogen in group 5) has the diamond structure and will have the electron structure shown in Fig. 6.12, which can be represented as either covalently bonded or ionically bonded. These bonds (the covalent and ionic bonds) therefore are modelled as **two extremes** of the same phenomenon of **electron sharing and partition** (the electrons from different atoms are depicted differently in the diagram).

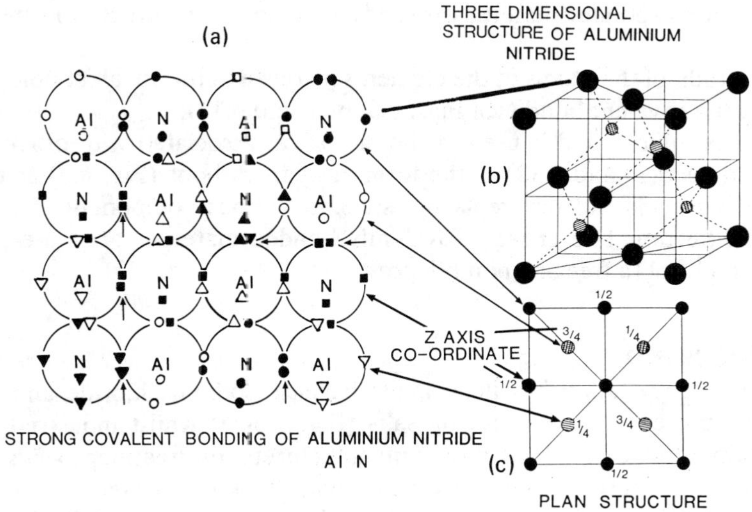

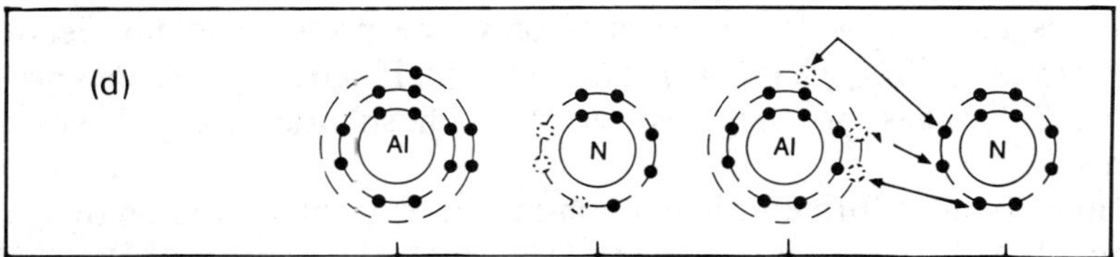

NUMBER OF PROTONS	13	7	13	7
NUMBER OF ELECTRONS	13	7	10	10
RESULTANT CHARGE	ZERO NEUTRAL	ZERO NEUTRAL	3 POSITIVE	3 NEGATIVE
DIAMETER OF EQUIVALENT SPHERE /nm	0.286	0.142	0.114	0.28
SIZE RELATED TO NEUTRAL ATOM			MUCH SMALLER	MUCH BIGGER

Fig. 6.12 — AlN can be modelled as an ionically bonded material or as a covalently bonded substance. The model chosen as the representation would depend on the properties which were being discussed. The ionic bond is considered as one extreme of the covalent bond. (a) AlN is modelled as covalently bonded and each atom is given different electron representations to show where the electrons have originated. (b), (c) AlN is ionically bonded, with the same lattice positions as shown in (a), (d) However, another consideration has to be involved. This is the packing of these positively and negatively charged units which have different diameters. No two like charged ions can be adjacent, and there must be the same numbers of positive and negative ions.

6.10.4 Summary of the covalent bond

(1) It is a directional bond.

(2) Wholly covalently bonded materials are mechanically very strong and non-deformable.

(3) The bond is relatively strong and therefore **entirely** covalently bonded materials have very high melting points.

(4) Covalently bonded materials are bad conductors of electricity and are used as insulators.

(5) The radii of the atoms of the elements involved in the covalent bond are **not** affected by the covalent bond (see Figs. 6.9, 6.10 and 6.12).

(6) The covalent bond is used in the motor car associated with other weaker bonds (Sections 6.10 and 6.12) in the form of hydrocarbons (Fig. 6.7) as plastics. More and more plastics are replacing steel in various components and this is discussed in Section 11.1 *et seq*. Covalently bonded materials are water, rubber alcohol and, glycol (a freezing point depressant).

6.11 IONIC BOND

The choice of illustration has been made because sodium chloride and calcium fluoride are present in road de-icing salts (Fig. 1.5(a)) whilst industrial fall out contains sulphuric acid and the motor manufacturers' pretreatment acids contain phosphoric acid. These acids are covalently bonded, but in water they 'ionize' to form hydrogen (H^+) ions. The hydrogen has left its electron on the SO_4^{2-} unit and is said to be ionically bonded. Fig. 6.13 shows the acids covalently bonded. As the ionization is the loss of [H^+] which is only one proton and no electrons, the remaining entity (HSO_4^-) has one negative charge. If two hydrogen ions are lost, the remaining SO_4^{2-} has two negative charges. Phosphoric acid behaves similarly (Section 6.18).

The ionic bond is formed when a non-metal 'X' from group 5, 6 or 7 combines with a metal 'A' from group 1, 2 or 3 to form an 'AX' structure. The metal loses its outer electron(s) depending on the group, and the non-metal gains the electron(s) depending on the formula 8 – N. Thus if a group 2 metallic atom combines with a group 7 atom, then the metallic atom loses 2 electrons while the non-metallic element

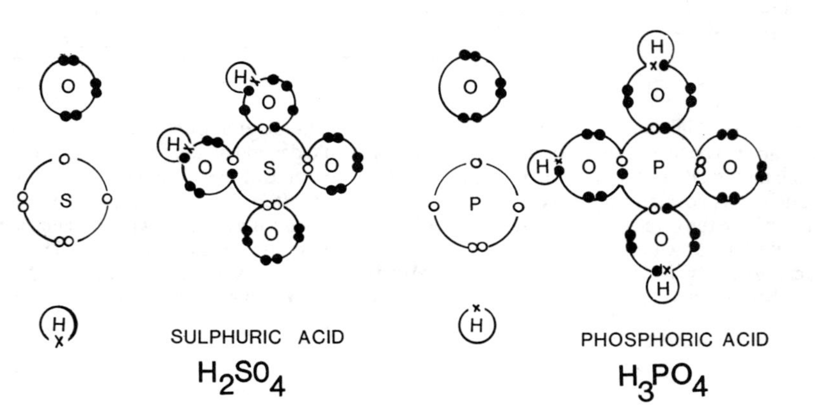

Fig. 6.13 — Some compounds can be modelled as showing a mixture of ionic and covalent bonds. Examples of this are H_2SO_4 and H_3PO_4 in which the hydrogen is modelled as ionically bonded and so can be detected in aqueous solutions, whereas the sulphur is never detected without the accompanying oxygen. When sulphuric acid H_2SO_4 (used as a 12 V battery acid) ionizes and loses the two H^+ ions, the resultant SO_4 unit becomes charged (2–). Similarly, with phosphoric acid the PO_4 unit becomes charged (3–). Phosphoric acid is discussed in Section 6.18 as a pre-treatment acid.

gains only one $(8-7=1)$. Therefore two of these non-metallic atoms 'X' are required to satisfy one divalent group 2 metallic atom and so the chemical shorthand formula will be AX_2. A condition of the ionic bond is that there must be no **overall electric charge** in the structure as a whole (Fig. 6.14). The net result is that the metallic atoms becomes charged positively (A^{2+}) (and is called a **cation**) while the two non-metallic atom becomes negatively charged (X^-)(called an **anion**). The ionic bond therefore produces a three-dimensional structure of positive ions and negative ions. Twice as many X^- ions must be packed into the available space as the A^{2+} ions as shown for calcium fluoride CaF_2 (calcium is in group 2, fluorine in group 7). An ion consists therefore of electrically charged atoms or groups of atoms. The sulphate ion is an example of the group of atoms existing as a charged unit (see Fig. 6.13).

The total positive charge must equal the total negative charge and so there must be twice as many anions (negative) as cations (positive) when a group 2 metal combines with a group 7 atom. The combination results in a compound called a **molecule** which, in general terms, is called a **salt**. The acids of the salts to be used in later sections (Section 6.18) are shown in Fig. 6.13 together with their chemical formulae which express the number of hydrogen ions forming that compound.

6.11.1 Ionic structures

There are essentially four simple A–X structures which concern us at this elementary stage (note that A refers to the metallic ion (positive charge) and the X to the non-metallic ion (negative charge). The numbers in square brackets in what follows refer to the number of nearest neighbours in the structural unit for both ions.

6.11.1.1 *The* [8]:[8] *structure*

The [8]:[8] structure is exemplified by the salt caesium chloride CsCl which can only be formed within a certain limit according to the 'free space' within the structure. This structure can be packed together only if the positive and negative ions lie within a certain band of radius sizes. This is expressed as

$$\frac{\text{cation radius}}{\text{anion radius}} = 1.0\text{–}0.73$$

6.11.1.2 *The* [6]:[6] *structure*

This is exemplified by the salt (sodium chloride NaCl) structure, and can only be formed by positive and negative ions lying within another band of radius sizes:

$$\frac{\text{cation radius}}{\text{anion radius}} = 0.73\text{–}0.414$$

6.11.1.3 *The* [4]:[4] *structure*

This is exemplified by zinc sulphide (ZnS). Again, this structure can only be formed by positive and negative ions lying within a certain size range:

$$\frac{\text{cation radius}}{\text{anion radius}} = 0.414\text{–}0.225$$

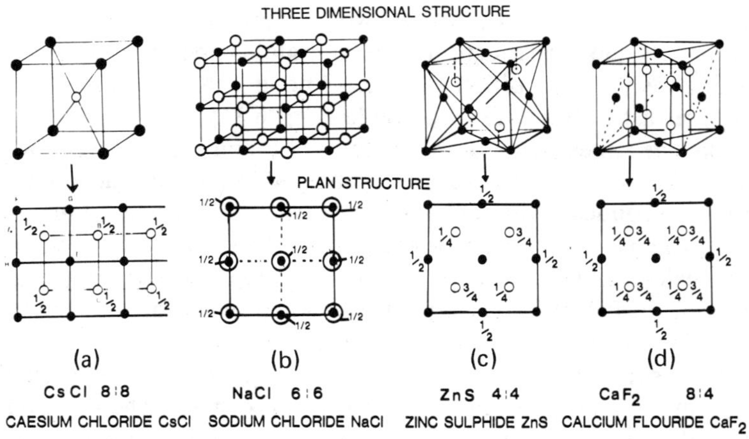

Fig. 6.14 — (a) The ionically bonded salt of CsCl has eight nearest neighbours to each ion and is therefore abbreviated as an 8:8 structure. (b) For similar reasons, sodium chloride (NaCl) and zinc sulplhide (ZnS) are represented as 6:6 and 4:4 structures respectively. These structures are formed from the same number of positive and negative charged ions. $MgCl_2$ and (d) CaF_2 structures require twice as many negative ions as positive ions and these must be packed in this ratio in a three-dimensional array. Each Ca^{2+} ion has eight nearest F^- neighbours, while each F^- ion has only four nearest Ca^{2+} ions, and so this is represented as an 8:4 structure. There are twice as many F^- ions as Ca^{2+} ions.

The structures are shown in Fig. 6.14 and, because the ions are of different chemistry (metals and non-metals), they are shown differently. The chemical formulae of the compounds are NaCl, CsCl and ZnS and because they have the same number of ions in the formula unit they must have the same number of charged ions (positive and negative) in the structural unit. The dependence of the structural unit on radius or radius ratio has a profound effect on durability, stability and melting points of the compounds formed between metals and non-metals. This will be examined with these elementary substances initially as examples. The size of the cation affects the number of nearest neighbours in the structure as shown in the CsCl, and NaCl structures. The charge of the cation affects the void space, e.g. a 2+ ion replacing a 1+ ion in these A–X structures must cause a vacant (1+) site for every 2+ ion substituted if overall neutrality is to be achieved Fig. 6.15. This produces a defect structure.

6.11.1.4 The [8]:[4] structure

This is exemplified by the salt calcium fluoride (CaF_2). This structure can only be formed by positive and negative ions lying within a certain size range. Note that twice as many negative ions are required as positive ions in this structure.

6.11.2 Summary of the ionic bonds

(1) The bond is non-directional and very dependent on the number of near neighbours, which is **dictated by the radius ratio**.

(2) The ionic bond is mechanically very weak and elastically non-deformable (there is a possibility of two like charges coming together if the ions are 'slid' over

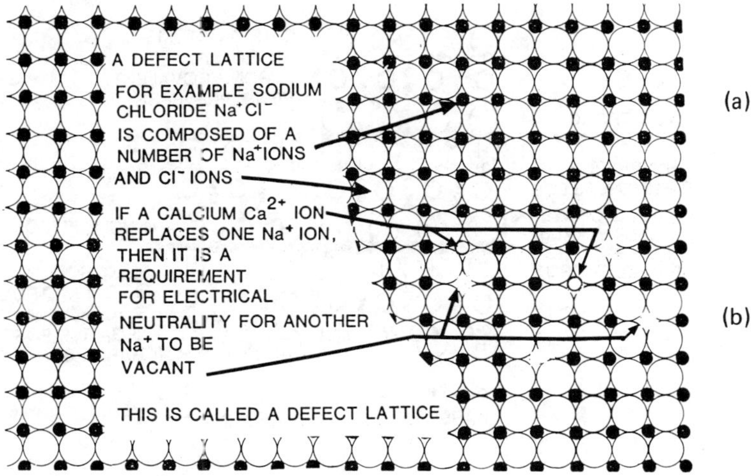

Fig. 6.15 — (a) A pure ionic substance is modelled as an array of positive and negative ions packed together so that no two negative ions are adjacent to each other and vice versa. (b) If mixed valency ions are considered, as shown with Ca^{2+} ions replacing Na^+ ions in the sodium chloride lattice, then for every 2+ ion which substitutes for a 1+ ion, another 1+ ion must be lost to keep overall electrical neutrality. This produces a void space in the ionic structure and the structure is termed a defect structure. The multivalent ferrous ion produces a defect structure which prevents the iron oxide film from being protective to steel. There are two states of iron in the ionic form: one is Fe^{2+} (ferrous) and the other is Fe^{3+} (ferric).

each other: this will result in a mutual repulsion). Hence ionically bonded structures are usually brittle and fracture without warning.

(3) The ionic bond is a medium to strong bond and therefore ionically bonded materials have average melting points.

(4) Ionically bonded materials are bad conductors of electricity in the solid state but very good conductors in the liquid or molten state, and are used in electroplating baths etc. In the electrocoating of motor vehicle bodies, it is a requirement to make the covalently bonded paint film have an ionically bonded unit on it so that it ionizes in water and can carry a charge and be plated out (Sections 9.7 and 9.8). The paint molecule can therefore be a big macromolecule so that good electrical efficiency can be achieved in that a large mass of paint can be deposited for one unit of charge. See also Section 6.13(3) and 6.13(4) for the methods scientists use for expressing solubilities and conductivities and Section 7.20 for the laws of electrolysis.

(5) The ions forming the structure are of different sizes (radii) and these have to be packed to give the smallest void space with no overall electrical charge within the ionic structure. Two positive (or negative) ions cannot be adjacent.

(6) Examples of the ionic bond are road de-icing salts and the growth of oxide films (Section 8.5.1). The most important point of understanding is in corrosion and the accelerating effect that the chloride ion has in corrosion (Sections 6.13, 6.14, 10.2 and 10.14).

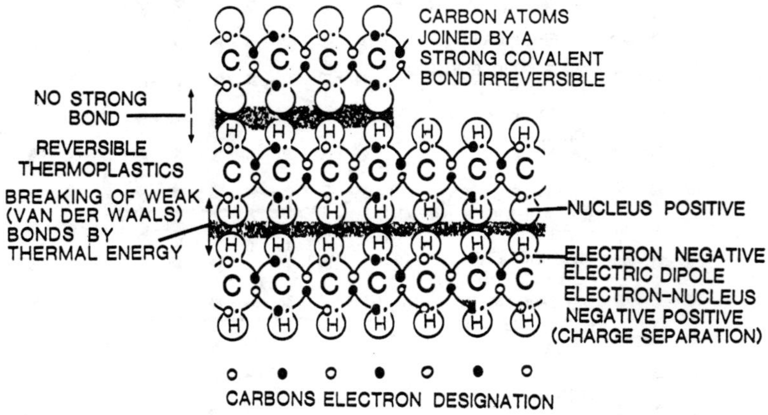

(a)

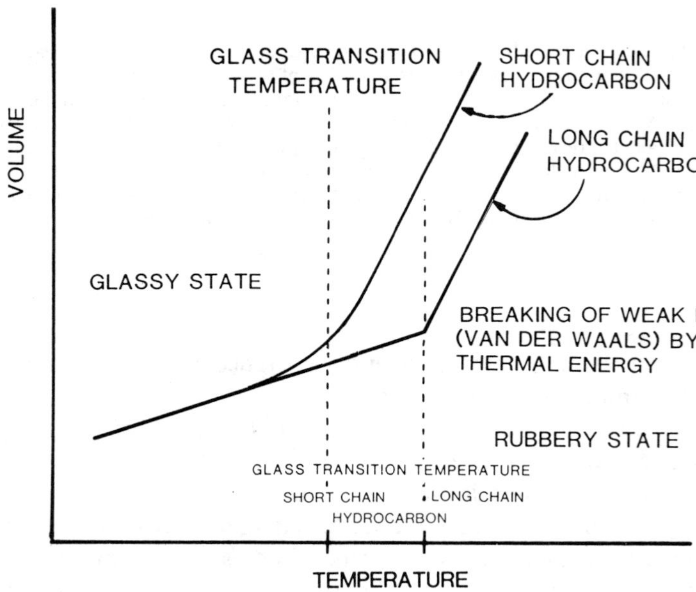

(b)

Fig. 6.16 — The electrical dipole (Fig. 6.8) formed by the directional bonding in the covalent bond leads to an electrical separation of charge called a dipole. This dipole is present in oxygen (Fig. 6.8) and hydrocarbons such as plastics. (a) The structural bonding of a plastic is shown here: the centre of each hydrogen atom is positively charged and the negative charge is fixed between the carbon and hydrogen atoms. It is the breaking of this dipole or van der Waals bond, by thermal energy, which makes the plastics thermoplastic. (b) The temperature at which this readily happens is often referred to as the glass transition temperature. The volume of the plastic is plotted against temperature. If the plastic were bonded by three-dimensional covalent bonds as a vulcanized patch on a tyre is, there will be no softening of the adhesive bonds. Hence rubber patches stuck onto a tube using an adhesive based on the evaporation of solvents (i.e. the weak van der Waals bond) will become 'unstuck' if the temperature rises appreciably. The use of adhesives which do not bond in three dimensions is illegal for the repair of motor vehicle tubes and tubeless tyres. Their use is still legal on pedal cycles.

6.12 VAN DER WAALS BONDS

This is a residual bond resulting from the **separation** of electrical charge. If the gases hydrogen, oxygen or chlorine are considered (Figs. 6.7 and 6.8) or the bond between molecules of plastic (Fig. 6.16(a)), it will be seen that the electrons in the covalent bond are now concentrated within the 'bond direction'. This localization produces a separation between the positive and negative charges because the nucleus is positively charged and the electrons are negatively charged. This charge separation is called an **electric dipole**.

The positive charge comes from the protons within the nucleus of each atom, and the electrons from the electron concentration in the bond. The van der Waals bond is a very weak bond and is responsible for the low melting points and boiling points of the gases and for the poor mechanical properties in manufactured, extruded or formed **thermoplastics** (Sections 11.1 and 11.5). The van der Waals bond is illustrated with reference to the case of plastics. The bond within the plastic chain is the covalent bond. The bond between the plastic chain molecules is the van der Waals bond.

6.12.1 Summary of the van der Waals bond

(1) The bond is directional.

(2) It is extremely weak.

(3) The weak bond means that materials which contain the van der Waals bond (e.g. plastics) have **non-specific softening points**. The onset of melting is called the glass transition temperature (Fig. 6.16(b)). These temperatures are generally low and give rise to a class of plastics called thermoplastics (Section 11.5). These plastics are very useful as they require less thermal energy to break the van der Waals bond and hence it is relatively cheap to make articles out of thermoplastics by moulding and cooling. Because of the poor bond strength some of the shorter chains or volatiles are lost from vinyl seats and these seats then become brittle. Plastic ash-trays, in contrast, are made from **thermosetting** resins which are bonded in three dimensions by the covalent bond whose strength overides the weak van der Waals bond (Section 11.14). They are strong and not melted by high temperatures (some ash-trays with metal stubs for extinguishing the cigarette are made from Nylon 66).

(4) Materials made from van der Waals bonds are bad conductors of electricity in both the solid and the liquid states. Hence plastics are used for electrical insulation. Small molecular weight chains are usually gases, liquids or low-melting solids at normal temperatures. The size of the covalently bonded chain associated with the van der Waals bond usually overrides the mechanical properties of the van der Waals bond, Fig. 6.16(b).

6.13 CHEMICAL SHORTHAND NOTATION — SOLUBLE AND INSOLUBLE NOTATION

The chemists' shorthand way of writing a reaction is by using the shortened abbreviations of the elements constituting the material, and representing the valency and structural proportions required for the compound as follows.

(1) In the form of a metal iron is represented as Fe. The letter e of the Fe has no relationship to the symbol e used to show the presence of an electron. Iron exists in the divalent (ferrous) state (Fe^{2+}) and in the trivalent (ferric) state i.e. (Fe^{3+}).

(2) In the form of the metal in solution it is represented as Fe^{2+} or Fe^{3+} and is called an 'ion'.

(3) An insoluble salt, or a salt which is not dissolved, is represented as $Fe(OH)_2$, Fe_2O_3 (iron oxide) or NaCl. These 'units' or the constituents such as NaCl in the solid state are not available for the carriage of electrical charge when written as NaCl. They are available as charge carriers when they are shown in the ionic form, e.g. as Na^+, Cl^-, and so the solution will have high conductivity (Section 10.1) or a low electrical resistance or, to use a common term, which is scientifically incorrect, a low 'insulation value'. Chapter 10 discusses resistance and conductivity. It is sufficient for our purposes at this point in time to consider conductivity as inversely proportional to the resistance.

(4) In the form of a soluble salt, or a salt that is capable of dissolving, or has dissolved,symbols such as Na^+ and Cl^- are used. These 'units' Na^+ and Cl^- called ions are available for the carriage of electrical charge and so the solution will have a high conductivity or offer a low electrical resistance.

(5) For a soluble acid symbols such as H^+ and Cl^- are used. In other words, the characteristic of an acid is that it gives an excess of H^+ ions compared with water (see Section 6.15).

(6) For a soluble alkali symbols such as Na^+ and OH^- are used. In other words, a characteristic of an alkali is the formation of excess OH^- ions. These units are available for the carriage of electrical charge and so the solution will have a low electrical resistance and act as a good conductor of electricity in liquid media.

(7) In writing equations to represent a chemical reaction, the scientists use arrows. The arrow is verbally translated as 'goes to', but in practice it should be translated as 'in equilibrium with'. This term is used to represent the fact that, in any equilibrium, a disturbance of the equilibrium can be obtained by increasing any of the participating units on the right-hand side or the left-hand side of the equation. A torch battery (Fig. 2.5) stops running down because the electrical energy (the electrons) is not consumed and therefore electrons appear in excess on the right-hand side to push the equation (see below) to the left because the torch has been switched off. This could be looked as a balance between the left-hand side and the right-hand side of the equation. Hence in some equations where the 'goes to' translation is required, the arrow $\rightleftharpoons$ will be used. A double arrow $\rightleftharpoons$ indicates that the equilibrium translation is required. Both translations can be considered simultaneously. The electrons may be are consumed by local cells (Fig. 7.10, Section 7.3) which would arise from impurities in the zinc. These electrons in the zinc electrode must be consumed by a cathodic reaction which takes place on the zinc surface. This is very slow if the battery is properly constructed, and the battery will have a fairly long shelf life.

Hence we can write the dissolving reaction of the zinc electrode on a switched–off battery electrode is represented as follows (an accompanying 'English language' translation of the equation is also shown):

Zn	$\rightleftharpoons$	Zn^{2+}	$+$	$2e^-$
zinc as the metal	goes to or is in equilibrium with	zinc into solution	plus	two electrons

There is an excess of electrons on the right-hand side and so these tend to stop the zinc from dissolving. The equation is thermodynamically balanced, and so the symbol $\rightleftharpoons$ is used. Some equations do not balance. The notation $\rightleftharpoons$ is used in these cases. The following example is unbalanced because the product formed on the right-hand side is nearly insoluble (this means that the ions Fe^{2+} and OH^- are taken out of solution):

$$Fe^{2+}_{aq} \quad + \quad 2OH^-_{aq} \quad \rightleftharpoons \quad Fe(OH)_{2\ solid}$$

The terms aqueous (aq) and solid are often implied and omitted. The right-hand direction is represented by $\rightharpoonup$ and the left-hand direction by $\leftharpoondown$ to indicate that the right-hand direction is predominant. See also Section 6.14(7) *et seq*.

6.14 CORROSION SCIENCE: THE CORROSION REACTION USING THIS NOTATION

Thus in science we tend to have a shorthand way of representing the various states of a metal, the metal in solution and the direction and equilibrium of reactions. Hence steel, in the form of a metal, is represented as Fe and the metal in solution as Fe^{2+} which are called ions (Section 6.11), and the electrical current as e^-, which indicates an electron. Hence we can write the dissolving reaction more correctly called the **anodic** reaction as eq. (6.1), with the 'English language' translation of the anodic reaction as eq. (6.2):

Fe	$\rightleftharpoons$	Fe^{2+}	+	$2e^-$	(6.1)
steel as the metal	goes to or is in equilbrium with	steel into solution	plus	two electrons	(6.2)

Some explanatory notes on this equation follow.

(1) The 'units' such as Fe^{2+} in solution are called ions.

(2) The electrons **cannot** go into the liquid environment and must always stay in the metal conductor. This has some important consequences for corrosion protection, one of which is that the protective film must be a non-conductor of electrons–electricity, i.e. an 'insulator'. Note that the common usage term insulator should not be used as it implies that the protective coating, e.g. paint film, should have a high dielectric constant. In fact a paint film acts by offering a high resistance to the passage of charged ions (Section 8.11).

(3) The arrow has been indicated in the translation as 'goes to or is in equilibrium with' but in practice it should be translated as 'in equilibrium with'. This term is used to represent the fact that in any equilibrium a disturbance of the equilibrium can be obtained by increasing any of the participating units on the right-hand side or the left-hand side of the equation, as in the case of the torch battery which stops running down because the electrical energy (the electrons) are not consumed and therefore electrons appear in excess on the right-hand side to push the equation in (Section 6.13) to the left. A similar effect is produced if the Fe^{2+} (or Zn^{2+}) ions are not removed from the metal environment. See (6) below. This could be looked on as a balance between the left-hand side and right-hand side of the equation.

There are several conditions which must be satisfied for corrosion to proceed.

(4a) The electrons in the metal must be 'consumed'. They **cannot be stored**.

(4b) The iron ions Fe^{2+} or Fe^{3+} in solution must be taken away from the liquid around the metal, otherwise the metal will stop dissolving (the equilibrium concept is discussed in Section 6.13(7)). See the discussion on battery rejuvenation Section 7.17.

(5) The current is carried by the electrons within the metal. In the electrolyte, the current is carried by both positive and negative ions. The current carried by the ions will depend on the **mobility** of the charge carrier and the electric charge, i.e. the valency. The total current carried by the positive and negative ions must always equal the current carried by the electrons within the metals. Ohms Law

$$V = i \times R$$

(where V is the voltage, i is the current and R is the resistance) will apply to both the flow of the electronic charge in the metal as well as to the ionic charge in the electrolytes.

These electrons, which must remain within the metal, are consumed at the other electrode (called the cathode) as in eq. (6.3) with the 'English language translation' as eq. (6.4). (This practice is followed throughout in the early chapters, where possible, to make it easy for the non-scientist).

$$O_2 \quad + \quad 2H_2O \quad + \quad 4e^- \quad \rightleftharpoons \quad 4OH^- \qquad (6.3)$$

oxygen as air   plus   water   plus   4 electrons   goes to or is in equilibrium with   4 hydroxide units (6.4)

The result is that the charge of the electron has **transferred** to the liquid environment: electrons can only exist in a metal.

(6) Hence corrosion will proceed providing that both the products formed when these electrons are 'consumed', i.e. the OH^- ions (eq.(6.3)) and the iron ions Fe^{2+} (eq. (6.1)) in solution, are **taken away** from the liquid **adjacent** to the metal. If this does not happen the corrosion (process or) current will stop. In practice we use this phenomenon in order to arrest or, more properly, to slow down corrosion. This is done by using a protective paint film, which stops the flow of these ions Fe^{2+} and OH^-. The paint film also keeps the environment away from the bare metal, but it should be pointed out that the paint film will allow sufficient water and oxygen to diffuse through to allow corrosion to proceed (Section 8.11). The prime function of the paint films, therefore, is to hinder the movement of these Fe^{2+} and OH^- ions from the anodic and cathodic sites, electrodes, or regions where they are formed. Corrosion cannot, therefore, be stopped completely, because there will always be a 'leakage current'. (This immediately should make you think of the 'shelf life' of the ordinary toy or 1.5 V transistor battery).

These ions react with each other and produce rust which has a much larger volume than the metal which has disappeared. Hence the rust tends to push the spot welded panels apart (Fig. 1.9) and to form excrescent mounds beneath the surface of paint films as shown in Fig. 8.18. The formation of rust is described by eq. (6.5) with the 'English language' translation in eq. (6.5b) and (6.6a).

$$Fe^{2+} + 2OH^- \rightleftharpoons Fe(OH)_2 \qquad (6.5a)$$

product of anode + product of cathode ⇌ iron hydroxide (6.5b)

The overall reaction is

iron + hydroxide + moist oxygen → rust + eight-fold increase in volume (6.6a)

(7) Rust is a complex mixture of iron oxides. These oxides are usually combined with water. This is why they cannot be painted over. If they are over-painted severe blistering will result (Sections 8.12 and 8.13). They are usually represented as the hydroxide $Fe(OH)_2$, but can also be written as the oxide FeO mixed with water H_2O. Thus $Fe(OH)_2$ is $FeO{\cdot}H_2O$. There are the same number of iron, oxygen and hydrogen atoms in both representations (rusting is a term which is reserved for iron and iron-based alloys that form predominantly iron oxides with variable water contents: see note (10) below. Other metals **corrode** rather than '**rust**' because they do not form iron hydroxides

(8) Eq. (6.5) is not represented by a double arrow ($\rightleftharpoons$) because iron hydroxide is insoluble and the shorthand way of depicting this fact is that insoluble compounds (i.e. undissociated) do not form ions.

(9) When rust is deposited on the steel surface it tends to exclude oxygen (Fig. 6.17). The cathodic reaction (eq. (6.3)) is thus transferred to the periphery of the rust layer and this leads to a high pH (Sections 6.16 and 6.17) and a stripping of the paint (Sections 8.11 and 8.13).

(10) Iron hydroxide $Fe(OH)_2$ or $FeO{\cdot}nH_2O$, *n* being used to represent the variable water content, is white in the pure form and is the diffusion barrier through which oxygen must diffuse for corrosion to continue (Fig. 6.17). The pH (Sections

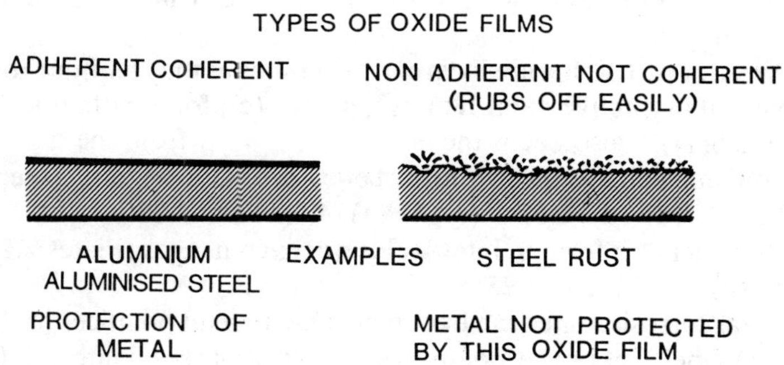

Fig. 6.17 — In an idealized theoretical situation, the corrosion of the steel would be accompanied by a corrosion product which would protect the steel from its environment, rather like the oxide formed on aluminium cooking utensils. In the ideal state this corrosion product would be ferrous hydroxide which, as explained in section 6.14(10), is not stable.

6.15–6.17) of saturated ferrous hydroxide $Fe(OH)_2$ is 9.5 and therefore is alkaline (Section 6.17). Iron will not corrode in a pH solution of 9.5 (Section 7.10) and so further corrosion can only proceed by this hydroxide reacting with the acidic gases SO_2 and CO_2, (sulphur dioxide and carbon dioxide (Section 10.5)). However, this divalent state of iron (Fe^{2+}) is not the most stable ionic state and the iron soon oxidizes to the trivalent state Fe^{3+}: this oxidation also produces an electron. This electron is consumed by the oxygen electrode (eq. (6.3)) so that the overall reaction as represented by eq.(6.6a) may be rewritten and represented as

$$4Fe(OH)_2 + 2H_2O + O_2 \rightarrow 4Fe(OH)_3 \qquad (6.6b)$$

The valency state of the iron is now +3 and this is termed **ferric** hydroxide and has a pH of 7. The oxide is red in colour and can be written as $Fe_2O_3 \cdot 3H_2O$. Most owners will have noticed that the rust formed on their vehicle beneath this brown oxide is black and slightly magnetic. This black magnetic oxide is the oxide Fe_3O_4 which can be considered as a mixture of Fe^{2+} and Fe^{3+} ions in the oxide lattice and written as $FeO \cdot Fe_2O_3$. Hence the oxides may be written as

⟶ increasing oxygen content ⟶

FeO	Fe_3O_4 or $FeO \cdot Fe_2O_3$	Fe_2O_3
ferrous oxide	ferrous ferrite	ferric oxide

(Note that $FeO \cdot Fe_2O_3$ and Fe_3O_4 are equivalent forms of writing the black magnetic oxide). Hence rust films usually consist of three layers of oxide: FeO; Fe_3O_4; Fe_2O_3. Therefore the anodic corrosion reaction (eq. (6.1)) takes place where there is a deficiency of oxygen, while the cathodic reaction (eq. (6.3)) takes place where oxygen and its replenishment is guaranteed (see Section 2.4). The paint film, a non-electron conductor (see note (6) above) prevents the flow–movement or diffusion of ions along the surface of the metal by a reaction which is equivalent to the throwing away of the electrolyte within a car battery. Hence the corrosion reaction is a simple 'battery' which cannot be turned off by a 'switch'. These are termed 'equivalent batteries' in this book. They are local corrosion sites which can be formed at the following locations.

(11) Areas of **bad design** are local corrosion sites as they harbour road dirt which restricts natural drying (Plates 2.1(a)–(g), 4.5(a)–(c); for a definition of bad design see Section 2.5(1)).

(12) Equivalent batteries can form beneath road dust 'dirt' (see Fig. 2.6 and Plates 2.2(a), 2.4(a)–(d), 2.6(a)–(c), 4.9(a), 4.9(b) and 4.13(b)).

(13) They can also form in inter-welds or clinch flanges (Plates 2.2(a), 2.3, 2.6, 4.1, 4.4, 4.7(a), 4.7(b), and 4.23.

(14) Local corrosion sites can exist just beneath mud poultices or dirt layers and at the 'water line' where the poultice is just wetted: see Plate 2.1 for the ledge beneath the wing–fender.

(15) They can also exist at the corrosion front or 'pit' which will explain the aggressive nature of 'pitting' corrosion or 'rust-through' corrosion. It also explains why pitting corrosion, internally nucleated, cannot be rectified from the outside

surface. This is often referred to as one or more of the following: rust penetration, metal perforation, inside-out corrosion or as pitting (Sections 7.8 and 7.11). The corrosion reaction will be dealt with in greater detail in Chapter 7.

6.15 CHARACTERISTICS OF AN ACID

The aim of this section is to explain (1) the meaning of the statement made in Section 6.14(9) concerning high pH, (2) the very acidic environment which exists at the pit front during metal perforation (Sections 7.9.2g and 7.11) and (3) the pretreatment acid phosphoric acid H_3PO_4 (Section 6.18, Chapter 9) for those who have studied science and want to know something about it.

Acids are often (wrongly) considered as being corrosive liquids. Some are highly corrosive, but others can be eaten in unripe fruit (citric acid) or with fish and chips–french fries (acetic acid or vinegar). Some are gaseous, e.g. hydrochloric acid (HCl), whereas others have to be dissolved in water to become acids. Examples of these are the acidic industrial by-products (carbon dioxide and sulphur dioxide and trioxide) from burning fuels such as coal and oils.

6.15.1 Characteristics of an acid

(1) Acids taste sour.

(2) Acids produce an excess of hydrogen ions in water and so often react to produce hydrogen gas with a metal (see pre-treatment phosphates and other 'rust cures').

(3) Acids change the colour of some dyes and therefore these dyes are used as indicators for acids. The colour change, for particular dyes, is often dependent on the number of hydrogen ions in solution (see later).

(4) Acids react with hardened cementitious products to produce the gas carbon dioxide (this should be avoided on your garage floor). They also destroy clothing material with time: this is usually a slow process and only shows up when the article is subsequently washed.

(5) These properties listed above are destroyed when an acid is mixed with an alkali or a metal. The acid is said to be neutralised.

6.15.2 Summary

The main points about acids are as follows. Acids (ionize and) produce an excess of hydrogen ions in water. The number of ions produced is in excess of the ions which already exist in pure water. This excess is responsible for that substances being an acid.

6.16 CHARACTERISTICS OF ALKALIS OR BASES

Alkalis or bases are terms which mean the same. They are opposites to an acid. An alkali is an element or 'unit' which is capable of taking up an hydrogen ion from an acid to form a new 'unit' called a **compound** (which is a **salt**), and producing water:

$$\text{acid} + \text{base} \rightarrow \text{salt} + \text{water} \tag{6.7}$$

e.g.

$$HCl + NaOH \rightarrow NaCl + H_2O \tag{6.8}$$

Eq. (6.7) is the 'English language translation' of eq. (6.8). It is conventional to represent the alkalis with OH^- and the acids with H^+ because the reaction to form the salt and water can be easily modelled, by considering these characteristic units which form water as follows:

$$\underset{\text{alkali}}{OH^-} + \underset{\text{acid}}{H^+} \rightleftharpoons \underset{\text{water}}{H_2O} \quad (6.9a)\ (6.9b)$$

This indicates that water is composed of an 'acid' part and an 'alkali' part which neutralize each other to form a neutral unit. The symbol $\rightleftharpoons$ in eq. (6.9a) indicates that the reaction is predominantly over on the right-hand side. Alternatively,

$$\underset{\text{covalently bonded representation (Section 6.11)}}{H_2O} \rightleftharpoons \underset{\text{ionically bonded representation (Section 6.10) (very low concentrations: see Section 6.17)}}{OH^- + H^+} \quad (6.10)$$

6.17 DEGREE OF ACIDITY AND ALKALINITY: HOW ACID IS ACID?

The degree of acidity is expressed as the reciprocal of the hydrogen ion concentration or minus the logarithm of the hydrogen ion concentration. This is given the abbreviation of **pH**. For newcomers to science it is best to consider the symbol p as a mathematical operator whose English language translation is to remind us to take minus the logarithm to the base 10 of the hydrogen ion concentration. Some examples are given in Table 6.9. Note that

$$[H^+] \times [OH^-] = 10^{-14} \quad (6.11)$$

(Square brackets [ ] means 'Concentration of'.)

Water is modelled as a covalently bonded 'unit' but also has a form where it forms ions (eq. (6.9a) and (6.9b)). It is estimated that only **one** water unit (molecule) in every **ten million** is bonded in this manner as ions. Hence this is often interpreted as stating that the water ionizes or dissociates to form H^+ ions and OH^- ions and this value is given the symbol K_w.

Table 6.9 — The modelling and nomenclature of pH

pH	$[H^+]$	$[OH^-]$	Nature
1	10^{-1}	10^{-13}	Very acidic
5	10^{-5}	10^{-9}	Fairly acidic
6	10^{-6}	10^{-8}	Acid rain
8	10^{-8}	10^{-6}	Alkalinity of some brackish waters
10	10^{-10}	10^{-4}	Alkalinity of some oven cleaners
12	10^{-12}	10^{-2}	Strong alkali

The hydrogen ion concentration of neutral solutions is determined from the ionization of water into H^+ ions and OH^- ions:

H_2O	$\rightleftharpoons$	$H^+ + OH^-$	
bad conductor of electricity: covalently bonded representation (Section 6.10)		good conductor of electricity: ionically bonded representation (Section 6.11)	(6.12)

$$K_w = \frac{[H^+] \times [OH^-]}{[H_2O]} \tag{6.13}$$

This reaction is reversible, (it goes in both directions), but is heavily biased towards the left-hand side (it is estimated that only one water unit (molecule) in every ten million is dissociated as represented above). This is is expressed as the ionisation constant K_w for water. The value of this constant is temperature dependent, but at 25°C it is 1.01×10^{-14}. Pure water at 25°C contains an equal number of H^+ ions and OH^- ions and so the pH is the square root of the ionization constant K_w. If the H^+ ions exceed the OH^- ions the medium is acidic and is represented as a pH from 1 to 7. If the OH^- ions exceed the H^+ ions the medium is alkaline. At temperatures above 25°C the ionization constant of water is greater and so the pH of pure water is temperature dependent (Table 6.10). Throughout this text the temperature of 25°C will be assumed.

Table 6.10 — pH of pure water at various temperatures

Temperature	$K_w \times 10^{-14}$	pH
0°C	0.115	7.47
10°C	0.29	7.27
25°C	1.008	7.00
40°C	2.916	6.77
60°C	9.614	6.51[a]

[a] The numerical value for neutrality is 6.51 for a reaction at 60°C.

The importance of this statement that one water unit (molecule) in every ten million is dissociated as represented above is as follows.

(1) Only a very few water units (molecules) are split up or dissociated to be able to conduct electricity and so pure water will only conduct electricity at high potentials. It therefore has a high resistance, or to use layman's language, a certain degree of 'insulation'. This is **only true** if there are no dissolved salts: if there are dissolved salts the water will have a 'low insulation value', owing to a low resistance to electrical currents or the passage of ions (see Table 10.2 for analysis of tap water).This explains the damaging and aggressive effect of the road de-icing salts in corrosion of the motor car.

(2) If an ion such as Fe^{2+}, removes some of these OH^- ions from water, more water will split up to balance the OH^- deficiency (eq. (6.9a) and (6.9b)) and so the anodic areas – as represented by eq. (6.1) become **acidic**. Hence crevices and inter-welds become acidic as corrosion proceeds, (some workers estimate a pH of 3). This acidic nature of the crevice or pit causes a rapid rise in the corrosion rate because the deeper the pit the more stagnant or denuded in oxygen is the pit front and the more acidic it becomes. The cathodic areas, in contrast, become very strongly alkaline and strips the paint off the metal substrate (Sections 8.12 and 8.13).

6.18 PRETREATMENT ACIDS — PHOSPHORIC ACID H_3PO_4

These are all based on the acid and the salts of the acid phosphoric acid ($H_3\ PO_4$) (Fig. 6.13) which will be abbreviated to H_3A. We can write an expression for the acid dissociating into the following negative ions:

Acid goes to ionic constituents

$$H_3A \rightarrow H_2A^- + H^+ \quad (6.14)$$
$$H_3A \rightarrow HA^{2-} + 2H^+ \quad (6.15)$$
$$H_3A \rightarrow A^{3-} + 3H^+ \quad (6.16)$$

For neutralization, each negative in the right-hand side will form a salt and these will have different characteristics. These different characteristics are best considered in terms of the following solutions made up of the same concentrations. The pH alters depending on the number of hydrogen ions in the solution. These hydrogen ions come from the salt's hydrolysing and contributing to the total number of hydrogen ions in the solution.

In N/10 solutions the acid phosphoric acid has a pH of 1.5, the sodium salt NaH_2PO_4 has a pH of 4.4, the sodium salt Na_2HPO_4 has a pH of 9.6 and the sodium salt Na_3PO_4 has a pH of 12.0. The molecular weight of H_3PO_4 ($H=1$; $P=30$; $O=16$) is $3\times1+30+4\times16=97$. An N/10 solution would therefore be 9.7 g of phosphoric acid dissolved in 1 litre of distilled water. Solutions are now expressed in molarities. This is the result of the reaction of the strong alkali (NaOH) in group 1 with a relatively weak acid to make the acidic salts NaH_2PO_4, and Na_2HPO_4 which both have replaceable hydrogen ions. As the hydrogen is replaced by the alkali so the salt becomes less acidic and more alkaline in nature.

Each one of these negative anions can react with a monovalent base as follows:

$$H_3A + BOH \rightarrow BH_2A + H_20 \quad (6.17)$$
$$H_3A + 2BOH \rightarrow B_2HA + 2H_20 \quad (6.18)$$
$$H_3A + 3BOH \rightarrow B_3A + 3H_20 \quad (6.19)$$

Note that for the car body, which is made essentially from iron (which gives a divalent alkali base, (i.e Fe^{2+})), the following would be the requirements for neutralization:

$H_3A \rightarrow H_2A^- + H^+$ (two H_2A^- ions need one Fe^{2+} ion for neutralization);
i.e. Salt is $Fe(H_2PO_4)_2$ (6.20a)

$H_3A \rightarrow HA^{2-} + 2H^+$ (one HA^{2-} ion needs one Fe^{2+} ion for neutralization);
i.e. Salt is $FeHPO_4$ (6.20b)

$H_3A \rightarrow A^{3-} \leftarrow 3H^+$ (two A^{3-} ions need three Fe^{2+} ions for neutralization);
i.e. Salt is $Fe_3(PO_4)_2$ (6.20c)

It is the salts of this acid with zinc which are used in many pretreatment processes. They have different solubilities and stabilities, and co-exist with each other in equilibrium, Plate 9.1.

6.19 PRETREATMENT PROCESSES

The car body is spot welded together to form an integral rigid unit, complete with doors, boot and bonnet (Plate 3.1). This is termed a 'body in white'. The problem is now to make a paint film adhere to the metal surface. This is difficult for the following reasons.

(1) The steel surface of the car body is covered with both the mill or pressing oils and the steel mill protective oil put on to stop the steel coil corroding in transit to the motor manufacturer. These pressing oils are used to lubricate the rolling and pressing operations to give the vehicle body panels a smooth polished surface which will reflect light uniformly (Fig. 6.18).

(2) The chemical bonding mechanism between the paint 'system' is completely different from the bonds which exist in the metal of the car body. Therefore it is possible for the corrosion process to strip the paint film off the 'polished' steel.

(3) A pretreatment degreasing process which is then followed by some 'process' which will provide a 'key' for the paint film is required. This can be a chemical or physical transformation process which must be applied to roughen the surface in order to compensate for the **absence of a chemical bond** between the paint and metal surfaces. (Figs. 3.2 and 8.18(a)).

(4) The process which applies the 'keying' must be cheap and be capable of application in mass production processes. Therefore a mechanical keying process (rubbing down) must be discounted (Fig. 8.18(a)).

The only cheap process involves a chemical acid attack (called an etch or a pickle), which leaves the crystalline salt of the acid and base metal as an insoluble residue upon the surface, Plate 9.1. The actual mechanism is discussed in Section 6.18 and Chapter 9.

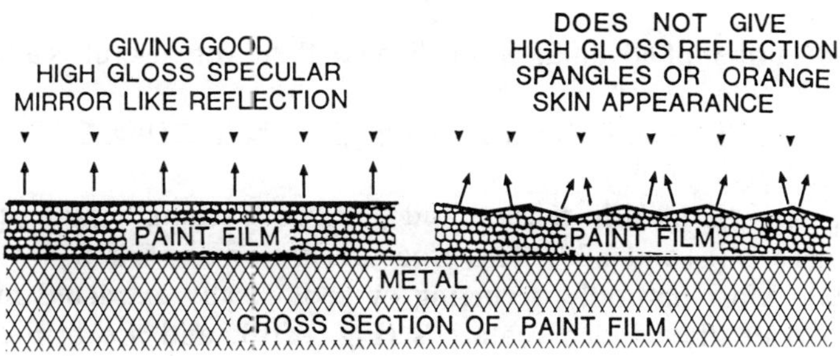

Fig. 6.18 — The show-room condition of the vehicle depends on how shiny the paint looks. Huge sums of money are spent on research to achieve this lustre. If the paint film 'cures' to an even coat, it follows that the metal surface and all the surface prepration techniques must not show through the paint film. Galvanized zinc gives rise to an uneven surface finish (termed 'spangles') which will also given an uneven paint finish. Orange peel is the most usual paint blemish that occurs in respraying. A cross-section through such an orange-peel surface is shown.

This pretreatment 'salt' process is known as the complex zinc iron phosphate film, whose function is two-fold: (1) to give a good 'key' for the subsequent paint film and (2) to contain any corrosion due to stone chipping and thus to prevent the '**undercutting**' of the paint film by corrosion and therefore to restrict the **lateral spread** of corrosion (Fig. 8.18; Sections 2.5, 7.12 and 7.13). This undercutting corrosion is usually called either **filiform** corrosion if the corrosion spreads out like a spider's web beneath the paint film (Fig. 8.21) or **scab** corrosion where the paint blisters up from the metal surface (Fig. 8.18; Section 8.12 *et seq.*).

Phosphoric acid was discussed in Section 6.18. It should be left until a full understanding of the book is attained. The phosphate coating is applied by an acidic phosphoric acid spray (eq. (6.21)). A spray process is used in preference to a dip process so that the mechanical impact of the acid droplets removes the hydrogen bubbles which are developed in the reaction. The metastable reaction (eq.(6.22)) can therefore take place once the hydrogen has been removed and deposit the complex iron and zinc phosphate crystals, which is the keying mechanism, Plate 9.1. Where double salts are used, the iron (Fe) and zinc (Zn) are represented by the single letter M (metal). The reactions can be represented as follows:

$$M + 2H_3PO_4 \rightarrow M(H_2PO_4)_2 + H_2\uparrow \tag{6.21}$$

The car body reacts with the acid to give a phosphate and hydrogen gas

$$\underset{\text{Soluble}}{3M(H_2PO_4)_2} \rightarrow \underset{\substack{\text{insoluble} \\ \text{keying crystals}}}{M_3(PO_4)_2} + \underset{\text{acid}}{H_3PO_4} \tag{6.22}$$

The phosphate salt $M(H_2PO_4)_2$ formed is unstable and breaks down to produce a new salt M_3PO_4 and acid H_3PO_4. The new salt produced acts as the mechanical key for the subsequent paint film. Unless a combined spray and immersion process is used, only the outer cosmetic panels are treated to the phosphate film and degreasing processes.

6.20 ALLOYS

Metals have a number of advantages which make them very useful as an engineering material.

(a) Metals may be readily fabricated (Fig. 3.1), bolted (Fig. 2.10), and welded (Fig. 8.7), (see Section 6.35).

(b) Metals can be used in tension and undergo a strengthening process when deformed (Figs 6.24(a) and 6.30). This process is known as **work** or **strain hardening** (Section 6.23). Most naturally occurring or geologically extracted materials are ionically or covalently bonded materials (Fig. 6.10), and are brittle and break without warning when stressed in tension. Wood is about the only exception.

(c) Metals are strong, conduct electricity and usually have a high strength:weight ratio. They can be readily shaped and they retain this new shape permanently. They may be stiffened by corrugating, curving, or making the metal into right-angle pieces, which increases the mechanical properties of the formed metallic product (Fig. 3.1). In this form they resist buckling and bending. Using this method of stiffening thin steel sheet can be made to have the mechanical properties of much thicker planar sections (without corrugating). This arises because metals have the same mechanical properties in compression as they do in tension. Perhaps the most important characteristic of the metals used in engineering is that the metals keep their shape even up to fairly high temperatures, whereas other materials such as plastics will revert to their original preformed shape. Metals acquire their 'new' shape under high stresses but with little 'spring-back' (Fig. 6.40), which is a beneficial characteristic for forming the complex panel shape.

The main problem with metals is that they corrode (Section 7.1).

The engineering strength of a metallic material is dependent on its structure, on both the microscopic and the macroscopic scales, and on composition (Section 6.36). The composition of the alloy is altered by the amount of alloy element added. The structure can be altered by several techniques used either singly or in combination with each other. These parameters affect corrosion. The strengthening mechanisms in metals are as follows:

(1) alloying by substitution (Fig. 6.4); the strengthening results from the **differently sized** atoms (Table 6.4)
(2) adding smaller atoms which go into the voids between the atoms: this is termed an '**interstitial**' hardening mechanism e.g. carbon or nitrogen can be used (Fig. 6.32).
(3) second-phase production (see Fig. 6.20(b))
(4) strain hardening (see Fig. 6.30)
(5) effect of grain size on yield stress and the alloy elements on grain size (Fig. 6.19(c))
(6) heat treatment (for the effect of grain size on the temperature of this heat treatment see Figs 6.28 and 6.35)

The metallic bond (Section 6.8) is indifferent to the type of metallic atom that is adjacent to it in its structure, provided that the following requirements are fulfilled. For simplicity two metals A (the solvent) and B (the solute) will be considered to form the alloy. **Substitution solid solution** occurs when the following conditions are fulfilled.

(7) The radii (Table 6.4) of the 'foreign' metal B should not be too different from that of the host metal A so that there is no large-scale distortion at the solute atomic sites or between the rows of atoms. Scientist tend to put an upper limit of 15% for the difference in radii. The distortion produced by this **15% difference** in radii is just sufficient to cause an **increase in the strength** of the alloy by preventing one plane of atoms from slipping over the adjacent plane.
(8) The structures of pure A and pure B should be the same, i.e. both f.c.c. or both h.c.p. or both b.c.c. (Section 6.9).
(9) The dissolution of metal A into metal B should result in a decrease in energy.

Thus the degree of miscibility is shown by a phase diagram (Fig. 6.20). In much the same way that water and alcohol are completely miscible, so are copper and nickel. Copper and nickel have radii within 15% of each other and have the same metallic structure in the pure state (Figs 6.2 and 6.6), and the solution of one in the

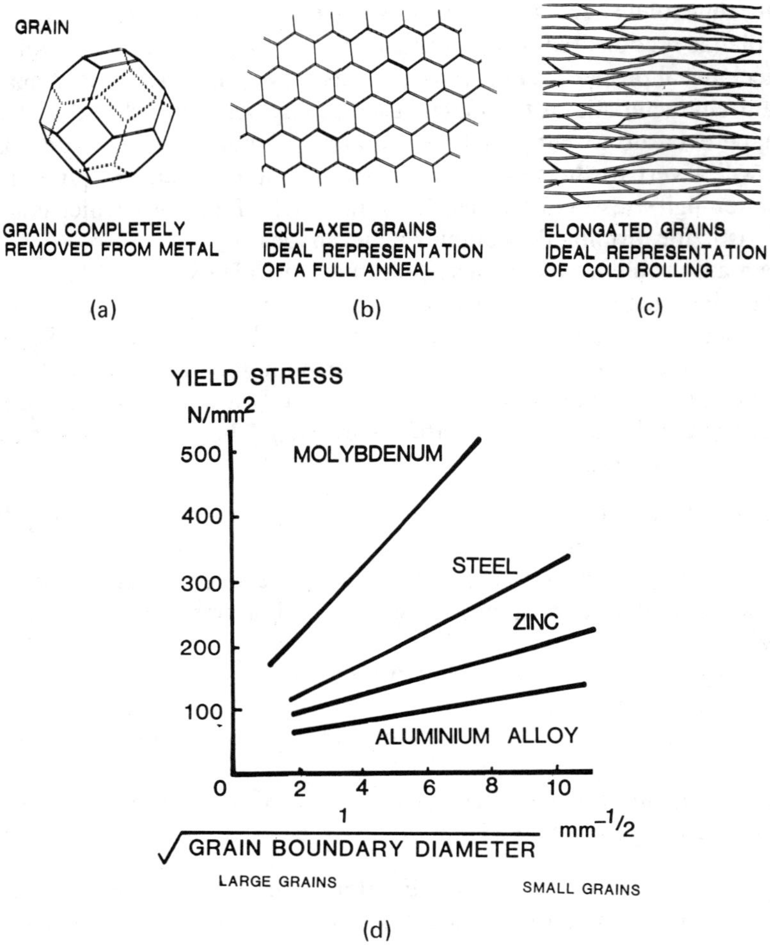

Fig. 6.19 — (a) If it were possible to remove a grain from a metal, then this is how an idealized grain would look.

(b) The grain boundaries within the metal from equiaxed equally sized grains packed together when sectioned and polished. Note that the point of intersection is a triple point. One never gets a meeting of four boundaries at a point.

(c) Cold rolling distorts these equiaxed grains into elongated grains. (d) The stress characteristic of a metal with different grain sizes. The resistance to deformation by an applied stress is much greater for a small-grain-sized material than a large-grain-sized material. The yield stress σ_y varies with grain diameter d according to the Petch equation:

$$\sigma_y = \sigma_1 + kd^{-1/2}$$

σ_1 and k are constants and are different for each metal.

other results in a decrease in energy. This is shown in Fig. 6.20(a), and it is termed **complete solid solution**.

If, however, copper and zinc are melted together there is very little solution of one in the other because the structures of the pure metals are different, i.e. copper is f.c.c., and zinc is h.c.p. Therefore there is **restricted solid solution**, and this is shown

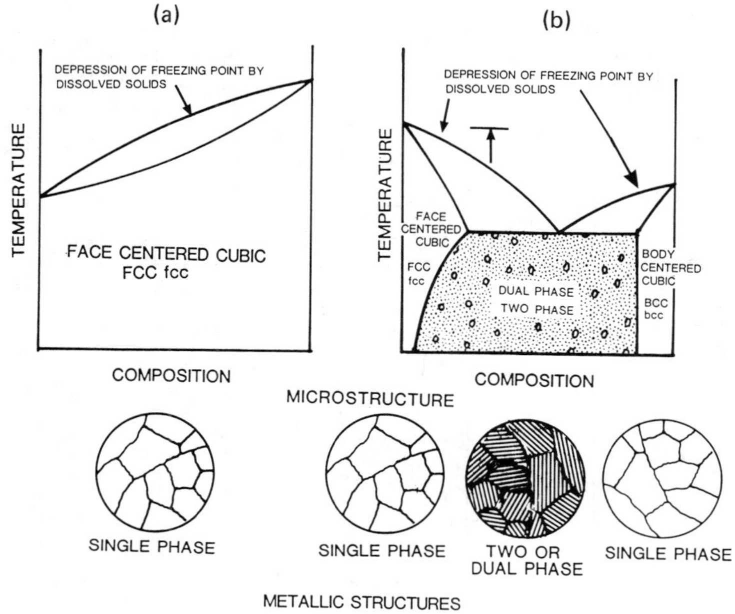

Fig. 6.20 — (a) An example of a phase diagram showing **complete solid solution**. The structure of the two metals is important as the solid solution region is represented as the α structure. This is the structure (f.c.c.) of pure copper and pure nickel because, to form a complete solid solution, both metals have to have the same structure. In the f.c.c. lattice of the pure metal there is random replacement of the copper atoms by atoms nickel (Fig. 6.4). The micro-structure is also shown for the range of compositions below the phase diagram. (b) An example of a phase diagram showing **restricted solid solution**. The structure of the two metals is important as the shaded region is represented as the two phase region (a mixture of α (f.c.c. structure) and β (b.c.c. structure). This results because the structures of the two metals are different and each alloy element wants its own structure, i.e. they are insoluble in each other. The micro-structures are also shown for specific regions below the phase diagram. In this case single phase and duplex (two-phase) structures are illustrated. The lines on the single-phase structures are grain boundaries (Section 6.21; Fig. 6.19). The parallel lines on the duplex structures are the α (white) and the β (black) lamellae.

in Fig. 6.21. In this case the two metals 'fight' over the structure to be achieved by this solution of zinc in copper. The result is that neither wins except at low concentrations of solid solutions, and a range of other competing crystal structures are produced which are dependent on the composition.

These single structural unit areas are shown by Greek characters and are referred to as **single-phase** regions. If the composition is not exactly that required to form these single-phase structural units, then there exists a **two-phase** region, rather like the two phases of oil and water which are immiscible. These two-phase regions are designated $\alpha + \beta$, $\beta + \gamma$ etc. and are very important because these second phases act as 'obstacles' or 'blockers' to the movement of crystal planes relative to each other. This relative movement is an important characteristic of metals and metallic deformation (Fig. 6.24). Other combinations of metals exhibit the simplest restricted solid solution phase diagram which is shown in Fig. 6.20(b) which also shows a

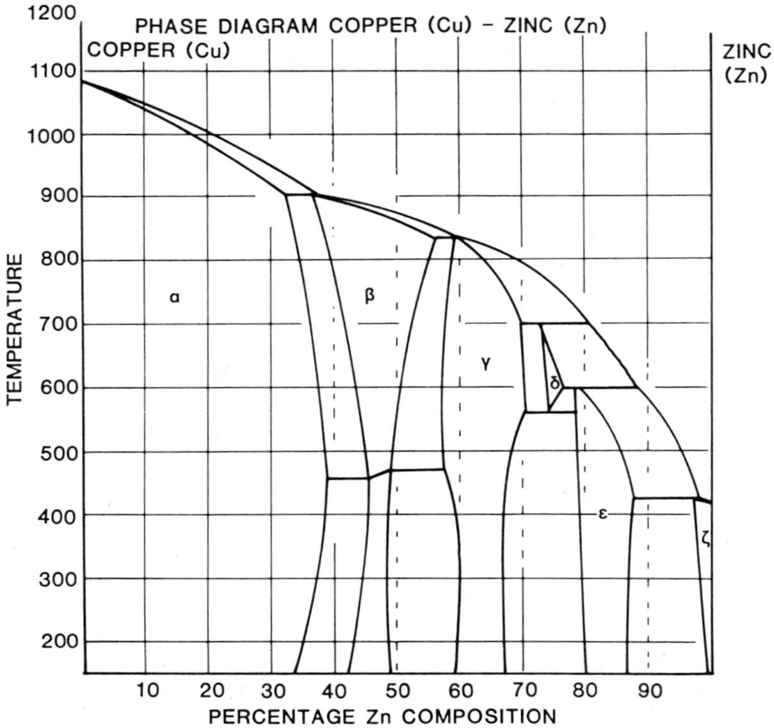

Fig. 6.21 — Brass is an example of a phase diagram showing restricted solid solution. The structures of the two metals copper (Cu) and zinc (Zn) are different (f.c.c. and h.c.p. respectively) (Note: this is only a difference in packing sequence (Fig. 6.2(d) and 6.5(c)). The single-phase structure regions are termed α and β. Compositions between these single-phase regions are two-phase regions. Note that there are several single phases throughout the compositional changes.

simplified artists impression of the micro-structure which accompanies each region of the phase diagram.

6.21 GRAIN BOUNDARIES, DEFORMATION AND DISLOCATIONS

Metals consist of regions of closely packed planes of atoms which all have a similar orientation to one another (Figs 6.2, 6.5 and 6.6). These are called **grains**, (Fig. 6.19). If the metal were all one orientation, such that the grain constituted the whole of the metal, it would be called a **single crystal** and as such would be very expensive as single crystals are rare and very difficult to produce. Normally, engineering metals consist of differently oriented regions or grains which are called **polycrystalline** because they consist of many grains, each having a different crystalline orientation. At the boundary between each of these grains there exists a defect called a **grain boundary**, (Figs 6.19(a)–6.19(c)). Metals deform in an elastic (recoverable) mode which gives them 'springiness' and by a permanent mode which allows them to be rolled and pressed to make complex shapes such as the panel shapes of the motor vehicle body. This is called plastic deformation. The grain boundaries act as an important obstacle to the continued deformation process (Fig. 6.24(a)).

The process of deformation is modelled as the movements of line defects called **dislocations** which cannot cross the grain boundary. These **dislocations are the boundary between slipped areas of a crystal and unslipped regions**. It is a one dimensional defect and is usually modelled in most textbooks as shown in (Fig. 6.22).

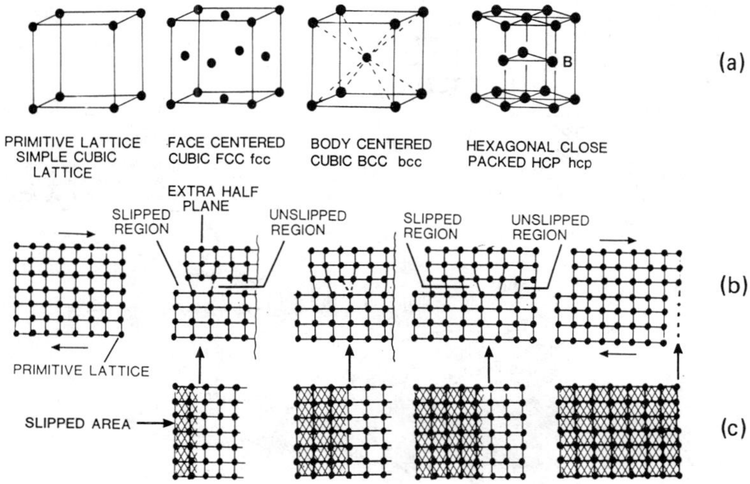

Fig. 6.22 — (a) The four possible structures of a metal lattice are primitive (P), (f.c.c.), (b.c.c.) and (h.c.p.). No metals have the primitive lattice structure, but the occurrence of the dislocation is always portrayed within this primitive lattice. (b) The edge dislocation in a primitive lattice as viewed from the side, which therefore makes students believe that the dislocation is a two-dimensional defect. The 'extra row' of atoms alters as the deformation passes across the crystal. (c) If we draw a 'bird's eye' view of the deformation, then the shaded area is the area where deformation has taken place. The **dislocation** should be defined as the boundary between slipped and unslipped regions, and therefore is a **linear defect** (one dimensional defect). Note. The extra half plane concept gives a two dimensional concept to a dislocation. A grain boundary is the **only** 2D defect known.

The slipped area is often referred to as the **slip plane**, and the magnitude of the deformation is called the **Burgers vector** (Fig. 6.23). (Fig. 6.23 shows how the Burgers vector is obtained and can be omitted on a first time reading as it is conceptually difficult to a non-scientist). For the characteristic deformation of an externally applied stress throughout a polycrystalline metal to continue, the initiation of simultaneous slip within several slip planes in the adjacent grains is required (Fig. 6.24). Hence, the ability of a metal with a smaller grain size, to withstand slip is greater as a result of the condition that several adjacent grains must also deform to allow just one grain to deform. The reason for this is as follows: it is not always possible to have this simultaneous slip, by dislocations, in so many slip planes at one time. Fig. 6.24 shows an artist's impression of a grain without any slip within it because of this fact.

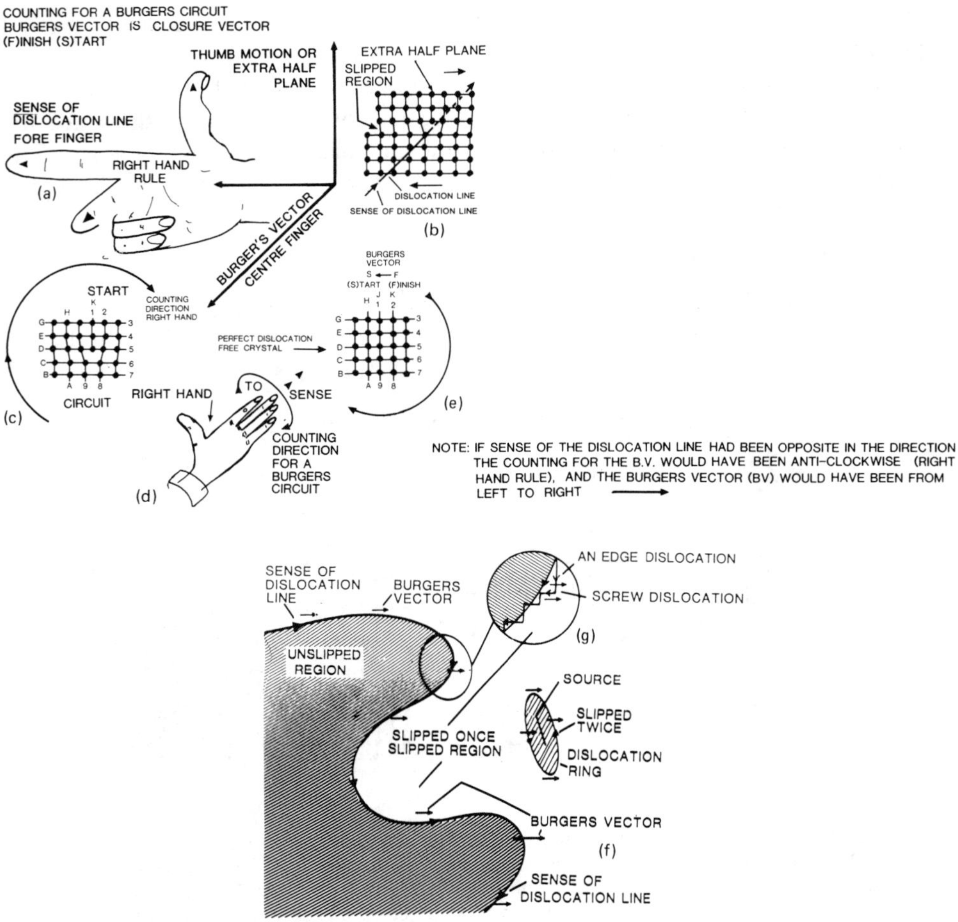

Fig. 6.23 — **The dislocation line, the boundary between slip and unslipped crystal**, is given a Burger's vector **b** and a sense rather like scientists give magnetic fields a sense. (a) The right hand is orientated as shown for edge dislocations so that each finger has a specific task. (b) The thumb represents the motion of the crystal due to the movement of the dislocation line, the centre finger, like the current in a conductor, is the sense of the dislocation line and the forefinger represents the Burger's vector. (c) The Burger's vector is determined by counting a **Burger's circuit** within the material containing the dislocation. (d) The direction of counting is specified by the right hand arranged with the fingers along the sense of the dislocation and the right hand is rotated clockwise. (e) The same Burger's circuit is then carried out in an ideal dislocation-free lattice. The **closure vector** FS between the finish and start (FS) is the **dislocation burger's vector** (BV). (f) A dislocation ring has a continuous sense, and the Burger's Vector represents the displacement of the whole crystal above the slip plane. From the diagram (g) there are several areas where the Burger's vector is parallel to the sense and other regions where it is at right angles to the sensed line. These are the only two conditions which can be modelled. These correspond to an edge (Figs. 6.22 and 6.25) and screw (Fig 6.26) dislocations. Any other condition is modelled as a rapid switch from regions where the Burger's vector is parallel to the sensed dislocation line to where the Burger's vector is perpendicular to the sensed line as shown in (f) for a dislocation ring spreading out from the centre of the page towards bottom left. Deformation towards top right has been omitted for clarity. The general rules for edges and screw dislocations are as follows: (1) an **edge** is with the Burger's vector at right-angles to the sense; (2) a **right-hand screw** is with the Burger's vector parallel to and in the same direction as the sense (shown in Fig. 6.26); (3) a **left hand screw** is with the Burger's vector parallel and in the opposite direction to the sense.

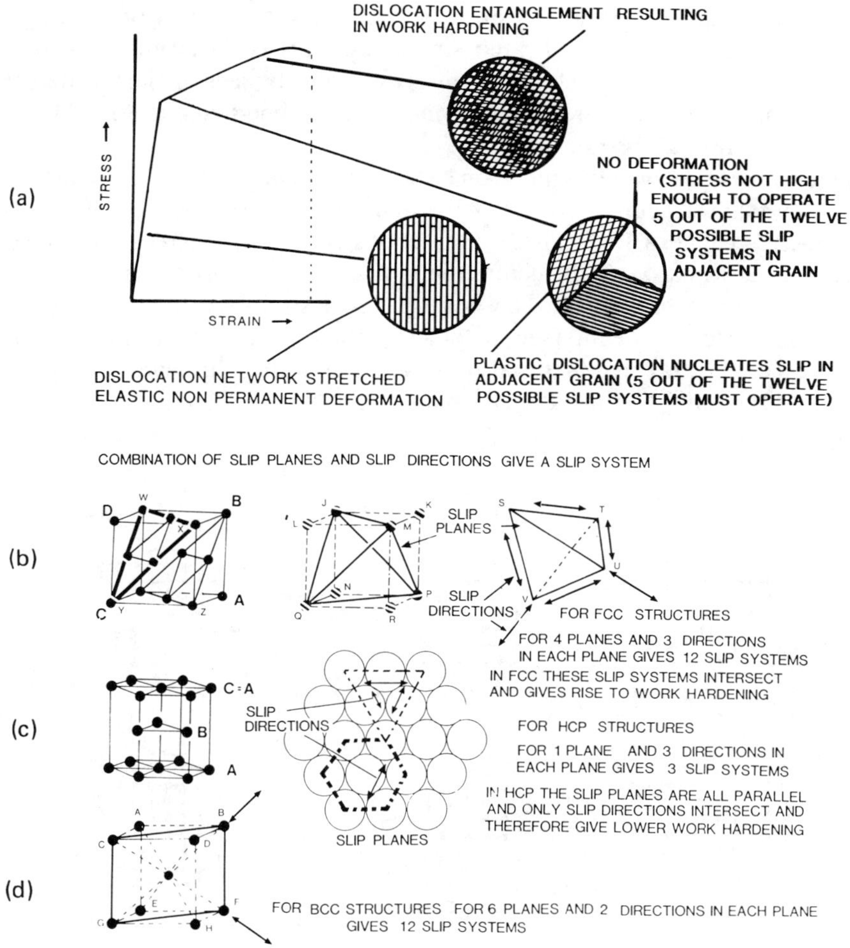

Fig. 6.24 — (a) Stress–strain curve showing the elastic and plastic regions of the metal. The diagram at the side shows the dislocation entanglement for each section of the stress–strain curve. As the stress increases, so does the degree of entanglement. A slip system is the combination of the slip plane (usually the most densely or close-packed plane) and the slip direction (Burger's vector). The slip planes and slip directions are shown as follows: b) for the f.c.c. structure, (c) for the h.c.p. structure and (d) for the b.c.c. structure. It should be pointed out that, in the modelling, only the slip systems which interact, i.e. intersect each other, can cause work hardening. Thus the hexagonal system has 12 slip systems, but there are no intersecting planes, because the slip planes are all parallel to each other. Thus only dislocations with different Burger's vectors can cause work (strain) hardening by intersecting on the same plane.

Conventionally the dislocation is drawn 'as an extra half plane' in simple three dimensional lattices which are not representative of the metallic structure. (Nature's metallic structure is either f.c.c., h.c.p. or b.c.c., while the simple lattice used in textbooks to draw the dislocation is a **primitive** cubic structure with atoms at the eight corners of the cube). There are **no** engineering metals with the primitive structure.

However this has to be scientifically excused owing to the difficulty of drawing the dislocation in these more complicated structures. This is an example of artistic licence in scientific modelling. This problem is overcome by **sensing** the line defect in the more complicated structures by applying the **right-hand rule** (Fig. 6.23(a)) to determine the extra half-plane.

In general these dislocations move on close-packed planes (Fig. 6.2(a) and 6.24) within the metallic structure and interact with each other to produce a dislocation 'entanglement' (called a **jog**). These interactions require an increase in stress to continue the deformation. This increase in stress as a result of the interactions is referred to as **strain hardening**. The occurrence of strain hardening is structure dependent as a result of the intersection of a slip plane with another slip plane (Fig. 6.24(b)–6.24(d)). Hexagonal materials have **no** intersecting slip planes, only intersecting slip directions, because the slip planes are all parallel to each other (Fig. 6.24(c)).

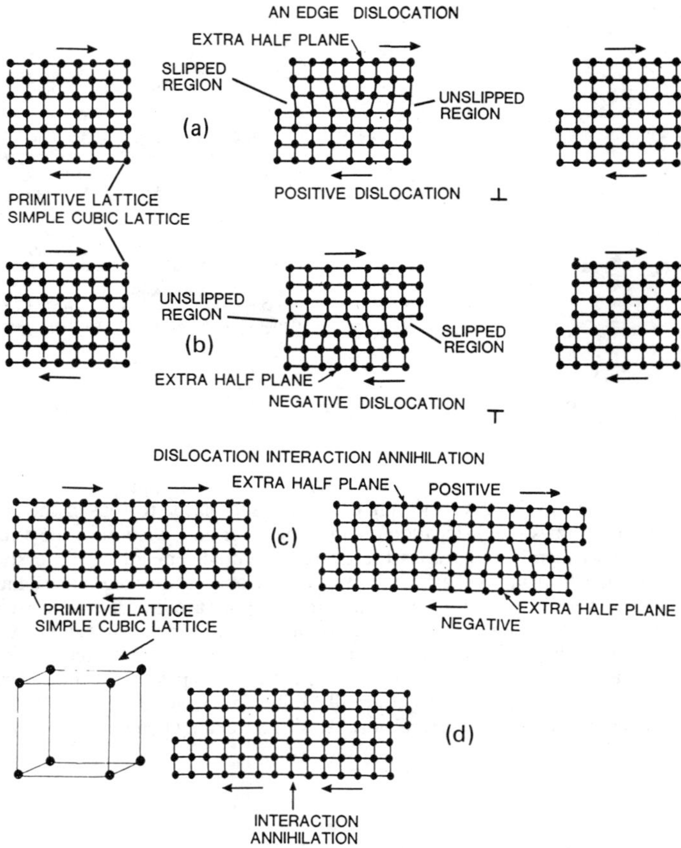

Fig. 6.25 — The (a) positive and (b) negative edge dislocations are shown in a primitive lattice. The passage of these dislocations gives the same result when the dislocation has passed through the crystal. (c), (d) If these dislocations are on the same slip plane they interact and annihilate each other. In metallic structures such as the f.c.c. and h.c.p. lattices, they are modelled on the direction of their Burger's vector with respect to the sensed dislocation line.

There are two types of dislocation which the scientist models for strain hardening. These are the **edge** and **screw dislocations**. These two dislocations are further subdivided. The edge dislocation is divided into **positive** and **negative** edge dislocations (Fig. 6.25) and the screw dislocation into **right-** and **left-handed** screw dislocations (Fig. 6.26). These dislocations are characterized by their sense and Burgers vector, as shown in Fig. 6.23 for the edge dislocation and Fig. 6.26 for the screw dislocation (both in primitive lattices).

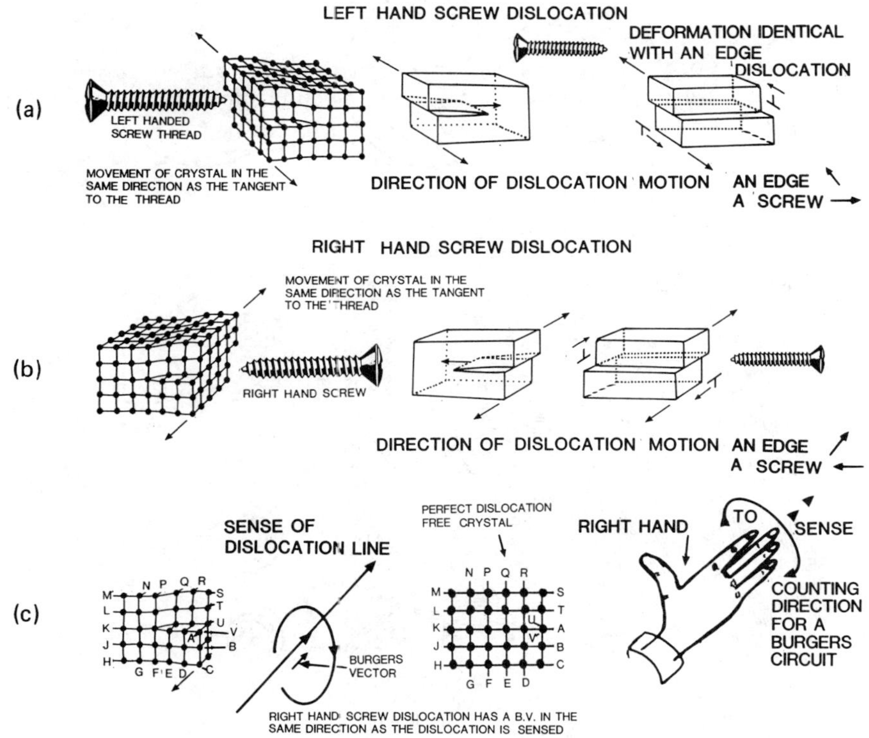

Fig. 6.26 — The two types of screw dislocation (left and right hand) are shown in a primitive lattice. (a) The **left-handed screw dislocation** gets its name from the fact that the left hand outstretched as shown in Fig. 6.23(e) would deform the crystal as follows: when the fingers push the dislocation line into undeformed material, the crystal would move as if the thumb was fixed to the crystal lattice, above and below the slip plane. (b) The **right-handed screw dislocation** gets its name from the fact that the right hand outstretched as shown in Fig. 6.23(d) would deform the crystal as follows; the fingers push the dislocation line into undeformed material, and the rotation of the right hand must be in the same direction as the corresponding engineering screw thread. The crystal would move as if the thumb was fixed to the crystal lattice, above and below the slip plane. Engineers who are more conversant with left- and right-hand screw threads (as illustrated) would appreciate that the crystal material moves at a tangent to the screw thread when the left- or right-handed screw thread is screwed into the crystal to take the dislocation line into undeformed material. (c) The Burger's vector is worked out for the right-handed screw only. Once again this is achieved by counting around the Burger's circuit (within the dislocated material) as defined by the outstretched right hand, and comparing this with the same circuit carried out in a dislocation free material. The Burger's vector is the closure vector FS in the ideal perfect dislocation-free crystal. In metallic structures such as the f.c.c. and h.c.p. lattices, dislocations are modelled on the direction of their Burger's vector with respect to the sensed dislocation line.

6.22 DISLOCATION INTERACTIONS

Dislocation interactions are best modelled by considering a trapped dislocation line anchored between two points A and B as shown in Fig. 6.27.

The deformation of the metallic lattice is modelled on the position of the dislocation line. Hence, if the deformation moves the dislocation out into unslipped

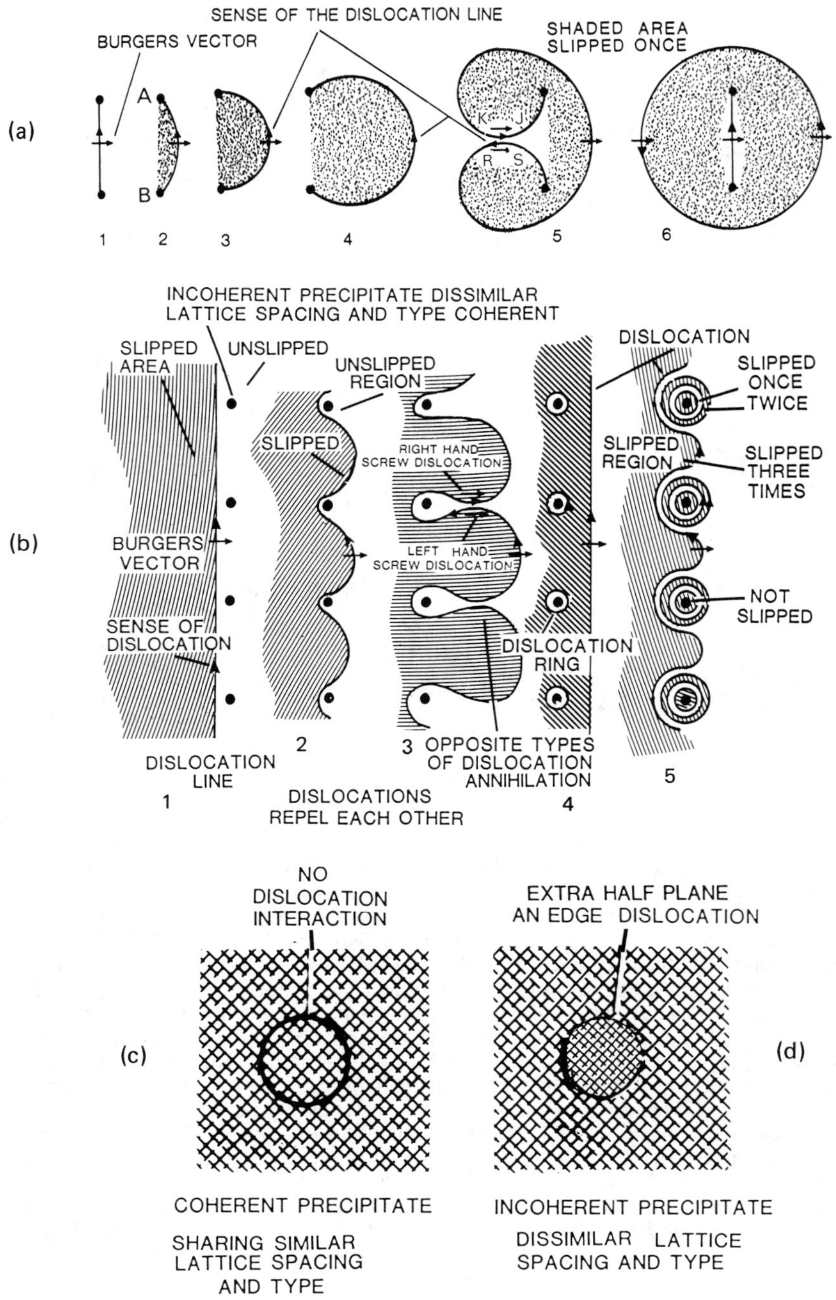

crystal, the length of the dislocation line increases. This is modelled as requiring a higher stress, rather like the stretching of a bow into a curve when the arrow is pulled back. The stress required is proportional to the reciprocal of the radius into which the dislocation line is curved. Fig. 6.27(a) shows the dislocation line at various stages of applied stress. Eventually the dislocation loops around the two anchoring points and produces a circle of deformation within the non deformed crystal. Slip can take place from here and move out and the second shaded area (Fig. 6.27(a), stage 6) becomes the area which has slipped twice. This is a Frank–Read dislocation source and was originally modelled to show why a crystal did not lose any dislocations from the crystal (note that in Fig. 6.27(a), stage 6), the single dislocation line shown in Fig. 6.27(a), stage 1, has now become two dislocations of the same sense and Burger's vector). If the crystal lost dislocations, then both the dislocation density and the probability of dislocation entanglement of the dislocation lines would decrease. On this modelling of the metal losing dislocations, the metal crystal would decrease in strength as a result of the loss of dislocations which would not be available for mutual interaction. The total linear length of dislocation lines would therefore decrease as the strained metal lost dislocations from the surface (Fig. 6.25(a)). This model can be extended to show how dislocations interact with coherent and non-coherent second-phase particles (Figs 6.27(c) and 6.27(d)). The interaction is shown in Fig. 6.27(b).

◀ Fig. 6.27 — The operation of a Frank–Read source (FRS). Originally this concept was laughed at by the scientific fraternity, and the standard joke about an FRS was that it was modelling 'beer froth' spilt on the bar. However, an FRS has been subsequently observed in alloys and now forms an important part of the modelling of dislocations. (a) Some sequential steps in the deformation process. Stage 1: a very high stress (σ) is required to make a straight dislocation bend into a curve of radius r. Mathematically it is expressed as $\sigma = k/r$, where k is a constant. The stress is always increasing until the radius of 'slip' is equal to the distance between the two anchor points, when the stress will decrease. (The same phenomenon is observed in blowing up rubber party balloons: they are difficult at first but when the radius becomes large, the increase in 'puff pressure' is smaller. The excess pressure is $4T/r$, where T is the 'elasticity' of the rubber. Stages 1–5: the bowing out of the dislocation line with the appropriate dislocation sense and Burger's vector (sense and Burger's vector described in Fig. 6.22–6.26). The two dislocations at KJ and RS are opposite types: KJ is right handed and RS is left handed (see rule in Figs. 6.25 and 6.26). Therefore dislocations KJ and RS will annihilate each other, forming a ring (stage 6), thus making a ring around the two anchor points and the original dislocation line is retained. (Compare this with the exit from the crystal of the dislocation shown in Fig. 6.25). The nearest analogy to this is the blowing of soap bubbles using a soap film suspended within a cicrular ring.) (b) This principal can be applied to a dislocation line meeting a row of precipitates or second phases provided that we model the second phases as 'non-deformable'. The scientist's interpretation of this is shown in (c) and (d), where there are two second phases: Fig. (c) the second phase has with the same crystal structure as the parent metal (termed a **coherent** phase); (d) the crystal structure and lattice spacing are different from those of the host metal (termed an **incoherent** phase). The model of the incoherent phase is that the crystal boundary is surrounded by a dislocation ring. This dislocation ring (or sphere as it surrounds the three-dimensional precipitate) cannot be broken into by an oncoming dislocation. Thus, in the sequence in (b), the dislocation is pushed through the randomly dispersed second phases and by the FRS mechanism given in (a), stages 1–6, leaves a dislocation ring around the second-phase incoherent phase. Subsequent dislocations arrive at the second-phase particle to find the distance between the particles reduced (by the new dislocation ring) and therefore a higher stress is required to push the new dislocation arrival through the smaller gap. Hence the second-phase precipitate is surrounded with new dislocation rings, which separate areas of the crystal which have slipped once, twice, three times etc. With each new dislocation which is forced through, the gap between the second phase particles becomes smaller, and the stress required for continued deformation becomes higher.

6.23 STRAIN HARDENING, RECOVERY, RECRYSTALLIZATION, AND ANNEALING

Strain hardening is the direct result of a dislocation source's requiring a higher stress to operate. This arises from the stress fields around the dislocation lines. These stress fields can be seen most easily in the case of the edge dislocation (Fig. 6.22) where the extra half-plane region is in compression. The region below the extra half-plane, is in tension(as shown by the inter-atomic distances). The complex interaction of dislocation lines places a back stress on the dislocation sources. This back-stress increases with both dislocation density (number of lines per unit volume) and the degree of dislocation entanglement. Thus the dislocation source, (as shown in Fig. 6.27(a), stage 1), must overcome the increasing back-stress for the dislocation sources to continue to operate and thus to produce further dislocations.

6.23.1 Hardness

Scientists often want to follow the increase in mechanical strength without the inconvenience of putting the metal in a stress–strain situation. Therefore the point of a diamond pyramid is used to make an indentation within the surface of the metal. The area supporting the load increases as the pyramid is forced into the metal surface. The imprint which the diamond pyramid makes under known loads gives a quantity termed the hardness. The hardness is used to determine the yield stress of the metal. There is a relationship between hardness and yield stress. There are several methods of determining the hardness. One of these is the **Vickers'** hardness number.

The hardness number is roughly 0.3 x σ_y where σ_y is the yield stress in meganewtons per square metre. Hardness is therefore a technical term in metallurgy and is used to follow the processes of strain hardening, recovery and recrystallization. A brittle material, although hard, will shatter under the loading of the diamond pyramid. These materials have another form of hardness based on the ability of one to scratch the other abrasively. A series of materials are arranged and given arbitrary hardness numbers from 10 (for diamond) down to 1 (for talc) and this is called the **Mohs** hardness scale (Glass is about 5.5 on this scale). These two forms of hardness should not be confused.

6.23.2 Recovery

Heating of a crystal above a certain temperature (0.3 T_m) where T_m is the melting point in kelvins (on the kelvin scale water freezes at 273 K)) produces a process known as recovery. In Recovery the dislocation entanglements take up a lower energy state. The dislocation entanglement therefore decreases and becomes 'coarser' as shown in Fig. 6.29, while the strength and hardness remain unaffected. The internal stress (as measured by such techniques as X-ray line broadening) are reduced. The grain size does not alter at all within the recovery process.

6.23.3 Recrystallization

Heating of a crystal above a certain temperature (0.5 T_m) (called the **recrystallization temperature**) causes a further reduction in this dislocation entanglement (or tangles) but more importantly an alteration in the grain boundary angles. Therefore the alloy 'softens' and its hardness decreases as atomic diffusion is now possible. The structure

recrystallizes using the intersection of three grains (triple point) as the nucleus for grain growth. Once recrystallization starts, (Fig. 6.28(b)), grain growth of this small new grain proceeds, its energy for growth being obtained by the reduction in total energy which results from the formation of a nearly dislocation-free new grain. Both the strength and the hardness decrease during this process because there are no obstacles to dislocation movement (Fig. 6.29).

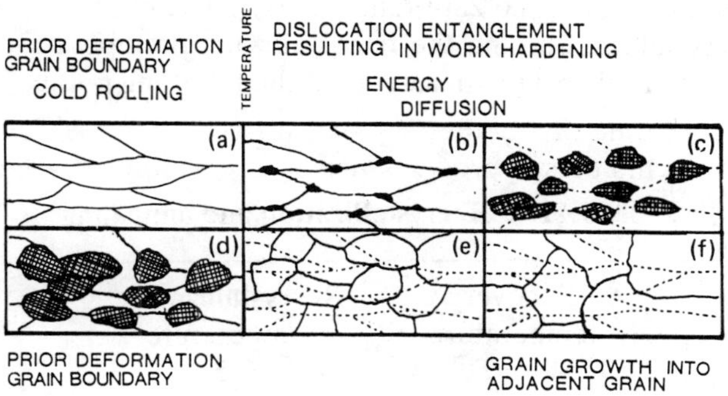

Fig. 6.28 — A schematic view of the micro-structure of a metal through various stages in (a)–(d) the recrystallization process and (d)–(f) grain growth due to the stresses around a dislocation. As the grain size increases, so the distance that the dislocations can travel before they are arrested at the grain boundaries increases. Thus a large-grain-sized metal is mechanically weaker than a fine-grained metal (Fig. 6.19). Both hot and cold rolling produce an elongation of the metal (and therefore of the internal structure) in the direction of rolling. The metal cannot slip sideways in the rolls owing to the high friction between metal and rolls and the high roll pressures.

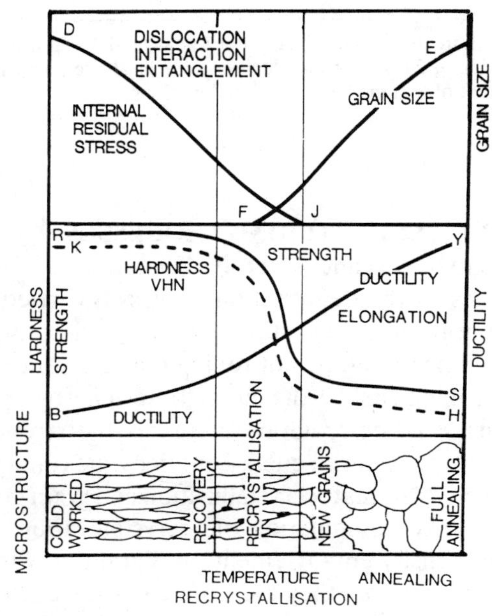

Fig. 6.29 — As the temperature of a work hardened metal is raised, several of the material parameters alter. This figure shows the decrease in dislocation entanglement, strength and hardness. The grain size, ductility and elongation to fracture increase with temperature.

6.23.4 Annealing

If the alloy is heated above this recrystallization temperature, a process of **annealing** takes place. Annealing is a process which markedly continues to softens the alloy and allows an increase in the rate of grain growth. These processes allow a complete redistribution of dislocations such that some of the grains grow bigger while the stresses necessary for deformation to take place become less. Annealing therefore completely obliterates the prior history of the metal (except where insoluble phases are deposited as shown in Figs 6.28(e) and 6.28(f)). The process of annealing is an extension of recrystallization, and the recrystallization temperatures for common metals are shown in Table 6.11 with respect to their melting points.

Table 6.11 — Recrystallization and annealing

Metal	Recovery temperature $\approx 0.3T_m$	Recrystallization temperature $\approx 0.5T_m$	Melting point (°K)
Tungsten	1200°C		3646
Nickel	600°C	700–900°C	1728
Pure iron	450°C	550–650°C	1808
Copper	190°C	350–650°C	1356
Aluminium	150°C	350–425°C	933
Zinc	20°C		692
Lead	<0°C		600

The recrystallization temperature of the alloy usually starts at around $0.5T_m$ where T_m (K) is the absolute temperature (degrees Celsius +273). If room temperature is above this onset of recrystallisation temperature the metals cannot obtain the benefits of cold working.

6.24 WORKING OR MANUFACTURING PROCESSES

Metals and alloys have to be cast and then rolled or extruded into shape. The energy required to do this is less if the deformation process is carried out at an elevated temperature than at say room temperature. Therefore deformation is easier (and energetically cheaper) if produced by hot rolling than if it were carried out by cold rolling processes. Hot rolling enables the metal to ‘soften’ (by a process of recovery) as the mechanical energy of deformation increases the stress required to cause further deformation. The mechanical energy put in by the rolling process also raises the temperature of the metal so that recrystallization and grain growth can take place during the hot rolling. The combination of recrystallization, which results in a decrease in dislocation density entanglement, and grain growth (which allows the dislocation to travel further before they are stopped by the grain boundary) reduces the mechanical energy required to deform the metal. All metals–alloys have a temperature range where they are best worked ‘hot’. Usually this is related to the phase diagram as the alloy should be in the **single-phase** region. However, this high temperature process produces an inferior surface finish owing to high temperature oxidation (Fig. 6.18 and 8.8(b)–8.8(d); Section 8.5.1). For a better surface finish and a uniform thickness of the steel sheet, which are required for motor vehicle panels, cold rolling processes are employed after hot working. This cold working process is more energy intensive because the rapid increase in dislocation density causes a high degree of work hardening, the metal–alloy usually being rolled below its recrystallization temperature (Figs 6.28 and 6.29 and Table 6.11). Cold working therefore is used as a finishing process in the production of metallic components which have previously been hot rolled. The cold deformation process produces a highly stressed material with elongated grains in the direction of working (Fig. 6.19(c)), which is accompanied by an increase in the yield and tensile strengths, but a decrease in the energy of absorption which is the energy beneath the stress–strain curve (Fig. 6.30).

Motor vehicle steel is categorized by 'CR' numbers which stand for 'cold reduced' followed by a number (e.g. CR 3) which relates to BS1449:Part 1, 1983. Other abbreviations are HR for hot rolled, HT for hardened and tempered, A for annealed and N for normalized. The types of steel used in automobile body manufacture are considered in Section 6.36 and Fig. 6.41.

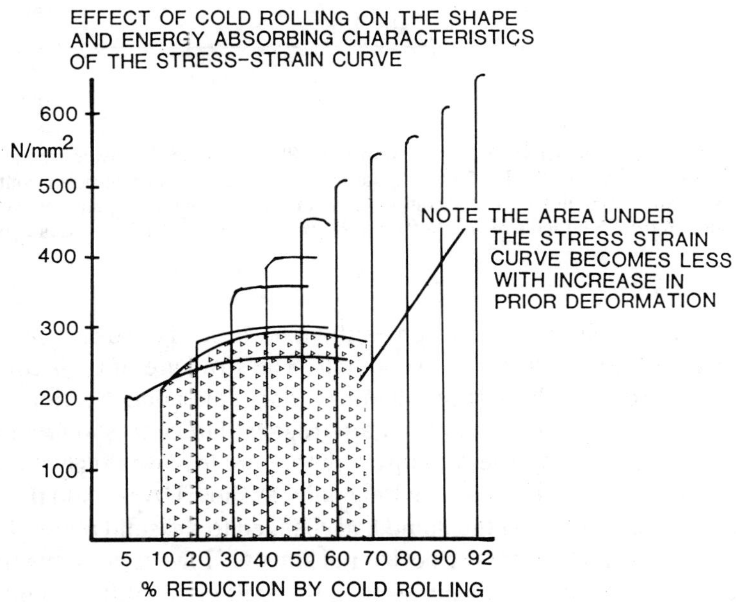

Fig. 6.30 — Effect of continuously rolling steel on the shape of the stress–strain curve. The percentage of cold rolling, determined by the thickness reduction, is plotted against stress and strain (the strain scale is not included, but is the same for each sample). The area beneath the curve is also altered with increase in cold rolling. Hence the energy absorption becomes less as the work hardening introduced by cold deformation increases. (This is applicable to all metals.) (After R. L. Kenton and R. S. Burns, Autographic stress–strain curves of deep drawing sheets, ASST, **21,** (1953) 577.)

6.25 READING PHASE DIAGRAMS

The most important part of studying a metal is to be able to interpret the phase diagram. The phase diagram for solder, an alloy formed from lead (Pb) and tin (Sn), is given in Fig. 6.31(a) together with the cooling curve for composition X.

The phase diagram is of a restricted solid solution type, and shows two solid solution single-phase regions, which are designated as α and β, with a long two-phase region designated α + β between compositions R and S at 100°C. As the temperature decreases, the solubilities of both tin in lead, and lead in tin decrease. (Note: in reading a phase diagram letters have been used. The melting point of pure lead is 327°C which on Fig. 6.31 is represented by the horizontal line uv and so can be further abbreviated as T_u, T_v or T_{uv}. This aids for clarity on the diagram.)

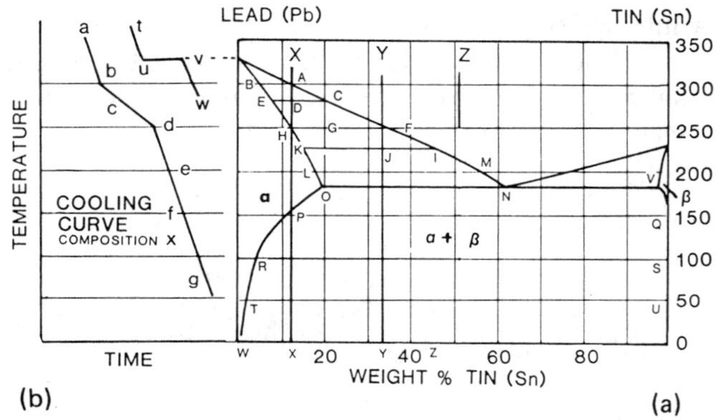

Fig. 6.31 — The restricted solid solution of lead (Pb) and tin (Sn) provides the alloy solder. Solder of composition 50% Pb, 50% Sn is plumber's solder, while an electrician would use 63% Pb, 37% Sn: (a) cooling curve for composition X in (b) the phase diagram. ((a) Redrawn by permission from Smithell's *Metals Reference Book*, Butterworths, London, 5th edn., p. 750.)

The restricted phase diagram comes about because of the difference in both radii (Table 6.4) and structure of the two metals. The structure of tin is complex having two structures, one a cubic lattice rather like that of diamond (Section 6.10, Fig. 6.9) and the other a distorted cube called a **tetragonal** lattice as shown in Fig. 6.32(f). Lead has an f.c.c. lattice. The cooling curve for pure lead would give (Fig. 6.31(b)) a temperature arrest at the metal's melting point as shown by uv, and the temperature would not decrease until all the liquid had changed to the solid phase. It is the **latent heat** evolved which produces the horizontal line uv. However, if some tin is added to the lead, there is first of all a **depression** in the melting point from temperature u to temperature b. The next important point to notice is that the temperature arrest uv has **disappeared**. The solder of composition X now cools slowly while solidifying over the temperature range bcd, and giving out **latent heat** (it can do this because the composition of the solid being formed is altering all the time as we shall see later on). At d the liquid is all solid and so no more latent heat is given out and therefore the rate at which the temperature falls alters. This is shown by the alteration of the slope bcd to defg. The phase diagram is constructed by recording the cooling curves of a

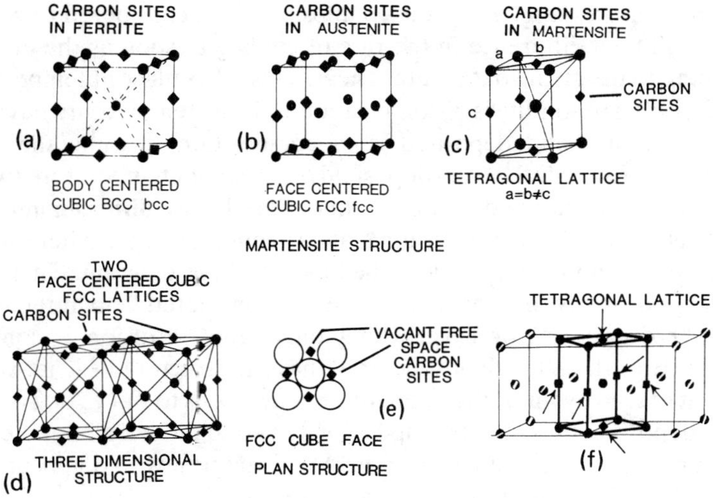

Fig. 6.32 — The solubility of carbon in steel is complex and structure dependent. The carbon goes into the 'free space' between the atoms as shown in (e). This is called **interstitial** solid solution. The interstitial sites are shown for the three structures of iron: (a) b.c.c., (b) f.c.c. and (c) martensite (Section 6.29). (d) The martensite lattice can be considered as two f.c.c. lattices joined together. Each f.c.c. lattice is composed of hard non-deformable spheres packed together which produce a small network of voids. (e) The face of one f.c.c. lattice. These voids are not large enough to take in water, but they are large enough to accommodate carbon (and nitrogen) atoms. In the iron–carbon phase diagram, the iron has two structural units: f.c.c. above 723°C and b.c.c. below 723°C. This diagram will also be used to show the structure of martensite (Section 6.29) where (d) two f.c.c. lattices (the interstitial carbon atoms are depicted, undergo a shear deformation and form a body-centred tetragonal lattice (distorted b.c.c. lattice), outlined in black and shown in (f) with the position of these carbon atoms in the face diagonal and along the edges of the lattice in one preferred direction. It is the carbon atoms along the *c* axis edges which give rise to the distorted lattice. The b.c.c. lattice of ferrite (Figs. 6.33 and 6.34), would have carbon atoms along all edges and none in the face diagonal as shown in (a).

series of alloys of different compositions and placing the points such as b and d on the temperature–composition plot. For the composition considered this gives points A and H. The temperatures at which solidification starts are joined up to give the **liquidus** line ACFIMN. The temperatures at which all the liquid has been changed to solid are joined up to give the **solidus** curve BEHKLO.

In order to examine how the phase diagram may be used to describe the cooling and subsequent solidification of a metal, I shall consider the following: Composition × (12 wt.% Sn, 88 wt.% Pb) will be cooled from the melt to room temperature in specific temperature drops, i.e. from 300° to 250°C, which will written as T_a or T_b for 300°C and T_h, T_g, or T_f for the temperature of 250°C. These happen to be the same temperatures, because the line AB is horizontal (Fig. 6.31). Similarly, we can express the composition by dropping the suffixes onto the composition axis, such that composition X can be described by a, d, h, or p. Thus 20 wt.% Sn is represented by points C or G or O. As capital (upper case) letters have been used on the phase diagram and lower case on the cooling curve, it will be customary to use lower case to refer to temperatures, e.g. T_b. If the composition X is considered to cool from the melt there will be a difference in the rate of cooling as soon as the metal starts to solidify owing to the giving out of latent heat. This takes place at temperature T_a (or T_b). The first solid to solidify is of composition B. If the temperature now drops to T_e (or T_d or T_c), the solid deposited is of composition E while the liquid is of composition C. However the solid deposited at the higher temperature (which was of composition B must alter and become richer in tin. Hence **diffusion** must take place and scientists describe the equilibrium state as being that state where the solid and liquid have the compositions as described by the phase diagram. Therefore **equilibrium conditions** imply that the 'temperature is considered held at the temperature points under discussion for a very long time' until diffusion is complete. The composition E of the alloy deposited at T_e has the structure of lead, with random replacement of lead by tin. If the temperature now drops to T_h (T_g or T_f), all the solid previously deposited must be of composition H and so diffusion has taken place. All the remaining liquid is of composition F. At this point it would appear that we are getting 'extra' tin from somewhere, because now the liquid is richer in tin than when it started. This apparent anomaly is explained by the fact that the solid is LESS rich in tin than the liquid with which we started. The ratio of liquid to solid at temperature T_E is given by

$$\text{liquid:solid} = \text{length of ED:DC} \tag{6.24}$$

This is called the **lever rule**. If E = 8%, D = 12%, C = 20 then the ratio of liquid:solid is

$$\mathbf{(12-8):(20-12)} = 4:8 \text{ or } 1:2 \tag{6.25}$$

At temperatures below T_h (250°C), the liquid has all solidified. The solid cools over the temperature range HP as a single phase, (f.c.c. structure with random replacement of lead atoms by tin atoms). However, at the temperature T_p (150°C), there exists a problem as the decrease in temperature is showing up the lattice or crystal distortions due to the different radii of lead Pb and tin Sn, and their different structures. Below this temperature T_p, the amount of tin which can exist in the lead's f.c.c. structure decreases, and so the α phase follows the composition line PRT. The excess tin is thrown out, not as pure tin, but as the β phase, (composition Q) that is the tin's structure with random replacement of tin by lead. The β phase must follow the composition curve QSU and so throws out lead not as pure lead, but as the α phase (composition R, at temperature T_r), that is lead's structure with random replacement of lead by tin. Therefore there exists a two-phase structure.

The cooling curve for composition Y is not shown. A discussion is given of the cooling curve and transformation processes, and it is suggested that a pencil cooling curve is made adjacent to the phase diagram on the tin side for this cooling curve. If the composition Y is considered to cool from the melt there will be a difference in the rate of cooling as soon as the metal starts to solidify owing to the giving out of latent heat. This takes place at temperature T_f. The first solid to solidify is of composition H. If the temperature now drops to T_k, the solid deposited is of composition K while the liquid is of composition I. However, the solid deposited at the higher temperature (which was of composition H) must alter and become richer in tin. Hence **diffusion** must take place. The composition K of the alloy deposited at T_k has lead's structure with random replacement of lead by tin. The liquid therefore follows the liquidus line FIMN, and the solid follows the solidus curve HKLO. If the temperature now drops to T_o, the solid deposited is of composition O while the liquid is of composition N, and there is a **temperature arrest** at this point (this is equivalent to uv for the pure lead). This composition (N) (60 wt.% Sn) of liquid is given a special name: **eutectic**. The eutectic liquid solidifies as follows:

$$\underset{\text{composition}}{\text{liquid}} \rightarrow \underset{\text{composition O}}{\text{solid } \alpha} + \underset{\text{composition V}}{\text{solid } \beta} \qquad (6.26)$$

However, the solid deposited at the higher temperature (which was of composition (L) must alter and become richer in tin. Hence diffusion must take place. The composition (O) of the alloy deposited at T_o has lead's structure with random replacement of lead by tin.

The solid at temperature T_o is a complex structure of the α and β phases which cools over the temperature range from T_o to room temperature by following the composition line OPRT. The excess tin is thrown out , not as pure tin, but as the β phase (composition Q) that is tin's structure with random replacement of tin by lead. The β phase must follow the composition curve VQSU and so throws out lead not as pure lead, but as the α phase, (composition R, T, and W: R > T > W) that is lead's structure with random replacement of lead by tin and thus follows the line OPRT.

Therefore there exists a two-phase structure as shown in the circles representing the structure on the micro-scale in Fig. 6.20. The second phases are the '**insoluble**' phases in the host matrix (in this case the α phase), which act as agents which increase the strength of the alloy as depicted in Fig. 6.27(d) by forming second-phase solids which do not have the structure (or atomic spacing) of the host metallic lattice (in this case the α phase). Note: the host phase would be the β phase if the composition of the liquid was to the right of N, i.e. a 80 wt.% to Sn–20 wt.% Pb will have a β phase host matrix.

6.25.1 Precipitation hardening alloys

The principle that a second phase, with a **different lattice spacing and structure**, is deposited within the host matrix solid is made use of in the strengthening of aluminium alloys. Aluminium alloys are being used in the automotive industry in experimental energy conservation vehicles. Aluminium alloys have been used by Land Rover for some time for non load-bearing-panels. When scientist's understood the phase diagram and the modelling of strength, the search was on for an element which dissolved in aluminium at high temperatures but was nearly insoluble in aluminium at room temperatures. Thus the left-hand side of the phase diagram depicted in Fig. 6.20 is the type of phase diagram required, but the right-hand side would not give the hardening requirements. The element copper (Cu) has this solubility property. The maximum amount of copper allowed to form a good precipitation hardening alloy is 4.5%. This adds to the complications of manufacture. In order to explain why, the principles of precipitation-hardening alloys will first be described. The first important point to appreciate is that the hardening process is not undertaken until all the rolling and forming processes have taken place. Therefore the hardening processes are undertaken with metal in the form of sheets. If the alloy is heated up to obtain the total amount of copper in solid solution, with no copper rich phase remaining as a second phase, a temperature is required which takes the alloy very near to the solid plus liquid phases. Hence if the temperature control of the furnace is not very good, or there is an uneven temperature distribution, the thin panels could melt. Therefore lower alloy contents are used for safety. Aluminium alloy with 2–4% Cu is often used as this now gives a much larger temperature range into which the sheet can be heated in order to obtain all the copper in solid solution, ie the sheet is all in the α phase. From this temperature, the sheet is quenched. (Quenching is the rapid cooling of the solid solution into water so that the copper does not have a chance to precipitate out and equilibrium conditions are not achieved. The quenching process therefore freezes-in the metallic structure of the higher temperatures.) The sheets can at this time be rolled in order to increase the dislocation density. Any shaping or cutting is then performed and the sheet is then aged. This is a process whereby the copper is deposited as the 'intermetallic' phase $CuAl_2$ by heating it to a temperature so that the alloy is still in the two phase region. The nuclei for this deposition are the dislocation entanglements (if there were some undissolved $CuAl_2$ particles prior to quenching, these would act as the nuclei and the full benefit of the precipitation-hardening process would not be obtained). The precipitation process is followed by modification of the alloys hardness (Section 6.23.1) and when the maximum hardness is achieved the sheet is cooled to room temperature. The materials technologist tries to get the second-phase particles to the stage where they are as small as possible and incoherent but distributed uniformly throughout the host matrix (Fig. 6.27(d)). These second-phase incoherent particles act as obstacles to dislocation and each dislocation which passes them creates a **dislocation ring** around the precipitate, making it appear bigger to the next dislocation. The spacing between particles thus decreases as more dislocations pass and so the material rapidly achieves good tensile strengths with only a small percentage of alloy element.

6.26 IRON–CARBON MILD STEEL

Steel is a very popular engineering material. Its popularity arises from the very wide range of strengths which can be introduced simply by alloying and by its heat treatment (Table 6.12(a); see Fig. 6.36).

In Section 6.20 it was a general rule that substitutional alloys were formed by substitution of the solvent atom by the solute atom. This was termed substitution solid solution and there were certain size considerations imposed for complete solid solution. In the iron–carbon system there is another form of solution because of the very small radius of the carbon atom (Table 6.4). This is termed **interstitial** solution hardening. The carbon atom is only 60% of the size of the iron atom and so is capable of filling the vacant sites between the atoms which are considered as hard spheres (Fig. 6.32).

Alloy steels, in contrast, such as 18/8 stainless steel (18% Cr, 8% Ni) are formed by substitution with the nickel (Ni) and chromium (Cr) replacing some of the iron atoms in the lattice structure. The carbon additions, however, are always interstitial, but their major contribution to the hardening and strengthening process is by the formation of **insoluble carbides**. These insoluble carbides are formed with both iron and the alloy additions. Hence the iron–carbon system is rather a misnomer, as it is the interaction of the iron with the relevant carbide (as opposed to carbon). The iron carbide normally used is called **cementite** (Fe_3C). The iron–carbon phase diagrams are given in Figs 6.33 and 6.34.

Iron is used in several forms. It is perhaps best to start the introduction to the iron–carbon (graphite or cementite) system by considering the various terms which exist in the ferrous alloys. All alloys which contain a large percentage of iron in their make up are called **ferrous** alloys while other alloys with little iron in their composition are termed **non-ferrous**. Hence brass is a non-ferrous alloy. Ferrous materials are the following.

6.26.1 Mild steel

Mild steel is iron with up to 2% C added which occupies the interstitial sites in the lattice structure Figure 6.32. The function of the carbon is to improve the mechanical properties of the steel by forming an insoluble second phase with the iron. These are known as carbides, and they act as obstacles to dislocations. The heat treatment of mild steel will be considered in Section 6.28 *et seq*. Mild steel is used for motor vehicle panels and the various types are considered in Sections 6.30 and 6.36.

6.26.2 Wrought iron

Wrought iron is almost pure iron with less than 0.03% C. It is used for ornamental work (e.g. scrolls) mainly and is found as garden gates. Wrought alloys are those which have a high melting point and do not flow well into moulds and are therefore

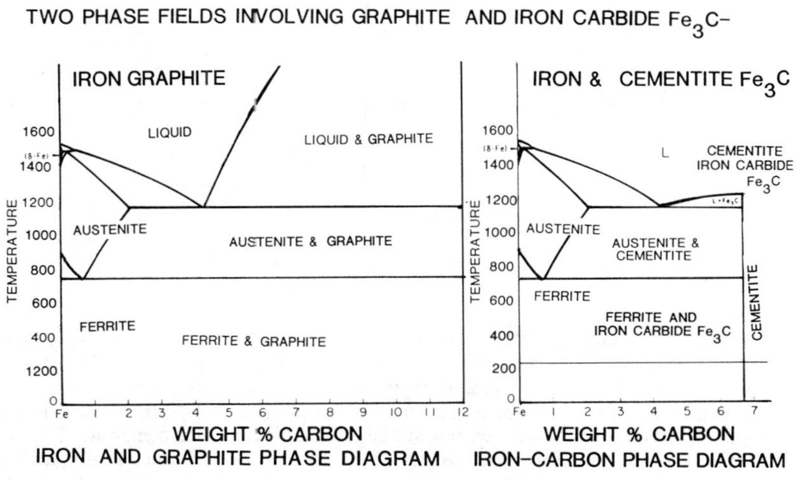

Fig. 6.33 — The iron–carbon phase diagram is complicated by the different structures which can be given to the carbon-rich side. Carbon can exist as the carbide cementite (Fe_3C) or as graphite. (Redrawn by permission from the *Metals Handbook*, Vol. 8, American Society for Metals, Metals Park, OH, 8th edn., p. 277.)

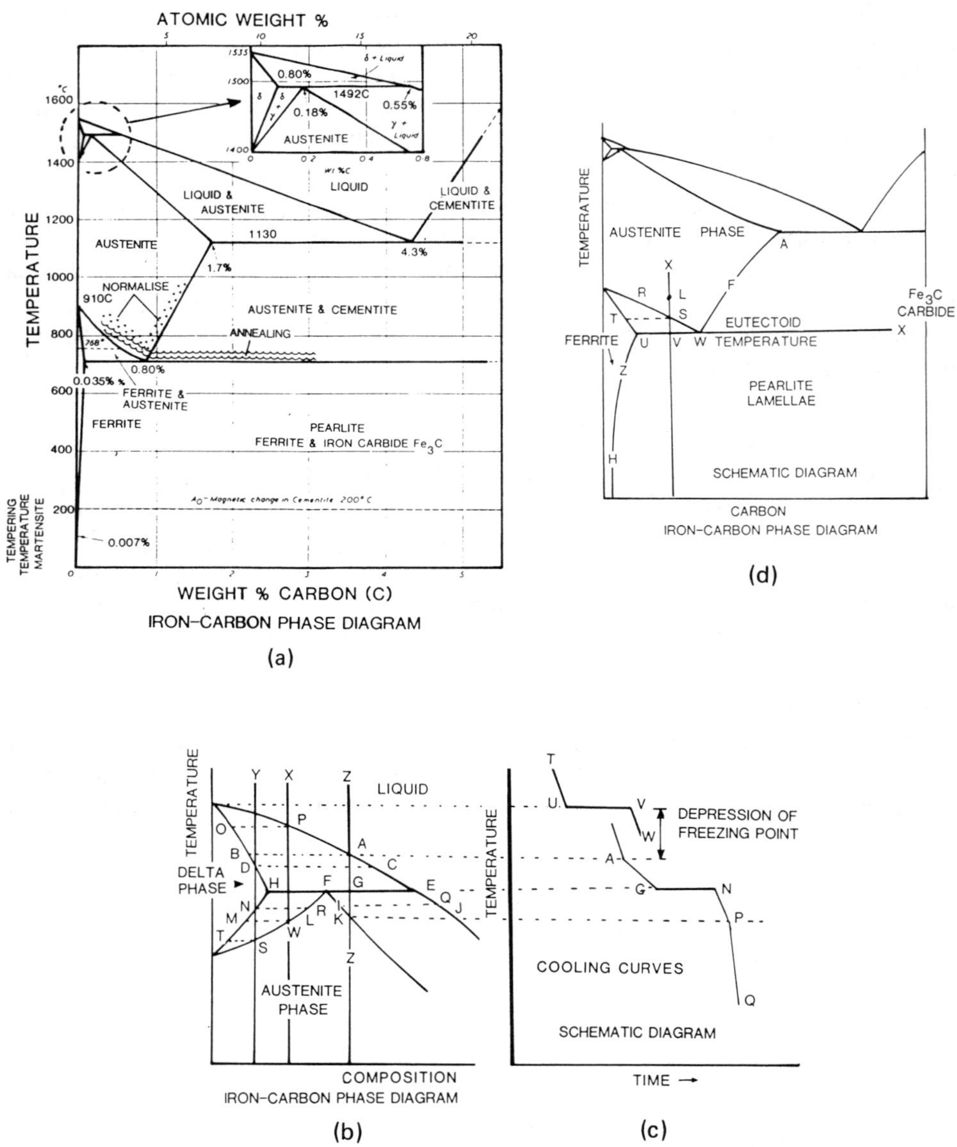

Fig. 6.34 — The iron–carbon system has been extensively studied. Each of the single-phases is represented by a Greek letter. Each single phase region is given a name, the main phases being cementite, austenite and ferrite. In these single phases carbon is in 'solution' in the interstitial positions (Fig. 6.32). In the two-phase regions, the carbon is no longer in solution and forms second-phase carbides. In the iron–carbon system the carbide is called cementite (Fe_3C). The mixture of ferrite and cementite phases is given the name pearlite because it resembles mother of pearl after being attacked by certain acids. The phase diagram shows that austenite transforms to pearlite below 723°C. As discussed earlier, therefore, it is less energy consuming to roll steel while it is in the single austenite phase (above 723°C) than when it is in the two-phase Pearlitic structure. In the two-phase state very rapid hardening will take place. (a) Reproduced by permission from Smithell's *Metals Reference Book*, Butterworths, London, 5th edn., p. 510.)

shaped by forging. The forged article is usually stronger than cast iron and does not suffer from brittle fracture (Section 6.31). Couplings on articulated trailers are made of wrought iron.

6.26.3 Cast iron

Cast iron usually has more than 2% C in it. This carbon exists in two forms, as **graphite** giving rise to the term 'grey' cast iron, and as the **carbide** Fe_3C, giving rise to the term 'white' cast iron. The form which the carbon takes is governed by the cooling rate of the cast. Cementite is produced if the cooling rate is fast. The terms white and grey arise from the appearance of the fracture surface. White cast iron is used for car cylinder heads and grey cast iron for piston rings (where the graphite has lubricating properties).

6.26.4 Bright drawn steel

Cold drawing of two-phase carbon steels produces a good finish to the steel as atmospheric oxidation is eliminated. The mechanical properties are significantly increased by the cold working. It is only really used where its tolerance and strength offers a cost saving. It would not (normally) be specified for subsequent machining operations unless mechanical properties were desired and the high cost was immaterial.

6.26.5 High tensile steel

This is a high carbon steel normally used in concrete prestressed products (Chapter 12).

6.26.6 Alloy steels

When certain alloy elements are added (e.g. chromium or nickel) they substitute for the iron (Fe) in the iron's lattice and modify the phase diagram. The addition of other elements (e.g. niobium (Nb)) may make carbides different from the Fe_3C cementite carbide as they have a greater affinity for carbon than does iron. Alloy steels are usually tougher than the equivalent carbon steel and have different corrosion characteristics (Tables 7.2 and 7.3).

The effect of alloy additions can be discussed in relationship to their effect on the iron–carbon phase diagram. Heat treatment also affects the strength properties of the steel. This heat treatment has to be related to the phase diagram. Hence the cooling of iron–carbon (steel) will be considered. The term steel is used to refer to various alloys based on carbon and iron. The carbon content is usually in the range 0.1–1.7%. The alloy element added is dependent on the mechanical and corrosion-resistant properties required. Where no alloy additions are added the iron–carbon system is referred to as mild steel. It is incorrect to drop the (mild) from mild steel, although it is often done in engineering and other trades.

6.27 COOLING OF AN IRON–CARBON (MILD) STEEL

The most important aspect of the study of mild steel is the interpretation of the iron–carbon phase diagram. The phase diagram for mild steel is given in Fig. 6.34.

The phase diagram is complicated, but it is hoped to show that a lot of the phase diagram can be ignored. It is complicated by the different structures which the iron may take at different temperatures. These are represented again by Greek letters α for the single-phase structure which is stable between room temperature and 723°C when there is less than 0.025% C by weight dissolved in this structure. The α phase is called the **ferrite** phase and has the b.c.c. structure (Fig. 6.6). The interstitial carbon sites within the lattice are shown in Fig. 6.32(a). The **austenite** phase is an f.c.c. structure and is only stable between 723 and 1490°C and can dissolve up to 1.7% C by weight at 1130°C. The interstitial carbon sites within the lattice are shown in Fig. 6.32(b). The δ phase is stable above 1490°C to the melting point and is again b.c.c.

The austenite region is the largest single phase region in the phase diagram: this is because of the difference in the amount of 'void space' in the structure with elevated temperature: the interstitial void space of the b.c.c. structure is 32%; the interstitial void space of the f.c.c. structure is 26%.

The structure of the carbide cementite is complex and does not really concern us except that it forms a **non-coherent structure** in both the austenite and the ferrite phases. Cementite is therefore an obstacle to the movement of dislocations and could be considered as having a fixed ratio of Fe:C as 3:1 as the structural formula suggests. When alloyed other carbides are formed and these have the general formulae $M_{23}C_6$ and M_6C (Section 6.34) where M is the alloy element.

6.27.1 The peritectic reaction

The peritectic reaction occurs in many phase diagrams and a materials technologist needs to understand how to read them. In the case of the iron–carbon system, the peritectic reaction can be ignored as all compositions give the austenite phase. It is the austenite phase which transforms to produce the obstacles to dislocations. In order to show the peritectic reaction, the iron–carbon system has been split up into sections.

The cooling curve for pure iron would give a temperature arrest at the metal's melting point as shown by UV in (Figs. 6.34(b) and 6.34(c)) and the temperature would not decrease until all the liquid had changed to the solid phase. It is the same reason as discussed for pure lead. However, if some carbon (composition Z) is added to the iron, there is , like solder, a depression in the melting point from temperature U to temperature A, even though the carbon is not going into substitutional solid solution. The iron–carbon of composition Z now cools slowly while solidifying over the temperature range AG, and giving out latent heat. The liquid follows the composition curve ACE while the δ phase follows the composition curve BDH. The first iron to be deposited will be of composition B% C. On cooling, diffusion will have to take place so that the carbon content can increase and follow the solidus line BDH. At temperature HE the liquid undergoes a **peritectic reaction.** This reaction is accompanied by a temperature arrest GN. The peritectic reaction can be written as:

$$\text{liquid E} + \text{solid H} \rightarrow \text{Solid F} + \text{excess of liquid E or solid H}$$

The austenite phase is of composition F and the δ phase is composition H. There will be an excess of liquid E from the peritectic reaction if the composition lies between F and E or an excess of solid H if the composition line lies within the region H–F where F is the composition of the **peritectic**. The amount of excess can be calculated from the lever rule. If the composition line fell on the point F there would be NO excess of either solid or liquid and the austenite phase of composition F would be formed. If the composition line fell to the right of F as shown by composition Z, then

liquid:solid = GH:GE

Thus there would be an excess of liquid which would mean that the austenite phase would cool and follow the composition line FIK (Fig. 6.34(b)) while the remaining or excess liquid would follow the composition line EQJ. At temperature K the last remaining liquid of composition J would have transformed to the austenite phase of composition K. This would require diffusion of carbon. Therefore the liquid steel ends up as the austenite phase of composition K even though the first austenite deposited was of composition F. (Note: the first solid phase to solidify out was the δ phase of composition B. Similarly, if the composition line X was between H and F, the austenite of composition F deposited from the peritectic reaction would transform by reacting with the excess delta phase of composition H to form austenite of such a composition that it would follow the composition line FRLW.)

If the composition line were to the right of H (say composition line X), then there would be an excess of solid in the above equation. Thus the excess solid would follow the composition line HNM and the austenite of composition F would follow the composition line FLW.

If the liquid was of composition Y, there would be no liquid phase left at the peritectic temperature T_e. For the composition Y the last remaining liquid which was of composition C disappeared at temperature T_{dc} to produce the δ phase of composition D. The δ phase of composition D will cool over the temperature range DN without undergoing any phase transformation. At the temperature T_n, the δ phase would have too much carbon in solution and so would deposit the austenite phase of composition R which would follow the composition curve RLWS, while the δ phase follows the curve NMT. When the composition lines FLS (or FK) intersects the original composition line, the austenite phase stops altering in composition and cools as a single-phase structure. This means that steels in the (single) austenite phase are very easy to work as there are no second phases which obstruct the movement of dislocations. Many textbooks leave this upper part of the iron carbon system off as all steels can be considered from the simple phase diagram which concerns the **eutectoid** reaction only. This means that the interpretation is now similar to that for the lead tin system except that the liquid phase (**eutectic**) of the lead–tin system is now the solid austenite phase (**eutectoid**) of the iron–carbon system.

The cooling curve of the system for composition Z is shown in Fig. 6.34(c). Latent heat is given out over a temperature range and so the rate of temperature drop alters when there is a two phase structure (liquid + solid) within the temperature range NP (Fig. 6.34(c)) compared to the cooling of a single phase below temperature P, where no latent heat is evolved.

6.27.2 The eutectoid reaction

If the austenite composition X (Fig. 6.34(d)) is considered to cool from these temperatures there will not be a difference in the rate of cooling when the austenite composition line meets the line RSW and starts to transform to the ferrite (α phase) which contains less carbon (T%) than the solid austenite phase which has S% C. This takes place at temperature T_s (or T_t) for the composition illustrated. The first ferrite to be thrown out of the solid austenite phase is of composition T and, as the temperature drops, this ferrite phase must follow the composition line TU, thereby becoming richer in interstitial carbon, while the austenite phase follows the composition line SW, also becoming richer in carbon. The solid deposited at the higher temperature (which was of composition T) must alter and become richer in carbon as does the austenite phase. Hence **diffusion in the solid state** must take place. At the temperature of 723°C (termed the eutectoid temperature) the ratio of untransformed austenite (now of composition W) to ferrite (now of composition U) is given by the lever rule:

$$\text{austenite:ferrite} = \text{length of UV:VW} \tag{6.28a}$$

At this temperature the **eutectoid** reaction takes place and there is a temperature arrest. The eutectoid reaction is:

$$\begin{array}{ccccc} \text{Eutectoid W} & \rightarrow & \text{solid U} & + & \text{solid X} \\ \text{austenite} & \rightarrow & \text{ferrite} & + & \text{cementite} \end{array} \tag{6.28b}$$

This eutectoid formation of ferrite and cementite is called **pearlite**. If the temperature now drops to T_z, all the solid ferrite previously deposited must be of composition Z and so diffusion must take place. All the cementite is of composition X. As the temperature decreases so does the diffusion rate. For diffusion to take place the cooling rate must be slow to allow for equilibrium conditions to be maintained. The phase diagram (Fig. 6.34(d)) only refers to equilibrium conditions.

There is no cooling curve for the cooling of this solid part as no latent is evolved. The composition line are determined by changes in volume because there is a phase change involving a f.c.c. structure with 26% 'void space' to a b.c.c. structure, the ferrite, with 32% void space.

6.27.3 The critical points

Often it is easier to talk of the steel's being heated up into the austenite phase. In order to generalize for all carbon steels whatever their carbon content, the phase diagram has specific **critical lines**. These are the lines RSW for low carbon steels ($<0.8\%$ C) and AFW for $>0.8\%$ C. Compare Figs. 6.34(a) and 6.34(d).

6.28 NON EQUILIBRIUM PHASES: MARTENSITE

The phase diagram only models and tells us what phases are present if **equilibrium conditions** are achieved. This means that the temperature drop is small and the temperature is not allowed to fall until all the necessary diffusion has taken place. This allows the metal to follow equilibrium conditions. Unfortunately, equilibrium conditions are only encountered under 'ideal' laboratory conditions where the steel can be cooled very slowly. In the normal manufacture of steel components, these equilibrium conditions may not be followed, simply because the rate of cooling is much faster than those where equilibrium is obtained. As the cooling rate is increased, so other phases are produced which are not represented on the phase diagram. These are martensite and bainite. Martensite (Fig. 6.32) is often described as a distorted b.c.c. lattice. This distorted b.c.c. lattice is called a **tetragonal** lattice. A tetragonal lattice is a rectangular lattice based on a square such that the lengths of the three sides are a, b and c where $a=b$, but $c \neq a$. The carbon is often described as occupying the c axis in a non–interstitial position. The dimensions of a and c axis interstitial sites depend on the carbon content as shown in Fig. 6.32(f). This is why the normally cubic lattice is distorted. Martensite is an iron–carbon phase which is not really stable under theoretical considerations. Martensite owes its stability or existence to the fact that the diffusion of carbon is very slow at low temperatures and the formation of martensite is rapid (this is really another example of the rates of formation against stability: see the discussion of Fe_3C and graphite in Section 6.10). Scientists tend to call these semi-theoretical but naturally produced phases **metastable**. Martensite is a very brittle phase and as such is not advantageous in the metallic structure. However the metastable phase can be heat treated (called **tempering** Section 6.30.2) to produce a very strong second-phase carbide. These carbides act as obstacles to dislocations. The presence of certain alloy elements promotes the existence of martensite. Some of these elements produce a completely new phase diagram. The effect these alloy elements have on the stability of the carbides and austenite phase is discussed in the next section.

6.29 NON EQUILIBRIUM COOLING

The difference between the structures formed from equilibrium and non-equilibrium cooling of steel is essential to our understanding of ferrous metallurgy. Many mistakes have been made in the past which have resulted in loss of life. We now know of the factors and explanations which gave rise to many sudden failures. However, these principles are not yet understood by many engineers in the repair industry. Examples are numerous, but because end users do not understand the science they are in no position to start litigation. Therefore this non-equilibrium cooling must be represented by scientists in an easily conceivable way. After all, this is what science is doing for us all the time, even though some find the conceptual modelling more difficult than others. Science puts or models 'nature's laws' into human understanding which later becomes 'our commonsense'. The scientist therefore interprets these non-equilibrium cools by temperature–time–transformation (**TTT**) curves, which are shown in Fig. 6.35. The difference between the structures formed from **equilibrium** and **non-equilibrium** cooling of steel is shown in Fig. 6.36.

The curves in Fig. 6.35 are built up on the premise that the completely austenite phase is quenched very quickly to a new temperature at which it is held for a variable time and the percentage of the austenite transforming at the variable times is calculated. These are represented on the figure so as to designate the time at which the austenite starts transforming to either martensite or pearlite depending on the temperature and the time of the finish of this transformation. The martensite lines are shown dotted as they are not time dependent. The TTT curves are really a 'fudging' of the clear differences between equilibrium and non-equilibrium conditions. The TTT curves are not really amenable to the super-position of the cooling curve, which is the usual method of interpretation. The important part of the TTT curve is the effect alloying elements have on the transformations and at what

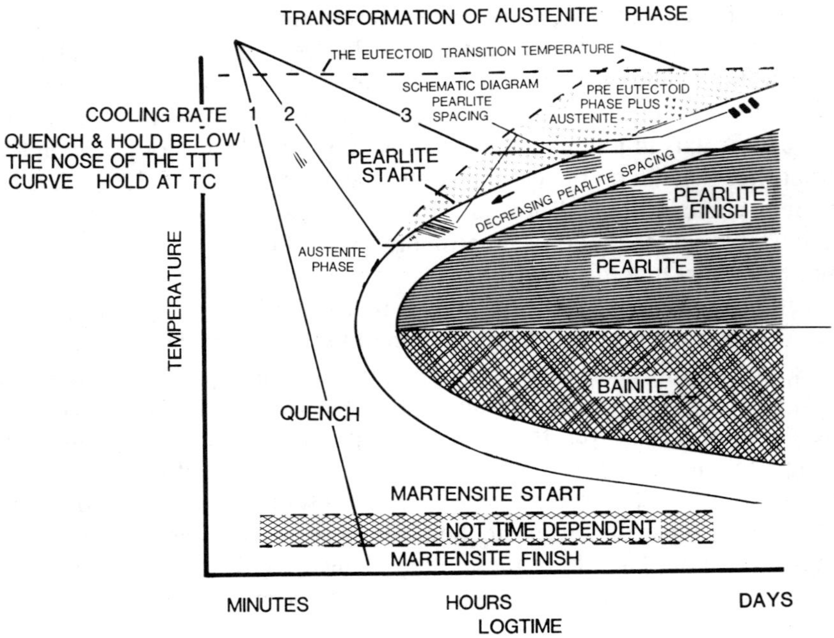

Fig. 6.35 — The structure of the iron–carbon system is very sensitive to the rate of cooling through the eutectoid temperature of 723°C. The diagram is often referred to as the time temperature transformation (TTT) diagram. The diagram models the transformation products dependent on cooling rate. The diagram is constructed from experiments in which the steel is rapidly cooled to a particular temperature and then held at the temperature for a long time (hence the horizontal axis is the logarithm of time). In practice the 'industrial rate of cooling' will not approach this 'laboratory idealised form', and so the TTT curves must be interpreted with this in mind. The spacing of the pearlite (α ferrite and cementite Fe_3C as lamellae) is also shown, as is the pre-eutectoid phase which would exist above 723°C.

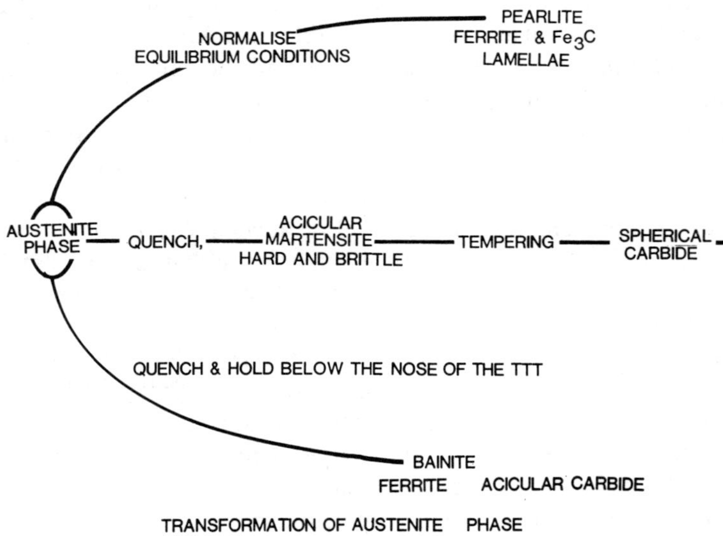

Fig. 6.36 — The various transformations which the austenite phase can undergo depending on the rate of cooling. See also Table 6.12.

temperature the '**nose**' happens to be. For normal steels the nose is at 535–550°C. A cooling curve puts the TTT curve into perspective for new readers to the subject. Three cooling curves have been used: 1, 2 and 3. The slopes of these cooling curves give an idea of the rate of cooling. However, a cooling curve does not level out and become horizontal as shown in Fig. 6.35 unless heat energy is added, as can be undertaken in laboratory conditions. These transformations and interpretations of engineering structures must be made with caution.

(1) Curve 1: the structure is fully austenitic at the start of quenching (i.e. point L on the simplified diagram Fig. 6.34(d). Cooling curve 1 indicates that the transformation of austenite to martensite takes place without the formation of any pearlite.

(2) Curve 2: the austenite does not start transforming until it has been at this lower temperature for a long time. This can sometimes be hours as the horizontal axis is logarithmic in time which has the effect of foreshortening the axis. After this time at this elevated temperature the austenite will start to transform to pearlite. The spacing between the pearlite lamellae is very small, and this can be nearly unresolvable by optical microscopy. This will produce good mechanical properties.

(3) Curve 3: The austenite takes a very long time to transform to pearlite (much longer than for cooling curve 2). The austenite transforms to a coarse pearlite. The mechanical properties in this condition are not very good, and the steel would be very soft as measured by the Vickers' hardness number (Section 6.23.1).

All the metal samples, when they have crossed the pearlite finish lines can then safely cross the martensite start and finish lines without undergoing any further transformation.

Several different metallurgical processes are used which produce superior strength steels. These will be discussed in the next section on the heat treatment of steels.

6.30 HEAT TREATMENT OF STEELS TEMPERING, AUSTEMPERING AND MARTEMPERING

Steel is capable of attaining very high strengths by the combination of alloying and the subsequent heat treatment of the steel. The increase in strengths available is shown in Table 6.12(a). The processes which are undertaken for these various heat treatments are best understood in relationship to the phase diagram or TTT curve.

Table 6.12(a) — The hardnesses produced by various heat treatments

Heat treatments	Hardness (VHN)	Microstructure
Quenched from γ phase	860	Martensite
As above and tempered 1 hr at 500°C	410	Precipitates of carbides in ferrite
Air cool from γ phase (normalise)	270	Pearlite
Slow or furnace cool from γ phase	210	Coarse pearlite
Heated for hours at 690°C	160	Carbides completely spheroidised

Table 6.12(b) — Methods of hardening for steels

Solid solution hardening	Transformation hardening	Ferrite grain refining	Precipitation hardening	Carbide promotors hardening
Phosphorus Silicon Manganese Nickel Chromium Molybdenum Vanadium Copper	Carbon Manganese Chromium Molybdenum	Aluminium Vanadium	Copper	Chromium

Table 6.12(c) — Relative merits of high strength steel grades

Material	Advantages	Disadvantages
HSLA	Inexpensive (mild steel + 5%) Readily available. Wide strength range (300–600 N/mm^2 YS)	Limited ductility YS variability (+ 125 N/mm^2) Increased spring-back + buckling compared with MS
Rephosphorized	Inexpensive (MS + 5%) Readily available Moderate formability	Contaminates baled scrap: segregation required. Some variability in YS Limited strength (255–280 N/mm^2 YS)
Bake hardening	Good cold formability, therefore less spring-back than rephosphorized Lower phosphorus level Develops strength during paint stoving	Limited strength (255 N/mm^2) On-cost unknown
Dual phase	High UTS:YS ratio; therefore high potential energy absorption Moderate–good formability	High on-cost (mild steel + 20%)

Figure contributed by G. Davies, Gaydon Technology Ltd. Reproduced by permission from G. Davies and R. A. Easterlow, 'Automotive design and materials selection, *Journal of Metals and Materials*, (January 1985) 20.

Table 6.12(d) — price of steel compared with that of other engineering materials (quoted with reference to mild steel)

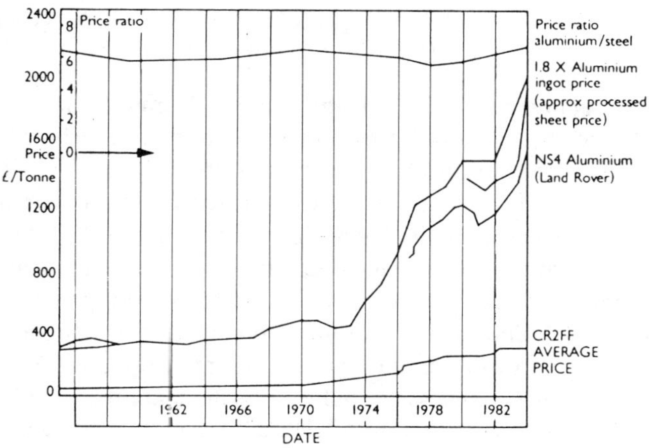

Figure contributed by G. Davies, Gaydon Technology Ltd. Reproduced by permission G. Davies and R. A. Easterlow, 'Automotive design and materials section', *Journal of Metals and Materials*, (January 1985) 20.

Table 6.13 — Current uses of cold reduces steels CR1–CR4 with costs. CR1 is usually an aluminium-killed steel, CR2 is a semi-killed steel, and CR4 is a rimming steel (cost increases in the order HR > CR4 > CR3 > CR2 > CR1)

Grade	Properties	Utilization (%)
CR1	Excellent formability, no bake hardening	26
CR2	Very good formability, bake hardens	36
CR3	Fair forability, bake hardens	26
CR4	Simple bending, bake hardens	5
HR	Fair formability	4
Coated	Fair–good formability (dependent on type)	3

Figure contributed by G. Davies, Gaydon Technology Ltd. Reproduced by permission G. Davies and R. A. Easterlow, 'Automotive design and materials section', *Journal of Metals and Materials*, (January 1985) 20.

6.30.1 Annealing

Annealing is the process where the steel is heat treated to obtain maximum softness. The steel must be heated for a long time above the eutectoid temperature and slowly cooled through this temperature. The result is that there are large grains and a coarse pearlitic structure.

6.30.2 Normalizing

Normalizing is the heating of steel up into the austenite phase above the critical point (Fig. 6.34(a); Section 6.27.3). This is usually about 50°C above the line RSW for steels with a carbon content of less than 0.8% (point W, Fig. 6.34(d) or above the line AFW for steels with a carbon content greater than 0.8%. The component steel is then removed and allowed to cool in still air. See also Fig. 6.34(a).

6.30.3 Hardening and tempering (Quenching and Tempering)

When steel is quenched into water or oil it transforms to martensite. The internal stresses caused by the carbon in the preferred crystallographic direction cause a marked increase in the internal stresses. These internal stresses can be relieved by a process of **tempering** which involves heating the steel up to 100–200°C, where carbon can diffuse and cause the precipitation of small carbides. Steels which have been correctly quenched and tempered (hardened and tempered) give the best combined properties of strength and ductility. This hardening and quenching can take place unintentionally, e.g. screwdrivers cannot have their tips remade on a 'dry' grindstone because the grindstone produces a lot of heat which makes the tip glow red and takes the tip up into the austenite field. Although the tip is allowed to cool in air, in fact the large mass of steel has quenched the tip to produce a martensitic structure. The tip will shear off very readily. Similar brittleness arises when the shafts to large armatures of large electrical machines are rebuilt. The heat-affected zone (Section 6.32) is very brittle.

6.30.4 Austempering

This is the rapid cooling of the steel to below the 'nose' of the TTT curve and transformation of the austenite to bainite. This has distinct advantages in that the risk of cracking, which can occur in a martensitic structure, is reduced. The bainitic structure is more ductile than that of tempered martensite.

6.30.5 Martempering

This is the rapidly cooling of the austenite past the nose to just above the martensite start temperature. The steel component is allowed to stay at this temperature, a process called **soaking**, until the temperature gradients are minimal. The steel is then removed and allowed to cool to room temperature and the austenite can then transform.

6.31 BRITTLE FRACTURE

Most of the metallic elements which raise the yield strength significantly (Fig. 6.37; Table 6.12(b)), have a detrimental affect on the transformation products which form when the temperature of the steel crosses the austenite–ferrite eutectoid of 723°C. Fig. 6.37 shows the rise in yield strength with alloy composition. The most efficient elements are carbon and nitrogen. After carbon and nitrogen comes phosphorus. Much less efficient in raising the yield stress are silicon and manganese. However, some of these elements also have a deleterious effect on the ductility of the steel, and the way in which the steel behaves either under high stress rates or when tested at low temperatures. Some of these elements promote the formation of the b.c.c. structure (Fig. 6.38).

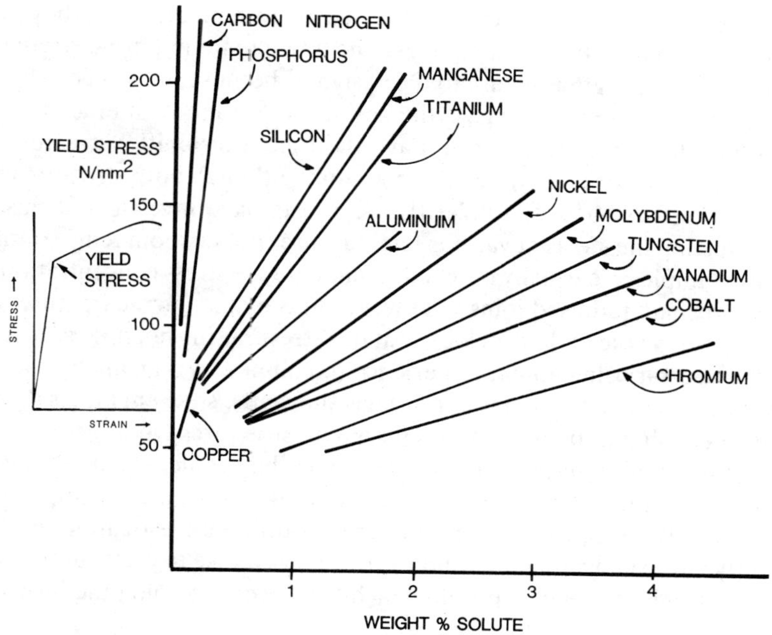

Fig. 6.37 — Different elements added to iron have different hardening effects because they are of different radii and/or different structure. Here the effect of the yield stress is depicted schematically for different alloy elements.

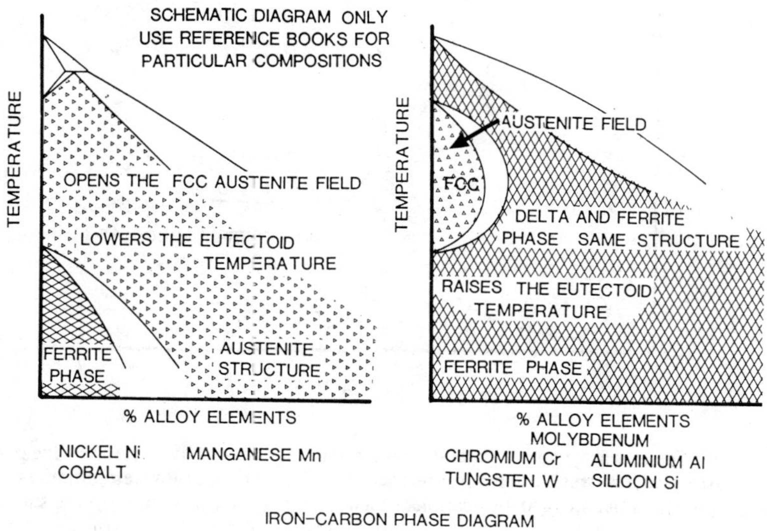

Fig. 6.38 — Different **alloy** elements in solid solution effect the iron-carbon phase diagram differently and this is summarized in these two diagrams which portray schematically the effect of alloy element on the austenite phase. Elements are subdivided into those which open up the austenite field, by lowering the eutectoid temperature, and give rise to the prefix 'austenitic' (e.g. austenitic stainless steels) and those which restrict the austenite field, by joining up the δ and ferrite phases, are referred to as 'ferritic' (e.g. ferritic stainless steels).

When steels with a **b.c.c.** structure are tested at **low** temperatures, they suddenly show a decrease in ductility and toughness and an increase in tensile strength. This increase in tensile strength is no use to the designer, because the steel can be broken in a brittle manner simply because the dislocations cannot interact to produce toughness and ductility. Steels which have the f.c.c. (austenitic) structure do not show this brittleness with decrease in temperature although both structures have 12 slip systems (Figs. 6.24(b) and 6.24(d)). Scientists have investigated this loss of ductility by testing the steels at various temperatures below room temperature (Fig. 6.39). These samples are usually notched to simulate the stress concentrations which may appear at small radiused joins or at weld joints etc. A ductile fracture depends on the steel's being able to deform locally at this stress raising notch so as to blunt the effect of the stress–raising notch. Experience (combined with an analysis of failures which has been handed down in tradition) has shown the steel engineers that a good margin of safety should be given when designing structures from ferritic steel (the b.c.c. structure) which have to be welded. This has come about by analysis of failures. For example, when riveting gave way to the welded steel ship, several of them mysteriously disappeared without trace or warning. Investigations undertaken subsequently have shown that the welding process gave rise to metallurgical changes within a very small region of the weld which has become termed the '**heat-affected zone**'.

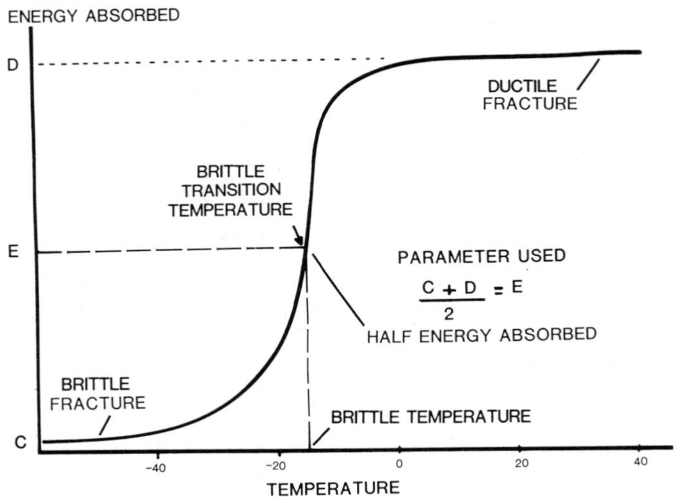

Fig. 6.39 — The characteristic of a metal is that it is neither drastically effected by the presence of notches forming a stress raiser nor influenced by deformation at low temperatures. This is explained by the movement of dislocations which work harden the region and the subsequent deformation which tends to 'blunt' the notch. This is not true for b.c.c. structures when the temperature at which the metal is tested decreases. The notch is not blunted at lower temperatures for steel, which undergoes a brittle transition at a temperature of -18°C. This temperature is called the brittle transition temperature and is modelled as the average energy of absorption. Some metallurgical processes tend to shift this curve to higher temperatures, i.e. the brittle transition temperature is shifted from − 18°C to + 5°C. This is very damaging for structural units made from such steels. The formation of a heat affected zone in welding is one such metallurgical process.

6.32 THE HEAT AFFECTED ZONE

When steels are welded, the bulk material acts as a **heat sink** and tends to conduct heat away from the fused or molten layer. This conduction of heat can make a metal go through the 723°C eutectoid region at a rate which promotes **martensite** (Section 6.28). Martensite is a very brittle material and occurs in a very narrow zone of the weld which is capable of undergoing a martensitic transformation as associated with cooling rate 1 in Fig. 6.35. This cooling rate is sufficient to cause brittleness in the steel. Adjacent areas will undergo different rates of cooling such that they do not miss the 'nose' of the TTT curve. The zone is called the **heat affected zone** and is usually abbreviated as HAZ. The HAZ plays host to both the initiation and the propagation of cracks. As the work hardening of the steel provides the designer with some element of margin of safety, this safety margin no longer exists as the deformation is accompanied by this brittle fracturing within the HAZ. Hence the first welded ships 'broke their backs' by this form of brittle crack propagation within the HAZ. Steps can be taken to eliminate this problem. These steps are often to pre-heat the metal prior to welding or to allow the welded steel to cool at a much lower rate through the temperature of 723°C (Figs 6.35 and 6.36). Some welded steels have to under-go a post-welding heat treatment in order to temper the martensite (Section 6.28). This would be an expensive method of overcoming the problem of the HAZ and would add to the vehicle price.

6.33 THE HEAT AFFECTED ZONE IN AUTOMOTIVE WELDING (See also Section 6.35.2)

In the case of spot welding, which is advantageous because it is fast, pre-heating of the vehicle body prior to welding could be used. The pre-heating process would cause added expense, and furthermore, if there were a stoppage in the line, there could be no guarantee that the 'still to be welded body' would go back into the pre-heating ovens. Therefore it is more practical to seek steels which can be welded without having to be pre-heated, or given the post-welding heat treatment process. In all cases of spot welding (Fig. 6.42), the region adjacent to the spot weld will always be quenched from its semi-plastic temperature (see the heat affected zone in Fig. 8.7(a)). The fusion process in spot welding is accompanied by very high electrode squeeze pressures.

6.34 LOW ALLOY HIGH TENSILE STEELS OR HIGH STRENGTH LOW ALLOY STEELS

Advances in design and the requirement for lighter more efficient vehicles have created a demand for structural steels which have a higher strength than structural mild steel. Mild steel which can be welded and spot welding (Fig. 8.7) is the preferred method of joining the vehicle panels together into the body in white (Section 3.1). This creates the demand that expensive and often impractical precautions should not apply. Therefore a weldable steel is one which can be welded like mild steel without losing ductility or toughness (in the heat-affected zone).

The least expensive way of strengthening a steel is by the additions of carbon (C), nitrogen (N), phosphorus (P), manganese (Mn) and silicon (Si). Some of the alloy elements act in different ways. (Section 6.20) gave the methods of strengthening an alloy. The alloy elements (Fig. 6.37) act in steels as shown in Table 6.12(b).

The disadvantages of these alloy elements are sometimes formidable. For example, the rephosphorized steel off-cuts cannot be mixed with normal scrap steel because phosphorus raises the brittle transition temperature of the steel. The brittle transition temperature is raised 5°C for every 0.01% P. Hence the scrap from this steel has very low scrap values as phosphorus is considered to be a very deleterious alloy element for other steels to contain.

6.35 REQUIREMENTS OF STEEL FOR FORMING THE STRESS–BEARING PANELS

The requirements made by the pressing operations are varied. This section will deal only with the effect of the metallurgical properties on the presssing.

6.35.1 Strain ageing and Lüders bands or stretch marks

There is a change in the mechanical properties of the steel after the final cold deformation (cold rolling: see Fig. 6.41). This is called **strain ageing**. Strain ageing occurs more rapidly at higher temperatures than at lower temperatures and manifests its presence by an increase in the tensile and yield strengths (Fig. 6.40). Although this may be beneficial in giving the designer a greater safety margin, it has a damaging effect on the final contoured surface of the bare unpainted steel, which will no longer reflect light like a mirror (Fig. 6.18). The surface is covered with 'stretch' marks, or **Lüders bands,** which are rather like the waves on deep-water seas when seen from above (it is difficult to photograph this micro-defect, but the effect can be readily noticed in the corners and lips of tobacco tins etc. which have been formed from one piece of steel). The application of subsequent high gloss paints accentuates and highlights this irregular defect.

In order to explain the defect, an elementary understanding of dislocations is required (Fig. 6.22). Strain ageing is best considered as the diffusion of the carbon and nitrogen elements to the 'bottom edge of the extra plane of the edge dislocation' where there is more room for the atom. This forms a local 'line of carbon and nitrogen atmospheres', (called **Cotterell atmospheres**) along the boundary between slipped and unslipped regions. Subsequent deformation, such as the pressing of a panel by the vehicle manufacturer, requires the dislocation line to move away from these 'atmospheres'. This requires a much higher stress because the dislocation is effectively anchored to the atmospheres, and therefore the material exhibits a higher yield stress. Once the dislocation line is moved away from its local atmosphere, the stress required to move the dislocation drops, until other dislocations in similar environments are required to move. This is shown on the stress–strain curve as a series of 'ripples' at the yield point. (Fig. 6.40). These ripples manifest themselves as deformation bands upon the surface of the steel pressing, and they are highlighted by high gloss paints. This is a serious disadvantage for external panels which are at eye level.

Strain ageing also has an effect on the brittle transition temperature of notched steels. The energy absorption curve (Fig. 6.39) is shifted to higher temperatures (Section 6.32).

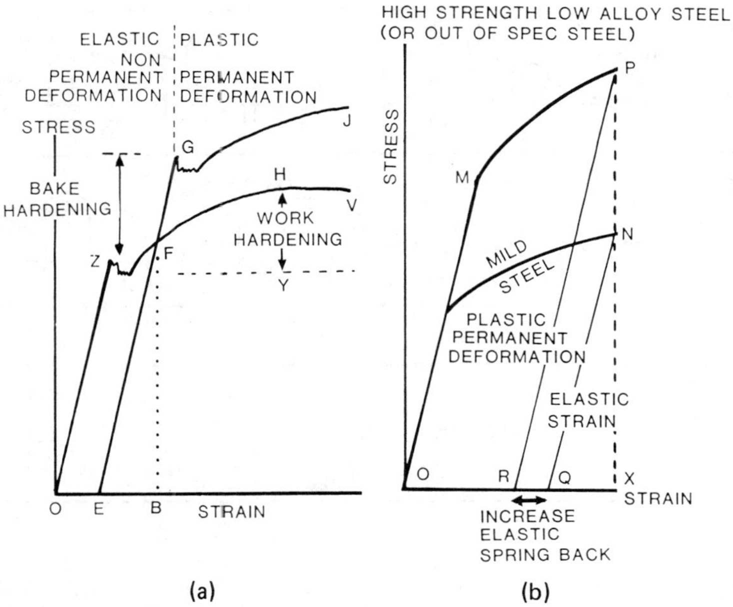

Fig. 6.40 — (a) Fig. 6.24(a) shows the dislocation interactions which occurred within the stress–strain curve. The yield phenomenon at Z, which is a characteristic of the b.c.c. metal, shows the stress variations which are required to strip the dislocations from their atmospheres. Hence strain ageing occurs when a steel sample is stressed to the point F (total strain is OB; the plastic strain component is OE) and the diffusion of solute carbon and nitrogen again takes place to the dislocations. If the sample is tested some time later, the yield point will be much higher (FG). Bake hardening takes place when the car body is heated–baked in order to fuse the electrodeposited powder into a continuous paint film. The metal is heated, which increases the diffusion rate of carbon to the dislocation sites and this increase in stress occurs in a shorter time. Note that this heating process will raise the whole stress–strain curve as shown here. (b) The stress–strain curve can be used to illustrate 'spring-back' and the manufacturer's requirement for uniform quality steels. Pressing requires a total strain of OX, which is composed of two components. For mild steel; these are an elastic strain (QX) and a plastic strain (OQ). For HSLA steels or steels with different yield strengths they are, an elastic strain (RX) and a plastic strain (OR). The increase in spring–back between the two steels is RQ.

Strain ageing can be reduced by the addition of aluminium (see Section 6.36.5) or by giving the sheet a temper roll (Fig. 6.41(b)), where an elongation of 1% is given before heat treating. Storage of temper-rolled steels at room temperature will allow the yield point phenomenon to reappear. (The rate of diffusion of carbon is much lower than that of nitrogen. Nitrogen will readily diffuse at room temperatures, whilst Carbon requires a temperature of 60°C to diffuse to these sites.)

6.35.2 Weldability

The traditional method of joining panels together is by spot welding. (Fig. 8.7(a)). In electrical resistance welding (spot welds) there is no inert atmosphere, no filler metal and no flux added to the weld as there is when MIG or MAG or electric arc welding is used (comparison made in Fig. 8.7(b) and 8.7(c)). The spot welding process is achieved by squeezing the two metal panels to be welded together and heating them

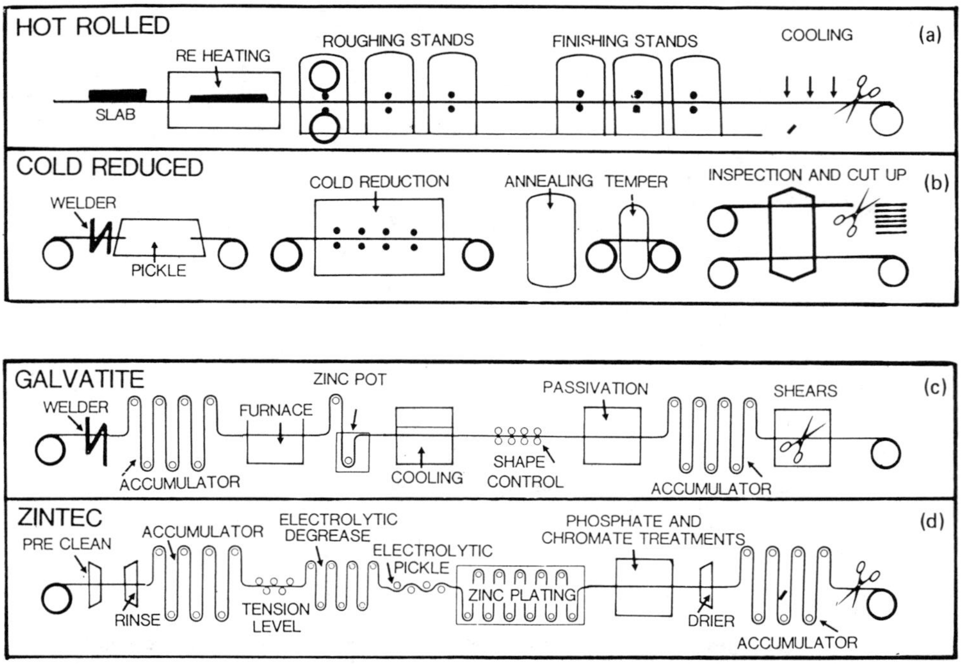

Fig. 6.41 — An abbreviated manufacturing route for British Steel's products. These routes are typical manufacturing processes for the steels considered in this book. (a) Hot rolled. As the name implies the steel is rolled at elevated temperatures. The mechanical energy put into the steel by the reduction process tends to keep the temperature high. If the temperature does fall the steel can be reheated. Usually the surface is rough owing to oxidation. (b) Cold reduced. Cold reduction reduces the thickness of the steel and provides a good surface finish. (c) Galvatite. Galvatite is a continuous zinc hot dipping process into molten zinc. Where an iron-zinc alloy is required an extra heating oven is used prior to the cooling stage. (d) Zintec. Zintec is British Steel's mild steel sheet coated with a thin film of electrodeposited zinc. The coating process is capable of producing steel sheet with either one side or both sides coated with zinc. Different thicknesses can be applied to the two opposite sides of the sheet. (Modified and reproduced by permission from those diagrams in British Steel Corporation publicity material on hot and cold reduced steels.

by a high current, and the weld 'nugget' is formed by the 'consolidation' of the 'plastic' metal under these very high pressures and squeeze forces. There is no physical melting of the area concerned as there is in the other techniques (Figs 8.7(b) and 8.7(c)). Fusion is by plastic deformation and pressure.

The heat supplied in joules is given by the formula

$$\text{Heat Supplied} = I^2Rt \tag{6.29}$$

where I is the current flowing, R is the electrical resistance across the junction and t is the time in seconds for which the current I flows. This heat is all supplied within the resistance path (Fig. 8.7(a)) between the metals and so is very localized. (Fig. 6.42). The vehicle manufacturer would like this process to be undertaken in the shortest possible time t in order to reduce distortion and costs by heat loss. Therefore to reduce t very high currents are used (500 A for light gauge work to 10 000 A for

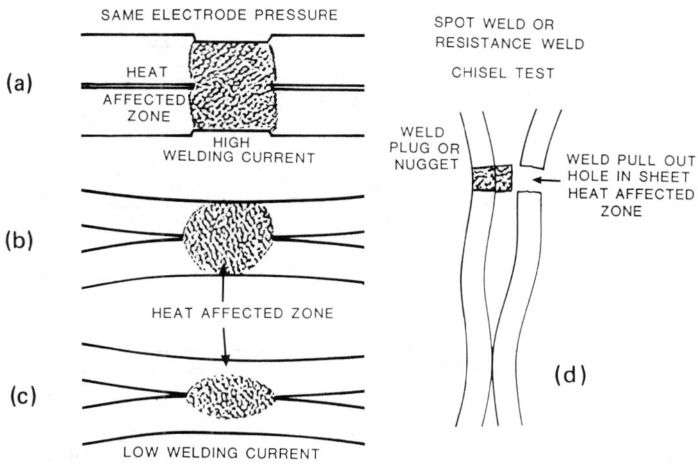

Fig. 6.42 — (a)–(c) The effect of using different currents to weld identical steel panels together. The region is also shown which is quenched through the 723°C phase change. This region becomes the heat-affected zone in those steels which are subjected to the austenite→ martensite phase transformation. (d) The weld should be able to resist peeling forces without fracture of the weld nugget across the welded interface. A satisfactory weld is one where the weld nugget is pulled out from either panel as shown in (d). Failure occurs by fracturing within the heat-affected zone. Industry uses a chisel to test for this. As a wide variety of different steels become used in the manufacture of the vehicle body, the repair industry will have to consider their policy of using the range of welding and 'cutting' equipment which is at present available.

heavier gauge steel). Metallurgically these factors work to produce a quenched heat-affected zone because the unheated panels conduct the heat away (see Section 6.32).

The weldability of the steel is often expressed as the carbon equivalent CE by the following expression:

$$CE = C + \frac{Mn}{6} + \frac{Cr + M + V}{5} + \frac{Ni + Cu}{15} \tag{6.30}$$

where all the elements are expressed as percentages. For a steel to be considered as weldable (Sections 6.31, 6.32 and 6.35.2), CE must be less than 0.45.

The weldability of a steel is important because most cracks are initiated within the heat-affected zone (Section 6.32) and this is one of the parameters which leads to a lack of subsequent toughness and ductility in the finished product.

6.35.3 Well–defined fatigue point

A well-defined fatigue point (Fig. 6.43) is imperative when designing with steel. Plastics and composites will never completely replace steel for the simple reason that they do not have a well-defined fatigue limit.

A cyclic stress S results in a cyclic strain ε. The number of cycles N which the metal can withstand before breaking is termed the fatigue life. The S–N curve is derived from a number of samples all fatigued to failure by an alternating stress. Unlike the stress–strain curve which is derived from one sample, the S–N curve is derived from

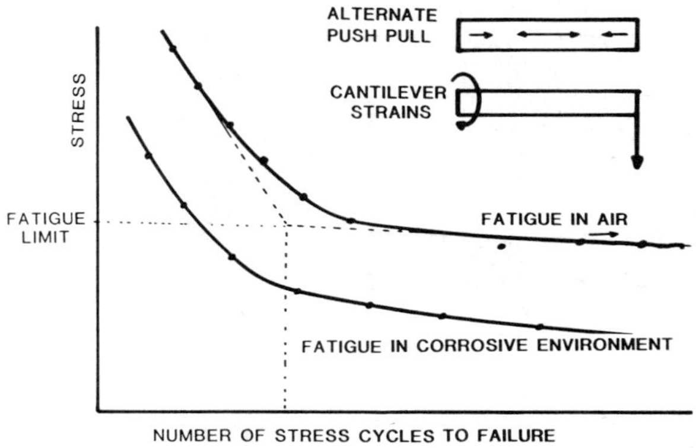

Fig. 6.43 — Fatigue curves are made from several samples by stressing in push–pull or by hanging a load on the end of a rotating bar which puts the surface into alternate stress cycles. The number of stress cycles required to cause fracture is plotted as one result (point). The points are joined up to give the curve. Each curve can take a year to obtain (whereas a stress–strain curve takes 10 min). Cyclic stressing in a corrosive environment lowers the fatigue resistance. Samples which do not fail are shown by a horizontal arrow.

several samples and is therefore very expensive to obtain. The magnitude of the strain gives rise to the terminology of **high strain–low cycle fatigue** for specimens failing within 10^3 cycles and **low strain–high cycle** fatigue for those which fail around 10^7 cycles. In order to classify steels (and other metals) with regard to their ability to resist cyclic deformation, it is often best to give the fatigue limit of the material. The fatigue limit is the stress required to cause failure within 10^7 cycles. Fatigue results from surface-initiated cracks which propagate and grow without any macroscopic evidence of their presence. Therefore fatigue failure is always sudden and the results are rather dramatic. The fatigue crack can be removed by continuously electropolishing away the surface layers (and with the layers the surface cracks). Hence, in weldable structures, not only has the steel to withstand notch sensitivity and phase transformations on cooling within the heat-affected zone but also, the weld has to withstand the micro-cracks within the surface layers which are initiated by fatigue. Fatigue resistance therefore has to be coupled to weldability.

6.35.4 Corrosion fatigue

Corrosion fatigue resistance is always less than the fatigue resistance of the metal tested in air. The fatigue limit in a corrosive environment is therefore lower than the fatigue limit in air.

6.35.5 Cheapness and availability

Steel is a readily available material with a well-known and understood manufacture. It is the cheapest material for its strength by some 6 times (Tables 6.12(c) and 6.12(d)).

6.35.6 Combination of strength and ductility

Steel possesses the required toughness, a term used to describe the combination of strength and ductility. The parameters causing deviations from toughness are well known: ageing (Section 6.35.1) weldability, (Section 6.35.2) and brittle transition temperature (Section 6.31).

6.35.7 Properties and paint protection and finishing technology

The properties of steel and the requirements for its protection combined with the problems of paint adhesion are clearly understood having undergone many years of development. Chapter 9 deals with the finishing technology. However, some elementary precautions as regards design against corrosion are often overlooked (Chapters 2–5).

6.35.8 Compatibility with existing production technology and knowledge

The steel required for the mass produced vehicle is compatible with existing production technology and the knowledge of the parameters which affect formability are understood.

6.35.9 Formability

Formability is the collection of properties required for a cold rolled sheet to take up a different shape permanently in order to produce an increase in stiffness. (A flat sheet of paper has no rigidity in one direction perpendicular to the plane of paper. If this piece of paper is made into a cylinder it has much greater rigidity and stiffness in three dimensions. Corrugating the paper will further enhance the stiffness.) Some of these new shapes for motor vehicle panels can be quite complex, and therefore the series of stamping operations must not produce; (a) rupture, (b) undue local thinning of one area relative to an adjacent area, (c) edge cracking or (d) localized stress points from which macroscopic deformation can be nucleated which would result in fatigue failure (Fig. 6.43; Section 6.35.3). If the panel is to be used for 'eye-catching panels', the steel must not produce 'rippling' either as a result of stressing aged steels (Section 6.35.1) or from the flow of steel within the stamping presses. Steel is much easier to deform in tension than it is in compression, and this fact often produces rippling where large contraflow distortions are required for body floors. These ripples are acceptable in out-of-sight panels (e.g. around transmission tunnels and floor pans).

6.35.10 Formability tests

The steel industry has produced its own set of formability tests. These involve tensile, bending and hardness tests. These tests are to tell the fabricator how the properties vary in the two principal directions: along the rolling direction (major axis) and at right angles to this direction (minor axis). The difference in deformations in these two directions results because the dislocation movement is restricted to particular planes (Section 6.22; Figs. 6.24(b) and 6.24(d)) and the condition put on the material deformation that 5 out of the 12 slip systems must operate in the adjacent grain (Fig. 6.24(a)). The result of these parameters is that the grains within the steel become preferentially oriented by rotation and slip with respect to the principal (major and minor) deformation directions. These are along the rolling direction (major axis), where all the deformation takes place, and parallel to the rolls where friction between the rolls and the metal means that the metal does not stretch in this direction (minor axis) (Fig. 6.44).

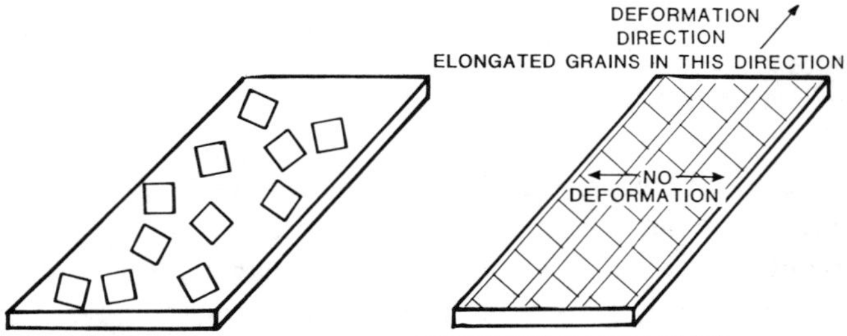

Fig. 6.44 — In order to reduce the thickness of a piece of sheet steel by rolling, there can only be deformation in one direction. This produces an elongated grain structure (Fig. 6.19), with all the slip systems (planes and directions) aligned. This alignment of slip systems requires grain rotation within the metal so as to accommodate the imposed deformation within their slip systems. The subsequent panel pressing involves plastic deformation in all directions. The number of grains with slip systems to accommodate this complex deformation are therefore restricted in the direction at right-angles to the rolling direction and so complex 'wrinkles' can be produced.

In order for the reader to appreciate this problem, woven wire has this preferred orientation. If a circle is cut out of the woven wire and deformed to make a tumbler, the result will be a series of ears on the rim of the tumbler. In the pressing of body panels this 'earing' would be unacceptable, and therefore the industry usually determines the stresses in the major and minor axes of the deformation directions by the following tests.

(1) In the Erichsen test a spherical plunger is forced against the steel plate and the deformations in two mutually perpendicular directions are determined.

(2) Special steel plates are chemically etched with circles and lines representing the major and minor axes of deformation after rolling. The diameters of the circles are measured at 45° to the two major and minor axes. The complex body pressing can be undertaken on these sheets, and the distortions of these circles can be measured. Different pressing lubricants are also used in order to determine which parameters give the best results. The deep drawing properties of the steel are represented by an r value (V in some countries) as defined by the equation

$$4r_{\text{mean}} = r_{\text{major}} + r_{45^\circ} + r_{\text{minor}} \tag{6.31}$$

The higher are the values of r, the better the properties (Table 6.14).

(3) Similarly, a stretch-forming index n is obtained for complex pressing operations and this gives a value of the work-hardening index as deformation takes place.

(4) Special tests on deep-drawn cylinders can be carried out.

6.35.11 Fittability and alignment prior to spot welding

The steels required by the vehicle body builder must be all of the same strength, strain ageing and ductility for a particular body pressing operation otherwise the amount of **spring-back** is different for each batch of steel. Hence a panel pressing which was originally designed to use a CR4 grade of rimming steel cannot be switched to use a high strength low alloy steel. The components would have a different form from the same pressing mould. This is shown in Fig. 6.40(b) which compares high strength low alloy and CR4 stress–strain curves. The degree of strain OX that each panel would be subjected to would be the same from the mould contour. However, on releasing the panel from the mould there would be spring-back. The plastic strain in each steel is shown as follows:

Steel	Plastic Strain	Elastic Strain
CR4	OQ	QX
CR4 out-of-specification steel	OR	RX
High strength low alloy	OR	RX

As the elastic strain RX is greater than QX, it follows that the high strength low alloy steel will have more **spring-back** and so will not fit the space provided for the panel in the jig for the spot-welding robot. Hence, if the same grade of steel varies in its tensile strength, the same degree of ill-fit will result. This example also shows that materials cannot be switched readily without producing problems on assembly. The corollary of this is: the steel and press are interrelated entities and the vehicle manufacturer requires consistent properties (in particular yield strength) from the steel producer. Fig. 6.40(b) is rather exaggerated for an out-of-specification CR steel. This has to be accepted in order to show up the increased spring-back.

6.36 STEELS FOR THE BODY STRUCTURE

The steels used for the automotive construction of a body are undergoing major review and developments as a result of the Energy Protection Act. This act requires the motor manufacturers to achieve specific fuel consumption figures as shown in Fig. 5.1, which has necessitated the reduction in body weight (Fig. 5.2). In order to achieve the same strength in the thinner body panels, higher strength steels have to be used. In the use of new steels corrosion protection must also be considered. The main alternatives to mild steel are the HSLA steels, polymer mouldings and aluminium sheet. The stress–strain curve for the contenders are shown in Fig. 6.45 (plastics and aluminium are not shown as they cannot be used structurally). Their use is for the non-load-bearing (cladding or cosmetic) panels. A fender–wing is a cosmetic panel as it is non-load bearing.

6.36.1 HSLA

This is high strength low alloy steels, (HSLA) e.g. British Steel's Tenform XK and XF formable steels, where XK stands for normal formability and XF for extra formability. In the case of the Tenform XK and XF grades the numbers refer to the yield strength in newtons per square millimetre. Hence the difference between Tenform XK 350 and Tenform XF 350 is the difference in formability, both having a yield strength of 350 N/mm^2. These steels have been developed for weldability (Section 6.35.2) and for formability (Sections 6.35.9 and 6.35.10).

Table 6.14 — As finishing technology improves, so different steels and coatings can be used for corrosion protection. The table compares the application and properties of materials used at present with those likely to be used in the future. Low bake electropaints will mean that bake-hardening steels may be used without the problem of over-ageing

			Required properties (minima)					
Panel	Current choice	Preferred choice	Yield stress (N/mm^2)	UTS (n/mm^2)	Elongation (% on 50 m)	r drawing index)	n (stretch-forming index)	Future alternatives
Roof	Mild steel	Mild steel	180	310	50 (typical)	1.8	0.23	Aluminium
Bonnet: Inner	Rephosphorized	Bake hardening	210	355	366	1.5	0.20	Plastic
Bonnet: outer	Rephosphorized	Bake hardening						Plastic
Trunk: inner	Rephosphorized	Bake hardening						Plastic
Trunk: outer	Rephosphorized	Bake hardening						Plastic
Doors: inner	Low YS rephosphorized	Bake hardening	255 (2%	360	—	—	—	Plastic
Doors: outer	Rephosphorized	Bake hardening	deformation					Plastic
Panel: front lower	Rephosphorized	Bake hardening	+ 20 min at					Plastic
Panel: rear lower	Rephosphorized	Bake hardening	170°C)					Plastic
Sills	Rephosphorized	Bake hardening						Aluminium
Fenders	Low YS rephos	Bake hardening						Plastic
Monoside	Low YS rephos	Low YS rephos	235[a]	350	35	1.7	0.20	Aluminium
Dash	Low YS rephos	Low YS rephos						Aluminium
Valance–wheelhouse	Low YS rephos	Low YS rephos						Aluminium
Wheel arches	Low YS rephos	Low YS rephos						Aluminium
Main floor	Rephosphorized	Rephosphorized	255[a]	360	32	1.6	0.18	Aluminium Composite
X member — complex	Rephosphorized	Rephosphorized						Aluminium
A post	Rephosphorized	Rephosphorized						Aluminium
B–C: outer	Rephosphorized	Rephosphorized						Aluminium
B–C: closing	Rephosphorized	Rephosphorized						Aluminium

Table 6.14 — As finishing technology improves, so different steels and coatings can be used for corrosion protection. The table compares the application and properties of materials used at present with those likely to be used in the future. Low bake electropaints will mean that bake-hardening steels may be used without the problem of over-ageing

			Required properties (minima)					
Panel	Current choice	Preferred choice	Yield stress (N/mm^2)	UTS (n/mm^2)	Elongation (% on 50 m)	r drawing index)	n (stretch-forming index)	Future alternatives
Roof	Mild steel	Mild steel	180	310	50 (typical)	1.8	0.23	Aluminium
Rear quarters	Rephosphorized	Low YS dual phase	220	440	38	1.2	0.24	Plastic
Rear floor	Rephosphorized	Dual phase	320	600	26	1.0	0.22	Aluminium
Front longitudinals	Rephosphorized	Dual phase						Aluminium
Rear longitudinals	Rephosphorized	Dual phase						Aluminium
Door seams	HSLA	High YS dual phase	550	950	19	0.8	0.22	Aluminium
Cross-members — simple	HSLA	HSLA	350[b]	450	25	1.1	0.19	Aluminium
Suspension	HSLA	HSLA						Aluminium
Jacking brackets	HSLA	HSLA						Aluminium
Small brackets —reinforced	HSLA	HSLA						Aluminium

[a]Maximum deviation = 40 N/mm^2 above specified YS.
[b]Maximum deviation = 75 N/mm^2 above specified YS.
Surface texture of all steels to conform to BLS 2 MP22] interim specification: CLA = 1.1 ± 0.2 μmm; min peak count = 250 ppi.
Table contributed by G. Davies, Gaydon Technology Ltd. Reproduced by permission from G. Davies and R. A. Easterlow, Automotive design and materials section, *Journal of Metals and Materials*, (January 1985) 20.

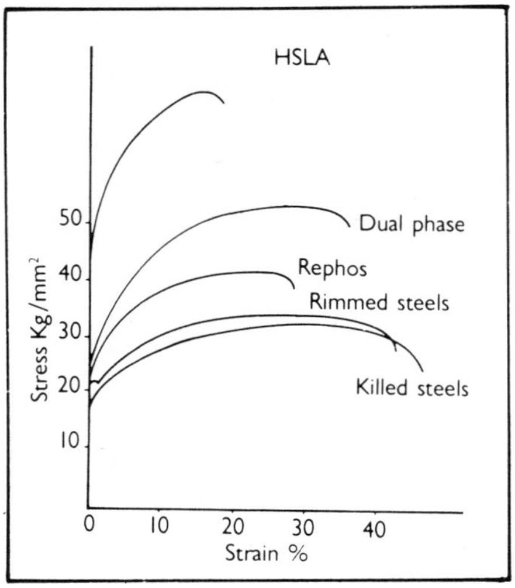

Fig. 6.45 — The stress–strain curve for a number of practical and experimental steels which could be used by the motor vehicle body builder as they satisfy many of the conditions discussed in this chapter. Cost is not considered here as research and development may reduce the cost of the steels. For a brief description of the steels refer to the sections as follows; HSLA, Section 6.36.1; Dual-phase steel, Section 6.36.2; Rephos steels, Section 6.36.6; rimming steels, Section 6.36.4; killed and semi-killed steels, Section 6.36.5; bake-hardening steels, Section 6.36.6. (Figure contributed by G. Davies, Gaydon Technology Ltd. Reprinted from G. Davies and R. A. Easterlow, 'Automotive design and materials selection', *Journal of Metals and Materials*, (January 1985) 20.

6.36.2 Dual phase steel

This steel is an experimental steel and at the present time high cost has not made it a viable proposition. These steels are made by heating the steel to above 723°C, into the ferrite and austenite region. With only a small amount of austenite present the steels are quenched to produce a martensitic structure within a ferritic matrix. The ferrite provides ductility and on heating the martensite will transform, producing carbides and hence a strong steel.

6.36.3 Rephos steels or rephosphorized steels

Steel is usually made with a very low phosphorus content. These steels have added phosphorus in order to increase the strength of the ferrite (b.c.c.) phase by solid solution hardening (Fig. 6.37). The brittle temperature (Fig. 6.39) is raised by 5°C for every 0.01% of phosphorus added and therefore for a weldable structural load bearing steel only a small amount of phosphorus can be added. The failure of the spot weld can occur when the welded area or 'nugget' is pulled out (Fig. 6.42(e); see Sections 6.30–6.33). The use of these steels tends to be restricted to bolt-on items where no added (unintentional) heat-treatment such as welding is performed.

6.36.4 Rimmed steels or rimming steels

The steel is poured before all the iron oxide (FeO) has been reduced by carbon. The steel metal solidifies on the mould wall in such a manner that a pure (gas-free, low in both carbon and sulphur) metal layer is formed without any gas holes within the chilled region. This is called the '**rim-in**'. Subsequently within the unsolidified mould the carbon reduces the iron oxide to produce carbon monoxide according to the equation:

$$FeO + C \rightarrow Fe + CO(gas)\uparrow \qquad (6.32)$$

This reaction produces 'blow holes'. The reaction is beneficial as the gas given off counteracts the contraction when the liquid phase solidifies. A rimming steel ingot therefore has a pure metal rim and an internal structure containing blow holes and occluded gas (carbon monoxide), iron oxide and sulphides. On rolling, this rim-in is maintained as a thin covering to the whole sheet steel. The art of making this rimming steel is to get the depth of rim sufficient so that the internal impurity-laden core or centre does not 'show through' the outer rim. Carbon contents of about 0.15% are normal.

In the HR and CR grades, as the number following the CR or HR designation increases so do the hardness, yield strength and ultimate tensile strength. Hence CR4 (a rimming steel) is a much stronger steel than CR1 (an aluminium-killed steel). The amount of inclusions and deleterious impurities present is also allowed to rise.

6.36.5 Killed steels and semi-killed steels

In order to reduce the above reaction (eqn. (6.32)), certain de-oxidizing agents (silicon (Si) and aluminium (Al) are added to the ladle. This completely stops the evolution of the carbon monoxide (CO) (hence the term killed), and so the ingot shows shrinkage. Semi-killed steels are formed when insufficient de-oxidizing agents are added and there is some evolution of the carbon monoxide. Killed steels (carbon contents of about 0.15%) are usually of higher quality than semi-killed (carbon contents of about 0.12–0.25%) or rimming steels.

6.36.6 Bake hardening steels

These steels (Fig. 6.40) have advantages in that strain ageing is induced after body build. Therefore the manufacturer has had the advantage of low yield strength and less spring back on forming a panel (Fig. 6.40) but the use of stronger steels in service life. These steels are formed by heating the mild steel to temperatures above 723°C in order to dissolve the carbon into the austenite and ferrite phases. The strip is then sprayed with water in order to retain this level of solubility. These sheet steels are then made into panels. The necessary deformation processes produce a large number of dislocations. The strain ageing process (Section 6.35.1) is then accelerated when the body in white is heated to 170°C (after electrocoating) in order to fuse the electrocoat powder into a coherent paint film. A fine dispersion of iron carbides is produced within the panels which increases the yield stress and ultimate tensile strength so that the panels offer greater dent resistance and strength in service life.

6.36.7 The effect of oxygen and nitrogen

Some steels are made by blowing air into the molten metal. This means that both oxygen and nitrogen can be dissolved in the metal.

(1) Oxygen exists in steel combined with iron and causes embrittlement problems as the solubility of iron oxide (FeO) in iron (0.10% at 715°C) decreases to 0.01% at room temperature. The embrittlement is caused by the precipitation of the iron oxide phase.

(2) Nitrogen is soluble in molten steel and is removed by aluminium as an insoluble alumium nitride (Fig. 6.12). If nitrogen is not fixed by aluminium, there is a danger that the steel will suffer from notch sensitivity and embrittlement with time as Iron nitride needles are formed. These needles act as stress raisers.

6.37 SUMMARY

There are many high strength steels which can be used for various panels of the vehicle body (Figs. 6.45 and 6.46) and their advantages and disadvantages are well known (Table 6.12(c)).

However, there is a trade off to be obtained, because variable yield strength steels or higher strength steels will automatically lead to a higher spring-back of the component in the panel stamping and forming process (e.g. RQ can mean 1.6° in a 90° bend, which is greater for HSLA steels than mild steel's spring-back; Fig. 6.40(b)). This would cause the panels to be ill-fitting prior to spot welding, which could aggravate the crevice phenomenon (Figs. 3.1, 3.9 and 3.11). Hence the requirement for spot welding the panels into a car body shell (Fig. 3.1) requires good '**shape fixability and fittability**'. As the panels become more contoured and shaped they increase in stiffness. This increase in stiffness is required to make thinner panels more dent resistant and to reduce the resonance within the panel. (Resonance is a mass phenomenon and at present it is sufficient to stick on small bitumastic panels on the inside of the resonating panel: see for example Plate 2.2(a)). This requirement

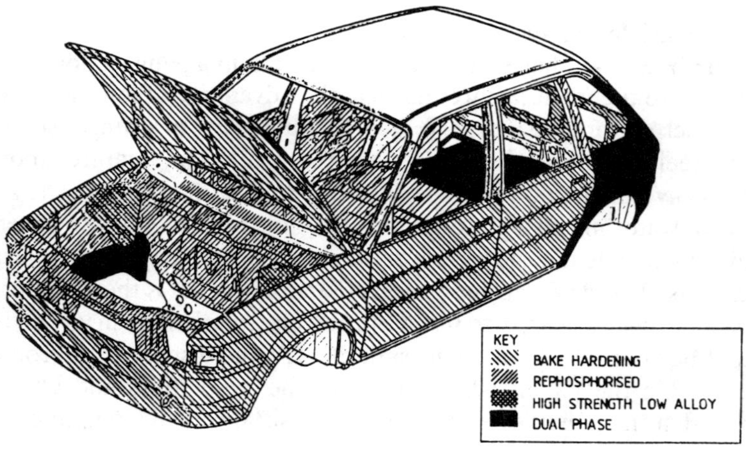

Fig. 6.46 — Figure showing how in the future some of the panels within a vehicle body will be made from different steels. (Figure contributed by G. Davies, Gaydon Technology Ltd. Reprinted from G. Davies and R. A. Easterlow, 'Automotive design and materials selection', *Journal of Metals and Materials*, (January 1985) 20.

for an increase in stiffness is why some vehicles have reversed curvatures in their side panels. Fig. 6.40(b) shows that it is not practical to change steels and to use the same stamping tools which were used with mild steel. The steels currently available for body pressings are shown in Tables 6.13 and 6.14.

6.38 COATED STEELS

As the thickness of steel decreases, the need to protect the steel by both sacrifical metals and design increases. Anodic coatings to steel are considered in greater detail in Section 8.3 *et seq*. For the sake of completeness their characteristics (as affecting body build) are summarized in Table 6.15. There are several manufacturers of coated steel products. Techniques for applying these anodic coats differ from manufacturer to manufacturer as shown in Table 6.16.

Zincro-metal (trademark of Diamond Shamrock Co.) is a material in which sacrifical zinc is obtained by a two-stage combined coating of 'Dacromet' which is a corrosion inhibitor (zinc dust and chromic acid) followed by a 0.102 mm thick coating of Zincromet which is a zinc-rich epoxy paint. These coatings are applied by a paint roller process.

6.39 ALUMINIUM ALLOYS

Aluminium alloys cannot be alloyed or heat treated to obtain the same strength as steel components. Strong aluminium alloys do not have the same formability as steel. However, their strength-to-weight ratio is superior to that of steel. (Alluminium-bodied commercial vehicles are a commercial viability. The reduction in weight obtained by using aluminium can be used to increase the carrying capacity of the vehicle. See Section 1.18.) The cost of aluminium alloys continues to increase, mainly because of the high cost of extracting the metal from its ores. Once formed, an aluminium alloy panel weighs half that of the steel component with a considerable penalty as to cost (Table 6.12(d)). At present it would not be economic to replace steel by aluminium. If 300 lb could be saved off the weight of the car by using aluminium, only half of the extra energy used to extract the aluminium alloy would be saved on petrol throughout the lifetime of the vehicle. The initial purchaser would not get the extra cost back at a fast enough rate to warrant the high initial purchase price (See Table 1.13).

At the present time aluminium alloys are being used in experimental energy conservation vehicles. The alloys considered for British Leyland's Energy conservation Vehicle are shown in Tables 6.17 and 6.18 (Plate 6.1).

6.39.1 Precipitation-hardening alloys

Aluminium alloys do not form martensitic-type structures when they are quenched. The alloy solute remains in supersaturated solution and is deposited as a very finely dispersed phase when the alloy is aged (e.g. at 195°C (Table 6.18)). Hence, to achieve the best precipitation-hardening alloy, scientists have been examining the phase diagram of aluminium with various additions in order to find one which has high solid solubility at an elevated temperature and low solid solubility at room

Table 6.15 — As the competition increases between motor manufacturers concerning corrosion protection, more research work is put into coatings which can be placed on steel without being damaged in subsequent forming. The coatings available for steel are considered in the context of the manufacture and durability of motor vehicle bodies

Product category	High strength coat ing capability	Formability	Weldability	Adaptation to surface treatment (phosphating)	Adeshion of paint	Anti-corrosion performance	Post-painted appearance
Organometallic	Appropriate for annealed steels in tight coils, problematic for continuously annealed steels	Average to good depending on coating adhesion	Average: problems arise owing to zinc and resin fumes	Very good	Very good	Good, but no cathodic protection	Very good
Galvanized sheet double-side minimized	Good: may require adaptation of metal substrate compo sition for galvaniz ing process	Average to good according to substrate composition and processing cycle	Average: problems arise owing to fumes and electrode water	Poor in conven tional baths; correct in baths adapted for zinc	Poor to good depending on surface treatment	Very good + cathodic protection	Poor to good depending on substrate and painting programme
Double-side alloy coating	Good: may require adaptation of metal substrate composition for galviniz-ing process	Poor to good depending on drawing operation; Powder formation on pressing recovery	Quite good	Good	Very good	Good + cathodic protection	Good
One-side	Good: may require adaptation of metal substrate composition for galvanizing process	Poor to good depending on drawing operation; power formation on pressing recovery	Average: problems arise owing to fumes and electrode wear	Performs like double-side galvanized on zinc side, good on exposed side	Poor to good on zinc side: very good on exposed side	Good protection against perforation- + cathodic protection	Poor to good on zinc side: good on exposed side
One-side 1/2	Good: may require adaption of metal at substrate composition for galvanizing process	Good to very good, taking into account ductility of ferro-zinc alloy	Average: problems arise owing to fumes and electrode wear	Performs like double-side galvanized on zinc side, good on zinc alloy surface	Poor to good on zinc side: very good on zinc alloy surface	Good protection against perforation and cosmetic corrosion + cathodic protection	Poor to good on zinc side: very good on alloy surface
One-side electrozinc coated sheet; pure zinc	Good	Good: certain powder formation problems are possible	Average: problems arise owing to fumes and electrode wear	Quite good on zinc side: good on exposed side	Good	Good protection against perforation + cathodic protection	Good
Zinc alloys	Good	Poor to good depending on components	Average: problems arise owing to fumes and electrode wear	Quite good on zinc side; good on exposed side	Good	Good protection against perforation- + cathodic protection	Good
Pure zinc + passivation	Good	Good	Improvement in electrode life according to Teksid	Quite good on zinc side; good on exposed side	Good	Good protection against perforation + cathodic protection	Good

Table contributed by G. Davies, Gaydon Technology Ltd. Reproduced by permission from G. Davies and R. A. Easterlow, Automotive design and materials section, *Journal of Metals and Materials*, (January 1985) 20. Table based on results from an article by Quinchon *et al*. Galvano Organo Treatments de Surface Journal, number 528, and acknowledged with thanks.

Table 6.16 — Summary of coated steels currently available

Product type	Commercial designation (manufacturer's trade name)	Features	Typical (automotive)
Organometallic coating	Zincrometal (26 producers worldwide, of which 9 in Europe: Usinor, Sollac, etc.): Immozinc (Thyssen)	Cold-rolled sheet continuously coated with zinc-rich primer, usually one skin	Doors, wings, radiator valences, wheel arches
Dip-galvanizes sheet	—	Continuously dip-galvanized products	—
Minimized coating (finish A) Minimized coating (finish B)	Ultrasmooth (Armco) Extrasmooth (Dofasco) Ultralisse (Usinor)	Zinc spangle can be eliminated with aid of steam (Armco) or by powder zinc spray (Usinor–Heurtey) These products are available in a coating thickness range of between 100 and 350 g/m^2 double-side	Cross members, sills, wheel arches, dash and trunk panel, etc.
Double-side alloy coating	Galvanneald (USA–Japan) Galvallia (Ziegler–Sollac) Galvamat (Usinor)	The zinc coating is modified by passage through an annealing furnace after the zinc bath to form a ferro-zinc alloy	Bodywork components in coaches, visible parts in certain vehicles of Japanese manufacture
Zn–Al coating (55% Al)	Galvalume (Bethlehem) Zincalume (Lysaght) Aluzinc (Svensk Stal)	Sheet is galvanized in a zinc–aluminium–silicon composite bath	Development stage
Zn–Al coating (5% Al)	Galfan (Phenix, Ziegler, Usinor, Stalco)	Sheet is galvanized in bath containing 5% aluminium and Mischmetal, remainder zinc	Development stage
One-sided: Pure zinc 10–14 μm	Zincgrip OS (Armco) (GIE, Phenix, Usinor–Heurtey) (Republic Steel) (Nippon Kokan) (Nippon Steel) Unizing (Italsider)	Zinc coating coating applied on one side via a meniscus Zinc is applied to one side by roller applicator Uncoated side is masked off prior to galvanizing Zinc applied to one side with aid of pump Uncoated side obtained by brushing galvanized sheet Unizing is produced using IFS steel	Wings rear panel

Table 6.16 — Summary of coated steels currently available

Product type	Commercial designation (manufacturer's trade name)	Features	Typical (automotive)
One-side 1/2	Paint Tite B (Inland) Jet Weld Ab (National Steel) Monogal (Usinor)	The light coated side of the differentially galvanized sheet is treated to form a ferro-zinc alloy with a coating layer weight of ± 30g/m^2. The zinc is brushed after galvanizing to leave a ferrozinc coating layer of some 2–10 g/m^2	Doors, wings, boot lid, bonnet (hood) liner, cross members
Electrozinc; pure zinc	(US Steel–Kawasaki) (Beautor)	Carosel process with soluble anode in vertical chamber Radiacol process with insoluble anode (US Steel patent) in vertical chamber Jet Flow process with horizontal insoluble anode	Doors, wings Boot lid
One-side	(Nippon Steel) In 1983 (Sollac–Usinor)		
Zinc alloys	Zn–Fe^3 Zn–Ni (Kobe Steel) NK FZ (Cr–Co) (Nippon Kokan) EZA (Nippon Kokan)	Electrodeposited alloys to enhance anti-corrosion at lighter coating weight	Development stage
Pure zinc + passivation layer	Zincrox (Teksid)	Electrodeposited zinc coated with chromium and a surface layer of chromium oxide	Development stage
Aluminized sheet	(Sollac) (Thyssen)	Sheet with aluminium coating applied on galvanizing line with ceramic bath to withstand temperature of 700°C	Exhaust silencer
Leaded sheet (Terne)	Usitomrne (Usinor)	Coating applied by passage of cold roller. Skin passed sheet through a molten lead bath solution containing 8–15% tin	Fuel tanks, air filters, radiators

Table contributed by G. Davies, Gaydon Technology Ltd. Reproduced by permission from G. Davies and R. A. Easterlow, 'Automative design and materials selection', *Journal of Metals and Materials*, (January, 1985) 20. Table based on results from an article by Quichon *et al.*, Galvano Organo Treatments de Surface Journal, number 528, and acknowledged with thanks.

Table 6.17 — Properties of aluminium available for car body construction

Alloy	Metallurgy	Advantages	Disadvantages
5XXX	Wrought alloys containing 2.5–3.5% magnesium	Relatively low cost (£1600–1800 per tonne) Good drawability Scrap more 'acceptable' to manufacturers	Prone to stretcher-strain markings Low strength Limited stretching performance
2XXX	Precipitation-hardening copper (2%+) alloy	Strength approaching that of mild steel Relatively good forming characteristics High and low YS version can be used for outer and inner (generally more complex) panels respectively, therefore no scrap segregation required Stretcher-strain free	High cost £1800+ per tonne Softens slightly on paint stoving Corrosion resistance marginally inferior to those of 5XXX and 6XXX in salt environment, unpainted
6XXX	Precipitation-hardening Mg–Si alloys	Same formability and strength characteristics as for Al–Cu alloys, also mixable scrap Some strengthening on paint stoving if high processing temperature used	Limited availablity in UK Requires carefull rocess control–continuous annealing for property control, therefore significant plant investment required

Table contributed by G. Davies, Gaydon Technology Ltd. Reproduced by permission from G. Davies and R. A. Easterlow, 'Automotive Design and Materials Section, *Journal of Metals and Materials* (January, 1985) 20.

Table 6.18 — By using a combination of spot welding and adhesive bonding, aluminium alloys may be used as structural parts of a lightweight vehicle. The table compares the strengths of aluminium alloys

Alloy	YS (N/mm^2)	UTS (N/mm^2)
5251 (5XXX)	85	180
2117 (2XXX)	157	291
CP609 (6XXX) as received	145	260
CP609 (6XXX) after 30 min at 195°C	250	300

Table contributed by G. Davies, Gaydon Technology Ltd. Reproduced by permission from G. Davies and R. A. Easterlow, 'Automotive design and materials selection', *Journal of Metals and Materials*, (January 1985) 20.

Plate 6.1 — Austen Rover's energy conservation vehicle. Plate supplied by Austen Rover and Gaydon Technology Ltd, and acknowledged with thanks.

temperature. Fig. 6.20 (restricted solid solution (left-hand side of the dual-phase compositions)) shows the form of phase diagram required. Phase diagrams where the solute line is nearly vertical will not form precipitation-hardening alloys (Section 6.25.1).

6.40 GLASS–REINFORCED PLASTICS AND POLYMERS

Polymers are discussed in greater detail in Chapter 11. Glass fibres used for reinforcement are discussed in Section 6.41.

6.41 GLASS

Glass is a hard brittle substance, made from silica sand (Fig. 6.10), which forms our glazings (windows). In order to be transparent to light, the glass must be **amorphous** or non-crystalline. If the molten liquid is allowed to crystallize, the resulting glass is opaque to light. This is the same phenomenon of water in the liquid state being 'transparent' to light but the ice formed on the cold window surface being opaque. This is because ice is the crystalline state, which makes the solid 'opaque'. Molten glass must be cooled quickly enough for it not to form a crystalline solid. Glass to the scientist is really a **supercooled liquid** which becomes a solid simply because its viscosity has increased so much with the fall in temperature that it has become a rigid substance which we loosely term a 'solid'. The silicates (Section 6.10), tars and bitumens ((Table 11.1, Section 11.4), are the only materials to show this large viscosity variation with temperature. Glasses are a common by-product of volcanoes.

Silica sand (SiO_2) (Fig. 6.10) is the main component of glass. Molten silica has a very high melting point (1700°C) and is a very viscous liquid in the molten state (Table 6.19(a)). Tar and bitumen products are similarly supercooled liquids where the solid state is discussed in terms of the 'liquid's' viscosity.

Table 6.19(a) — The viscosity of glass at various temperature at which the manufacturing processes are performed

Process	Temperature	Viscosity ($N\,s\,m^{-2}$)
Melting	1400°C	100
Working	1200°C	$1000 \rightarrow 10^6$
Softening	1000°C	$10^7 \rightarrow 10^8$
Annealing	500°C	10^{13}

The addition of metal oxides (sodium and calcium oxides) to sand at elevated temperatures depresses the melting point of sand to some extent (down to 800°C for sodium oxide). The manufacturer of glass uses this depression of the freezing point by alkali oxides as the lower melting point makes the moulding of glass artefacts cheaper. However, there is a limit to the amount of metallic oxides which can be added, because the glass starts to become soluble in water. Water glass is a glass-like substance which is soluble in water and is used to pickle eggs as the soluble water glass does not allow oxygen into the egg shell. Calcium oxide is added to reverse this solubility in water.

Glass is a mixture of the oxides of silica (SiO_2) (structure shown in Fig. 6.10) with sodium and calcium oxides (Na_2O, CaO) which on cooling do not crystallize. The solid glass has the structure of the liquid state. Glass owes its transparency and mechanical properties to the fact that the atoms constituting its chemical make-up are not ordered into a crystalline array. If the metallic oxides components of glass are kept low, the glass is insoluble in water and does form a very good corrosion-resistant material. Indeed, glass is used for the storage of many aggressive acids. Glass is also used to package food and drink because it does not impart a 'taste' or emit any foreign substance (toxic or non-toxic) into liquids. Glass is chemically attacked by acidic fluorides and acids of this type are is used for engraving, for security engraving where number plates are etched onto vehicle windows. Glass does not discolour and has very good dimensional stability.

6.42 STRESS RAISERS

Glass is susceptible to brittle fracture because, being an amorphous structure, there are no dislocations (1) to blunt a small micro-crack, (2) to reduce the stress-raising characteristics of a scratch or (3) to work harden the area where this deformation is concentrated. Surface imperfections therefore reduce the strength and ultimate usefulness of the amorphous material's properties. The tensile strength of glass produced without any surface imperfections is 7000 MN/m^2. This has to be specially produced. Non–specifically produced glass has a wide scatter in the mechanical properties and the tensile strength can vary between about 25 and 70 MN/m^2. Scientists, by modelling the fracture process, have been able to improve the mechanical properties of glass by heat treatment. These processes give rise to either '**laminated**' glass or '**toughened**' glass as used in motor vehicle glazings. Laminated glass is generally used for the windscreens and toughened glass for the side glazings and rear windows. These forms of glass are referred to as 'safety glass', which is defined by British Standard 857 as glass which when fractured is less likely to cause severe cuts or serious physical injury than ordinary glass. Ordinary glass fractures to give long razor-sharp 'splines' which are not restrained (see Section 6.44). The Road Traffic Act of 1930 made compulsory the fitting of safety glass in 1932. The two types of glass (laminated and toughened) are manufactured completely differently, but they both start from the same raw material, i.e. **float glass** (ordinary domestic glass).

6.43 MECHANICAL PROPERTIES OF GLASS

Glass has poor thermal conductivity and so unequal cooling rates set up thermal gradients whose expansion forces can exceed the mechanical properties and cause breakage. The ability of glass to withstand thermal gradients is dependent on its thickness. Glass undergoes elastic deformation up to the point of fracture (Fig. 6.47(a)). Both the fracture stress and the strain to fracture are low. The energy absorbed in fracture (the area under the stress–strain curve (Fig. 1.9)) is very low as a result of the shape of the stress–strain curve, which is a direct result of glass's being an amorphous supercooled liquid which does not contain any dislocations (Figs. 6.22

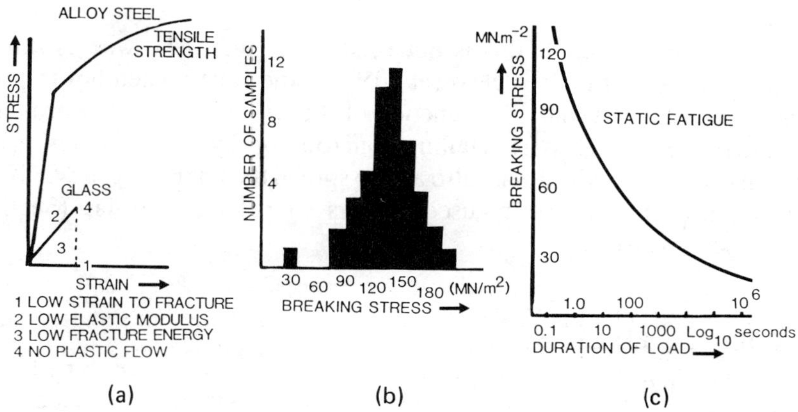

Fig. 6.47 — (a) The stress–strain curves for glass and steel. Glass has a low strain to fracture (1) and deforms elastically only (4). The elastic modulus of glass (2) is lower than steel and these two aspects mean that glass has a low energy of fracture (3). (b) The results from testing glass in tension have a wide scatter. The result obtained is dependent on the surface stress raisers. (c) The breaking stress of glass is dependent on the duration of the applied load. This is often referred to as 'static fatigue'. If glass is loaded rapidly, it will withstand a much higher load than if it were loaded slowly. (Drawn from information supplied by courtesy of Pilkington Glass, Ormskirk, and acknowledged with thanks.)

and 6.23). The absence of dislocations means that the glass lacks ductility (Fig. 6.24) and cannot be permanently deformed by an applied load. All permanent deformation must be carried out at an elevated temperature. The resultant glass product must be annealed otherwise the differences in cooling rates between the centre and external regions will freeze-in stresses which will cause the glass to shatter unexpectedly. The strength of glass is very dependent on the rate of deformation and the length of time that the glass is under stress (Fig. 6.47(c)). Glass has a history of failing under '**static fatigue**', where sudden unexpected failure arises at stresses which the component has withstood before. Glass gives no warning as to when it is about to break and does not withstand thermal or mechanical shocks unless certain additions are made to the melt process which confer on the molten liquid a much lower thermal expansion coefficient.

Table 6.19(b) — Thermal conductivities of two different glasses, one which is very resistant to thermal shock, the other not resistant to thermal shock

Name of glass	Use	Additions	Thermal expansion
Borosilicate	Ovenware	B_2O_3	3×10^{-6}
Window glass	Domestic	Na_2O	8×10^{-6}

6.44 WIRED GLASS

Contrary to popular belief, this is **not reinforced glass**. The wire is sandwiched between two sheets of molten glass (at 1050°C) and rolled when hot to produce a uniform composite material. The function of the wire is to hold any broken pieces together to prevent the glass from falling from roofs or skylights and causing injury to pedestrian traffic. The wired glass also acts to spread the temperature evenly through the glass in the event of a fire. It is used in doors where laminated glass (Section 6.46) would be too expensive.

6.45 TOUGHENED GLASS

The weakness of glass lies in the defects at the air–glass surface. Small scratches, which the eye cannot resolve, act as stress raisers. Glass is cut with a diamond scribe which causes a deep scratch. This scratch acts as a stress raiser, and for the glass to break along the scribe line the stress concentration has to exceed the stress raiser of the normal inherent defects for crack propagation. If the stress raiser caused by the inherent defects is greater than the stress raiser of the scribe line, the glass fracture will run off the scribe line. The scribe line will also contain many small cracks at right-angles to the major line. In toughened glass, the air–glass interface is placed into compression so that the inherent flaws and scratches do not act as stress raisers until this compressive stress has been overcome by externally applied tensile forces. Thus these external forces in the outer layers of glass are balanced by tensile forces within the central region of the glass which has no glass–air interface (Fig. 6.48) and so cannot initiate fracture. The surface compressive forces can be some 2–2.5 times the central tensile forces. In some cases these compressive forces can be up to 3.0–4.0 times the central tensile forces. For most applications the stress profile is a parabolic distribution as shown in Fig. 6.48. Toughened glass can be produced by **thermal** processes (e.g. for vehicle glazings, because the thickness is uniform) or by **chemical** processes. Chemical processes are used where the thickness is not uniform, e.g. for milk bottles. Chemical toughening involves the substitution of sodium or lithium ions by the much bulkier potassium ion in the surface layers. The toughening arises because the potassium ion (0.133 nm in size) is larger than the sodium ion (0.098 nm) and the lithium ion (0.78 nm) (Table 6.4).

These thermally induced compressive forces are produced after all the cutting, forming and bending processes have been undertaken. The preformed unit is heated to a temperature of about 600–670°C where the viscosity of the glass is just sufficient to relax most of the manufacturing stresses but not high enough to allow any shape deformation. The glass is then cooled quickly by cold air jets which are played onto both surfaces of the glass. This cools the surface layers more than the central region which remains hot as a direct result of glass's low thermal conductivity (0.8–1.3 W/m°C). After the external surfaces have cooled and stiffened, the central hot region cools to room temperature, putting the outer precooled layers in compression. This makes the toughened glass very strong (some three times stronger than annealed glass of the same thickness). In order to break the glass by propagating a small surface flaw, the outer surface compressive stresses have to be exceeded, in conjunction with the bulk internal central stress distribution which is free of surface stress-raising defects. When the toughened glass does break, the tensile 'locked-in'

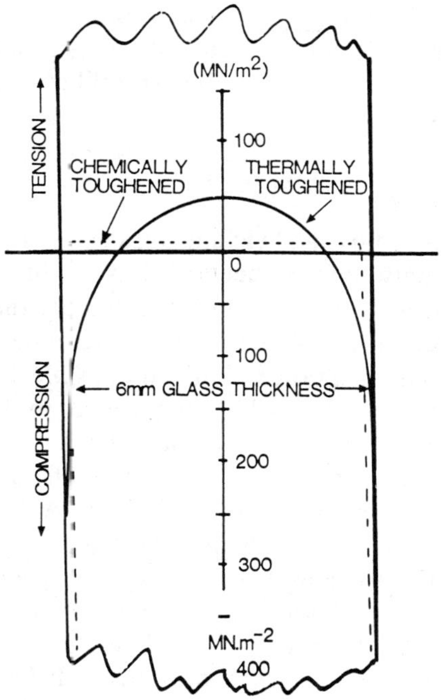

Fig. 6.48 — Toughened glass is made with the surfaces in compression which is attained by cooling the surface while the centre is hot. When the centre cools to room temperature, the associated contraction places the surface layers in compression, thus closing up the surface micro-cracks. The stress distribution is parabolic. Glass can also be toughened by chemical means. The stress distribution is flat topped as shown. Chemical toughening can be undertaken on glass of varying wall thickness. Thermal toughening can only be undertaken on uniform thicknesses of glass. (Drawn from information supplied by courtesy of Pilkington Glass, Ormskirk, and acknowledged with thanks).

stresses within the central region initiate multiple cracks. Since the fracture surfaces are perpendicular to the surface, the razor-sharp dagger-like splines are not produced. The toughened glass breaks into fairly small cubes as the complex locked-in stresses are relieved.

When car windscreens were first glazed, they were flat and nearly upright. The cracking of toughened screens into small cubes did not present a great problem in obscuring the driver's vision of the road ahead. Almost all car glazing is now curved and raked backwards. This causes problems as the crack interfaces reflect light so that the drivers vision is obscured. This can be overcome using curved laminated screens.

6.45.1 Producing curved windscreens

There are two methods of producing the curvature required for the modern raked windscreen. These are **die bending** and **sag bending**. The more modern process is sag bending where the sheet of glass is heated to its softening point of 580–670°C (viscosity, $10^{6.6}$ N s m^{-2}) in a furnace supported on a cushion of air which prevents the glass from contacting the support bed. The glass at this temperature is quite 'floppy' and sags under gravity. The glass is then chill cooled in order to lock in the thermally induced stresses.

6.46 LAMINATED GLASS

Laminated glass is produced by sandwiching a transparent plastic sheet (poly-vinyl butyral; 0.76 mm thick) between two sheets of glass and firing the resultant matrix at 150°C to an adherent sheet. At this very low temperature there is no formability left in the two sheets of glass. The two sheets of glass have to be made as a pair. They are bent together as a pair in the sag process with an infusible dust layer to keep the two sheets apart. These two sheets have different stresses induced (Fig. 6.49(c)). The outer sheet must be low stress toughened glass. The inner glass is toughened to a higher extent. Hence either (or both) of the two sheets of glass can break without producing complete disintegration of the windscreen. As the outer surfaces are not toughened to the same extent as toughened glass, there is no shattering into small pieces on impact with road stones as there is with the toughened glass screens. Hence only limited cracking is obtained within the outer sheet, and the driver's vision is not impaired. The inner sheet of the pair is toughened to a higher degree so that it shatters into small pieces if the head is thrown forward. This reduces the amount of cutting and head injuries. Advertisements often show the laminated glass being

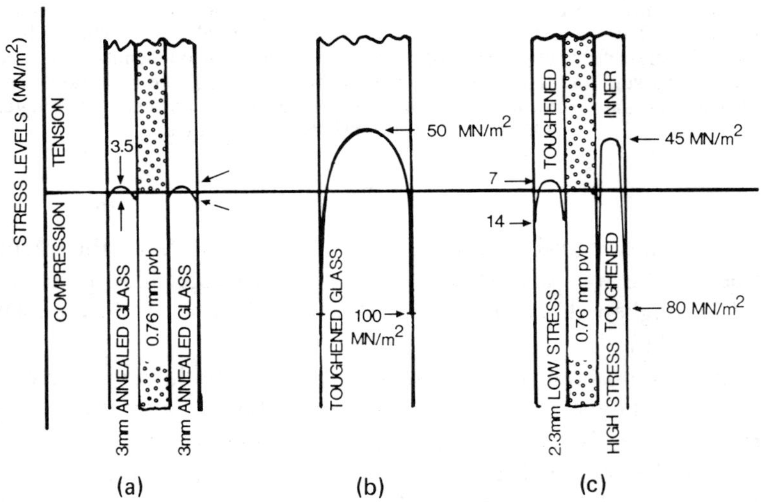

Fig. 6.49 — Laminated and wired glass are composite rolled units. (a) indicates a conventional laminate between annealed float glass which are used in domestic patio doors. The laminate material is polyvinyl butyral (PVB)interlayer. (b) Ordinary toughened glass with the compression (100 MN/m^2) and tensile (50 MN/m^2) stresses shown. (c) Modern laminated windscreens are made from low stress toughened glass on the outside, with high stress toughened glass towards the occupants. The low stress toughened glass fractures producing single cracks, while the inner higher stressed glass fractures to produce small 'cubes'. (Redrawn from information supplied by courtesy of Pilkington Glass, Ormskirk, from D. S. Oliver and G. J. Copley, Glass for Construction Purposes.)

attacked with hammers to no detriment. The middle sheet of plastic (plasticized polyvinyl butyral) can undergo an elongation of 200–350% before fracture. The outer sheets of glass, being a good corrosion-resistant coating, prevent the degradation of the plastic and ensure complete transparency over the service life of the laminated glass.

6.47 FIBRE GLASS REINFORCEMENTS

Glass fibres may be used as a reinforcement. In this technology the surfaces of the glass fibres are protected and cracks only propagate through a single fibre. The fibres are arranged so that their outer surfaces are in compression by chilling the fibres from a high temperature (compare toughened glass). These glass fibres can also be treated with a tin compound which diffuses into the surface layers of the glass fibre. The mechanical strength and resistance to abrasion and the formation of micro-cracks within the surface layers are therefore superior to those of ordinary glass. Thermoplastics (Section 11.5) can be strengthened by the addition of the glass fibre (Table 6.20).

Table 6.20 — Reinforcement of plastics: strengths of thermoplastics and thermosets

	Young's modulus ($\times$ 10 NM^{-2})		Tensile strength ($\times$ 10 NM^{-2})	
Polymer	Unfilled	Filled	Unfilled	Filled
GRP (thermoset)	—	30	—	1200
Polypropylene	≈1.2	4	35	50
ABS + Glass	2	6	40	100
Polycarbonate	2	4	70	60

The stress σ and elasticity of a fibre composite are given by the relationship

$$\sigma = V_f\sigma_f + (1 - V_f)\,\sigma_m$$

Young's modulus is given by

$$E = V_f E_f + (1 - V_f)\,E_m$$

In these equations, V_f is the volume fraction of fibre, and σ_f the strength of the fibres, σ_m the strength of the matrix and E_m the Young's modulus of the matrix.

Glass fibre is some 30 times stiffer than the polymer and volume replacement glass fibre is much cheaper than the plastic polymer. A panel made of steel is some 10 times stiffer and 2.5 times stronger in tension than the corresponding panel made from glass-reinforced polymer. Hence, for a glass-reinforced polymer panel to have the same stiffness as the steel panel, it would have to be increased in thickness or designed with double or complex curvatures to increase the stiffness. Often other additional (inert) materials are also used to increase the stiffness of plastics. These are referred to as fillers (e.g. Chalk, fuller's earth etc.: see Section 11.17).

6.48 SURFACE COATINGS

Glass is transparent to light and heat (diathermous). Larger glazings increase the solar energy gain of the vehicle interior. This causes problems with the plastic components within the glazed area. Surface coatings of dielectric metal oxides are applied to glass in order to control the transmission properties of the glass to particular wavelengths from the solar spectrum (Fig. 6.50). The oxides often produce a colour tint to the glass. The reduction in total heat admitted to the vehicle interior as shown in Fig. 6.51 is obtained without reducing the rigorous standards required by the Road Traffic Acts for undistorted vision.

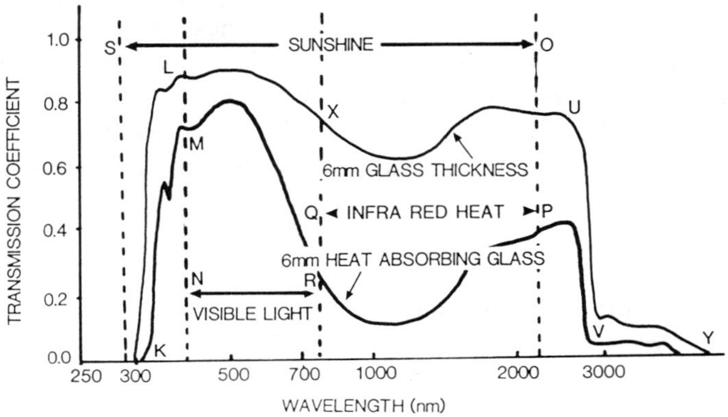

Fig. 6.50 — The energy radiated from the sun is of many wavelengths, but only a small fraction of this radiation (NR) produces the sensation of vision. Amongst much of the radiation is the infrared radiation (QP). Only a small fraction (KY) of the total radiation is transmitted by glass. The heat-absorbing glass absorbs the infrared (QP) and red (XR) end of the visible spectrum and therefore appears with a 'blue' or grey colour. This absorption heats up the glass and therefore the glass reradiates more heat than the conventional toughened glass (Fig. 6.51). (Redrawn from information supplied by courtesy of Pilkington Glass, Ormskirk, from D. S. Oliver and G. J. Copley, Glass for Construction Purposes.)

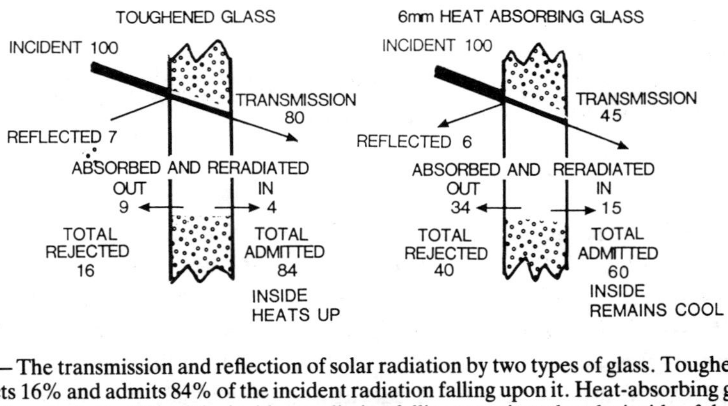

Fig. 6.51 — The transmission and reflection of solar radiation by two types of glass. Toughened glass rejects 16% and admits 84% of the incident radiation falling upon it. Heat-absorbing glass rejects 40% and admits 60% of the incident radiation falling upon it so that the inside of the car remains cool. (Redrawn from information supplied by courtesy of Pilkington Glass, Ormskirk, from D. S. Oliver and G. J. Copley, Glass for Construction Purposes.)

6.49 FUTURE DEVELOPMENTS

Over the last 20 years there has been an annual 3% growth in the car's glazing area. This increase in area gives better visibility to the driver and provides a feeling of spaciousness to the occupants. Another trend has been started by the Energy Protection Act (Section 5.1) which requires the manufacturer to obtain better fuel consumption and reduce the weight of the vehicle. The vehicle glazing adds considerably to the weight of the vehicle and, if thinner glazings could be developed, not only would the weight of the vehicle be reduced but so would its centre of gravity. Approximately 22% of the weight of glass can be saved by reducing the glazing glass from 5 to 3 mm toughened glass and 6.8 to 5.0 mm laminated glass. If this were carried out across Europe some £150 million of petrol could be saved per year (Pickard, 1979). Pickard compared a 1969 model with a 1975 model where the glazed area went up from 2.35 m^2 to 3.24 m^2. The use of the thinner glazing would have given a weight saving of 22%.

Furthermore, the width of the A pillars can be reduced with the adhesively bonded laminated windscreen which can now strengthen the body structure. Modern adhesives (Section 11.26) can insulate the glass from any mechanical strains and provide mechanical rigidity to the unitized structure.

6.50 GLASS CERAMICS

Ceramics are fundamentally the same as glass in all aspects of working and manufacture. Ceramics are usually high in aluminium oxide (Al_2O_3) which promotes a crystalline structure on cooling. The fine crystallite size gives an opaque structure. Ceramics have much higher strength and shock resistance properties (both thermal and mechanical) than glass owing to their much smaller crystallite size (compare Fig. 6.19). Ceramics therefore find greatest use in sparking plug insulations, where good electrical and thermal properties are required. They are also used domestically in the 'oven to table' kitchen utensils.

6.51 CRACK STOPPING AND CRACK INITIATORS

6.51.1 Crack stopping

There are several ways of stopping a crack from propagating. The surface treatment of toughened glass shows the main principle which the construction industry used in the building of railway bridges, aquaducts etc. with their majestic arches. Cathedral domes and their associated buttresses are another example where the whole structure is in compression. The answer is: design for the structure to be in compression. The greatest disadvantage that an engineering material can have is a lack of toughness which implies a poor resistance to the ability to stop or blunt cracks. Dislocations are a good method of blunting a crack but, as every person undertaking DIY engineering knows, it is very easy to break a metal by subjecting it to very large stress cycles. These reverse cycle stress cycles tend to sharpen up the crack when the blunted crack is placed in the compression cycle of the alternate bending programme. This is called 'high strain fatigue' (Section 6.35.3). Fatigue is damaging to metals and exceedingly damaging to composites.

6.52 CRACK INITIATORS

Scientists have been able to model what happens to the crack tip. Griffith's concept was that the stress σ_t at the tip of a crack of length l, and of radius r magnifies the applied stress σ_a according to the formula

$$\sigma_t = \sigma_a \left(\frac{l}{r}\right)^{1/2} \tag{6.33}$$

Hence a small surface defect whose length is 1000 times the interatomic bond distance and that has been sharpened up by being in compression so that the radius of curvature is of the order of the interatomic bond gives a concentration factor of some 300. This technology is used in many engineering design functions where sharp 90° are replaced by radiused fillets. Cracks may be stopped from spreading in Perspex by drilling out the front of the crack with a small drill, which effectively blunts the crack as a stress raiser.

Griffiths also modelled the fracture process by correlating the relationship between the fracture stress σ_f required to form a crack in non–crystalline solids which behave elastically in that the interatomic bonds are stretched to their breaking point. In breaking the interatomic bonds, a new surface is created. This new surface requires energy to form.

$$\begin{matrix}\text{stress required}\\ \text{to form a crack}\end{matrix} = \sqrt{\left(\frac{\text{surface energy} \times \text{Young's modulus}}{\text{crack length}}\right)} \tag{6.34}$$

Putting in typical values of surface energy and Young's modulus for glass fibres shows that a critical crack length of 100 times the interatomic distance is required to initiate fracture.

REFERENCE

Pickard, J. (1979). Automotive Engineer, **4** (3) 54–56.

7

Basic corrosion theory

7.1 CORROSION SCIENCE — EQUILIBRIUM

This chapter develops the ideas behind corrosion so that the reader becomes conversant with the many half-truths and myths which surround the multidisciplinary subject of corrosion. By taking the scientific approach, it is hoped that the reader will be able to overcome the elementary mistakes that are made by having little or no knowledge of science.

The shorthand chemical formulae always act as a stopper to the gain of knowledge, and yet their use is a very good way of putting down knowledge in a concise way. It can be compared with a shorthand typist's notes, which are completely unintelligible to the person without a knowledge of shorthand.

Batteries use the corrosion process to produce a voltage. The prime function of a battery is to turn **chemical energy** into **electrical energy**. This chemical energy is the energy, in the form of say heat energy, which is required to produce (or win) a metal from its oxide or ore in which it exists in the earth's crust.

$$\underset{\text{Requires energy to go in this direction}}{\text{Metal-containing ore as the oxide} \rightarrow \text{metal}} \tag{7.1}$$

$$\underset{\text{Gives out energy to go in this direction}}{\text{Metal} \rightarrow \text{oxide}} \tag{7.2}$$

Reaction (7.2) is used commercially in the production of pretty fireworks, which use aluminium and magnesium metallic flakes for this reaction.

In the case of a battery the corrosion process is controlled by stopping the flow of electrons in the metallic part of the circuit in order to preserve the chemical state of the metal which is zinc. A section through a popular 1.5 V battery and its associated external circuit are shown in Fig. 2.5. We shall return to the commercial battery later (Section 7.16). However, for the purposes of this exercise consideration will be given to both the dry cell battery and the lead acid 12 V car battery shown in Fig. 7.1.

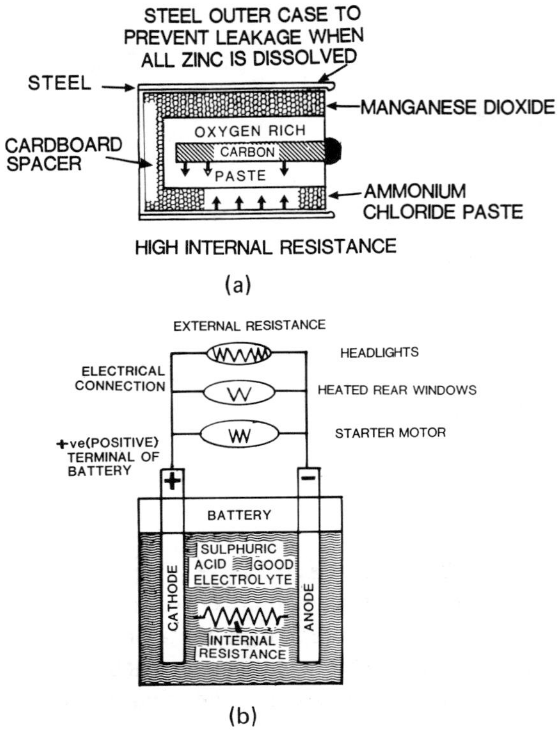

Fig. 7.1 — (a) A cross-section through a 1.5 V dry cell battery. The battery produces electrons at the negative terminal as the zinc outer case dissolves. In order to make 'leak-proof' batteries, the zinc case is given an external layer of steel. The oxygen is provided from a chemical compound in the ammonium chloride paste which acts as the electrolyte. Several ($8\times1.5=12$) of these dry cell batteries can be joined up in series to produce 12 V. This will not start the car because there is a large potential drop within the paste electrolyte owing to the 'high internal resistance' of the battery. The 12 V car battery, in contrast, has a very low internal resistance which allows the battery to produce the very high starting current required to turn the starter motor over. If other items are also run off the battery at the same time as the starter is activated, the 12 V supply available for the battery drops to sometimes unacceptable levels (most motorists will have noticed the brightness of the head-lights dropping as the starter motor is engaged).

As shown in the introduction to Chapter 6, there are a number of metallic elements in the periodic table. The metallic elements all have a different tendency to ionize. This is shown in Table 7.1.

$$\text{A Metal in the solid state} \rightleftharpoons \text{the ionic form dissolved in water plus electrons} \quad (7.3a)$$

$$\text{i.e. } M \rightleftharpoons M^{n+}+ne^{-1} \quad (7.3b)$$

where n is the natural valency of the metal (Sections 6.7–6.9) and M is any metal.

If the reaction proceeds to the right it is a **corrosion reaction**, while if it proceeds to the left it would be an **electroplating process**. If the rate of the reaction to the right equals the rate of the reaction to the left, there exists a state of **equilibrium**.

Table 7.1 — Part of the electrochemical series

'Noble' end	Half-cell reaction	Voltage (V)
Gold	$Au^{3+}+3e^{-}\rightarrow Au$	+1.42
Oxygen electrode	$O_2+2H_2O+4e^{-}\rightarrow 4OH^{-}$	About +1.40
Silver	$Ag^{+}+e^{-}\rightarrow Ag$	+0.7995
Copper	$Cu^{2+}+2e^{-}\rightarrow Cu$	+0.3448
Hydrogen (1 atm)	$H^{+}+H^{+}+2e\rightarrow H_2$	0.0000 (by definition)
Lead	$Pb^{2+}+2e^{-}\rightarrow Pb$	−0.126
Tin	$Sn^{2+}+2e^{-}\rightarrow Sn$	−0.186
Nickel	$Ni^{2+}+2e^{-}\rightarrow Ni$	−0.250
Cadmium	$Cd^{2+}+2e^{-}\rightarrow Cd$	−0.4020
Iron	$Fe^{2+}+2e^{-}\rightarrow Fe$	−0.440
Chromium	$Cr^{3+}+3e^{-}\rightarrow Cr$	−0.71
Zinc	$Zn^{2+}+2e^{-}\rightarrow Zn$	−0.7620
Aluminium	$Al^{3+}+3e^{-}\rightarrow Al$	−1.67
Sodium	$Na^{+}+e^{-}\rightarrow Na$	−2.712
Magnesium	$Mg^{2+}+2e^{-}\rightarrow Mg$	−2.875
'Base' end		

The 'half-cells' for most of the common metals in combination with their own ions is tabulated above. Each reaction has been written so as to consume electrons. The oxygen electrode reaction is strictly not part of the electrochemical series; it has, however, been included here as it is the main reaction for the corrosion of the motor vehicle. The more negative the electrode potential, the more stable is the ionic form, or the more unstable is the metallic state. An example of this would be sodium, which will react with water with the evolution of hydrogen. The reaction is so vigorous that the hydrogen and sodium spontaneously ignite. If the two reactions could be separated, a voltage of −2.7 V would be produced.

Therefore an electrode is no longer at equilibrium when a net current flows to or from its surface. The measured potential of this electrode–ion system deviates from the equilibrium potential depending on the magnitude of the current flowing. Therefore in using the equilibrium potentials of a metal in its environment, we are only able to obtain a corrosion '**tendency**' and we cannot express a corrosion rate. As corrosion rates are the main requirements for the designer and end user, it is particularly important that these equilibrium potentials are used with caution. Aluminium, for example, from its equilibrium potential shows a pronounced tendency to react nearly as vigorously as sodium, but in actual fact aluminium shows a greater stability in a lot of environments. It is used for kitchen utensils where it is more resistant than some of the metallic elements which show, by virtue of their potentials, less tendency to react. These other metallic elements (e.g. uncoated steels) are in fact worse than aluminium.

Each corrosion reaction consists of two reactions: an anodic half-cell and cathodic reaction or half-cell. These two **half-cells** (anodic and cathodic) make up the battery or corrosion cell. There is no difference between a corrosion cell and a battery. **A battery is a corrosion process which we want to make more efficient. A corrosion cell is an unwanted reaction which we want to stop or make less efficient.**

The anodic reaction (the dissolving electrode) rises in potential when a current is drawn from the equilibrium potential (Fig. 7.2). The cathodic potential falls from its equilibrium potential when a current is drawn from the equilibrium potential (Fig. 7.2). The extent of this potential difference when measured in volts is referred to as **polarisation** (Section 7.18).

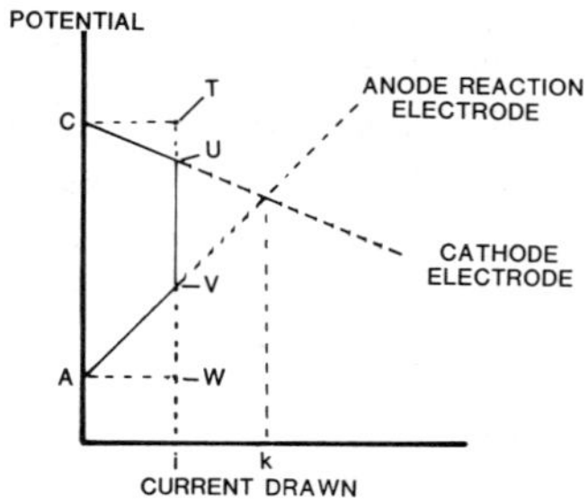

Fig. 7.2 — The potential difference between the electrodes of a battery or a corrosion cell falls with respect to the current drawn from the cell. This is modelled by considering a voltage drop across an internal resistance within the cells shown in Fig. 7.1(a), which arises because the products of each electrode reaction remain adjacent to the electrode. Hence representing the anodic potential and cathodic potential by A and C respectively, the external potential difference measured for a current i flowing is UV. The open circuit potential AC has dropped owing to a voltage drop across the cathode environment of 'TU' and anode environment of 'VW', making a total potential drop across the internal resistance of 'TU+VW'. These potential drops are referred to as polarizations: thus 'TU' represents the cathodic polarization, and 'VW' the anodic polarization.

If the metal concerned were iron (Fe) the reaction would be written:

$$Fe \rightleftharpoons Fe^{2+} + 2e^- \tag{7.4}$$

Note: The chemist's shorthand symbol for iron is Fe and it should be remembered that the letter e in Fe has nothing to do with the shorthand representation for the electron.

These anodic and cathodic reactions at the electrodes are referred to as half-cells, in that they soon stop with the build-up of either or both ions and electrons on the right-hand side of the equation (in practice this means a build-up of electrons in the metal and a high concentration of ions adjacent to the metal environment). There is a voltage associated with these half-cells, but to measure voltage or potential difference we have to make a connection to another half-cell (Fig. 7.6) or have some elaborate method of measuring the potential of the ion in solution. The 'other half-cell' is usually given the position of being an **arbitrary zero** (in other words potential is measured with respect to this other electrode). This arbitrary zero must be fairly reproducible so that potentials can be measured in England, America, Australia etc. and the results be the same. A good analogy of this process is the ability to measure a distance. Historically, distances were initially measured using the human being's anatomy, an inch being the distance of the thumb finger nail to first joint, and the yard being the average person's step. Obviously these 'reference units' changed from person to person and, after several revisions throughout our scientific history, the standard unit of distance is the metre. The metre is defined as the 1 650 763.73 of the wavelengths, *in vacua* of the orange light emitted by Kr^{36}_{86} in the electron transition $2p^{10}$ to $5d^7$. (Note that 36 is the number of protons, or the Atomic Number (Section 6.4), and therefore automatically identifies the substance as krypton (Kr), and 86 is the atomic weight, 2p and 5d refer to the electron levels discussed in Section 6.6.)

In corrosion, this main reference electrode was taken as the **hydrogen electrode** which was given the status of being the reference arbitrary zero. Today it is not favoured as the reference electrode, although all voltages of the half-cells are still given with respect to the hydrogen electrode. Other more practical electrodes are used. These more practical **reference electrodes** are copper in contact with copper sulphate, silver in contact with silver chloride or mercury in contact with mercury chloride (Fig. 7.3). Each one will give a different half-cell potential.

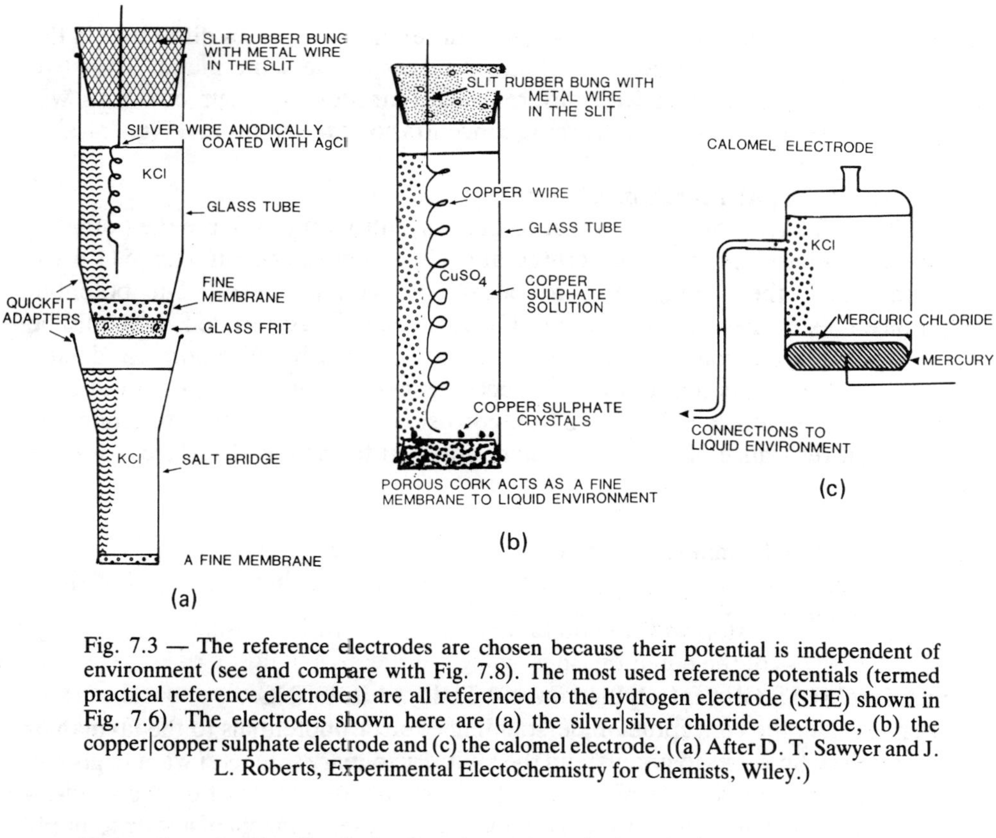

Fig. 7.3 — The reference electrodes are chosen because their potential is independent of environment (see and compare with Fig. 7.8). The most used reference potentials (termed practical reference electrodes) are all referenced to the hydrogen electrode (SHE) shown in Fig. 7.6). The electrodes shown here are (a) the silver|silver chloride electrode, (b) the copper|copper sulphate electrode and (c) the calomel electrode. ((a) After D. T. Sawyer and J. L. Roberts, Experimental Electochemistry for Chemists, Wiley.)

Using these half-cells all the metals can be tabulated with respect to the arbitrary zero and the results are given in Table 7.1. Each corrosion reaction consists of two half-cells. There are three factors which affect the voltage of these half-cells which take part in a corrosion reaction. These half-cells can also form the basis of a battery which would be composed of two half-cells. Differences in potential can be produced by the following.

7.1.1 Dissimilar electrode cells

These cells include the combination of two different metals acting in conjunction with each other and this form of battery–corrosion cell will be considered in Sections 7.3, 7.6 and 7.15.

7.1.2 Deformation cells

If the metal is cold worked or two identical metals with different grain sizes (Fig. 6.19 and Section 6.21), or degrees of work hardening (Figs 6.24 and 6.30) are combined (Section 6.22), a difference in potentials will be exhibited (Fig. 7.4(b)). This means that the same metal will have a different tendency to corrode depending upon its prior history and 'state' of deformation. This is a small effect and sometimes gives rise to the myths about the corrosion phenomena and the metallic state. For newcomers to science it is best to consider this at a later date.

7.1.3 Concentration cells

These are cells which will produce a potential difference if there is a difference in the concentration of ions between two identical pieces of the same metal. Thus two pieces of copper in contact with different concentrations of their own ions will corrode until both metals are in the same concentration of copper ions (Fig. 7.4(a)).

7.1.4 The differential aeration cell

The differential aeration cell is similar to the concentration cell. Here the potential difference between the two differently aerated electrodes can be fairly large (1.4 V). This voltage is the driving force for most corrosion processes. The potential difference falls as the two pieces of steel become equally oxygenated. It should be pointed out that the voltage cannot be detected normally because both anode and cathode are short circuited in everyday occurrences. The voltage is demonstrated using the U-tube shown in Fig. 2.4. The (corrosion) current is therefore limited by the external resistance of the environment adjacent to the metal surfaces (Section 10.1).

7.1.5 Differential temperature cells

A metallic component immersed in the same liquid will exhibit corrosion if there exists a temperature gradient along its length.

7.2 CONCEPTS OF POTENTIALS AND THEIR MEANING

To help explain the conceptual understanding of electropotentials to the layman or non-scientist who wants to understand the multidisciplinary subject, I want to give an analogy of potential (and its effect) with temperature and its effect on the solid-to-liquid transition. This comparison is not meant to compare mechanisms or to imply there are similarities in the fundamentals between corrosion and melting. The fundamentals of corrosion and melting are **completely different**.

Temperature is a scale which does not require another medium or reference point to get a reading with a thermometer. This referencing of the thermometer has already been undertaken when the thermometer was made. Mercury-in-glass thermometers are made by placing a known volume of mercury in a glass bulb to which

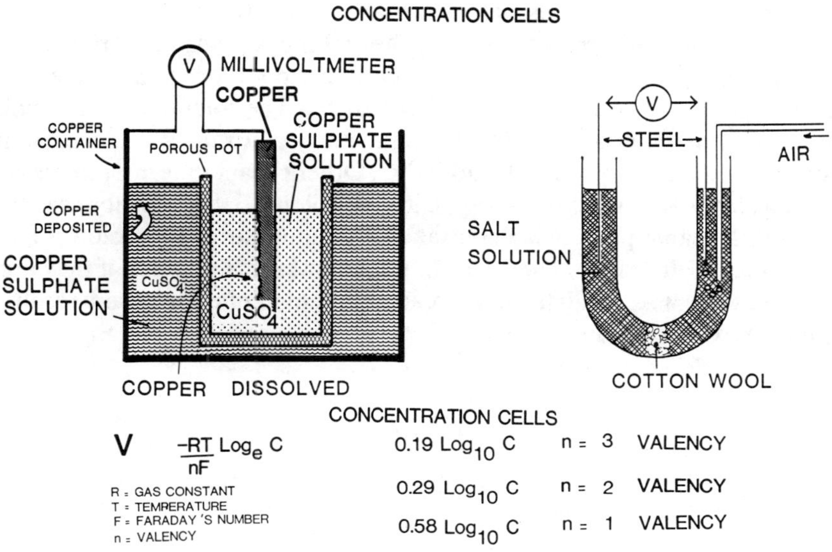

(a)

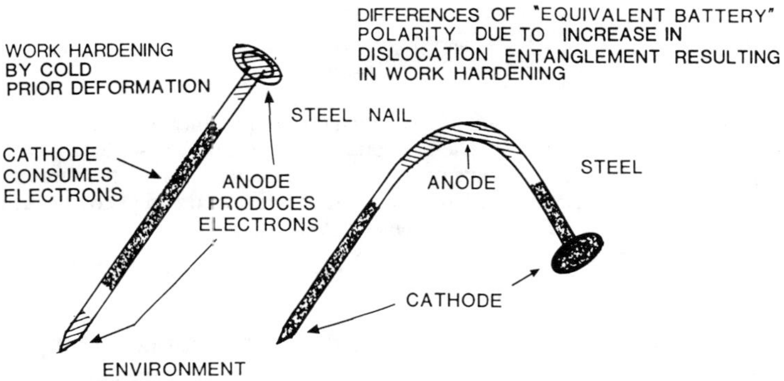

(b)

Fig. 7.4 — (a) If copper is immersed in two different concentrations of copper sulphate, a concentration potential will be produced. This can easily be demonstrated in the laboratory. The corrosion of the motor vehicle is due to a concentration cell. It is the concentrations brought about by the differences in oxygen replacement within certain areas (crevices, dirt deposits and areas which have restricted drying). (b) Differences in the amount of deformation or work hardening (Section 6.24, Figs 6.22–6.27) cause differences in electrochemical potentials. A nail is made from cold rolled steel. The head is subsequently made by an 'up-setting' process which involves deforming the head with a 'hammer'. The point is made by 'flying shears'. A nail placed in water will therefore corrode with the head and point as the anodes. The less deformed shank will be the cathode. It is possible to reverse these potentials by bending the shank of the nail, which produces more work hardening in the shank than in the head and point. The newly deformed shank becomes anodic. There are thus two means of reversing the polarity of corroding steel: differences in deformation and in oxygen replenishment.

there is a small capillary. Heat energy causes the mercury to expand and the length of this column of mercury is measured at two arbitrary reference points, the melting point of ice and the boiling point of water. The melting point of ice is arbitrarily given the 'honourable' position of 0°C and to the boiling point of water at the standard atmosphere is ascribed the arbitrary value of 100°C. The column is then subdivided and marked into 100 equal divisions. The mercury is *assumed* to have a uniform coefficient of expansion between 0 and 100°C. Other cheaper thermometers are also made using alcohol (which is dyed to give it a red colour so that it is more easily seen) in glass and the same process is undertaken. Again uniform expansion is assumed over the temperature range. However, this assumption is incorrect: if they are both placed in the same water bath they will 'read' different temperatures as a result of this non-uniform expansion (Fig. 7.5).

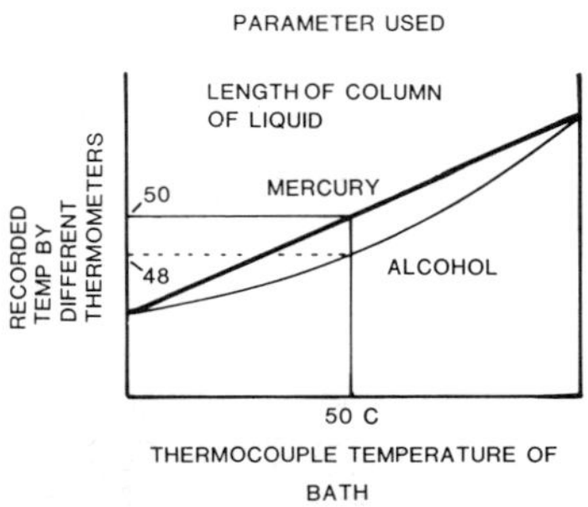

Fig. 7.5 — The figure shows the result of measuring the temperatures using three temperature measuring devices (thermometers). The temperatures are measured by alcohol and mercury (both liquids contained in glass) thermometers and are plotted against the temperatures as measured by the third thermometer. The temperatures measured at 50°C by the alcohol and mercury thermometers will disagree.

With potential differences this is not done. Each reading is compared with the artificial zero as shown in Fig. 7.6 and recorded. Thus in the presence of its own ions a metal (e.g. zinc in a solution of zinc sulphate) has a characteristic potential relative to the solution. If the potential of the metal is artificially raised by some outside source, then more of the metal will dissolve so as to increase the concentration of its ions in solution. Thus at elevated potentials the ions are more stable whereas if the potential is artificially lowered, then the metallic state becomes more stable and some of the metallic ions will tend* to plate out (electroplating). Similarly, if we consider the melting point of a metal to be analogous to the electrode potentials, then a raising of the temperature means that more metal will melt and the liquid state is more stable. In contrast, if the temperature is lowered the metallic state is more stable. This is

*There is a certain activation energy or energy hill which has to be overcome before plating out proceeds. The everyday analogy to this is a roller coaster. The carriages have to go up an incline to race down the other side. If the inclune (equivalent to the activaton energy) were any higher, the carriages would not reach the top and would roll backwards.

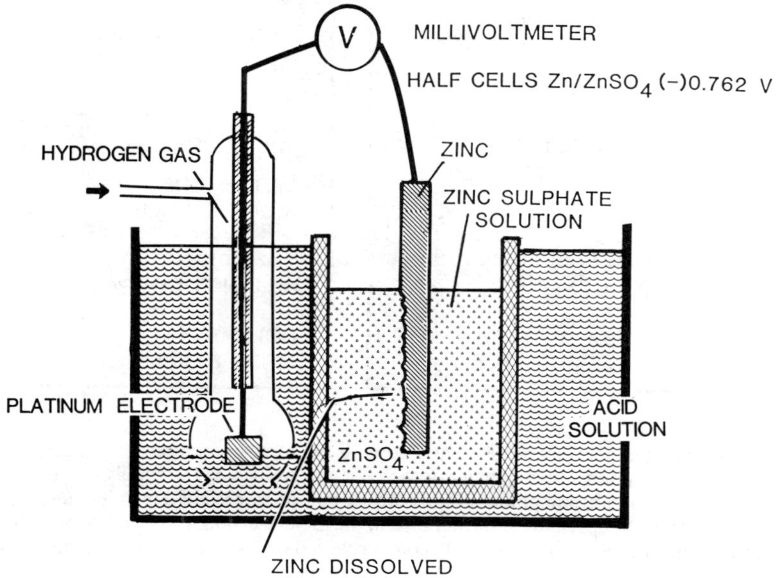

Fig. 7.6 — Schematic view of the potential of zinc, in contact with its own ions determined against the hydrogen electrode. This can be repeated for all metals to produce the electrochemical series. This series gives no reference to the corrosion rate of the metal concerned. Note: equilibrium means that the rate of zinc dissolving is equal to the rate at which the zinc is plated out.

shown diagrammatically in Fig. 7.7. The analogy therefore is that, if a metal has a melting point of 800°C, then its liquid form will be more stable at a temperature of 801°C than the solid form, while at 799°C its metallic solid state will be more stable because the equilibrium

$$\text{metal solid} \rightarrow \text{liquid state} \tag{7.5}$$

takes place at 800°C. If a metal has its potential raised above the equilibrium E_m potential given in Table 7.1, then it will dissolve and prefer to be in the ionic state whereas, if the potential is lowered, the metallic state is more stable than the ionic state. This is shown diagrammatically in Fig. 7.8 where the environment is also considered. Environment (considered in greater detail in Section 7.10) affects the ease with which a metal dissolves but does not influence the melting point of a metal. A few examples will be given to illustrate the point and then its application to the motor industry will be considered.

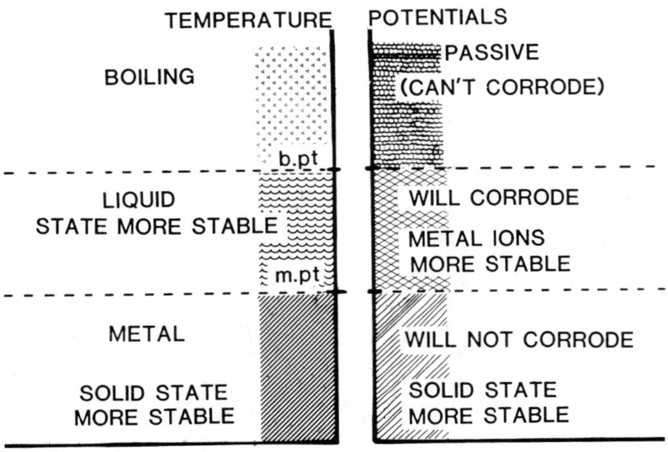

Fig. 7.7 — The temperature scale compared with the electropotentials of a metal. A metal above the melting point will melt at a rate which is determined by the supply of latent heat. A metal above the equilibrium potential will corrode but the rate is not dependent on latent heat or potential. If the temperature is further raised, the metal will boil. If the potential of the metal is raised, it often ceases to dissolve and becomes covered with an oxide film. The metal is said to be passive.

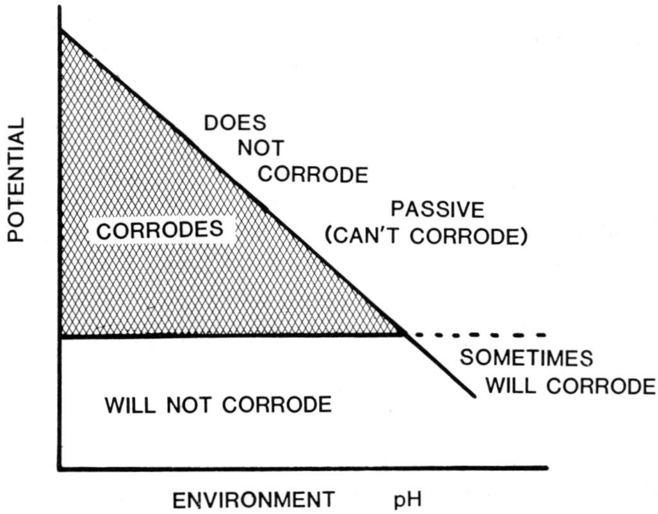

Fig. 7.8 — The melting point of a metal is independent of the environment. The ability of a metal to corrode is dependent on the pH (Section 6.17) of the environment. In this case the region of potential above the equilibrium potential means that the metal will corrode. A potential below the equilibrium potential means that the metal will not corrode because it cannot. These areas are dependent on pH, and there is an area for some metals where the metal cannot corrode even above the equilibrium potential. This is due to the formation of a passive oxide film. This is shown diagrammatically here: see Section 7.10.

7.3 EARLY BATTERIES

One of the earliest batteries (1836) (Fig. 7.9) was made from copper in copper sulphate (+0.34 V with respect to the hydrogen electrode (Table 7.1)) coupled to zinc in zinc sulphate (-0.76 V with respect to the hydrogen electrode). This

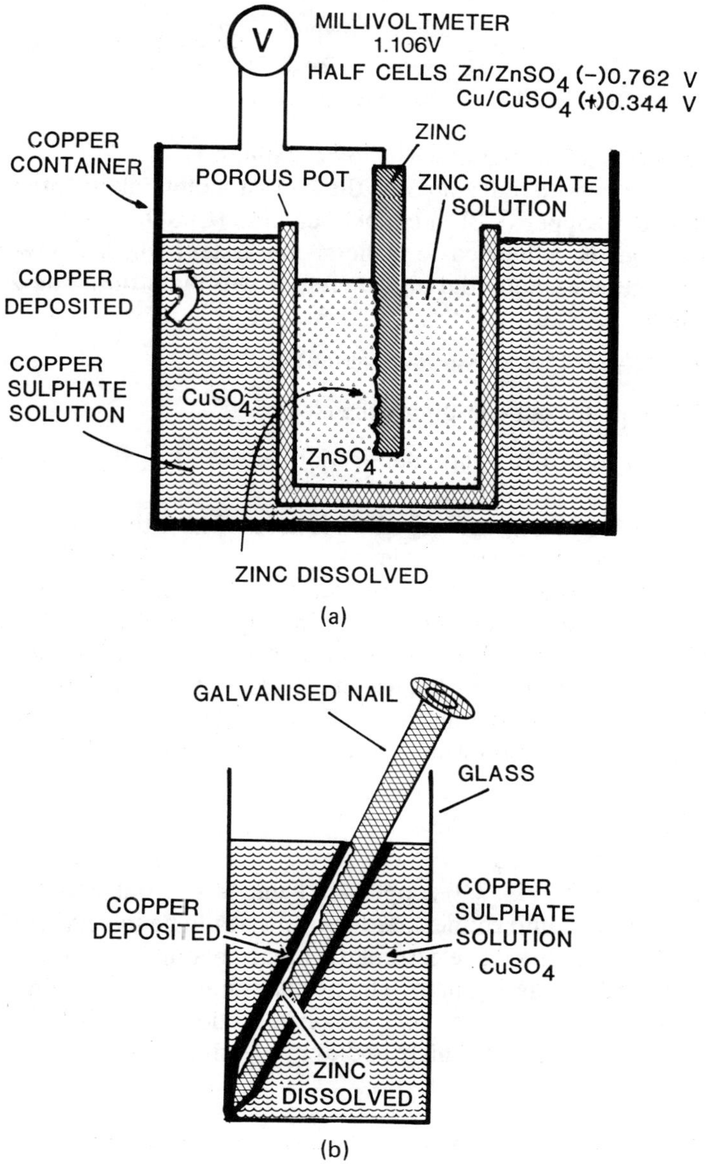

Fig. 7.9 — (a) An early battery attributable to Daniells (1850). Zinc is in contact with its own ions (zinc sulphate). This is contained in a porous pot which allows hydrogen (H^+) ions to pass through to the solution of copper sulphate, which is in contact with copper. An 'open circuit' (meaning no current is drawn) voltage of 1.106 V is produced. (b) The porous pot is required because, if zinc (e.g. a galvanized nail) is immersed in copper sulphate, a layer of copper is deposited at the same time as the zinc dissolves.

combination gave a potential difference of 1.1 V. (This can be deduced if one considers the potentials to be like mileages between three towns A, B and C. All the mileages are measured from the central town B (the arbitrary reference point):

town A is 34 miles west of town B
town C is 76 miles east of town B

● town A — west / 34 miles — town B ● — east / 76 miles — town C ●

← 110 miles →

The distance between A and C is 34+76=110 miles.

In the case considered zinc has the lowest potential and therefore, when connected to the copper (Cu) in copper sulphate ($CuSO_4$), the zinc will dissolve, while the copper is connected to a metal–ion system which is lower than its its equilibrium reference potential and therefore will plate out. The Zinc dissolves according to

$$Zn \rightarrow Zn^{2+} + 2e^- \quad \text{(Anode reaction)} \tag{7.6a}$$

and the copper plates out consuming the electrons:

$$Cu^{2+} + 2e^- \rightarrow Cu \quad \text{(Cathode reaction)} \tag{7.6b}$$

The zinc is represented as the negative electrode whilst the copper is the positive electrode of the battery which produces 1.1 V. Thus the *Chemical poential energy* of the zinc is being turned into *electrical energy*. This can be achieved because the liquids are separated and the battery can be switched off. Sometimes impurities prevent a 'battery' from being switched off. Local 'cells' shown in Fig. 7.10 can cause the consumption of the zinc by the use of electrons at the impurity site. This is why distilled water has to be used in the 12 V car battery system. Domestic tap water would provide impurity sites which would effectively short circuit the electrode.

7.4 ELECTROPLATING

In the above example (Fig. 7.9) copper was being 'plated out' and this is electroplating of copper (albeit onto copper). However, it would be too expensive to use Zinc every time industry wanted to electroplate. Electroplating is carried out by supplying a DC source across a plating bath. The normal supply of DC is obtained from AC rectified power sources. This is expanded on in Section 7.20 *et seq.* and forms part of the story for the painting of vehicle bodies by electrophoresis.

7.5 CATHODIC PROTECTION OF UNDERGROUND PIPES

Cathodic protection of underground pipes is carried out by a process which is not too dissimilar to the above example of electroplating. However, there is no electroplating because of the high over-voltages formed at the electrodes. In cathodic protection, the potential of the structure requiring protection is forced down below its reference electrode potential by an external source of direct current, usually from

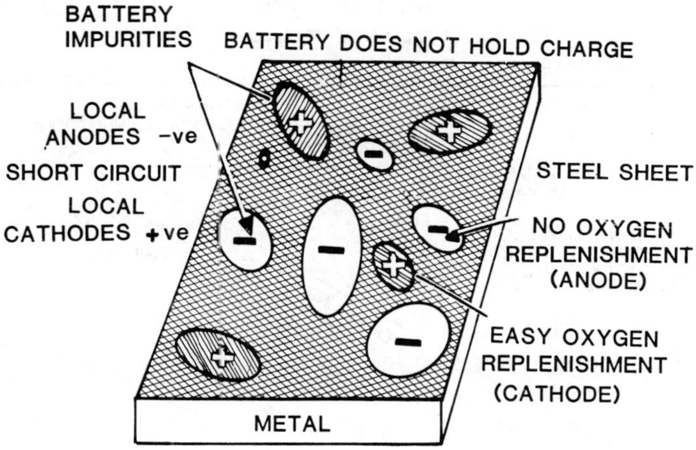

Fig. 7.10 — A commercial battery does not have an infinite shelf life unless the metal and electrolyte are pure. Impurities can cause local cells to operate which discharge the zinc. This happens with the lead acid battery if ordinary tap water is used for topping up. These local cells arise because the impurities are deposited on the electrode surface and effectively 'run down' the electrode. The zinc for example will dissolve by a series of short circuits between the centres of impurity and the unadulterated zinc surface. The only way to prevent this run-down is to drain the electrolyte or liquid environment away. These local cells operate upon a steel surface where there are differences in oxygen availability. The positive areas (terminals of the local cell) are the areas with adequate oxygen replenishment, while the negative areas are those where the oxygen is restricted. The only way to stop these cells is to remove either the difference in oxygen availability or the electrolyte.

mains rectification. This is shown in Fig. 7.11. This implies that any Fe^{2+} ions in the ground waters will be 'plated out' onto the steel pipe but, using the analogy developed earlier (Section 7.2), there will be no Fe^{2+} ions from the steel pipe going into the ground waters. Hence the pipe will not corrode. In most industrial uses, the presence of these large DC generators would be a severe limitation to their use, in addition to the maintenance required on the non-consumable electrode, and so other forms of cathodic protection are used. In these systems the cathodic protection is achieved using chemical energy. The sacrificial metals are usually zinc, aluminium and its alloys, or magnesium. Plate 7.1 shows the sacrificial anode as applied to ships, whilst Fig. 7.12 shows the application of impressed current cathodic protection of ships from rectified alternating current.

7.6 CATHODIC PROTECTION IN THE MARINE INDUSTRY

The marine industry makes use of sacrificial magnesium metal which is placed near the propellers (Plate 7.1). In this case the magnesium dissolves and depresses the steel below the −0.76 V (−760 mV) equilibrium point. Using a similar principle to the sacrificial metal on the ships hull, some sacrificial metals are being sold as special add-on attachments for the motor vehicle. The claim is that these solitary sacrificial anodes will reduce corrosion of the total vehicle. These will not work, and this section will attempt to outline the theory for the protection of the ship. The

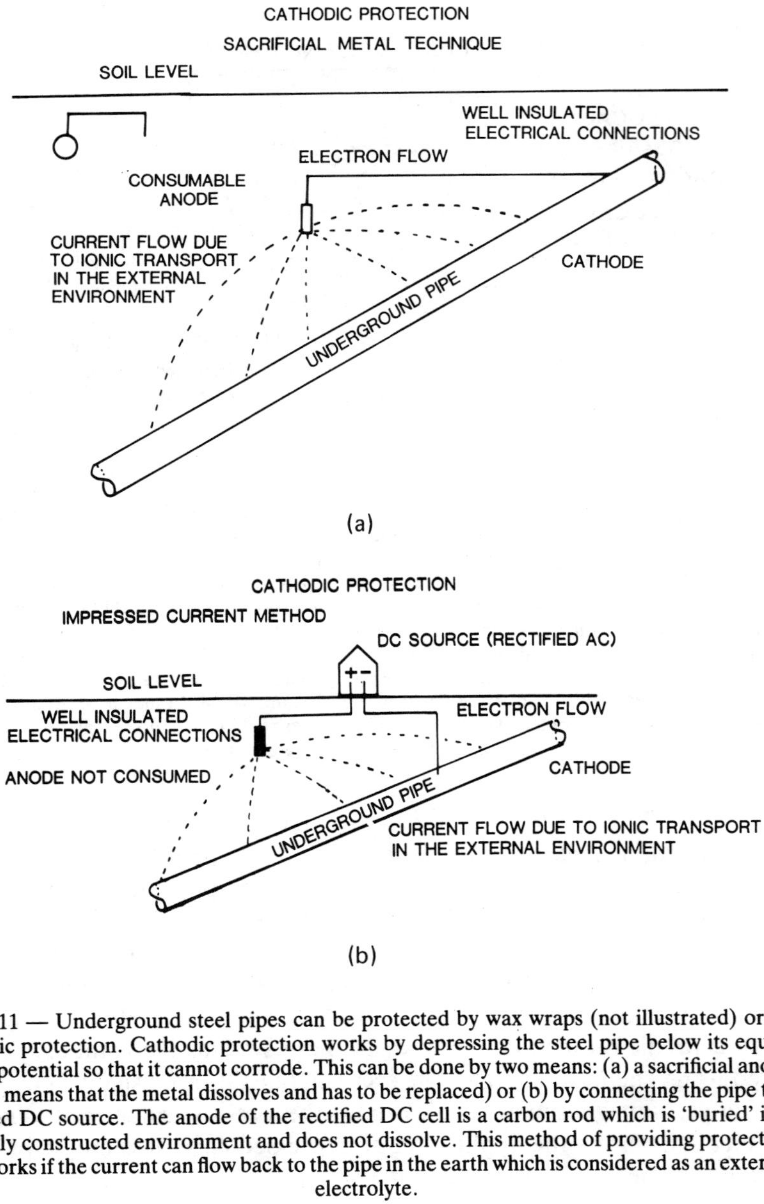

Fig. 7.11 — Underground steel pipes can be protected by wax wraps (not illustrated) or by cathodic protection. Cathodic protection works by depressing the steel pipe below its equilibrium potential so that it cannot corrode. This can be done by two means: (a) a sacrificial anode (which means that the metal dissolves and has to be replaced) or (b) by connecting the pipe to a rectified DC source. The anode of the rectified DC cell is a carbon rod which is 'buried' in a specially constructed environment and does not dissolve. This method of providing protection only works if the current can flow back to the pipe in the earth which is considered as an external electrolyte.

mechanism for this protection lies in the fact that the ship is immersed in a saline solution which is a good electrical conductor of ions (Section 6.11). Therefore the electrical current flowing in the sea water can flow out in all directions from the sacrificial anode (magnesium slugs in Plate 7.1) to all parts of the ship. The sacrificial magnesium will not protect the structure which is out of the sea water even though there will be a thin layer of salt water on the superstructure (Fig. 7.12). The reason is

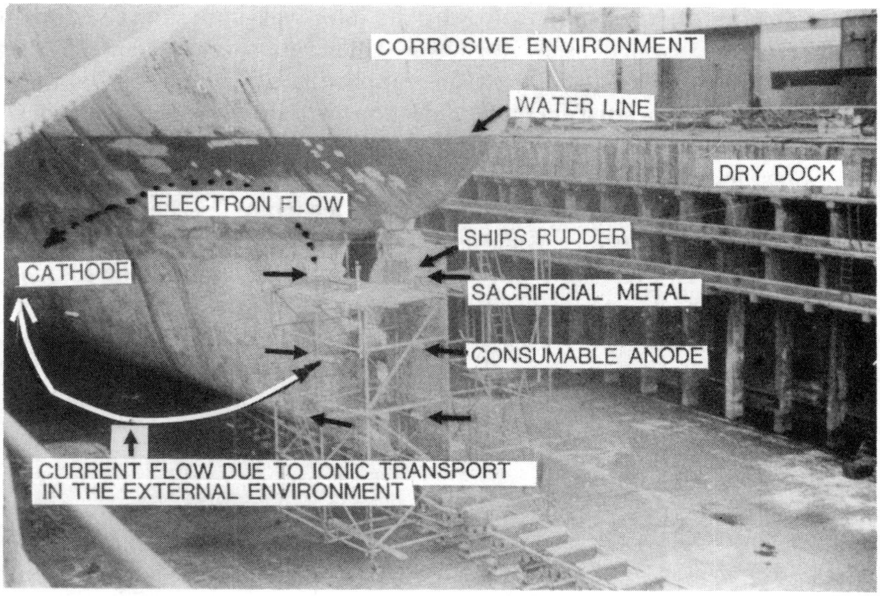

Plate 7.1

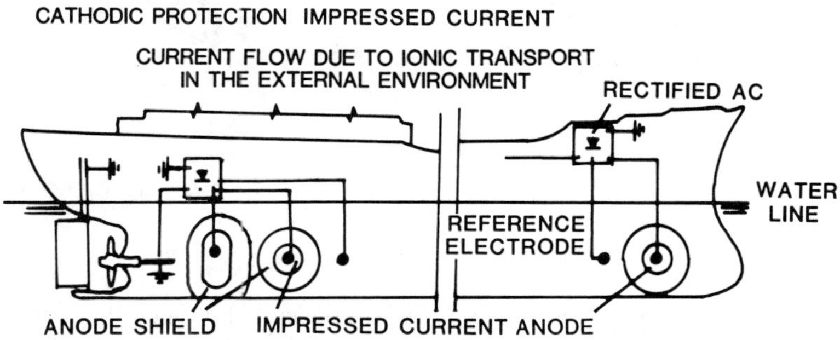

Fig. 7.12 — Cathodic protection of ships is achieved with metals below steel in the electrochemical series (Table 7.1). These bolt-on sacrificial anodes are shown in Plate 7.1. The dissolving metal is able to 'throw' its current over the whole of the submerged steel. It cannot protect the super-structure as the potential drop through the thinly adsorbed sea spray is too great. Again impressed current cathodic protection can be used as shown here. It cannot be used on motor vehicles as the resistance of the paint film and absorbed water is about 50 MΩ (After Ashworth and Booker, Cathodic Protection 1985, Ellis Horwood, Chichester, Fig. 7.6(b)).

that most of the derived potential from the dissolving Magnesium will be 'lost' through the high resistance of the this thin absorbed film of water. The degree of protection of the superstructure above the water-line is affected by the current density:

$$\text{Current density} = \frac{\text{total current}}{\text{cross-sectional area}}$$

This is similar to the reasoning as to why the inter-welds are not electropainted (Section 3.3.) The current density is too low within the inter-weld (Chapter 9 *et seq.*).

Hence, when these sacrificial metals are applied to the motor vehicle, they will only protect a small area adjacent to their location owing to the very thin film of water on the underside of the vehicle. This is a good example of the correct scientific principles (cathodic protection) but incorrect justification with the complete theory and the associated requirements. This form of technology is being used by the automotive manufacturers on specific panels where corrosion has been a problem in the past models. Fig. 7.13 shows the application of the sacrificial zinc to one–side coated steels. Just as the sacrificial anode does not protect the superstructure of the ship, so the zinc will not protect the area of the car when the zinc is not in aqueous contact with the steel through the electrolyte. This fact is not understood by the motor manufacturers as they tend to rely too heavily on the zinc-coated steels without providing adequate protection for the zinc.

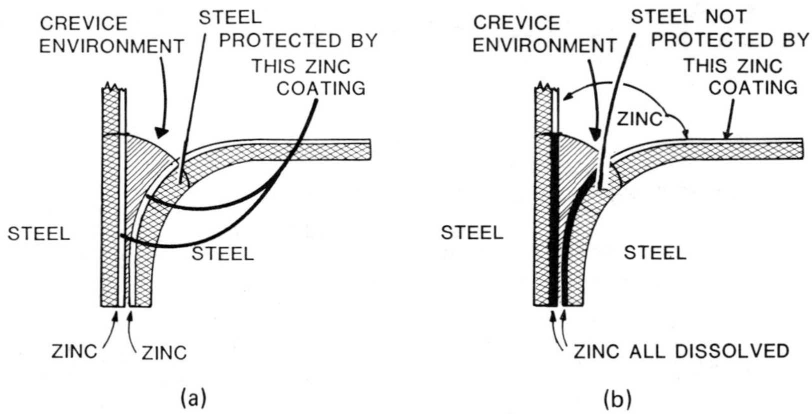

Fig. 7.13 — The same factors described in Fig. 7.12 help to protect the internal panels of vehicle bodies which are made from one-side-zinc coated steels. (a) The zinc only protects the adjacent metal if there is good aqueous contact. (b) zinc which is not in the aqueous environment cannot offer (sacrificial) protection. The steel within the crevice environment will corrode once the zinc within the liquid environment is dissolved. If the crevice liquid environment grew to make contact with the undissolved zinc, there could be once again sacrificial protection. As soon as contact with the zinc is lost, the steel will continue to corrode.

7.7 CORROSION SCIENCE: GALVANIC CORROSION; ANODES AND CATHODES

So far in the corrosion process, there has been constant mention of batteries and the fact that electrons are produced at the surface of one electrode. The terminals in battery technology are called the **positive** and **negative** terminals. In corrosion technology the negative terminal is called the **anode**. The positive electrode, which consumes the electrons, is called the **cathode**. There are three reactions which represent the process of corrosion. These were (with the English language translation) (where IEW means 'in equilibrium with' (Section 6.14)). The anodic reaction is

Fe	$\rightleftharpoons$	Fe^{2+}	+	$2e^-$	(7.7a)
steel as the metal	goes to or is IEW	steel into solution	plus	2 electrons	(7.7b)

These electrons are consumed at the other electrode called the cathode as follows in the cathodic reaction

O_2	+	$2H_2O$	+	$4e^-$	$\rightleftharpoons$	$4OH^-$	(7.8)
Oxygen as air	+	Water	+	4 electrons	goes to or is IEW	4 hydroxide units	(7.9)

7.7.1 The removal of the ions at anode and cathodic sites

These ions (Fe^{2+} and OH^-) react with each other and produce rust (Section 6.14(10)), which has a much larger volume than the metal which has disappeared. Hence the rust tends to push spot-welded panels apart (Fig. 1.8 and 1.9), and to form excrescent mounds beneath the surface of paint films (Figs 8.18(b) and 8.18(c)). The reaction is

Fe^{2+}	+	OH^-	$\rightarrow$	$Fe(OH)_2$ (insoluble)	(7.10)
Product of anode	+	Product of cathode	$\rightarrow$	Iron hydroxide	(7.11)

Iron hydroxide+more oxygen→RUST+eight fold increase in volume

The importance of these reactions is that they are all dependent on the electron, which can only exist as a free electron within the metal. Hence all the reactions (except those represented by eq. (7.10) and (7.11) must take place on the metal surface.

The important points of these equations to the understanding of corrosion are as follows.

(a) The 'units' such as Fe^{2+} in solution are called **ions** and are capable of moving only in a liquid environment (Sections 6.6, 6.11 and 6.13).

(b) The electrons cannot go into the liquid environment and must always stay in the metal (Section 6.14). This has some important consequences for corrosion protection.

(c) In the metal the current is carried only by the electrons. In the electrolyte the current is carried by both positive and negative ions and not by the free movement of the electrons. The current carried by the ions will depend on the mobility of the charge carrier and the electric charge, i.e. the valency. The total current carried by the positive and negative ions must always equal the current carried by the electrons within the metals. Ohm's law is

$$V=i \times R$$

where V is the voltage, i the current and R the Resistance. The current is the electronic charge in the metal which must equal the ionic charge which flows in the electrolytes.

(d) The product of the cathodic reaction as represented by eqs. (7.8) and (7.9) is an alkali (Section 6.16) as represented by the OH^- units. This OH^- unit has a softening action (called saponification (Section 8.16)), on the automotive paint system. The result of the saponification of paint is that the corrosion process will undercut the paint film and cause the paint film to 'bubble up' (Fig. 8.18 *et seq.*). The anodic reaction (eqs. (7.7a) and (7.7b)) will then take place beneath the paint film and severe loss of paint adhesion will follow over a large area (Fig. 3.10, Section 3.6, Fig. 8.18 *et seq.* and Plate 8.1). To prevent this from happening an acid pretreatment process (Section 6.18) is used. (The effect of these hydroxyl ions on the one-side zinc-coated steels is detrimental: see Sections 7.9 and 7.15).

(e) If the anode and cathode reactions are too close to each other (e.g. a stone chip causing loss of paint), the rate of corrosion is slow. The reasons are the following:
   (1) The anode and cathode have an equal supply of oxygen and therefore the corrosion battery is not 'fully charged'.
   (2) An insoluble rust $Fe(OH)_2$ (section 6.14) tends to form, which prevents the product of eqs. (7.8) and (7.9) and eqns. (7.6) and (7.7) moving, and therefore the corrosion rate is slowed down.

Hence a stone chip is very slow at corroding through the metal from one side to the other. Stone chips very rarely cause metal penetration. Paint stripping is the usual result as shown in Fig. 8.18(b), Plate 8.1, and Sections 7.12 and 8.14.

7.8 TYPES OF CORROSION ATTACK

There are five types of corrosion attack. These are classified and shown in Fig. 7.14 as follows.

7.8.1 Uniform attack

This as the name suggests is the uniform loss of metal with time and is an idealized form of corrosion of steel. Uniform corrosion exists in the tarnishing of silver and the formation of the green patina on copper which is so admired on buildings. This form of corrosion is not usually observed on steels.

7.8.2 Pitting

This is the result of highly localized attack and is the usual method for steel corrosion. Pitting usually results from a small anode–large cathode combination and is the major cause of body corrosion. The comparison with uniform attack is shown in Fig. 7.14. Other forms of pitting result from impingement attack, fretting (usually due to small amplitude movement which rubs away the oxide layer) and cavitation erosion (usually due to the collapse of vapour bubbles onto the surface of the metal, e.g. the impeller on the vehicle water pump).

7.8.3 Dezincification of brasses

The local loss of zinc out of the alloy brass (which is an alloy of copper and zinc) (Section 6.20, Fig. 6.21) is shown in motor vehicle radiators if the alloy brass is not stabilised with the element arsenic. This leads to very small holes within the radiator core and a loss of coolant which is very hard to track down. This is because the loss takes place under pressure, and there is rapid evaporation from the core surface due to the temperature.

7.8.4 Intergranular corrosion

Intergranular corrosion is the localized corrosion of the grain boundaries of the metal. In this case the grains act as cathodes while the grain boundary is the anode. This is not observed in motor vehicle body corrosion.

7.8.5 Cracking due to stress corrosion or corrosion fatigue

A metal which is continuously bent into an arc is said to be stress or strain 'cycled' (Fig. 6.43). The failure of the metal is by fatigue. If these cycles are carried out in a corrosive environment, the failure is said to be by corrosion fatigue, e.g. failure of main journals.

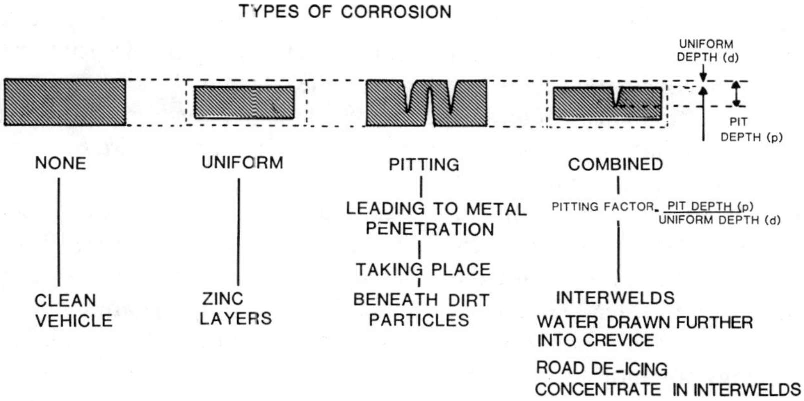

Fig. 7.14 — A comparison of the forms of uniform corrosion for the same amount of steel dissolved.

7.9 ALKALINITY OF THE CATHODIC AREAS AND ACIDITY OF THE ANODIC AREAS

7.9.1 Alkalinity of the cathodic areas

Water is modelled as a covalently (Section 6.10) bonded molecule but also has a form where it forms ions as if it were ionically bonded (Sections 6.11 and 6.16; eqns. (6.10) and (6.12)). It is estimated that only one water unit (molecule) in every ten million is bonded in this manner to give ions. Thus the following reaction is very biased towards the covalently bonded water molecule and is therefore indicated with a triple-headed arrow:

$$H_2O \leftrightharpoons H^+ + OH^- \qquad (7.12)$$

Hence this is often interpreted as stating that the water ionizes or dissociates to form ions H^+ ions and OH^- ions and is given a constant K_w (Section 6.17). The hydrogen ion concentration of neutral solutions is determined from the ionization of water into H^+ ions and OH^- ions given below:

$$H_2O \rightleftharpoons H^+ + OH^-$$

Bad conductor of electricity — Good conductor of electricity

$$K_w = \frac{[H^+] \times [OH^-]}{[H_2O]} = 10^{-14} \tag{7.13}$$

This reaction is reversible (it goes in both directions), but is heavily biased towards the covalently bonded left-hand side (because it is estimated that only one water molecule in every ten million is dissociated as represented by the ionic bond). Therefore only a very few water molecules are split up or dissociated to be able to conduct electricity and so pure water will only conduct electricity at high potentials. However, if the cathodic reaction of eqn. (7.8) which produces hydroxyl ions OH^- is considered, the following is the result. The excess OH^- ions produced by the cathodic reaction must mop up some of the H^+ ions from the dissociation of water and make the covalently bonded water molecule. As the H^+ ions are mopped up, their concentration must decrease and therefore the pH must rise (Section 6.17). A high pH means that the region is alkaline. Thus cathodic areas produce an alkaline environment in the cathodic neighbourhood.

Note: these alkaline regions of the cathode can sometimes cause 'secondary' corrosion reactions. This is pertinent to the metals which occupy groups 3 and 4 of the periodic table (Sections 6.5 and 6.6). These metals (e.g. aluminium (Al, group 3), zinc (Zn, group 2), lead (Pb, group 4) and tin (Sn, group 4)) are amphoteric and dissolve in both acids and alkalis. Of these only the zinc and aluminium are used commercially for motor vehicle manufacture. Zinc will therefore corrode in the cathodic reaction products without the beneficial environment of carbon dioxide to form the basic carbonate film (Section 8.4) which is so protective. The basic carbonate will not form because the corrosion reaction is proceeding beneath the paint film.

7.9.2 Acidity of the anodic areas

If an ion (such as Fe^{2+}), removes some of these OH^- ions from water because its hydroxide ($Fe(OH)_2$) is insoluble, more water will split up to balance the $[OH^-]$ deficiency (eqn. (6.9(a) and 6.9(b)), and so the anodic areas (as represented by eqn. (6.1); Section 6.14) will become acidic owing to the dissociation of water to try to 'replace' the removed hydroxyl ions OH^-.

Hence the water reaction can be represented now as a displacement to the right:

$$H_2O \rightleftharpoons H^+ + OH^- \tag{7.14}$$

As more water dissociates, so there is an increase in the hydrogen ion concentration which means that the anodic regions become acidic. The rate of corrosion of steel in acidic media is much faster than in neutral solutions. Hence crevices and inter-welds which form natural anodes as a result of the differential aeration of the crevice become *acidic* as corrosion proceeds, while the cathodic areas become very strongly *alkaline* and strip the paint off the metal substrate. This is primarily due to the fact that paints are not stable in alkaline media (Section 8.16).

7.10 POTENTIAL-pH

The potential dependence of a metal's stability in environments of different pH was discussed in Section 7.2. The equilibrium potential of a steel electrode is affected by the concentration of Fe^{2+} ions in solution and by the pH of the environment. A scientist called Pourbaix first examined the relationship of pH and potential and plotted these relationships in an atlas of electrochemical equilibria in water. They are a summary of the thermodynamic data and so do not express a rate of corrosion but do enable the scientist to tell at a glance whether the metal is corrodible or immune (and does not corrode). The immunity of steel at high pH is attributed to the formation of an oxide film within an alkaline environment. The oxide is called a **passive oxide film**, and is chemically protective to the metal substrate. The oxide film is only stable between a pH of 9 and 13 in the **absence of chlorides** (Section 12.13).

The Pourbaix diagram is shown in Fig. 7.15 for iron, zinc, chromium, copper and aluminium with the regions of corrosion, passivity and immunity marked. The immunity area is where the metal will not corrode because the metal's potential is too low (Section 7.2; Figs. 7.7 and 7.8). The Pourbaix diagram essentially plots the equilibrium potential (vertical axis) against pH (horizontal axis). The metallic state of the metal is more stable at potentials below the line EFH. For this to be true, both the sacrificial anode and the steel must be in the same liquid environment. The potential of the steel is not reduced from 'A' to 'D' just by the attachment of the zinc metal.

Thus, when steel is galvanized, or a ship cathodically protected by say magnesium or zinc, the steel's potential (at the point say A for a pH of 7) is brought to a potential D below this line EFH. It is the dissolving zinc which produces this lowering of potential. If, however, the potential is raised, rapid corrosion would result, and then with further raising of the potential the steel would be protected by the passive oxide film Fe_2O_3 as the steel is taken into the region UKFH. The nature of this passive protective oxide film to a first approximation can be considered as equivalent to the anodizing process of aluminium (Section 8.20).

This passive film is only stable under high pH and is the reason why concrete protects steel even though the concrete contains micro-voids. The corrosion of embedded metals in concrete e.g. motorway bridges is discussed in Section 12.11 *et seq.*.

7.11 PITTING — THE EFFECT OF A LARGE CATHODE AND SMALL ANODE

The anodic reaction represented earlier (Section 7.7; eqns. (7.6) and (7.7)) can be written as

$$Fe \overset{\text{rate 1}}{\rightleftharpoons} Fe^{2+} + 2e^- \qquad (7.15)$$

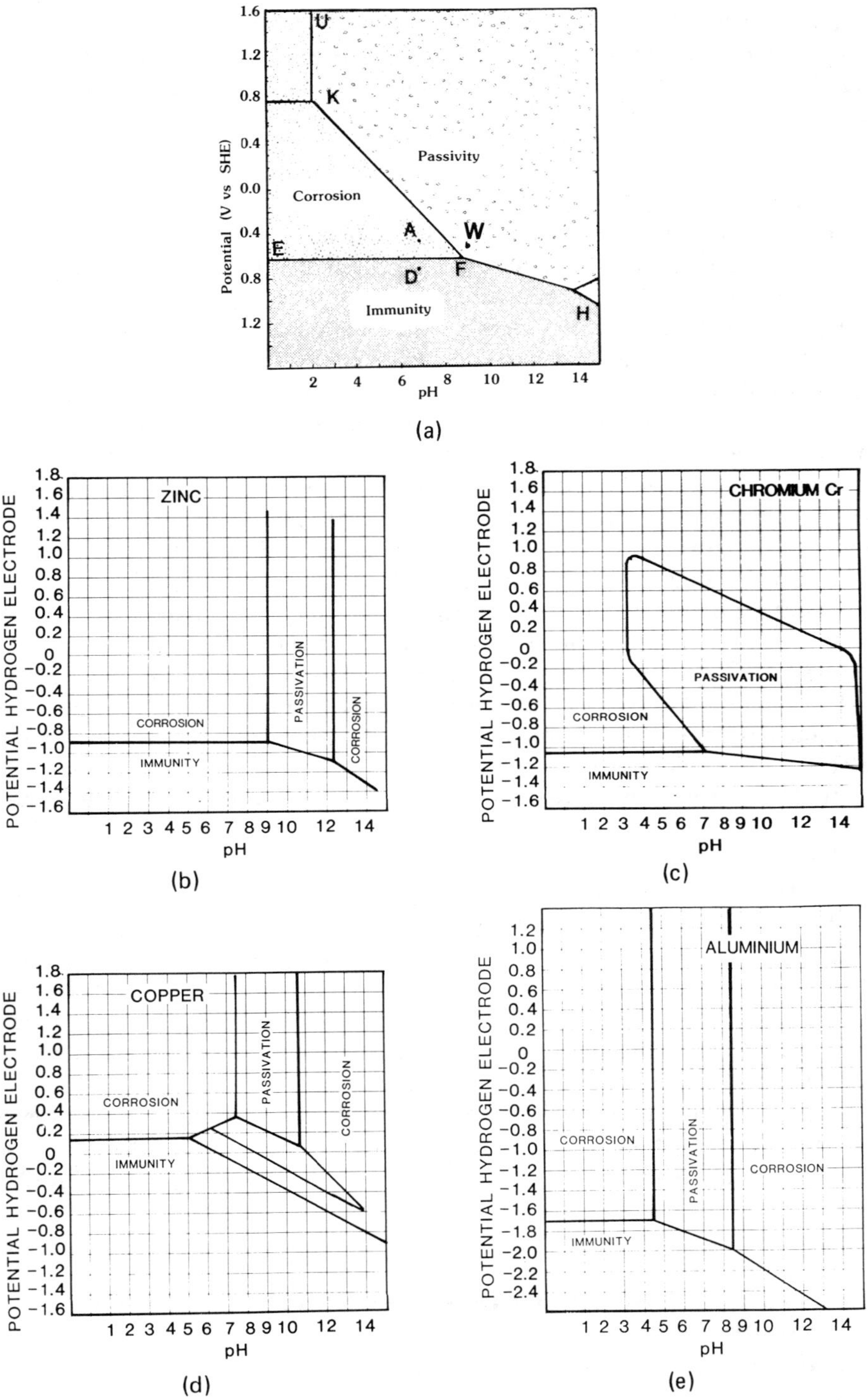

Fig. 7.15 — The potential–pH diagrams for (a) steel (iron; Fe), (b) zinc (Zn), (c) chromium (Cr), (d) copper (Cu) and (e) Aluminium (Al) as postulated by Pourbaix. The areas of corrosion, immunity and passivation are shown. See also Fig. 7.8

These electrons, which must remain within the metal, are consumed at the cathode as

$$O_2 + 2H_2O + 4e^- \overset{\text{rate 3}}{\rightleftharpoons} 4OH^- \tag{7.16}$$

Hence corrosion will proceed providing that both the products formed when these electrons are 'consumed' and 'produced': i.e. (the OH^- ions and the iron ions Fe^{2+} in solution) must be taken away from the liquid adjacent to the metal. If this does not happen the corrosion process or current will be reduced. In practice the rate of corrosion will be dependent on several rate-determining steps. We use this phenomenon of **rate determining steps** in order to arrest or, more properly described, to slow down corrosion. The possible rate-determining steps are listed below. In order to establish a fundamental model it is necessary to take a close-up view of the corrosion cell and to examine it with regard to the flow of charge (Fig. 7.16).

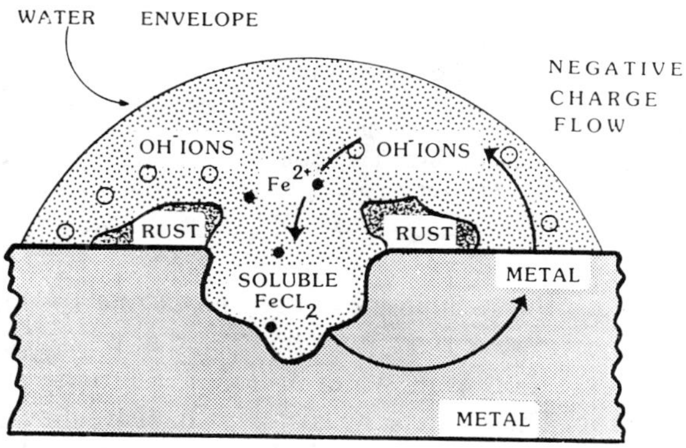

Fig. 7.16 — The circular transfer of negative charge from its production at the anode and its transference to the cathode area through the metal as electrons. At the cathode the negative charge enters the aqueous phase as the OH^- ions by taking part in the cathodic reaction. These OH^- ions migrate to the anodic area where they neutralise the positive ions at the metal/liquid interface. The negative charge has gone round the whole circuit: hence the analogy to the escalator of Fig. 7.17. The same is true for the positive charge transfer. This figure also shows the soluble $FeCl_2$ salt (ferrous chloride) being formed. This speeds up the diffusion of the Fe^{2+} ions away from the anodic site and therefore accelerates corrosion where large cathodes–small anodes are operating at, for example, corrosion pits.

The flow of the 'ions' Fe^{2+} and OH^- is shown, and one can see that that a negative charge has gone round a complete circuit. The electrons move from anode to cathode in the metal and the OH^- ions from cathode to anode in the liquid environment, rather like the steps on an escalator going round and round. In fact both negative and positive charges have gone round the corrosion circuit, although it is too difficult to show the positive charge transfer here. Thus the underside of the escalator represents the negative charge in the metal, which consists of the electrons e^- in the metal, and the passenger topside treads are the OH^- ions in the liquid environment (Fig. 7.17).

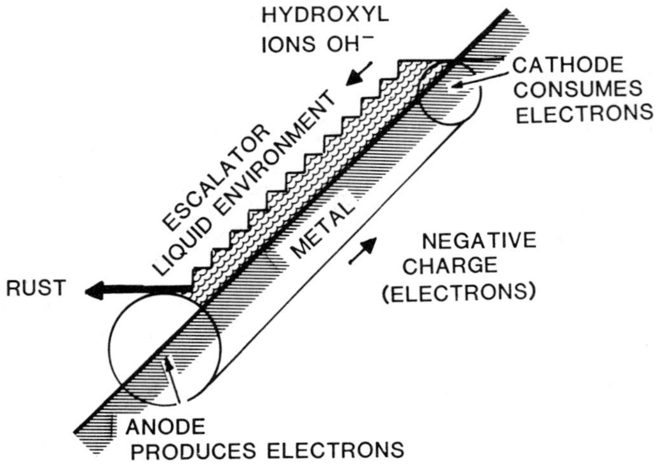

Fig. 7.17 — The escalator carries people from one floor to another. Its speed has no bearing on the pedestrian (traffic) density or how quickly they are served at the counter, pack their bags and depart from the store. The corrosion reaction is a 'sensible escalator' in that its speed is governed by the 'traffic' at all points and moves at the slowest rate to avoid 'human congestion' at any of the points mentioned above. The escalator itself, represents the flow of negative charge, and the speed of the escalator is governed by the slowest chemical reaction (see text).

Therefore, because the escalator can only go at one speed, it means that all processes in corrosion, however fast, must be slowed down to the slowest one — the escalator speed. Thus if there is no water or liquid environment outside the metal, the escalator must stop. There can be no corrosion because the flow of these charged entities will be the rate-determining step, and so they would build up at the electrode which produces them and stop all other processes. There would be no corrosion. Hence excluding, or even making the flow of, these ions more difficult in the environment outside the metal causes a slowing down of the whole corrosion reaction.

Thus the possible rate determining steps are as follows:

Rate 1 rate of production of Fe^{2+} ions
Rate 2 removal of Fe^{2+} ions from the anodic region
Rate 3 consumption of electrons at the cathode
Rate 4 diffusion of oxygen to the cathodic area
Rate 5 removal of OH^- ions from the cathodic region
Rate 6 ion combination to form iron hydroxide

The slowest of these will determine the rate of corrosion. In pure water (Section 10.1) the overall reaction is determined by rate 6.

The presence of chlorides causes other rates to govern the overall corrosion rate because a more soluble iron salt ($FeCl_2$) is produced which means that the Fe^{2+} ions may more readily diffuse away (Fig. 7.16).

$$Fe^{2+} + 2OH^- \xrightarrow{\text{rate 6}} Fe(OH)_2 \quad (7.17)$$

$$\text{Product of anode} + \text{Product of cathode} \rightarrow \text{insoluble iron hydroxide} \quad (7.18)$$

These rates will depend on the size of the anode or cathode. Hence, in discussing corrosion rates, it is nearly always best to discuss **current density**. A large cathodic area with a small anodic region will produce a very high current density at the anodic region if rate 5 is the rate-determining step in the overall corrosion reaction. This high current density at the anode will achieve the rapid dissolution of the metal provided that the products of the anodic reaction can diffuse away from the anodic site. Now this will make rate 2 the new rate-determining step. The mobility of the Fe^{2+} ion is very slow in pure water, and really does not exist, because the Fe^{2+} ion is 'insoluble' in water (Section 7.7 eqn. (7.10)), and is taken out of solution in the liquid environment by eqns. (7.10) and (7.11). The Fe^{2+} ion does, however, exist and is more soluble when in combination with the the chloride ion Cl^- which is present in the road de-icing salts (Section 10.2) when they dissolve in water. Thus, when considering the corrosion of steel by road de-icing salts, rate 2 is no longer the rate-determining step (Fig. 7.16). Other rate-determining steps take over control of the corrosion rate.

7.12 STONE CHIPS — LARGE ANODES AND SMALL CATHODES

A stone chip damages the paint and makes it non-adherent to the underlying metal. Often the non-adherent paint falls off and the bare metal is exposed equally to oxygen. There is no potential diffference produced when adjacent areas of steel are subjected to the same degree of oxygen replenishment (Section 7.1). Fig. 2.6(f) also shows diagrammatically the effect of a stone chip on the paintwork of a vehicle. Such paint chipping is relatively harmless because there is now a large unrealised anode and a small latent cathode. Both have oxygen readily available and so there is only a small potential developed. Because there is a small cathode, which must be oxygenated more than the bare metal latent anode, the cathodic rate-determining steps prevent the severe corrosion from manifesting itself in metal perforation.

However, as soon as the corrosion process starts to 'undercut' the paint film then the rate-determining steps alter as the anodic areas are realized. Hence corrosion is of little consequence where there is a good supply of oxygen at both the anode as well as the cathode and so the reactions would stop each other. This is called **polarisation**, (see Section 7.18). Note: Corrosion is also of no consequence if the steel is totally denuded of oxygen. Rate 6 now becoming the rate-determining step. **Corrosion only takes place if there is a different rate of replenishment or availability of oxygen**.

As soon as the corrosion process starts to undercut the paint film, the anode would increase in efficiency as it would enter a region where oxygen replenishment (beneath the paint film which is now the corrosion front), is difficult. The cathodic area is the bare metal exposed by the stone chip. As the vehicle paint surface dries these paint chips become devoid of the electrolyte, thus breaking the circuitous negative charge flow in the liquid environment. Using the escalator analogy, the escalator would stop. Corrosion therefore stops. However, the 'crevice' effect of the broken paint, and the possible undercutting of the paint film by corrosion, ensures that the metal remains wet for a longer period of time. (see the effect of water condensation within narrow capillaries given in Table 10.10). Hence given time, the interface between paint and metal at the fractured surface becomes the new crevice.

7.13 QUESTIONS AND ANSWERS from Section 2.5

In Chapter 2 a non technical explanation of corrosion was given. In that explanation the corrosion of the motor car (Fig. 1.4) was a result of the flow of these electrons from one region of the metal (of the car) which is not well oxygenated and so acted as the anode (negative electrode of the 'corrosion battery') to another region, the cathode, (which is the positive electrode of the corrosion battery), which is plentifully supplied with oxygen. This only took place if the two regions were connected by a liquid environment analogous to the escalator steps discussed in Section 7.11. It is the metal in contact with the oxygen-denuded regions which dissolved or corroded. The voltage, or magnitude of the electromotive force (emf), of the corrosion depended on the difference in oxygen availability or the rates of replenishment at the two sites (anode and cathode). If there was no difference in the degree of oxygen availability at the two regions there would be little or no voltage generated and therefore the rate at which the Fe^{2+} ions were generated would be low. Therefore the corrosion rate would be low. Hence, if the whole corrosion process is governed by the presence of the liquid environment analogous to the escalator step (Fig. 7.17), then regions which could maintain the liquid environment could act as voltage generators for some considerable time.

7.14 ANSWERS TO QUESTIONS in Section 2.5

These 'equivalent batteries' occur at all places where there is a difference in the availability of oxygen and the presence of a good electrolyte which will make the Fe^{2+} ions more mobile (Sections 7.11 and 10.2). The 'equivalent batteries' can be formed at the following locations, which are all characterized by the fact that they can maintain a liquid environment for some time **and** provide the metallic surfaces with different rates of oxygen replenishment.

(1) Areas of bad design (see Section 2.5(1) for definition): these harbour road dirt which form local dirt poultices which restricts natural drying and provide different amounts of oxygen replenishment at various points around and within the poultice.

(2) Beneath hygroscopic (Section 10.5) road dust 'dirt' (Fig. 2.6(a)) or other surface deposits.

(3) In inter-welds, clinch flanges or hemming folds. These also form stagnant liquid environments which present a small liquid surface to the atmosphere and so drying by evaporation is severely restricted. More important for the corrosion process, oxygen replenishment at the metal surface which is furthest from the liquid surface is restricted or denied. These crevices also aid condensation of water from the atmosphere as a result of their capillary size (Table 10.10; Section 10.7).

(4) Beneath larger mud poultices or layers.

(5) At the corrosion front or 'pit' which will explain the localized 'pitting' nature and speed of 'rust-through' corrosion often termed 'inside-out' corrosion because it

is initiated by the deposition of salts on the inner panels of box members which then corrode through by the localized pitting to the outer surface.

(6) Beneath non-adherent but coherent surface coatings.

In Section 2.5 several questions were posed with regard to Fig. 2.6 which gave a schematic view of the road de-icing salts concentrating at the base of the road aggregate, which helps to fix the aggressive saline environment upon vertical surfaces. The liquid envelope which is produced by the uptake of water is also shown. The absorption of atmospheric water on humid days by the road de-icing salts is a result of their hygroscopic nature (Section 10.5). The reader was asked to indicate which is the region or area which has (i) least oxygen replenishment and (ii) most oxygen replenishment.

(7) Answer: least replenishment of oxygen occurs beneath the dust particle (see also Fig. 10.5). The perimeter of the metal–droplet interface would have the most access to oxygen as this would be the shortest diffusion path for oxygen replenishment to the metal surface (rate 4 in Section 7.11).

Fig. 2.6 also shows diagrammatically the difference between internally initiated pitting corrosion causing metal penetration and scab corrosion (surface corrosion beneath a painted surface). The reader was asked to indicate which are the areas of oxygen denudation.

(8) Answer: the tip of the corrosion front would be the most oxygen-denuded area, as the oxygen would have a long way to diffuse to the corrosion front (rate 4). As one goes towards the outer surfaces of the metal from the corrosion front, the oxygen availability would increase. Therefore there is more propagation of the pit than there is a growth in the circumference of the pit.

Fig. 2.6 also shows diagrammatically the effect of a stone chip on the paintwork of a vehicle. Such paint chipping is relatively harmless. The reader was asked (a) why? and (b) what will happen if the corrosion process starts to 'undercut' the paint film?

(9) Answer: (a) because there is a good supply of oxygen at the anode as well as the cathode and so the reactions would stop each other. This is called **polarization** (Section 7.18) and is one mechanism of slowing down the corrosion process. (b) As soon as the corrosion process started to undercut the paint film, the anode would increase in efficiency as it would enter a region where oxygen replenishment, at the corrosion front, was difficult.

There are three options for stopping, or more correctly speaking slowing down, corrosion. These are as follows.

(10) Keep the vehicle dry, thus breaking the contact in the liquid environment. This would affect rates 1,2,3,4,5 and 6 in Section 7.16.

(11) Have a 'high electrical insulating' film on the metal to separate the metal from the environment. This is usually a paint film, but wax films (Section 8.8) are an alternative. Note: as the basis of corrosion theory is unravelled it is necessary to consider more parameters for this 'high insulating film', e.g. water penetration, ionic conduction and oxygen diffusion (see Section 8.12).

(12) Keep out road de-icing salts which are naturally 'hygroscopic' (Section 10.5) and absorb water from the atmosphere. This affects rate 2 but also affects the 'escalator' (the salt on a car dries white on a dry day: most of the time it is wet and cannot be seen except for a very thin film of moisture on the car body).

7.15 DISSIMILAR METAL CONTACTS

There are several different metals from which metallic parts can be fabricated in order to make a structural or decorative article. If all these different chemical elements, in the pure state, are categorized as regards their ability to resist corrosion, then we obtain the **electrochemical series** (Table 7.1). In this table the chemical elements are positioned according to their electrical potential with respect to an arbitrary reference electrode, called the hydrogen electrode, which is given the 'honourable' position of being the arbitrary zero. From this arbitrary zero other elements have an electrode potential with respect to the hydrogen electrode (Sections 7.1 and 7.2). Unfortunately these pure metals are not structurally strong and so for service conditions other metallic elements are added to form alloys (Section 6.20 *et seq.*). The addition of these 'foreign' metals, and the alloy's subsequent heat treatment process, alters the potential of the alloy from that of the pure metallic system. The potentials of the engineering components are shown in Tables 7.2 and 7.3, and are differentiated from the electrochemical series by being called the practical galvanic series of metals.

Another factor which alters the electrode potential of an engineering material is the environment. The galvanic series of metals is shown in Table 7.3 for both aerated sea water and de-aerated fresh water. It is worth noting that the aerated sea water, to a first approximation, is equivalent to a solution of road de-icing salts. The electrode potentials of some materials in de-aerated fresh water are considerably different from their electrode potential in aerated saline (sea water) solutions. For example, there is a difference of over 350 mV between the two environments for 18/8 stainless steels (in this case the potential is measured against a standard reference Calomel cell (Fig. 7.3(c)). The corrosion rate of engineering metals depends on the environment.

Table 7.2 — Potentials of metals and alloys

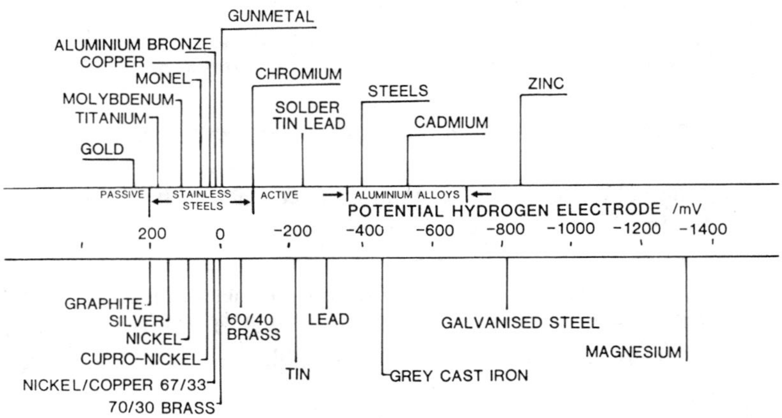

Pure metals cannot be used for engineering structures. Alloys are much stronger than the pure metals, and the formation of an alloy alters the chemical potential of both metals forming the alloy. These alloys are arranged in order of their potentials with respect to the hydrogen electrode.

Table 7.3 — Galvanic series of metals in (a) aerated sewater and (b) de-aerated fresh water

Contact of dissimilar metals may result in galvanic corrosion of the more active metal, that is, the metal with more negative potential.

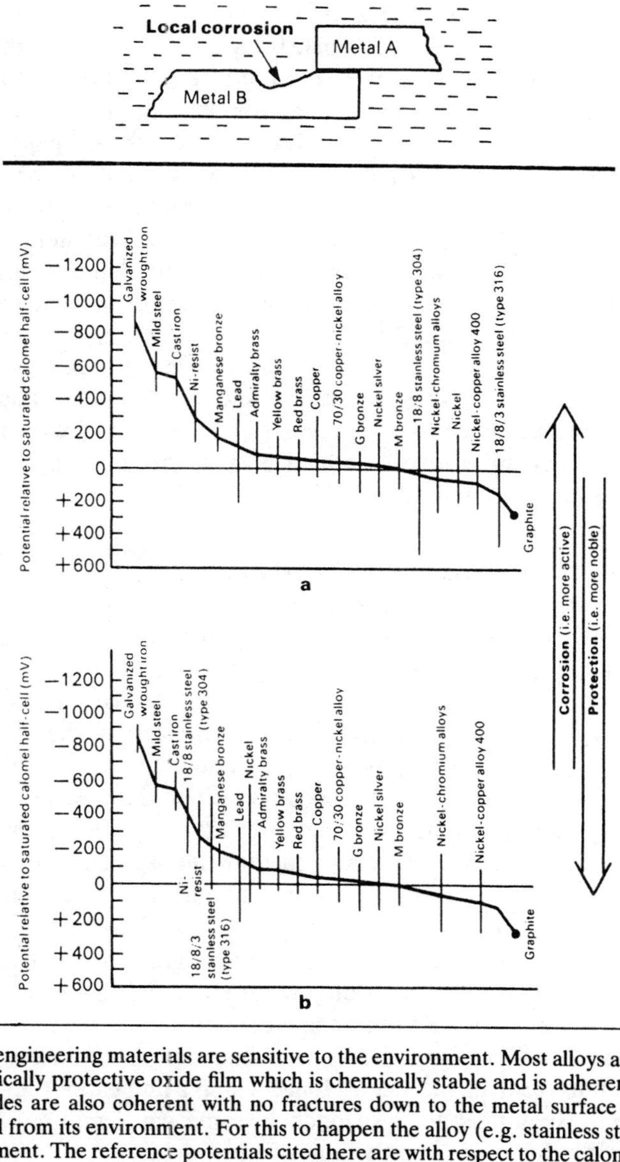

The potentials of engineering materials are sensitive to the environment. Most alloys are chosen because they form a chemically protective oxide film which is chemically stable and is adherent to the substrate metal. These oxides are also coherent with no fractures down to the metal surface and so physically separate the metal from its environment. For this to happen the alloy (e.g. stainless steel) must be in an oxidizing environment. The reference potentials cited here are with respect to the calomel electrode (Fig. 7.3(c)). Some engineering alloys suffer a potential shift under extremes of oxidation, e.g. As an engineering rule of thumb, 1 A is equivalent to 1 g h^{-1} corroded (for aluminum it is about $\frac{1}{3}$ g h^{-1}). 18/8 stainless steel type 304 suffers a shift of some 350 mV.

(Reproduced from Corrosion Control Checklist for Design Offices, Department of Industry, Contributed by T. K. Ross, courtesy of Inco Europe Ltd. and Metal Corrosion Engineering Design Guide 1977, Oxford University Press.

Comparative rates are shown in Fig. 1.6. The difference in oxygen or aeration in sea water is well shown in Fig. 1.6(c) which shows the corrosion rate of steel is greatest in the 'splash zone' where aeration is greatest. The motor vehicle could be said to be operating in the winter in the 'splash zone' (Fig. 1.6(c)). Fig. 1.6 also shows other relevant corrosion rates for zinc and aluminium in marine environments.

The importance of this galvanic series of metals is that any alloy more positive than mild steel from which the motor vehicle body is made will be 'cathodic' to the car body and so accelerate the corrosion of the mild steel. Hence the alloy decorative trim will act as the cathode (consuming electrons), while the car body dissolves to produce these electrons. Hence the car body dissolves at the expense of the car trim. Therefore this form of dissimilar metal contact causes localized breakdown of the paintwork. It is another example of a large cathode–small anode (Section 7.11) at the crevice junction, if the areas in contact with the liquid environment are considered.

This ability of one metal to protect another metal is used in steel ships in sea water (Plate 7.1; Section 7.6). In this case zinc or magnesium slugs are screwed to the hull of the ship below the waterline and these dissolve (producing electrons) whilst the hull of the ship is protected from corrosion because it is cathodic and is required to consume electrons. These slugs are called sacrificial anodes and the form of protection is called 'cathodic protection'. This is the reason why the motor industry has gone over to 'galvanized steel', which is steel protected by a sacrificial zinc coating (discussed in Section 8.3).

7.15.1 The misconceptions concerning sacrificial anodes

The sacrificial metals used in the shipping industry are not coatings (Plate 7.1), but they do protect the whole structure below the waterline because the ship is totally surrounded by a lot of water. The current within the electrolyte, the sea water, can spread out as shown in Plate 7.1. The similar spread of return current is shown in the case of the underground pipelines in Fig. 7.11. In the case of the car there is only a thin film of water, at best, and so a piece of zinc added as a sacrificial anode will not have the same effect. The zinc will corrode and protect the *adjacent* steel. It will not protect the whole car as it does the ship because of the high resistance of the thin film of water. However, there will be a small current (leakage current) which could be detected from the total circuit. The important question is whether this leakage current is high enough to provide cathodic protection. Similarly, zinc-coated steels (Fig. 7.13) will only protect a small area from the aggressive micro-climate and not the whole structure or panel which may be zinc coated. Zinc-coated steels only protect the steel at the crevice if the zinc makes contact with the environment that is in contact with the steel (Fig. 7.13). Any zinc not in contact with the steel's environment cannot protect the steel. Hence a ship's hull cannot be protected by a zinc slug connected to its masthead. In much the same way, a motor vehicle cannot be protected by a zinc slug placed on the inner wheel arches.

7.15.2 Disadvantages of zinc

Zinc is one of the few metals which dissolves in both alkaline and acidic environments (Fig. 7.18). In Section 7.9 it was stated that the cathodic reaction produced an alkali which would dissolve the zinc and so aid the undercutting of the paint film. Hence

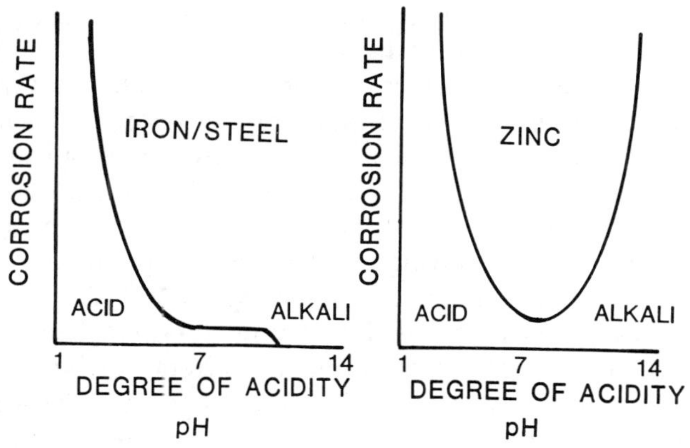

Fig. 7.18 — The corrosion rate of (a) steel with that of (b) zinc for different pH environments. The steel reaches a minimum (zero) rate at a pH of 9.5. This pH is hard to achieve for long periods owing to the acidic nature of the impurities in the air (Section 12.11). These acidic impurities are carbon dioxide (burning of carbonaceous fuels) and sulphur dioxide (an industrial pollutant from oils and manufacturing industries). Generally the scientist has combined them together to call them 'acid rain'. (After J. Frasch, *Corr. Anticorr.*, **9,** 1961, p. 225.

zinc-coated products soon shed their coat of paint as shown by galvanized pedestrian railings at controlled crossings and zinc-coated vehicle panels, etc. A similar effect may take place on the one-side zinc-coated steel used in the box sections where the environment becomes trapped at the inter-weld crevices as shown in Fig. 7.13.

7.16 DRY CELL BATTERIES — (PROPER) BATTERIES

The function of a battery is to turn chemical energy into electrical energy at a constant voltage. This chemical energy is the energy which the metal possesses as a result of the heat energy which is required to produce or win a metal from its oxide or ore in which it exists in the earth's crust. The 'corrosion process' in the battery is controlled by stopping the flow of electrons in the metallic part of the circuit so as to preserve the chemical state of the metal. A section through a popular 1.5 V battery and its associated external circuit shown in Fig. 2.5 (Section 2.2) and Fig. 7.1 (Section 7.1).

Hence we can write the dissolving reaction of the zinc electrode which is the negative electrode (anodic reaction) as follows:

Zn	$\rightleftharpoons$	Zn^{2+}	$+$	$2e^-$	(7.19)
zinc as the metal	goes to or is in equilibrium with	zinc into solution	plus	two electrons	(7.20)

The larger the current drawn from the battery, the more zinc has to go into solution as is required by Faraday's laws (Section 7.20). The electrons cannot go into the liquid environment and must always stay in the metal conductors. The double arrow should be translated as 'in equilibrium with' (Section 6.20). This term is used to represent the fact that, in any equilibrium, a disturbance of the equilibrium can be obtained by increasing any of the participating units on the right hand side or the left-hand side of the equation. In the case of a torch battery which is switched off, it will stop running down because the electrical energy (the electrons) is not consumed at the cathode. The switch (Fig. 2.5) in the external circuit prevents the electrons from getting from the anode to the cathode along the metallic conductors. These electrons and Zn^{2+} ions therefore appear in excess on the right hand side to push eqn. (7.19) to the left (i.e. the zinc stops dissolving and the battery is preserved).

The common method of stipulating a battery's output is by voltage. This is the open-circuit voltage, (i.e. when there is no current flowing). Thus a car battery is 12 V without a current being drawn from it. When a current is drawn from it, the voltage difference between the terminals drops. The advantage of the lead acid battery is that this voltage drop is not very high. Eight zinc dry cell batteries can be connected in series to produce 12 V, the same voltage of the car battery, (i.e. eight 1.5 V dry cell batteries in series: = 8 x 1.5 V = 12 V). About 100 A is required to start a car. The eight dry cells will perform very badly in their ability to start the car engine when compared with the lead acid battery. The eight dry cells will will not produce the 100 A required because there is a high voltage drop across the battery's internal electrolyte when a high current is drawn. Conversely there is a very small drop in voltage across the lead acid cells and so the battery can deliver the high starting current. The voltage drop across the dry cell internal electrolyte is due to polarisation and this will be discussed again in Section 7.18. Hence in the following discussions only a slight disturbance of the equilibrium conditions (of no current flow) will be considered so that the mechanism can be discussed.

To produce a voltage (of 1.5 V), these electrons (in the metal) must be 'consumed' and the zinc ions in solution must be taken away from the liquid around the metal, otherwise the metal will stop dissolving. This removal must be done rapidly if the output voltage is to remain at 1.5 V for the period of use, otherwise the voltage would fall as the concentration of zinc ions increased around the zinc electrode.

These electrons, which must remain within the metal, are consumed at the other electrode called the cathode, which is made of carbon, as quickly as they are produced to maintain the output potential:

O_2	+	$2H_2O$	+	$4e^-$	$\rightleftharpoons$	$4OH^-$	(7.21)
oxygen as an element	+	water	+	4 electrons	goes to or is in equilibrium with	4 hydroxide units	(7.22)

Hence a 1.5 V power source will be produced providing that both the products formed when these electrons are 'consumed' (the OH^- ions) and the zinc ions (Zn^{2+} (eqn. (7.19)) in solution must be taken away from the liquid adjacent to their electrodes:

Zn^{2+}	+	$2OH^-$	$\rightarrow$	$Zn(OH)_2$	(7.23)
product of anode	+	product of cathode	goes to	zinc hydroxide	(7.24)

Note: this equation is not an equilibrium because zinc hydroxide comes out of solution in an efficient battery as an insoluble product. The symbol → shows this.

If a high current is taken from the torch batteries, the output voltage will fall because the rate limiting step will be the diffusion of the OH^- and Zn^{2+} ions. In this case the battery will need a 'rest' in order to allow these ions to diffuse away. This is discussed in the next section. Conversely, if the cell is left on open circuit, it will not consume the zinc unless there are impurity particles within the zinc which act as local cathodes (i.e. just like the central carbon electrode). In this case there will be an internal short circuit within the zinc and the zinc will corrode away. These local cells are shown in Fig. 7.10. No useful current can be produced by batteries made up from local impurity sites and zinc.

7.17 BATTERY REJUVENATION

If the anodic and cathodic products are not removed from their corresponding electrodes quickly, the corrosion process or current will slowly stop. Therefore an efficient battery requires the addition of chemicals which will achieve this rapid ionic charge removal. When the battery becomes old, and the chemicals used to remove these 'corrosion' products become used up, the battery voltage falls very quickly when used to provide a current. Usually a rest and the application of heat will disperse these products to produce a 'renewed life', although this is somewhat temporary. This build-up of electrode products at the electrode at which they are produced is called **polarization**. A build up of metallic ions at the anode will arrest the further ionization of the metal to produce more metallic ions. (Another way of looking at this is that there is a local back emf as a result of a concentration cell (Fig. 7.4)). Use is made of this fact to protect metals from the aggressive environment by a paint film (Section 8.11 *et seq.*). Hence the paint film provides two functions: (1) a physical boundary between the metal and its environment, and (2) control of the movement of charged ions at the interface between metal and paint, which slows down the corrosion.

7.18 POLARIZATION APPLICATION TO PAINT FILMS

The potentials recorded in the electrochemical series are in fact equilibrium potentials which means that there is no current flowing and the circuit is open circuit. The half-cell electrodes are not at equilibrium potential when a current flows either from them or away from them (in the metal circuit). The deviation from equilibrium is known as polarization. The degree of polarization depends on the current flowing and can be represented as in Fig. 7.19. The form of these curves was first discussed with regard to Fig. 7.2.

Thus the following cause polarization:

Anodic polarization	
rate of production of Fe^{2+} ions	rate 1
removal of Fe^{2+} or Zn^{2+} ions from the anodic region	rate 2
Cathodic polarisation	
consumption of electrons at the cathode	rate 3
diffusion of oxygen to the cathode area	rate 4
removal of OH^- ions from the cathodic region	rate 5
Mixed polarization	
ion combination to form iron hydroxide	rate 6

The lowest rate will determine the rate of corrosion.

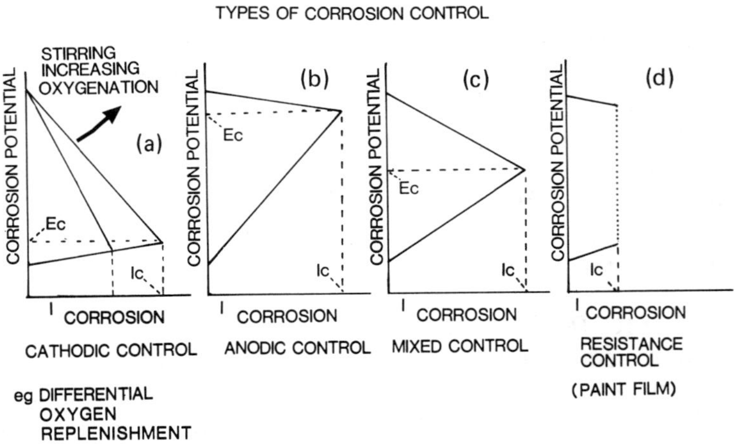

Fig. 7.19 — If reference is made to Fig. 7.2 the differing potential drops at the electrode interface can lead to different types of corrosion control. Fig. 7.2 developed the general concept of polarization. This figure represents the effects of anodic, cathodic and mixed polarization on the cathodic and anodic potentials. Resistive control is characteristic of paint film technology.

If we take as an example, a rod of steel corroding as a result of the differential aeration cell, then from the equilibrium conditions, with no current flowing, the cathode electrode will be at +1.40 V (Table 7.1) with respect to the hydrogen electrode. The anode electrode will be at −0.44 V (Table 7.1) with respect to the hydrogen electrode. Fig. 2.4 shows that we must separate these two electrodes. If no current is allowed to flow, in other words no corrosion takes place, the open-circuit voltage will be around 1.8 V. This is close to the open-circuit voltage of the dry cell battery of 1.5 V. When a current is allowed to flow, the voltage drops just as the voltage drops from the dry cell battery when the torch is switched on. In the normal corrosion reaction, there is no possibility of switching off the corrosion cell in an equivalent manner to switching off the torch. This is because the metal shown in Fig. 2.4 is not cut into two pieces in practice. The structure is one continuous metal path between anode and cathode (provided that the anode and cathode regions are linked by a continuous aqueous environment). The metal is a good conductor of electricity (of electrons) and so there can be no potential difference between the metal at the anodic site and at the cathodic site. The metal itself must assume a potential somewhere between +1.40 V and the −0.44 V. The rest potential will depend on the degree of polarization of each electrode. Thus Fig. 7.19 shows the effect of the three types of polarization considered in corrosion: cathodic polarization control in Fig. 7.19(a), anodic polarization control in Fig. 7.19(b), and mixed polarization control in Fig. 7.19(c). Resistance control is shown in Fig. 7.19(d).

These curves are a modelling proposed by Evans and thus bear his name. Their use will be shown by considering the voltages put out by a zinc dry cell battery. If the dry cell battery delivers a high current, a large voltage drop will be experienced between the battery terminals as a result of its own internal resistance r. Thus the dry cell battery can be modelled as shown in Fig. 7.19(d). Any current which flows must flow through the internal resistance and the external circuit resistor (for example the light bulb). It follows therefore that the sum of the voltages around the circuit must add up as shown in Fig. 7.1 and 7.2. If these voltages are now transferred to the polarization curves along the length of the line

TU is the potential drop across the cathodic electrode (Fig. 7.2)
UV is the potential drop across the external resistor (e.g. bulb)
VW is the potential drop across the anodic electrode for the current drawn

If the same analysis is carried out for a piece of steel corroding by the differential aeration cell, then there is no external resistor (equivalent to the electric light bulb) because the steel is a good electronic conductor. Hence the intersection of the anodic and cathodic curves is the corrosion current. The magnitude of the corrosion current is limited by the **degree of polarization**. Hence if Fig. 7.2 represents this situation, then the length of the line UV represents the voltage drop across the resistance of the aqueous medium. Thus, if the external resistance of this medium is altered by the addition of sodium chloride, which produces a lower resistance, then the length of line UV can only become smaller by moving out to the right with higher corrosion currents.

7.19 CORROSION RATES

Rates of corrosion are usually expressed as a mass loss after exposure of that metal to the corrosive environment. These are expressed in milligrams (mg) per square decimeter (dm) per day, i.e. mg dm^{-2} day^{-1}. If the loss of thickness is more important this is expressed in millimetres per year. However, these corrosion rates do not take account of 'pitting' which is localized at a site and cannot be expressed in these terms which assume uniform corrosion. Therefore corrosion rates should be considered with caution where localized pitting corrosion is a possibility.

The potential difference generated by the differential aeration mechanism is proportional to the **difference in oxygen replenishment**. If this fact is applied to steel, there will be a different corrosion rate for steel with time of corrosion because the rust formed will act as a barrier to the diffusion of oxygen to the metal surface. Hence initial corrosion rates can be as high as 100 mg dm^{-2} day^{-1} and reducing to a steady state corrosion rate of 25 mg dm^{-2} day^{-1}. There cannot be any relationship to a corrosion rate for pitting.

7.20 CORROSION RATES AND FARADAY'S LAWS

Faraday was investigating the deposition of metals from solution which we now know as electroplating. In this book it is a first approximation to consider corrosion as the reverse process of electroplating. As motor vehicle manufacturers use the deposition of paints by an electrical current, it is worth looking at Faraday's laws to see whether the electrocoating process may be made more efficient.

Faraday's laws state the following.

(1) The mass of a substance liberated during electrolysis (M) is directly proportional to the quantity of electricity passed (Q). The quantity of electricity (Q) is proportional to the current (i) and time (t). Conversely, for corrosion, therefore, the quantity of electricity produced is proportional to the amount of metal corroded.

Therefore

$$Q = i \times t$$

i.e. M is proportional to Q

(2) For a fixed amount of electricity the weight (W) of an element deposited is proportional to its chemical equivalent. The Faraday constant (F), which is equal to 96 493 C (where C is the symbol for a coulomb) liberates the gram-equivalent of an element of atomic weight (A) and charge n. Thus

$$W = \frac{i\,t\,A}{n\,F}$$

The equivalent of each metal is the atomic weight divided by the charge on the metallic ion (Section 6.4 *et seq.*). This is usually the group which the element occupies in the periodic table but there are exceptions to this rule. Consider the following metals and the results given in Table 7.4: copper, zinc, chromium, iron and aluminium.

Table 7.4 — The electrical efficiency of metal deposition from aqueous solutions

Metal	Atomic weight	Charge on ion	Equivalent weight
Copper	63.5	1	63.5
Zinc	65.5	2	32.7
Iron	55.8	2	27.9
Aluminium	27.0	3	9.0
Chromium	52	6	8.7
Chromium	52	3	17.3
Chromium	52	2	26.0

Chromium exists in the three valence states of +2 +3 and +6 and, from the point of electroplating chromium onto a metal, it is more economic and energy efficient to plate from the lowest, divalent, state than from either the trivalent (+3) or the hexavalent (+6) state. However, it is not always possible to achieve this and chromium is deposited from the hexavalent state.

With the electrophoretic paints used in the motor industry, the paint molecule is made very large and given a unit negative (or positive) charge (for the importance of charge (positive or negative) see the discussion of anodic and cathodic electrocoats in Chapter 9). This means that for the passage of one Faraday a large mass of paint will be deposited and so the process is very economic as well as having the other advantage of coating re-entrant angles.

7.21 ELECTROLESS PLATING

These are fairly expensive plating solutions where chemical energy is used to deposit the metal. The most common electroless solution is that for nickel, where a reducing agent is capable of reacting with the metallic ion and reducing it to the metallic state.

These solutions are 'modified' so that the reduction will only take place when there is the surface to be coated. This prevents the metal from being deposited as a sludge in the electroless nickel solution. These solutions are used where it is a requirement to plate onto non-metals, such as plastics. Electroless plating provides a metal conducting layer which can then be plated by conventional plating solutions. These metallized plastic components are replacing the zinc-based die castings for motor vehicle name plates and other such embellishments. The plastics, by virtue of their covalently bonded structure, are non-conductors of electricity (Section 6.10). In some cases adhesion failure is a common result as it difficult to give a good key onto the plastic surface for the first electroless plate. Mechanical keying can be obtained by attacking the plastic with a solvent, followed by chromic and sulphuric acid etches, but these add costs and only benefit the subsequent second, third or fourth owner (Section 1.4).

7.22 ELECTROPHORETIC DEPOSITION

In this process the articles are made an electrode within the cell. A DC voltage of about 50 V is applied between the electrodes which causes the charged species to migrate to the electrodes (Figs 9.5 and 9.6). The process is referred to as anodic electrophoretic deposition if the article is the anode or positive terminal of the DC supply, and cathodic electrophoretic deposition if the article is made the cathode or negative electrode. This phenomena is discussed in detail in Section 9.1.

8

Basic corrosion protection

8.1 METHODS OF CONTROLLING CORROSION

Fig. 5.4 shows some elementary design parameters suggested by the Department of Industry after the report of the UK Committee on Corrosion and Protection in 1971. There are several other ways of protecting a metal from corrosion. These available methods are best discussed under the following headings: design (which is the most important aspect; see Chapter 5 for a discussion relating to automotive conditions), materials selection, environment, and method of corrosion protection. Not all these methods are practical or suitable for motor vehicle protection. However, they are discussed here as they show how pernickety a corrosion technologist must be to eliminate unwanted corrosion. In Chapter 7, it was shown that corrosion is an unwanted battery, while a battery is a corrosion process which the battery manufacturer wants to make more efficient.

For any structure, corrosion protection must start with **design**. The choice of metal to be used will be governed by the parameters shown in Fig. 8.1. Hence the selection of metallic component in a particular environment is important. Usually the choice of material is governed by other factors, which will include 'prime' costs, ease of fabrication and the ease of the subsequent joining methods. Often in the design of any structure, the materials' corrosion properties are not fully considered at this juncture and this omission often turns out to be costly to some other party in terms of higher maintenance or repair costs. Most difficulties arise because it is usually very difficult to define the environment exactly because of the complex interaction with industrial pollution, e.g. sulphur dioxide and the level of oxygen replenishment resulting from small geometrical factors. Likely temperature ranges are often omitted and the likelihood of condensation is overlooked. Hence a full 24 h cycle (or plant operating cycle) should be considered for the four annual seasons.

In automobile corrosion, while these conditions are important, they are minor aspects of a much larger problem. One of the most difficult problems to overcome in the automobile environment is the turbulence which is caused by both the preceding vehicle and indeed the vehicle in question. This is shown in Fig. 5.6. In 'normal'

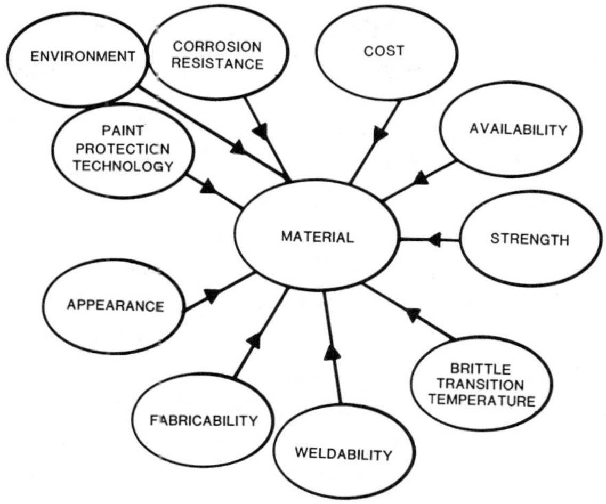

Fig. 8.1 — The parameters which should be considered before a designer designates the material from which the vehicle–part is to be constructed.

industrial corrosion dealing with static structures, the annual rainfall would be taken into account. Motor vehicles are now designed to move at speed, and Fig. 5.5 shows the effect of a vehicle collecting spray by small drainage holes. For this mechanism to operate, it does not have to be raining, as the effect is produced on a wet road. The vehicle's speed therefore '**recycles**' the rainfall and produces a much more difficult environment to design against unwanted corrosion. Hence the usual parameter of rainfall is not strictly relevant. Top speed and pressure variations caused by overtaking vehicles would be more relevant parameters to be considered.

Hence the multidisciplinary nature of designing against automotive corrosion is more complex, but it can be reduced to a few basic requirements. These have been discussed in Chapter 5. Normally a designer considering a structure or article would consider the following parameters.

8.1.1 Environment

The environment of the motor vehicle is considered in great detail in Chapter 10. In other industrial aspects of corrosion, the important question is: can the environment be modified? There are several ways of modifying the environment, some of which are not applicable to the vehicle body but do affect the other mechanical (spare) parts in transit.

(1) Can we remove oxygen and make all areas in contact with the same oxygen potential? This is performed by a paint film and works well provided that a micro-climate is not allowed to develop on top of the paint film. In other industries, this can be done by a chemical scavenger such as sodium sulphite or hydrazine, which is used for boiler-water treatment. Proprietary inhibitors are used for vehicle engine cooling water which flows through a radiator made of copper alloys and either an aluminium

or a cast iron engine block. Hence the interaction of metals and the coolant is complicated by the fact that one inhibitor will not suffice for all metals (see Table 8.6). Weld through inter-weld mastics would reduce the crevice corrosion at panel interfaces.

(2) Can we remove the aggressive contaminants? Road de-icing salts are the cheapest and most readily available freezing point depressant available country-wide. Some experimentation has been undertaken to add corrosion inhibitors to the salts, but this puts up the costs of road management. Hence, assuming that road de-icing salts will be spread on the road for the foreseeable future, horizontal surfaces and sharp right-angle bends should be designed out, and washing made easier.

(3) Can water as moisture vapour be excluded? In order to achieve this a range of temperatures must be specified for the article in question so that condensation is eliminated. This method is very effectively used with small articles (e.g. cameras) where sachets of silica gel are packed with the article in the same polythene bag. Corrosion is at a minimum at a relative humidity of below 70% (Fig. 10.5; Sections 10.5 and 10.6), but once again it should be mentioned that a drop in temperature will cause the relative humidity to rise even though no water has been added to the system. This method is not suitable in automotive corrosion.

(4) Can corrosion inhibitors be used? These are usually added to radiator cooling waters as they become de-oxygenated with rising temperatures. Flow velocities at various points become important in the corrosion mechanism. There are two types of corrosion inhibitors, and they are termed according to the reaction which they effect. Anodic corrosion inhibitors interfere with the anodic process and these are dangerous if the concentration is insufficient. Cathodic inhibitors interfere with the cathodic process. This means that the corrosion process is stopped by not using up the electrons produced in the anodic process (Section 6.14). Corrosion inhibitors are dealt with elsewhere (Section 8.7). Note: one could state that the environment was a form of inhibitor, but the absence of a micro–environment is the best way to arrest the corrosion process in motor vehicle bodies. Therefore it is best to look at barrier methods for corrosion protection.

8.1.2 Barrier methods of separating the metal from the environment

(1) Is the barrier coating to be a metal film? These are classified as cathodic to steel, e.g. chrome plating, or anodic (sacrificial) to steel, e.g. galvanized steel (a zinc coating on a steel component). Metals cathodic to steel are usually less corrodible and rely upon a continuous pin-hole-free coating to protect the underlying steel. The industrial methods of applying the coating are numerous and out of the scope of this book. They are briefly (1) spraying, (2) cementation (called sherardizing for zinc coatings, calorizing for aluminium coatings or chromatizing for chromium coatings), (3) electroplating, (4) hot dipping, (5) cladding (a metallurgical 'sandwich' of two or more metals or alloys for added corrosion resistance or (6) the metal as a pigment powder suspended and contained in a paint. Some of these techniques can also produce anodic–cathodic reversals and care must be exercised on manufacture. This subject is dealt with elsewhere (Fig. 8.3 and the following sections).

(2) Is the barrier film to be an organic coating such as a paint film (Section 8.11) or wax film (Section 8.8), the latter being correctly termed '**temporary protectives**'.

Temporary protectives were developed as a class of materials which are used for corrosion against atmospheric pollution whilst the metal is in storage or transit. These temporary protectives may contain vapour corrosion inhibitors (VCIs). They are called temporary protectives because they can be readily removed and are oils, greases or thick strippable coatings or the VCI impregnated into brown paper. For these processes to be effective, surface preparation is very important.

(3) Is the barrier film to be an inorganic coating put onto the metal by a manufacturing chemical process such an anodic film used on some aluminium trim products? Anodizing is the formation of a 'semi-porous oxide' but impermeable oxide film on aluminium alloys which can be dyed to different colours (see Fig. 8.2). A phosphating process (Fig. 3.2), where complex phosphates of zinc and cadmium and other metals are precipitated out of solution (Section 6.18, Plate 9.1), or chromating for aluminium alloys are both processes that provide a good 'key' for the subsequent paint process.

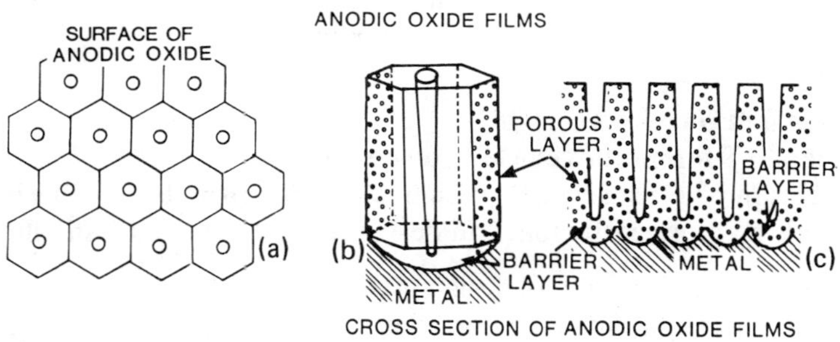

Fig. 8.2—Anodic oxide films are 'semi-porous' as a result of the central holes within each grain. These holes allow the oxide film to absorb a dye and so make coloured anodic oxide films. The holes also absorb ordinary household paint pigments should any paint accidentally come into contact with anodic aluminium window frames. (Fig. 8.2(c) redrawn by permission from L. L. Schrier, Corrosion, Vol. 2, Newnes Butterworth, London, 1977, Fig. 16.1).

(4) Is the barrier film to be an inorganic coating put onto the metal by a cementitious product which produces a chemically protective oxide film on the steel rather like the anodic film on aluminium (Section 8.21)?

8.1.3 Electrical methods

(1) Electrical insulation can be used to prevent electron flow from anode to cathode. This is a process which is frequently overlooked and should be used where any second surface is attached to another. It is cheap but very effective (see Fig. 2.9). Bimetallic couples can be successfully used in conjunction with each other if the two metals are *adhesively* stuck together (see Section 11.22), thus ensuring simultaneously a gap-filling and an electrically insulating role.

(2) Cathodic protection from either an external DC power source or a sacrificial anode (e.g. zinc on galvanized steel, magnesium anodes on ships (see Plate 7.1)). This is a very effective way of protecting a metal structure where repainting or any other form of protection system is expensive to maintain. It finds most of its uses in

buried pipelines (Fig. 7.11). As the vehicle electrical system is direct current, and there is a lot of misconception about applying sacrificial anodes to the automobile, it will be considered as a possible method (Section 7.15.1). Since the anodic reaction works by producing electrons, which must be consumed at the cathode site, it follows that the anodic reaction can be stopped by flooding the anodic site with 'spare' electrons from the DC source. A battery in a commercial torch or flashlight is stopped from running down by the same process: a break in the electrical circuit (the switch) prevents the electrons from getting to the cathodic site (see Fig. 2.5).

(3) Anodic protection such as galvanization is discussed in Section 7.15 and 8.3).

8.1.4 Vapour-deposited coatings

Bright lustrous metal coatings may be produced by vapour depositing the metallic coatings under a partial vacuum. This may be achieved by several techniques: (1) cathodic sputtering; (2) thermal decomposition of the vaporized halide or carbonyls; (3) vaporization of the metal by thermal means. Coatings from the vapour phase can be deposited upon non-metals and thermoplastic materials as well and so these are replacing the older chrome-plated small components.

8.1.5 Design and corrosion control

See also Chapter 5 for motor vehicle body design.

(1) Corrosion is best reduced by design. Good design at the drawing board in the first instance is no more costly that bad design. In fact the maxim should be

Good design is free, bad design is expensive

(2) Prototype design should be considered. No coating will have a lifetime in a continuous salt spray cabinet of over 1000 h unless it is very thick (in which case it will not penetrate crevices) or very good, in which case it will be expensive. Guarantees are often given against metal penetration (and not for the first signs of internally nucleated corrosion). From the results of Chapter 4 this surface nucleation takes place very early in the vehicle's life. This is shown in the appropriate photographs in Chapters 1–4. Fig. 1.6 shows the corrosion rates for steel (microns per/year averaged over 10 years) in several environments. These facts should be compared with the following

1 year is 8760 h
6 years, the body warranty period, is 52 560 h
8 years is 70 080 h
12 years median life is 105 120 h

In considering automobile corrosion the final question perhaps should be: does it take account of the factors which are all scientifically derived and discussed and listed in this book, e.g. temperature cycles leading to condensation within the boot–trunk, bonnet–hood, drainage positions, passenger air extraction mechanisms and efficiency, crevices, and their ability to 'switch on' a battery which is powered by differential aeration, arising from out-of-sight road dirt or mud deposits–compactions? Some of these are represented in Chapter 5 and associated figures for body design.

The available methods will now be discussed in further detail.

8.2 BARRIER METHODS OF SEPARATING METAL FROM ENVIRONMENT

Metal coatings may be applied by several processes (Tables 6.15 and 6.16). Each process or coating produces different properties when formed and welded. The vehicle body fabricator requires uniform well-known properties as shown in Section 6.35.11 *et seq*. The different properties as regards spot welding are shown in Table 8.1.

Table 8.1 — Properties of precoated steels sheet

Coated sheet product	Corrosion resistance ranking[(1)]	Painted appearance[(1)]	Weldability (number of spots)	Formability: available grades				Upper temperature limit for oxidation resistance (F)
				FH	CQ	DQ	DQSK	
Aluminium–zinc coated	10	7–8	700	X	X			1250
Aluminium-coated (Type 1)	9	7–8	500		X	X	X	1250
Zinc-coated								
(G 90)	5[(2)]	4–6	2000[(3)]	X	X	X	X	550
(A 60)	3[(2)]	7–8	3000[(3)]	X	X	X	X	550
Zincrometal	3	9	3000	X	X	X	X	Not applicable
Terne	2			X	X	X	X	Not applicable
Bare steel	0	10	9000	X	X	X	X	Not applicable

[(1)] Relative ranking on a scale of 0 to 10. [(2)] Coating life is proportional to coating weight. [(3)] Electrode life increases with decreasing coating thickness. Abbreviations; CQ: Commercial quality; DQ; Drawing quality; SK: Special killed; FH: Fully hardened; HR: Hot rolled.

Reproduced by permission from H. E. Townsend and L. Allegra, Precoated Sheet Steel for Automotive Applications, in R. Baboian (ed.), Automotive Corrosion by De-icing Salts, National Association of Corrosion Engineers, Houston, Tx, p. 343, Table 2.

The cheapest method of producing a coating is by immersion in a molten bath of the metal which is to be used as the coating. This is termed '**hot dipping**' and is shown in Fig. 6.41(c) for the manufacture of Galvanite (Galvanite is a registered trademark of British Steel Corporation). Other well-known methods are immersion in an electrolyte of the metallic salt to be deposited and 'electro-plated' as shown in Fig. 6.41(d) for Zintec steel (Zintec is a registered trademark of the British Steel Corporation).More obscure methods may be used for spraying the molten metal onto the surface from a high temperature gun. This process is generally used on large structures *in situ* and has little application to the automobile. Other methods of production include cementation where the metal article is heated and tumbled with a flux and the metal powder to form the coating. Aluminium and zinc coatings can be applied this way and are mainly used for small part production. Coatings of the metal nickel may also be formed from a solution of the salt by a chemical reduction process (Section 7.21) (as contrasted with the electrical reduction process of electroplating).

Such chemical processes are called 'electroless' to distinguish them from electroplated metals.

Metals cathodic to steel (copper, chromium, nickel, lead) are usually less corrodible and rely on a continuous pin-hole-free coating to protect the underlying steel. This is never achieved by electroplating techniques owing to the evolution of hydrogen which disrupts the film to produce a pin-hole termed a 'holiday'. This is shown with chrome plating in Fig. 8.3. The porosity of the chromium plate is shown in Fig. 8.4.

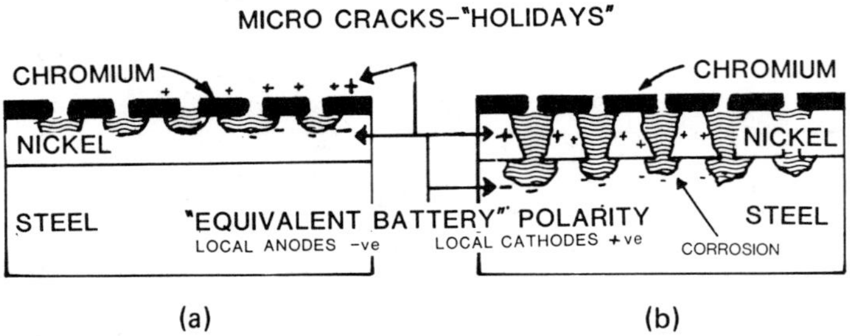

Fig. 8.3 — The electrochemical series used with these diagrams concerning the initiation of corrosion at a 'holiday' (or hole) in the chromium plating. (a) The nickel plate which is placed on the steel before chromium plating is anodic (negative) to the chromium layer which is cathodic (positive): hence the nickel is corroded preferentially. The function of the nickel is to obtain adhesion of the chromium layer to the steel layer. Often a copper flashing is also used. (b) When the corrosion pit reaches the steel substate, the steel is anodic to the nickel layer which now suffers an electrode polarity reversal and becomes a cathode to the steel. Hence the steel corrodes preferentially with respect to the nickel layer. Chromium plating therefore breaks down into small 'rust pits' which allow corrosion to continue beneath the outer cathodic chromium plate (cathodic to both nickel and steel). (After Electroplating and Metal Finishing, **12** (1) (1959), 3; redrawn by permission from L. L. Schrier, Corrosion, Vol. 2, Newnes Butterworth, London, 1977, Fig. 14.13).

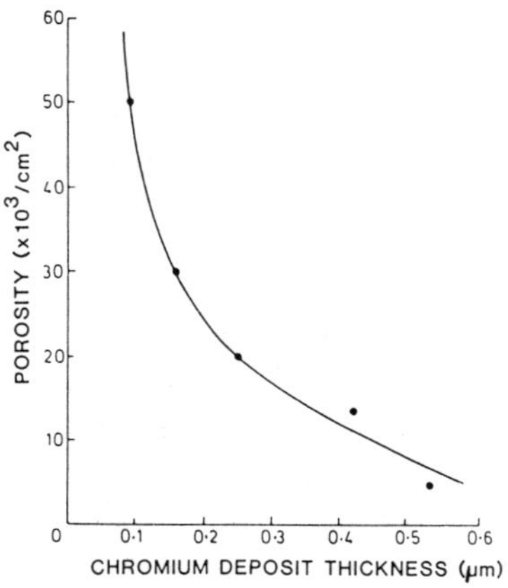

Fig. 8.4 — The number of holidays decreases with increase in chrome plate thickness as shown by the change in the micro-porous chrome plate with its thickness. (Redrawn by permission from L. L. Schrier, Corrosion, Vol. 2, Newnes Butterworth, London, 1977, Fig. 14.4.)

8.3 ANODIC COATINGS TO STEEL

An example of anodic coatings to steel is galvanized steel (e.g. Galvatite), which is a coating of zinc. The life of zinc-coated steels is dependent on the thickness of zinc as shown in Fig. 8.5. Another example of an anodic coating to steel is cadmium. Cadmium is used on some expensive bolts and screws. Table 8.2 compares the corrosion rates of these two metals.

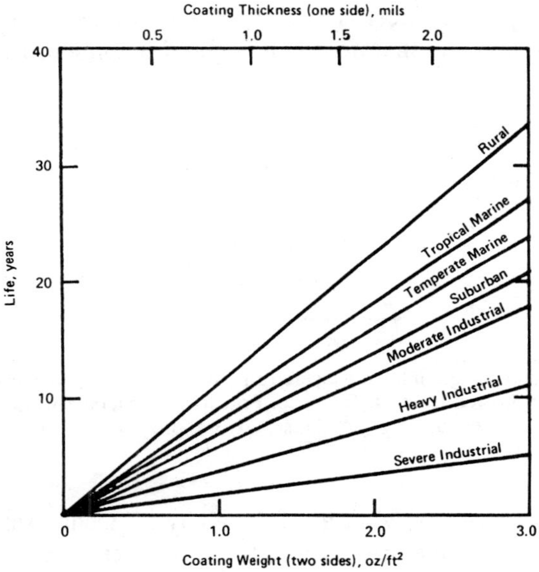

Fig. 8.5 — The life of zinc coatings on steel in different environments. The times are given for the first significant appearance of rust. (Reproduced by permission from H. E. Townsend and L. Allegra, Precoated Sheet Steel for Automotive Applications, in R. Baboian (ed.), Automotive Corrosion by De-icing Salts, National Association of Corrosion Engineers, Houston, Tx, 1981, p. 344, Fig. 1.).

Table 8.2 — Relative corrosion rates of cadmium and zinc in different environments

Atmosphere	Corrosion rate (mm/yr)	
	Zinc	Cadmium
Industrial	0.005	0.01
Suburban	0.0018	0.0023
Marine	0.0025	0.0013

Reproduced by permission from D. R. Gabe, Principles of Metal Surface Treatment and Protection, Pergamon, Oxford, p. 40, Table 2.4.

'Galvanizing' has come to be used as a general term for the deposition of the sacrificial metal zinc onto the surface of steel to provide corrosion protection. The term galvanizing should only be used where the metal is deposited from a molten bath as incorrect specifications may be written if this usage is debased. Sometimes a zinc–iron alloy finish is required. This is achieved by passing the hot dipped steel strip through another furnace which is placed between the zinc pot and the cooling stage of the process outlined in Fig. 6.41(c). The second phase formed is often brittle and polarity changes can occur. The micro-structure arising from the hot dipping (Fig. 8.6) is completely different from that of the zinc applied on the steel by the Zintec process. The Zintec process steel is used by the automobile manufacturer as it contains no second phase and therefore bending and shaping processes are easier.

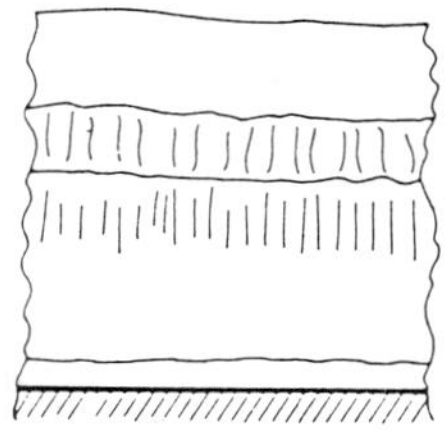

Fig. 8.6 — A schematic diagram for the layered structure of hot-dipped galvanized coatings. (Reproduced by permission from D. R. Gabe, Principles of Metal Surface Treatment and Protection, Pergamon Press, Oxford, p. 99, Fig. 4.7.)

The protection offered by zinc arises from two mechanisms: (1) physically protecting the steel from the atmosphere; (2) the sacrificial corrosion of the zinc in most environments to provide protection for the steel. The reason that this seems a contradiction in terms is that one protects one metal from corrosion by allowing another metal, with a more negative potential (Section 7.15; Table 7.1) to corrode. This apparently illogical step arises from the aesthetically displeasing deep brown colour of rust, which society hates. This results because the natural corrosion products from steel are incapable of physically separating the metal from the environment. (There is an exception to this rule. Specially alloyed steels called 'weathering steels' produce a protective film on the surface. These have no use in the automotive industry. Stainless steels produce a coherent adherent oxide film and these will be discussed in Section 8.5.2. The metallurgy is discussed in Section 6.20 *et seq.*)

This deep brown should be compared with the colourless (white) Zinc corrosion product. Hence the colourless result of the corrosion of zinc is more acceptable to society. (Note that the normal atmospheric corrosion product from Zinc is a colourless complex compound. The Zinc ions react with air, water and carbon dioxide to form a coherent and adherent film which protects the zinc from the environment.) Zinc coatings are more resistant to chloride marine environments than they are to industrial environments which contain sulphur dioxide (see Fig. 1.6 and Section 10.5). The micro-climate produced by corroding zinc is alkaline in nature which causes saponification of the subsequent paint coatings (Sections 8.11–8.16).

Galvanizing is therefore one of the most promising corrosion prevention measures that can be applied to steel because it can be carried out so easily, in some cases before the steel leaves the steel mill or manufacturers. The durability of the total coating is dependent on thickness and environment as shown in Fig. 8.5. There are several ways of putting this (zinc) protection onto steel, as discussed in the following sections.

8.3.1 Hot dip galvanizing

The article is dipped in molten zinc and the sheet is therefore coated on both sides. This technique is little used in the automotive industry because there are several cold rolling and deformation defects which when added to the galvanizing defects give rise to an inferior final paint coat. The final paint film is marred by Lüder's bands (deformation irregularities: Section 6.35.1), zinc **spangles** (large grain size of the hot-dipped zinc surface) and zinc **pimples** (which become very pronounced when the galvanized sheet is drawn). The adhesion of the zinc layer must be good to withstand this drawing process, but the second phase produced by the alloy at the zinc–steel interface is very brittle (Fig. 8.6). This second phase is also cathodic to steel (see Section 8.6 for the effect on steel). Another problem which arises if galvanized sheet steel is used is the need for higher currents for spot welding, which produces a shorter electrode life (Table 8.1). Some of the disadvantages of using zinc-coated steels are the production of zinc vapour and shorter electrode life due to the higher electrode forces needed to pinch the weld together (Table 8.1). Hot dip galvanized steel is used satisfacorily by Porsche for the load-carrying sections of the body.

8.3.2 Sherardizing

The steel is coated with zinc dust and heated to 400°C so that the zinc diffuses into the surface (the melting point of zinc is 419°C).

8.3.3 Electrolytic zinc

Electrolytic zinc-coated steel is free of zinc spangles. Here the zinc is electro-plated onto the steel, which allows the steel to be coated on one side only (Fig. 6.41(c)). This is the form of 'galvanized' panels which the motor industry now uses for sills–rockers and for inside box members which in the past have shown premature corrosion (Plates 2.2 and 2.4).

8.3.4 Painting

The metal as zinc dust is suspended in a paint and, provided that the zinc content is high enough, electrical contact will be made between the zinc particles and the steel to provide a sacrificial 'metallic coating'. These are often referred to as zinc-rich primers.

8.4 ADVANTAGES AND DISADVANTAGES OF ZINC

8.4.1 Advantages

The use of a metal lower in the electrochemical series than steel is somewhat at first surprising. Initially one would expect the zinc to have a faster corrosion rate than steel, simply because it is lower than steel in the series. However, it is important to realize that position does not govern the rate of reaction. This is shown dramatically by aluminium and its alloys, and their use in the kitchen as cooking utensils, where they stand up to high temperatures and fairly acidic solutions of various fruits. Zinc coatings on steel have gained in popularity because zinc has a much better corrosion resistance than does steel in most service conditions (Fig. 1.6) (for the exceptions see Section 8.4.2). The superior corrosion resistance of zinc is due to its ability to form a dense and adherent corrosion product which acts as a barrier to further corrosion. There is an additional advantage of zinc as a surface coating: at raw or sawn edges the zinc will sacrificially protect the steel provided that the two metals are in aqueous contact. As the zinc coating applied to the automotive steel is not heated no brittle phase is formed. This means that the sheared edges may also obtain a smearing of zinc on the exposed surfaces. This sacrificial protection is dependent on the conductivity (and resistance) of the external liquid environment which coats the edges. This is dramatically shown by painted Zintec steel surfaces which have been scribed such that the paint film is cut down to the metal (Zinc) substrate.

This protection is only produced when both the zinc and the steel are in contact with the *same* liquid environment as shown in Fig. 7.13 and discussed in Section 7.6. Zinc itself is protected by a basic zinc carbonate which is formed on zinc products subjected to non-aggressive environments. Chlorides and sulphur dioxide containing environments tend to promote the dissolution of this protective zinc carbonate. This basic zinc carbonates does not form in mildly acidic media. The environment has to be alkaline before it can form. Thus the corrosion rate of the zinc is retarded as the zinc is in general physically protected from the environment by its own corrosion product. This is different from the case of steel, whose corrosion rate is increased when exposed to the chloride ion and no protective film is formed. However, zinc coatings do not form this protective carbonate film when exposed to atmospheres containing small concentrations of sulphur dioxide (because the environment is too acidic to form the basic zinc carbonate $ZnCO_3{\cdot}3Zn(OH)_2$. There is also evidence to suggest that the corrosion rate of zinc in pure atmospherically condensed water is fairly high and so zinc products usually suffer more when 'facing the ground' than when they 'face the sky'. One factor in this anomaly is the rate of evaporation from condensed droplets which have formed on sheltered surfaces, such as those which 'face the ground'. Condensation in automative bonnets and clinch folds of doors would be one example where the corrosion rate of the zinc coating would be high owing to atmospheric condensation and slow evaporation of this 'pure' (but industrially contaminated with sulphur dioxide) water.

Zinc is also subject to differential aeration cells especially when the level of oxygen falls below that which is required for saturation. This is typified by the corrosion obtained (by these concentration cells; Fig. 7.4) in zinc alloy carburettors as a result of the stagnant pools of water formed beneath petrol.

The majority of the corrosion products formed by the corrosion of zinc alloys if freely exposed to air are white, very adherent to the substrate and coherent within themselves. When exposed to water under anaerobic conditions, the products are voluminous. These are often termed 'white rust'. This unwanted corrosion can be reduced by ageing the newly produced zinc sheet in dry air or by using a chromate treatment which provides temporary protection to the zinc (see Fig. 6.41(c) and 6.41(d)).

8.4.2 Disadvantages

Zinc is one of the few metals which dissolves in both alkaline (pH > 12.5) and acidic (pH < 5) environments (Fig. 7.18). In Section 7.9 it was stated that the cathodic reaction produced an alkali, which would dissolve the zinc and so aid the undercutting of the paint film. Hence zinc-coated products soon shed their coat of paint as shown by some of the earlier uses of double-sided zinc coated steels in the automotive industry. This paint shedding is also shown by some painted galvanized steel window frames, pedestrian railings at controlled crossings etc. It is very difficult to keep a paint film on zinc-coated (steel) products if there is little or no pretreatment (chromate, phosphate or preageing treatments). A similar effect takes place on the one-side zinc-coated steel used in the box sections if the pH rises above 12.5. This effect will not be noticed by the owner of the vehicle as the paint stripping will be proceeding in out-of-sight areas (e.g. see Fig. 7.13). The right choice of paint and the correct pretreatment will allow zinc-coated steels to keep their paint film.

8.4.3 Other zinc products

Alloys of 43% Zn and 55% Al, 1.6% are used to protect steel. This alloy is termed galvalume and is sacrificial to steel. The presence of aluminium provides increased protection (some fourfold increase) over galvanized products. These alloys are used in the automotive industry as a replacement for galvanized products. Galvalume products do not provide anodic protection to raw (cut) edges.

8.4.4 Cadmium coatings

Cadmium is a metal below zinc in the periodic table (Table 6.5) and is also used to plate small steel components. Cadmium has similar properties to zinc as shown in Table 8.2.

8.4.5 Weldability

Methods used for joining metals in automobile construction are welding, soldering and brazing. Coated steels must be capable of being joined by any of these methods (Table 8.1). Often only slight changes to general procedures need to be made. Resistance spot welding is used for joining sheet steel into body of the vehicle (Fig. 8.7(a)).

8.4.6 Formability

Formability is the main requirement for coatings used in the automotive industry. These include stamping, bending, stretching, drawing and rolling. Coatings such as zinc phosphate often reduce the frictional properties between metal and die. The coating must be resistant to any damage which these processes may cause.

8.4.7 Painted appearance

The final paint coat to any coated steel, which is seen by the consumer, must be free of imperfections (Fig. 6.18) such as Lüder's bands and spangles (see Section 8.3).

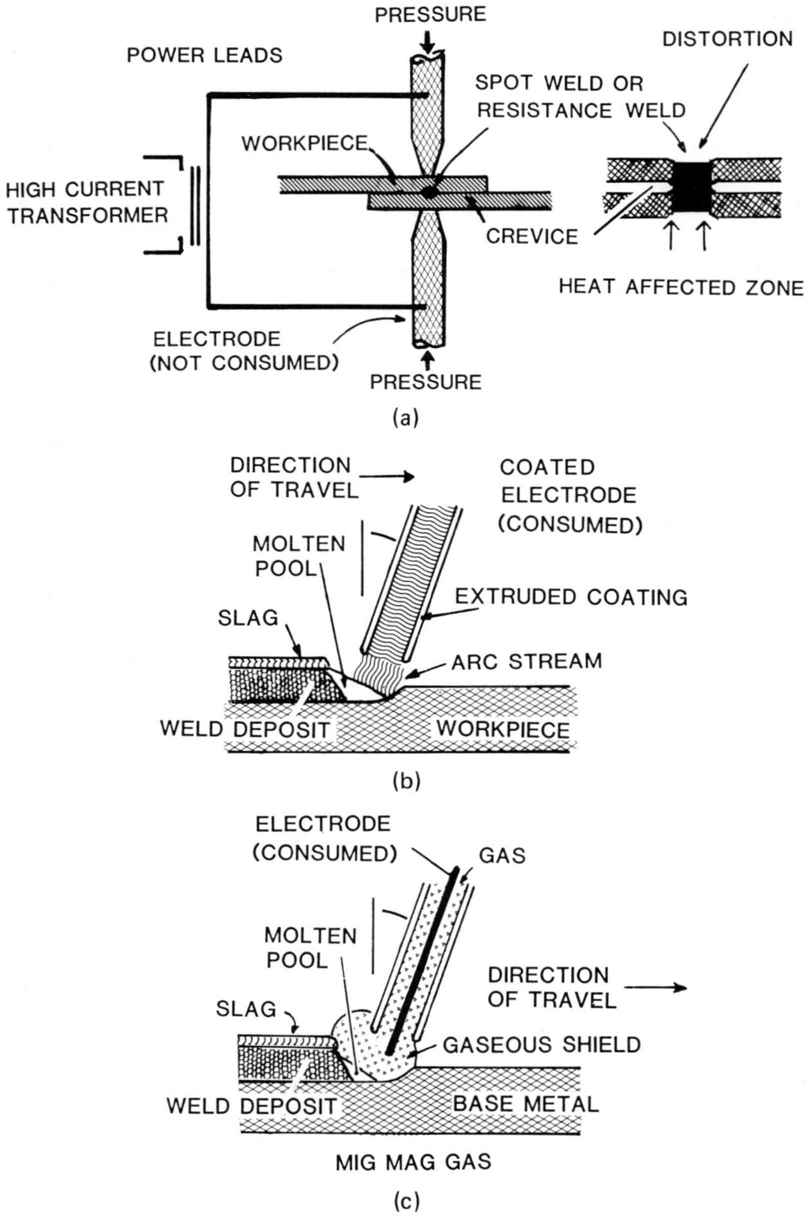

Fig. 8.7 — Three methods of joining metals together are shown. (a) The resistance spot weld requires no electrode. A high current is passed between two eletrodes (which are not consumed except by wear). The heat generated locally melts the metals over the area to which sufficient pressure has been applied to obtain contact. The fusion area is circular and forms a neat round contact are a termed a spot weld. (b) The conventional welding mechanism used by most other trades. A consumable electrode melts when a high energy electric arc supplies sufficient heat to the two metals. (c) Metal inert gas (MIG) and metal argon gas (MAG) welding are used in the repair of crashed vehicle bodies. These gases are passed down the arc gun which contains a coil of consumable wire electrode so as to shroud the fused metal and to prevent it from oxidizing rapidly. A heat-affected zone may be produced where the heat is rapidly conducted away from the weld area. See the discussion of brittle fracture problems in Section 6.32.

8.5 ALUMINIUM-COATED STEELS

The metal aluminium is anodic to steel. Aluminium coatings are therefore anodic to steel, and can be applied by a continuous hot-dip process by surrounding steel in a bath of molten aluminium with 5–11% Si beneath an inert atmosphere. The purpose of the silicon is twofold: (a) to minimize the growth of the intermetallic layer and (b) to increase the metallurgical bond between the steel and the aluminium. The advantage of this coating is that it offers good corrosion resistance to both exhaust/muffler condensates (Table 8.3) and chloride environments and so is used for exhaust/muffler systems, where the life is extended by a factor of two as shown in Table 8.4. The cheaper replacement exhaust systems are often made from mild steel. For exhaust–muffler conditions, corrosion resistance at elevated temperatures to acidic condensates is also important (Table 8.3 and Fig. 8.8). It is estimated that the condensate is equivalent to aqua regia (Table 8.3), which is very strong acid formed by mixing 1 part nitric acid with 3 parts hydrochloric acid.

Table 8.3 — The exact analysis of the condensate from exhaust gases produces a very corrosive mixture of acids. In order to examine the performance of various steels in the laboratory, the mixture described here was used to obtain the results depicted in Table 8.4

Acid	Concentration
HCl	1000 ppm Cl^-
H_2SO_4	470 ppm SO_4^{2-}
HNO_3	4800 ppm NO_3^-

Reproduced by permission from A. J. Cannings, in D. R. Holmes (ed.), Dewpoint Corrosion, Ellis Horwood, Chichester, 1985, Chap. 14, Table 1.

Table 8.4 — The relative corrosion resistance of various steels for exhaust manufacture

Material	Relative corrosion resistance
Mild steel	1
Galvanized mild steel	1.2
Galvalume on mild steel	1.3
Aluminized mild steel	1.5
3CR12 stainless steel	1.8
Aluminium	2.2
Hyform 409 stainless[1]	2.7

[1] Only tested for 7 days.

Reproduced by permission from A. J. Cannings, in D. R. Holmes (ed.), Dewpoint Corrosion, Ellis Horwood, Chichester, 1985, Chap. 14, Table 2.

Galvalume is the registered trademark of the Bethlehem Steel Company.

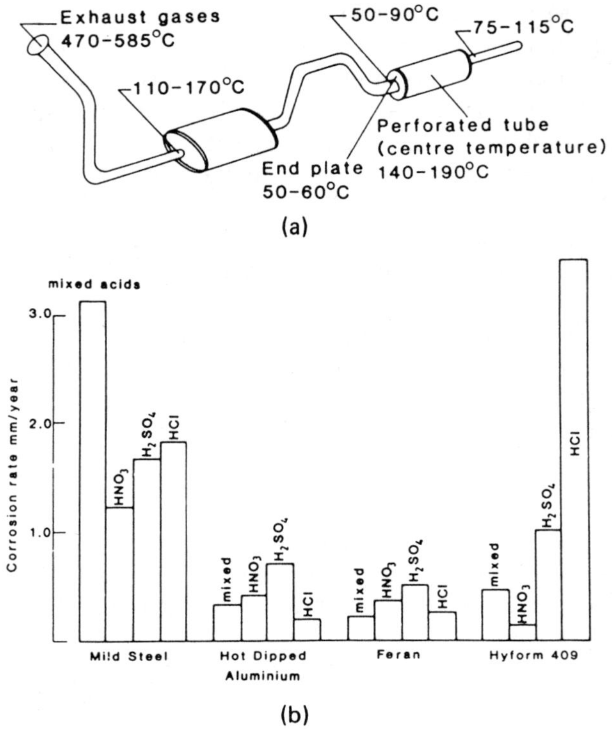

Fig. 8.8 — (a) The temperatures of the various sections of a vehicle exhaust–muffler system when the engine is running at 2000 rev/minute. There are three types of corrosion mechanisms here: high temperature oxidation in the region 470–585°C, internally nucleated corrosion resulting from the acidic nature of the condensation products and externally nucleated corrosion by virtue of the external road de-icing salts. (b) Corrosion rates of various metals in short-term laboratory tests at 75°C. (Reproduced by permission from P. J. Boden and S. J. Harris, in D. R. Holmes (ed.), Dewpoint Corrosion, Ellis Horwood, Chichester, 1985, Chap. 15, Fig. 1 and 15.)

Although aluminium coatings are anodic to steel, the aluminium does not dissolve readily as it is protected by its coherent and adherent oxide film. Temperature cycling, which causes fracturing of the aluminium oxide film, can be overcome by adding titanium to the coating. The titanium combines with the carbon (of the steel) and nitrogen from the air, thus decreasing the amount of austenite/ferrite phase change (at temperatures above 723°C (1270°F) and dependent on the percentage of carbon, i.e. line RSW of Fig. 6.34(d) as the exhaust metal is cycled through the temperature composition line RSW in Fig. 6.34(d) (Section 6.26). The phase change is from b.c.c. to f.c.c. and is accompanied by a volume change of some 8%. Thus warpage, distortion and cyclic oxidation properties are improved, and there is also the possibility of having a higher operating temperature (676–833°C (1250–1500°F)).

8.5.1 High temperature oxidation

The exhaust gases entering the exhaust pipe cause high temperature oxidation. Oxidation is corrosion process which occurs without an aqueous environment. In this process metallic ions and their electrons can diffuse through the oxide film where the electrons react to produce the oxide ion. Growth of the oxide film therefore takes place at the oxide/environment interface (Fig. 8.9(b)). Alternative mechanisms have been postulated where the oxygen ions formed on the surface as a result of the diffusion of electrons through the oxide film cause the film thickness growth to take place at the oxide–metal interface (Fig. 8.9(c)). Fig. 8.9(d) shows a combination of the two mechanisms. For high temperature corrosion resistance, the oxide film which is formed on the surface of the metal is required to protect the underlying metal from the environment. This would produce a thickness–time curve as shown in Fig. 8.9(a) parabolic curve. However, this is very rarely observed as the oxide film tends to fracture as a result of temperature fluctuations or volume differences as the dense metal becomes impregnated with large O^{2-} ions. As a result the oxides become non-adherent to the substrate. This causes different growth laws depending on the actual mechanism postulated.

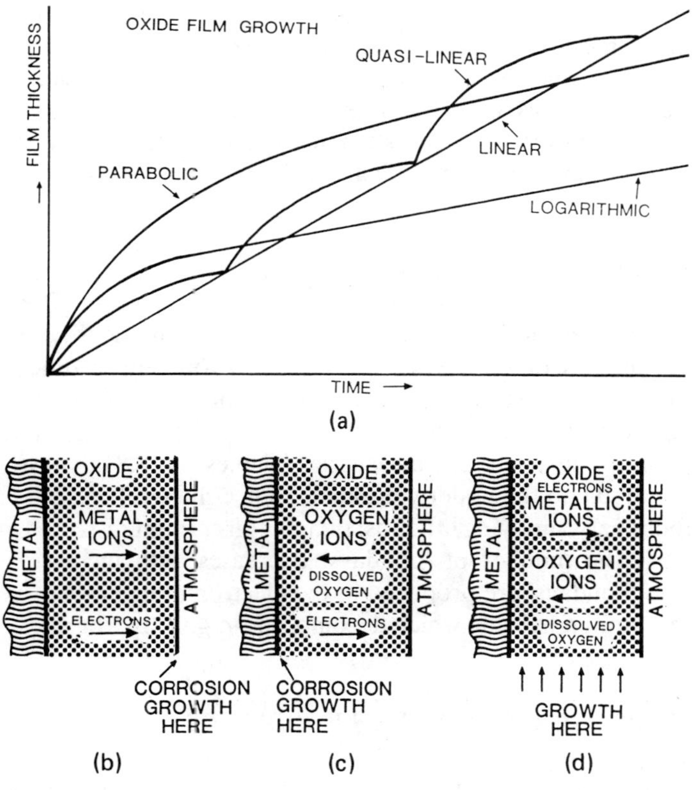

Fig. 8.9 — (a) The oxide film thickness (y) increase with time is modelled on certain mathematical equations; logarithmic, $y = \log_e(kt)$; linear, $y = kt$; parabolic, $dy/dt = k/t$, where dy is a small increase in thickness in a small interval of time dt; quasi-linear, a combined linear and other form depending on where the oxide films fracture owing to the increase in volume. (b)–(d) The growth of the oxide film is modelled upon which species diffuses: (b) diffusion of metallic ions, only giving growth at the oxide–air interface; (c) diffusion of oxygen ions, giving growth at the oxide–metal interface; (d) diffusion of both ions, giving growth within the oxide film.

8.5.2 Stainless steels

Stainless steels owe their corrosion resistance to the tenacious coherent oxide film which is developed under atmospheric conditions. In aqueous conditions stainless steel is not as good as the name implies because it requires oxygen-containing waters to remain "stainless". The potential is shifted considerably (approximately 330 mV for type 304), depending on the environment as shown in Table 7.3.

8.6 CATHODIC COATINGS TO STEELS

These coatings are classified as cathodic to steel and so promote or accelerate the corrosion of the underlying steel.

8.6.1 Lead alloy coatings

Lead alloy coatings (abbreviated as Pb–Sn in Fig. 8.10) are referred to as Terne-–coat, which is used for automotive brake pipes. These coatings are more corrosion resistant than zinc coatings, but do not provide sacrificial protection at 'holidays', pin-holes and breaks in the coating. It is difficult to produce a pin-hole-free coating and the current method of overcoming the severe pitting of these coatings is to interlace them with layers of other metals which are more corrosion resistant (Fig. 8.10(b)). The effect of this corrosion-resistant interlayer is to cause a lateral spread of corrosion rather than the characteristic metal penetration which causes the brake pipe to fail under heavy 'panic' breaking. The hydraulic bursting pressure is given in Fig. 8.11.

8.6.2 Chrome plating

These coatings of chromium are cathodic to both steel and nickel. The effect this has on the underlying steel is shown in Fig. 8.3. The chromium does not dissolve at all, but contains 'holidays' or voids which allow the environment to attack the substrate nickel coating. The nickel inter layer therefore dissolves, producing electrons which are consumed by the cathodic reaction which takes place at the chromium plate. The nickel is higher up the electrochemical series (Tables 7.1–7.3) than steel and so the nickel pit is cathodic to the underlying steel. Hence the steel corrodes galvanically, producing those very annoying little rust pimples on chromium-plated products. Fig. 8.4 shows that the number of 'holidays' decreases with increase in chromium thickness. The 'holidays' are produced in the electroplating process as a result of a metastable chromium hybride which decomposes to give hydrogen.

8.7 METHODS USING CORROSION INHIBITORS

Inhibitors are a class of material additives which can be added to the aqueous environment to undertake a variety of functions. Thus not all inhibitors are corrosion inhibitors. For example, inhibitors are added to lubricants to improve their resistance to oxidation, as well as reducing the effects of corrosion (Table 8.5).

Corrosion inhibitors are used to control the corrosion of motor vehicle parts, prior to manufacture and in shipment of spare parts. They are used in the radiator cooling water and anti-freeze solutions, engine lubricating oils, metal pretreatments, in paints and in cavity waxes. Most proposed inhibitors fall into one of the following

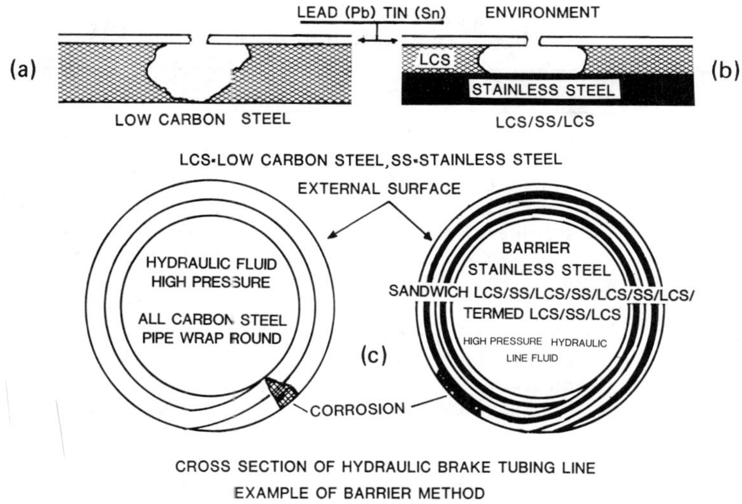

Fig. 8.10 — Coatings may be made to act as barriers in some cases, while in others they may aggravate the problems (see Section 8.2; Fig. 8.3). The beneficial example chosen here is that of hydraulic brake pipes where the stainless steel (SS) barrier band increases the life of the brake pipes. (a) At local sites where the Terne coating (Pb–Sn) is porous, the underlying steel will be anodic to the outer lead alloy. Pitting corrosion of the low carbon steel (LCS) pipe will therefore take place beneath what appears to be a good outer 'corrosion-free' coating. Penetration by pitting resulting in a sudden bursting under 'panic stop' conditions will be the ultimate result (see Fig. 8.11). (b) Low carbon steel (LCS) is made the sandwich in a triple barrier system with stainless steel (SS). The stainless steel is now cathodic to the low carbon steel and so pitting of the hydraulic brake pipes does not take place. The cathodic barrier to the sandwich will withstand better both the highly corrosive atmospheres and panic stop pressures within the hydraulic oil by a factor of three. The stainless steel barrier does not 'cathodically' corrode (if this barrier were zinc, it would preferentially corrode in both the anolyte and catholyte environments: Fig. 7.18). (c) The hydraulic brake pipe is made by wrapping the triple sandwich as shown in (c). (After R. Baboian, Clad Metal Systems for Automobiles, in Automotive Corrosion by De-icing salts, National Association of Corrosion Engineers, Houston, Tx,1981, p. 410 Fig. 12.)

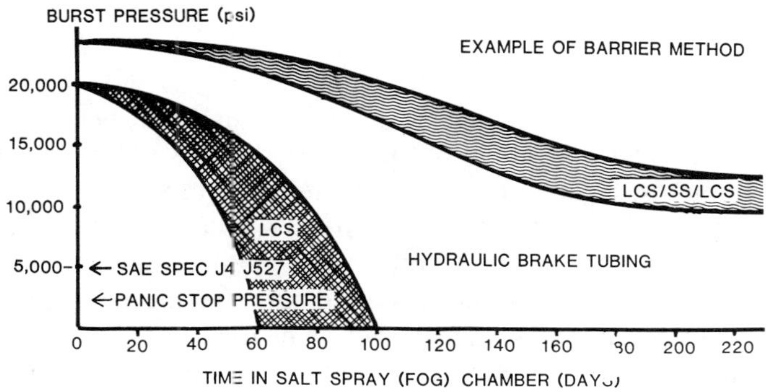

Fig. 8.11 — The bursting pressure of the hydraulic pipe is shown after a period of time in laboratory accelerated corrosion tests for the LCS and LCS/SS/LCS triple sandwich. (After R. Baboian, Clad Metal Systems for Automobiles, in Automotive Corrosion by De-icing Salts, National Association of Corrosion Engineers, Houston, Tx, 1981, p.410 Fig. 14.)

Table 8.5 — Inhibitors are added to lubricating oils to enhance the anti-corrosion behaviour of the lubricant

Additive	Function	Typical chemical types
Oxidation inhibitor (anti-oxidant)	To increase oxidation resistance of lubricants by interfering with the chain reactions which give rise to acid and asphaltene formation	Oil-soluble amine and phenol derivatives for temperatures up to about 120°C. For higher temperatures dialkyldithiophosphates and other compounds listed below as metal deactivators; also P_2S_5-olefin and -terpene reaction products
Metal deactivator	To form inactive protective films on metal surfaces which otherwise might catalyse oxidation and corrosion reations	Trialkyl and triaryl phosphites, organic dihydroxyphosphines, some active sulphur compounds, diamines; in lubricating greases, mercaptobenzothiazole and phosphites
Corrosion inhibitor	To protect metal surfaces, particularly bearing surfaces (copper, silver, lead) against corrosion	Zinc dialkyldithiophosphates, sulphurised olefins, sulphurised terpenes
Rust inhibitor	To eliminate rusting in presence of moisture	Polar compounds, such as metallic soaps, esters and derivatives of dibasic acids; barium and calcium sulphonates
Water repellant	To impart water-resistant properties, particularly in greases	Aliphatic amines, hydroxy fatty acids and some organic silicone polymers
Basic compound	To neutralise strong mineral acids (from fuels)	Barium or basic calcium salts of sulphonic acids and alkylsalicylic additives are basic salts with a large excess of base
Dispersants	To keep insoluble combustion and oxidation products in suspension and dispersed	Salts of phenolic derivatives; polymers containing barium, sulphur and phosphorus; calcium or barium soaps of petroleum sulphonic acids

Reproduced with permission from L. L. Schrier, Corrosion, Vols. 1 and 2, Newnes Butterworth, London, 1977, Table 2.20.

categories: natural plant or animal products, dyes, industrial by-products or pure chemical compounds. When the active chemical compounds are isolated and examined in detail, they are all found to contain a group 5 or 6 element, the most common being nitrogen or sulphur, through which they are thought to act. Inhibitors are thought to act by being physically adsorbed onto the surface. A group 5 element will want to form a compound with five or three nearest neighbours. The valency of an element was derived in Section 6.5(6) as being either the group to which it belonged or the group number subtracted from 8. For example, nitrogen in group 5 of the periodic table (Table 6.5) has $8-5 = 3$ as the valency.

A corrosion inhibitor is a substance which is added to the environment in order to reduce the corrosion rate. Most common inhibitors are a mixture and are proprietary compositions so that the constituents are not disclosed, while other solutions are well protected by patients and their compositions are well publicized.

Inhibitors are often specific towards a particular metal, so that an inhibitor for copper in a solution will often not be inhibitive towards steel in that same liquid (Table 8.6). Thus most practical systems will require several inhibitors for each of the metallic components present. Chromates and dichromates are the only exception to this rule. These salts provide excellent inhibition to most metals and alloys but suffer from a toxicity problem. Prolonged contact of the skin with chromates produces a rash, and stringent regulations are in force for their disposal in waste. The precise mechanism of corrosion inhibition is not fully understood and is the subject of much research and debate. However, the methods by which it is thought that the inhibitors operate give rise to their classification.

Table 8.6 — General guide to the effectiveness of various inhibitors in the near-neutral pH range

Metal	Chromates	Nitrites	Benzoates	Borates	Phosphates	Silicates	Tannins
Mild steel	Effective	Effective	Effective	Effective	Effective	Reasonably effective	Reasonably effective
Cast iron	Effective	Effective	Ineffective	Variable	Effective	Reasonably effective	Reasonably effective
Zinc and zinc alloys	Effective	Ineffective	Ineffective	Effective	—	Reasonably effective	Reasonably efective
Copper and copper alloys	Effective	Partially effective	Partially effective	Effective	Effective	Reasonably effective	Reasonably effective
Aluminium and alumi-nium alloys	Effective	Partially	Partially	Variable	Variable	Reasonably	Reasonably
Lead–tin soldered joints	—	Aggressive	Effective	—	—	Reasonably effective	Reasonably effective

Reproduced by permission from L. L. Schrier, Corrosion, Vols. 1 and 2, Newnes Buterworth, London, 1977, Table 18.1.

The methods by which the inhibitors may work are

(1) by altering the nature of the corrosion product, (2) by chemisorption onto the metal surface, (3) by rearranging the metal reference electrode potential of both anode and cathode and (4) by changing the electrical resistance of the electrolyte environment. Their classification can be as follows: absorption type; hydrogen evolution poisons; oxygen scavengers; oxidizers; vapour phase inhibitors.

8.7.1 Absorbtion type

These inhibitors absorb onto metal surface and affect both the cathodic and the anodic reactions. They could be considered as a monolayer of a 'paint'. Their molecular size is usually large so that the molecule can be likened to the fabric shade part of an umbrella. This acts like a very thin barrier coating. The molecule has an active part, usually a positively charged group 5 or 6 atom e.g. a nitrogen atom, which is equivalent to the 'umbrella handle' by which it attaches itself to the metal surface: compare Fig. 8.12(a) and 8.12(b). Typical of this class of inhibitor are the organic amines (Fig. 8.12). This absorption is electrostatic in nature and the whole absorbed molecule is thought to act as a barrier coating. Consequently a high molecular weight and suitable spatial geometry of the molecule (e.g. balance about this group 5 atom) are required to aid the 'packing' of the 'umbrella' at the metal surface (Fig. 8.12(d)). Balance is required so that the umbrella sits flat to the surface rather than tilted. Molecular compounds which are not spatially balanced have been found to be inferior to those which are balanced. Most inhibitors of this type are large oily macromolecules or ions.

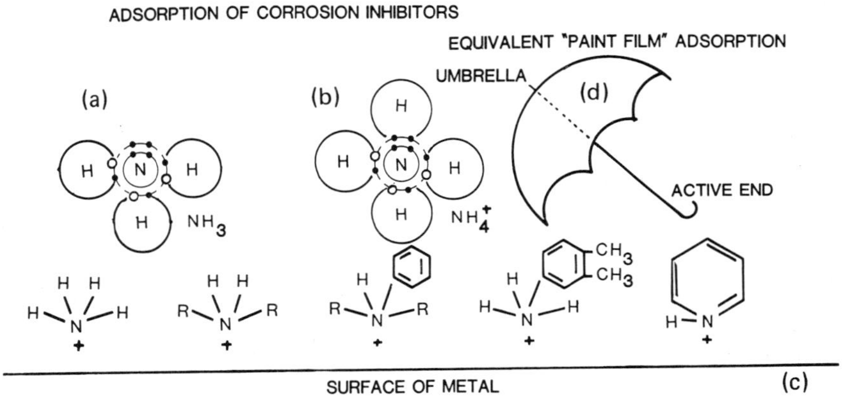

Fig. 8.12—Corrosion inhibitors based on the amines produce chemically absorbed monolayers adjacent to the metal surface. These amines depend on their physical size to absorb onto the substrate via the 'active end', which is through a 'positively' charged nitrogen atom. (The positive charge arises from the addition of one hydrogen proton as we go from (a) NH_3 to (b) NH_4^+. This will be the same for the other amines shown in (c).) The size of the amine unit is likened to (d) the shield provided by an umbrella during a rain storm. The barrier need be only one layer thick, and can be likened to a very thin film of paint.

8.7.2 Hydrogen evolution poisons

These inhibitors prevent the discharge of hydrogen on the cathodic areas of iron and thus reduce (eliminate) corrosion. They achieve this by being a 'negative catalyst', such as antimony or arsenic ions, or by forming a blanketing layer or film which prevents the evolution and discharge of hydrogen. In this case the inhibitors are of little use in oxygenated waters and cannot be used outside acidic environments (pH 1–6). These inhibitors are used in the motor trade for inhibiting phosphoric acid solutions where rusted steel is etched prior to painting, e.g. in relation to stone chips etc. (Chapter 9). Their use should be avoided if there are any problems with cyclic deformation where hydrogen embrittlement of the metal can cause brittle failure.

(Hydrogen embrittlement is where hydrogen is absorbed by the steel. This absorption results in the raising of the brittle transition temperature (Fig. 6.39; Section 6.31).

8.7.3 Oxygen scavengers

In cases where oxygen is involved in the cathodic reaction, e.g.

$$O_2 + 2H_2O + 4e^- \rightarrow 4OH^-$$

oxygen scavengers will be effective at controlling corrosion. Such an inhibitor is hydrazine which is oxidized to nitrogen and water. Such inhibitors are therefore consumed with time as shown below:

$$N_2H_4 + O_2 \rightarrow N_2 + 2H_2O$$

They are not suitable for acidic solutions.

8.7.4 Oxidizers

These are chromates, nitrates and other oxidizing salts are used to inhibit the corrosion of iron and its alloys. Chromate rinses are used by the motor vehicle manufacturer to confer better corrosion resistance after phosphating in the paint pretreatment process of the car body (Chapter 9).

8.7.5 Vapour phase inhibitors

These are substances which possess a very low vapour pressure and readily transfer from their liquid formulation to the metallic surface by a process termed 'sublimation'. Vapour phase inhibitors (VPIs) therefore are only effective in enclosed areas. They are similar to the organic absorption inhibitors in their mode of operation. These VPIs are usually found on oily, brown paper which has a distinctive smell and is used as a wrapping paper for metallic components purchased as spare parts in the automotive industry. The brown paper carries the VPI. VPIs have been used in wax formulations for the injection of cavity box sections. They are usually volatile nitrites such as dicyclohexylammonium and di-iospropylamine nitrites.

8.7.6 Concentration level on inhibitors

Corrosion inhibitors do not affect the environment as a whole by either pH or oxygen concentration. However, they do require to be present above a certain critical level for any degree of effectiveness.

8.7.7 Geometric design and inhibitors

Crevices which form differential aeration cells impose certain restrictions on inhibitors which must be considered. For example, the access of the inhibitor to the anode of such a cell is restricted by the same factors which prevented any dissolved oxygen gaining access to the anodic area which set up the cell in the first instance. Hence inhibitors acting on the anode will be more susceptible to concentration within the medium.

8.7.8 Some inhibitors and uses of corrosion inhibitors

(1) Nitrites are good for ferrous metals, cast iron and steel covered with scale. Nitrates are not very effective for copper and aluminium alloys. The optimum concentration is 2000 ppm at a pH of 8.5. They are dangerous inhibitors if lead and lead solder are present. This can occur in leaded overlap joints, where nitrites are used in the pretreatment salts as oxidizing accelerators.

(2) Phosphates give average performance for most metallic systems, but the active component, the phosphate unit, tends to be precipitated out if dissolved in hard waters.

(3) Polyphosphates are good for steel but less effective for other metals. Polyphosphates have been used as inhibitors in cooling waters for a long time. Additions of this inhibitor have been tried in water washes in an attempt to reduce the effect of corrosion caused by the road de-icing salts in the automotive industries. They were found to be too expensive when used on the scale necessary to reduce the aggressive effect of the chloride ion.

(4) Borax or tetraborates are good for steel and zinc but less effective for cast iron and copper. Mixed results are obtained with aluminium.

(5) Chromates are good for all metals, but their use should be controlled where they are mixed with organic compounds, e.g. ethylene glycol mixtures. Ethylene glycol is used as a freezing point depressant in anti-freezes. In these cases the chromate inhibitors can be reduced from the hexavalent chromium ion Cr^{6+} to the non-inhibiting trivalent chromium ion Cr^{3+} and corrosion will result. Chromates can give rise to dermatitis. They are used in car radiators where concentrations of 1000 ppm provide protection. Chromates are also used in the wash after the paint pretreatment process (Fig. 9.1(19)). These chromates prevent the formation of scab and filiform corrosion (Section 8.12 *et seq.*; Fig. 8.21; Fig. 2.6).

(6) Benzoates are good for steel and effective for copper and aluminium alloys. Sodium benzoate is classed as a 'safe' inhibitor in that it is not concentration dependent. Its efficiency is greatly influenced by the presence of chlorides. It was originally developed for use in radiators containing glycerol anti-freezes with which it is compatible (see the discussion on chromates above). It is used in waxed wraps and temporary protectives (Section 8.8) based on rubber latex or bentonite pastes because it does not cause dermatitis. Poor protection is obtained with cast iron and zinc (galvanized products tend to have an anode–cathode reversal, the zinc becomes cathodic to the steel and severe corrosion can result). Benzoates are not deleterious to lead and lead solders.

(7) Silicates are fairly good with most metals, particular nickel.

(8) Molybdates are good for steel. They are used in conjunction with other inhibitors to obtain a synergistic effect.

(9) Triazoles are good for copper and brass but not very effective with other metals.

(10) Thioureas are successful in pickling solutions, but can cause eye problems, leading to temporary blindness.

(11) Tannins are reasonably effective for most metals. These are used for 'conversion' coatings, where the rust is converted into a black magnetite oxide film, which acts as a base for subsequent paint films.

8.8 TEMPORARY PROTECTIVES (PROTECTIVE OILS AND WAXES)

The protective oils and waxes used in the automotive industry for corrosion protection of internal cavity box members are a derivation of the coatings used by the packaging industry. The packaging industry uses wax coatings as a cheap method of excluding moisture vapour from foodstuffs. Wax on its own or impregnated in paper is used for metal components, in storage and prior to assembly. The waxes are hydrophobic (dislike of water) substances which retard the ingress of water and oxygen to the metal surface (compare with paints Section 8.12). They are particularly useful in those industries where the components cannot be supplied in the painted condition, e.g. pistons or other steel components etc. In the packaging industry, these coatings have to be removed, and they became known as 'temporary protectives' because they were readily removed by a hydrocarbon solvent. The term 'temporary protectives' does not give any implication as to the length of time that the coatings are effective as a corrosion-preventative coating. Paint, in contrast, is not removed by a solvent, because the paint film undergoes a further chemical reaction with the oxygen of the atmosphere.

Protection of steel components by temporary protectives arises from both a 'tackiness' and a mechanical keying of the waxes/oils onto the surface as the volatiles evaporate. In some formulations there is also a polar attraction of the oils onto the metallic surface in the same way that the adsorption type inhibitors worked (Section 8.7). These polar compounds are usually metallic soaps of long chain fatty acids, and alkali salts of the the petrolatum sulphonates and oxidized petroleum residues. (A calcium salt of a weak organic acid is fairly alkaline owing to the position of calcium in the periodic table (Tables 6.2 and 6.5). Alkaline environments are protective towards steel as discussed in Section 7.10). The exact compositions of these polar additions and their derivatives are proprietary information. However, because these polar compounds have a greater tendency to be adsorbed onto steel than does water, these materials are very effective coatings for preventing corrosion. They are often called 'water displacing additives', because they displace water by a combined process of density and chemical polar group activity. This class of temporary protective is used to coat machined parts and to remove the machine 'oils' (a water–oil emulsion) to leave a thin coating of oil. They usually require a separate cage and water drain-off area as shown in Fig. 8.13. They should not be used as a substitute for proper drying where they are used in wax injection systems in automobile bodies where they cannot act as a water displacer, e.g. where drainage holes to doors are high as shown in Fig. 8.14. In cases where the water-displacing additive is incorporated into the waxes, the relative density between water and the wax will govern the film thickness. This is shown in Plate 4.2(b).

There are several types of temporary protectives (Table 8.7), which can be arranged in the following classes as follows:

(1) Class A: oil-type protectives. These are essentially mineral or vegetable oils to which corrosion inhibitors and anti-oxidants have been added. They are usually inferior to the other classes of protectives discussed below, but they do have the advantage that they are highly penetrating liquids with a hydrophobic surface. They are ideal for crevices. The main disadvantage with them is that they can be physically displaced by touch or abrasion and by water.

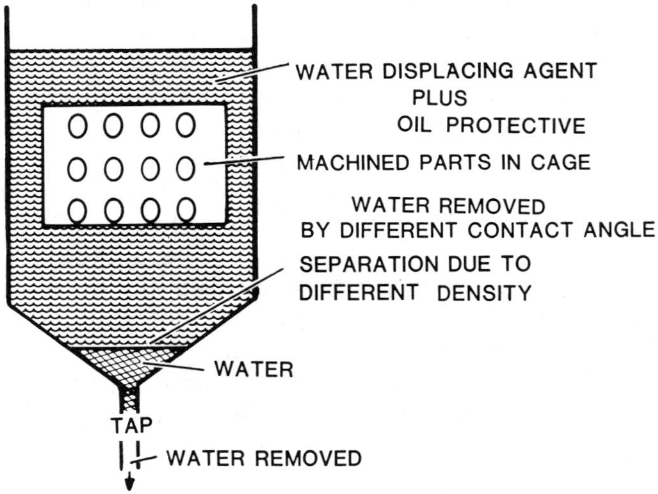

Fig. 8.13 — Schematic view of a water-displaced agent being used to coat newly machined steel parts with a temporary protective before storage or transport. Machining oils are a water–oil emulsion. Any water left on the steel component will cause rapid corrosion. The surface-active compounds have a higher affinity for the steel surface than does the water. As the water is also denser than the protective oil, it falls downwards owing to gravity where it may coalesce and form a pool at the bottom. From this point it can be periodically drained off. A machine part being suspended in the water–oil line will not be protected as the associated water will not be displaced.

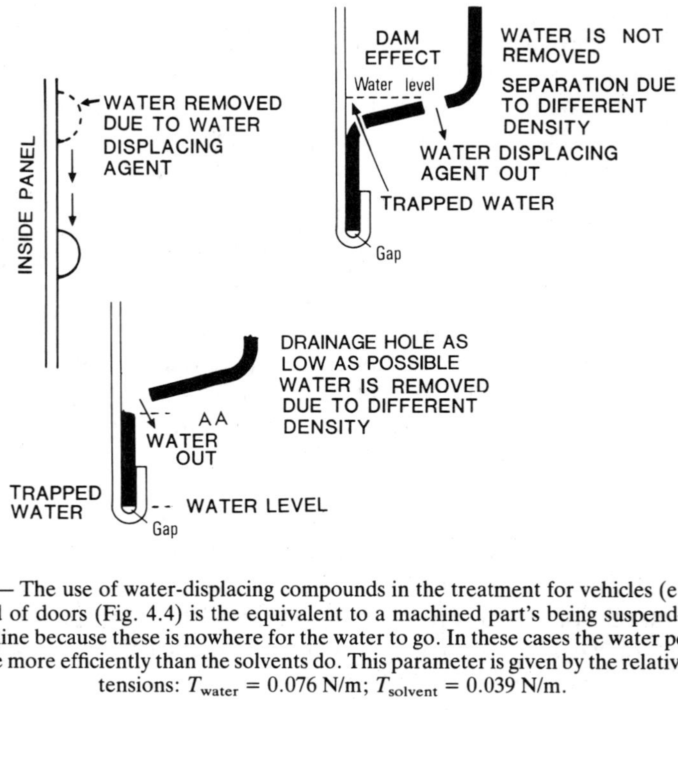

Fig. 8.14 — The use of water-displacing compounds in the treatment for vehicles (e.g. at the clinch fold of doors (Fig. 4.4) is the equivalent to a machined part's being suspended in the water–oil line because these is nowhere for the water to go. In these cases the water penetrates the crevice more efficiently than the solvents do. This parameter is given by the relative surface tensions: $T_{\text{water}} = 0.076$ N/m; $T_{\text{solvent}} = 0.039$ N/m.

Table 8.7 — Types of temporary protectives in general use

Type of protective	Typical ingredients[a]	Method of application	Properties of film
Solvent-deposited hard film		Dipping, spraying, brushing	Solid, thin, tough, non-sticky, removable by wiping with solvent
(a) ordinary grade	(a) Plasticized bitumens, plasticised resins, white spirit, coal tar naphtha, chlorinated solvents		
(b) water-displacing grade	(b) As (a) above together with water-displacing agents		
Solvent-deposited soft film		Dipping, spraying, brushing	Solid, thin, greasy, removable by wiping with solvent
(a) ordinary grade	(a) Lanolin, petroleum, with and without specific corrosion inhibitors and anti-oxidants, white spirit, coal tar naphtha, chlorinated solvents water-displacing grade together with water-displacing agents		
Hot-dipping soft film	Petroleum, lanolin, with and without specific corrosion inhibitors	Dipping in molten material	Solid, thick, waxy or greasy, removable by wiping with solvent or immersing in hot oil
Smearing	Metallic soap and mineral oil, soft petroleum, lanolin (castor oil/lead stearate for rubber-containing components)	Smearing, brushing	Solid, thick, greasy, removable by wiping with solvent
Slushing	Metallic soap and mineral oil, oil-softened petroleum, lanolin, small amounts of solvent	Smearing, brushing	As for smearing protective
Oil	Mineral oil, specific corrosion inhibitors and anti-oxidants	Dipping, rinsing, spraying	Liquid, thin, oily,
Strippable			
(a) hot-dipping grade	(a) Ethyl cellulose, cellulose acetate butyrate, mineral oil, plasticizer, resins, stabilizers	(a) Dipping in molten material	(a) Solid, tough, non-adherent, often leaves oily film with lubricating properties; film removed by stripping

Continued overleaf

Table 8.7 (cont.) — Types of temporary protectives in general use

Type of protective	Typical ingredients[a]	Method of application	Properties of film
(b) cold applied grade	Vinyl copolymer resins, plasticizers, stabilizers — flammable or non-flammable solvents	Spraying, dipping	Solid, tough, non-adherent film, removed by stripping
Volatile corrosion inhibitor (V.C.I.)	Organic amino salts (e.g. dicyclohexylamine nitrite, cyclohexylamine carbonate)	From solution by spraying, as a powder by sprinkling, by wrapping with V.C.I-impregnated paper	Adsorbed, non-visible film

[a] Some details of typical compositions, where these are available, are given in *Petroleum, Oils and Lubricants (POL) and Allied Products*. Defence Guide DG-12, Section IV, Ministry of Defence, H.M.S.O., London (1968).

Reproduced by permission from L. L. Schrier, Corrosion, Vols. 1 and 2, Newnes Butterworth, London, 1977, Table 17.5.

(2) Class B: soft, thin, non-drying films which are left when the solvent evaporates. The solvents are usually white spirit, or coal tar naphtha. These films are based on wool-fat derivatives (lanolin), petrolatum (Section 8.10.3), petroleum–oil mixtures or soap-thickened oils. They may or may not contain corrosion inhibitors. They are often very temperature-sensitive films.

(3) Class C: hard, thin, films which are left when the solvents evaporate. These usually consist of resin-hardened wool fat, plasticized bitumens and certain metallic soaps. These materials, when applied, tend to have very poor capillary creep.

(4) Class D: thick greasy films. These can be calcium- or barium-based greases from the wool grease stearin or from compounds based on petrolatum, with or without corrosion inhibitors.

(5) Class E: peelable or strippable coatings which surround gear components in transit. These are based on cellulose compounds.

(6) Class F: VPIs or volatile corrosion inhibitors. These are usually present in the wrapping paper of ferrous articles. Some non-ferrous metals are tarnished by the VPIs.

8.9 INDUSTRIAL OILS AND WAXES

The terms oil and wax refer to consistency and do not refer to a particular chemical composition. In general the oils and waxes can be represented by a saturated hydrocarbon where the general formula is $C_n\ H_{(2n+2)}$ (Fig. 6.7). Viscosity is the physical parameter which distinguishes an oil from a wax although temperature is the one agency which can make a wax from an 'oil'. A good example of this is the waxing of diesel fuel in the winter. Here the longer chain hydrocarbons solidify with decrease of temperature.

The general formula is $C_n\ H_{2n+x}$ (Fig. 6.7(b), 6.16(a). The value of x depends on the type of material:

normal paraffinic wax	$x = 2$	completely adductable in urea.
simple branching	$x = 2$	no change in the value of x
one saturated ring	$x = 0$	Fig. 8.15(c)
two saturated rings	$x = -2$	i.e. dicyclic compounds (Fig. 11.1)

The first classification of waxes is whether they are animal, vegetable, mineral, or synthetic in origin. A wax is group of **organic thermoplastic** substances which have a characteristic application and properties. They melt at about 50-90°C to a clear liquid which does not exhibit thread-spinning characteristics (e.g. like molten nylon). They form **gels** with organic non-polar solvents. Waxes have good water-repellant properties and are capable of imparting a high gloss (e.g. furniture and car polishes).

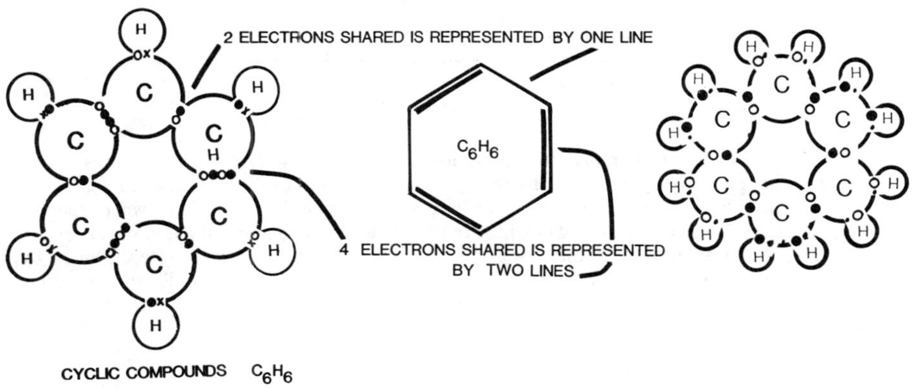

(a) (b) (c)

Fig. 8.15—(a) Benzene C_6H_5 is the most usual form of cyclic compound (Section 11.3). (b) The scientists shorthand representation of benzene. (c) The fully saturated (no double bonds) cyclic aliphatic compound, cyclohexane C_6H_{12} found in oils and to which the text applies. The hydrocarbons are all pigeon holed according to their general formulae and structure, Sections 11.2 and 11.3. Non scientists should consider reading Chapter 6, 10 and 11.

8.9.1 Transition temperatures

The waxes have one or more transition temperatures (Figs. 6.16 and 11.4) where the internal forces modify the crystalline structure with a release of energy (Fig. 8.16). For some waxes this anomalous expansion will take place at a temperature through which the panels are thermally cycled on standing out in the midday sun. The waxes will at the same time be losing their most volatile components with time. The net result is that the cavity-injected waxes show a marked tendency to crack and craze. This cracking takes place in the thickest waxes initially and the cracks propagate to the base of the crevice. This effect is more noticeable on the waxes applied by the motor manufacturers which must fulfil another requirement: **they must not drip**. It

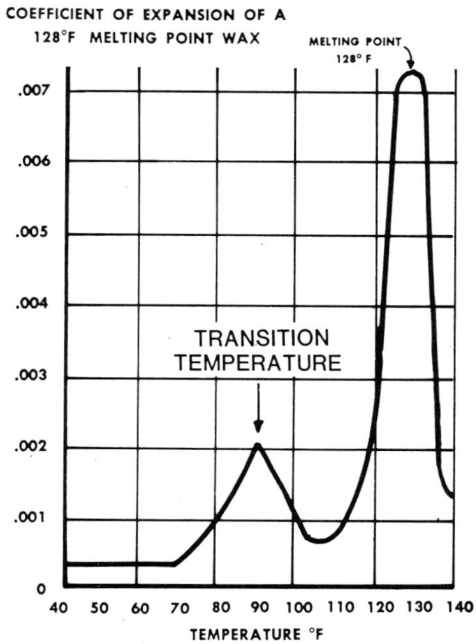

Fig. 8.16 — The various transition temperatures which characterize certain waxes. Each transition temperature is related to the thermal decomposition of the weakest van der Waals bonds (Section 6.12). This expansion of the wax is non-reversible and so some waxes tend to crack and craze. This cracking and crazing are usually to the base paint film and so, in certain locations, it is just as though there was no wax added to the structure. (Reproduced by permission from H. Bennett, Industrial Waxes, Vols. 1 and 2, Chemical Publishing Co., New York, 1975.)

was difficult to photograph this owing to the narrow apertures, but this crazing is shown in Plate 4.16.

The iso-paraffins do not have a transition temperature. Other waxes do not show a transition temperature unless the cooling rate is fairly rapid. This transition change leads to a contraction in crystal boundaries and this transition does not give the optimum properties where the waxes are used for moisture/water exclusion, e.g. in cavity injection (It is an advantage, however, if the wax is to be moulded as shown in the manufacturing of candles as the candle will shrink away from the mould sides). Waxes can occlude 7-15% of their volume of air at ordinary temperatures.

8.9.2 Wax types

The types of waxes on the market are a follows:

(1) Refined waxes are soluble in benzene, ether, chloroform, carbon disulphide, carbon tetrachloride, turpentine, petroleum and fixed oils. They are insoluble in water and cold alcohol. They have an oil content of less than <0.3%. They are tasteless and odourless.

(2) Semi-refined waxes have melting points in the range 104-116°F with an oil content of 0.6–3%. Wax and oil mixtures with melting point below 100°F are referred to as **cerols**.

(3) Crude scale waxes are the higher molecular weight waxes (large value of *n* (Fig. 6.7(b)) in the hydrocarbon chain) which may contain up to 3% oil, which makes the resultant material very crumbly and weakens its mechanical properties, giving rise to softer waxes. These are not odourless or colourless and are sold as 'white' or 'yellow' wax depending on the degree of refining.

8.9.3 Properties

The convention is to study waxes in their molten state with respect to their refractive index. The refractive index of the molten wax varies with their chain length and structure. This variation is represented on the 'wax triangle' as shown in Fig. 8.17.

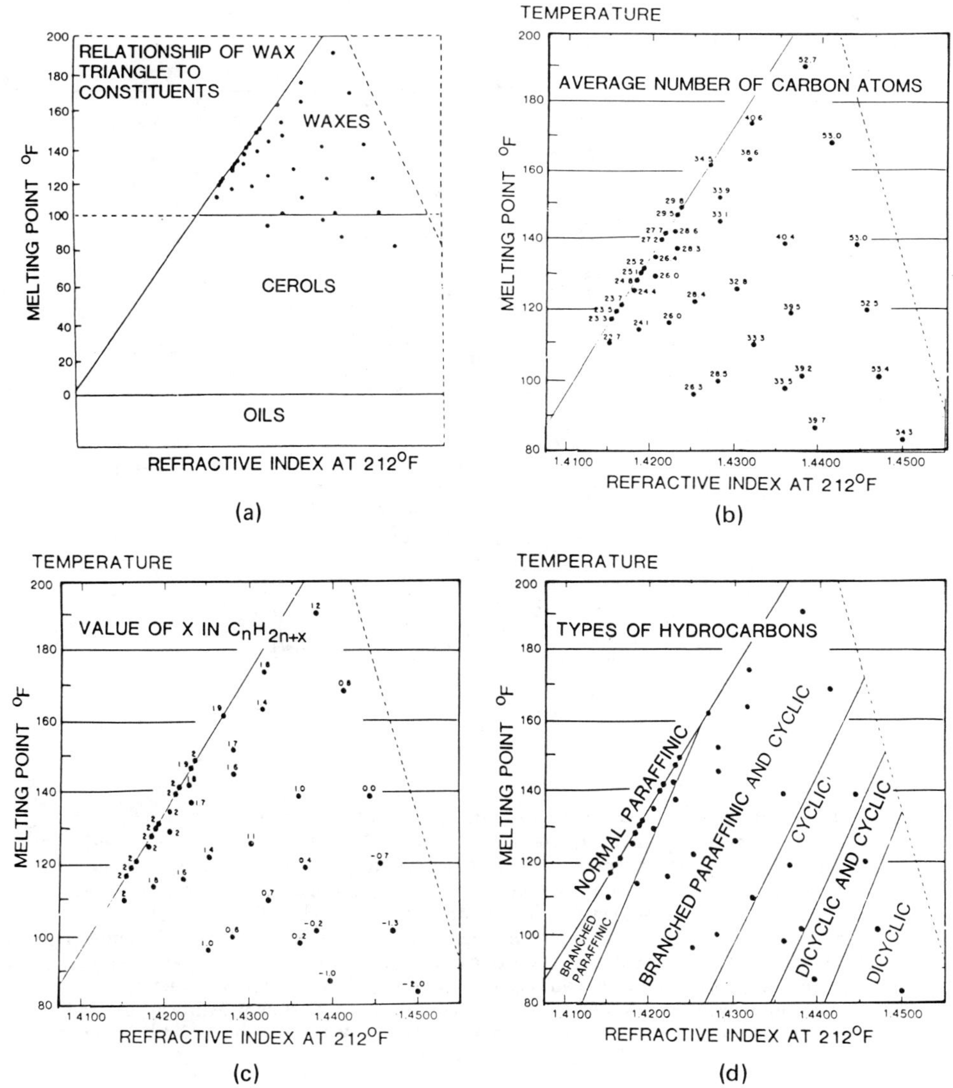

Fig. 8.17 — (a) (b) (c) (d).

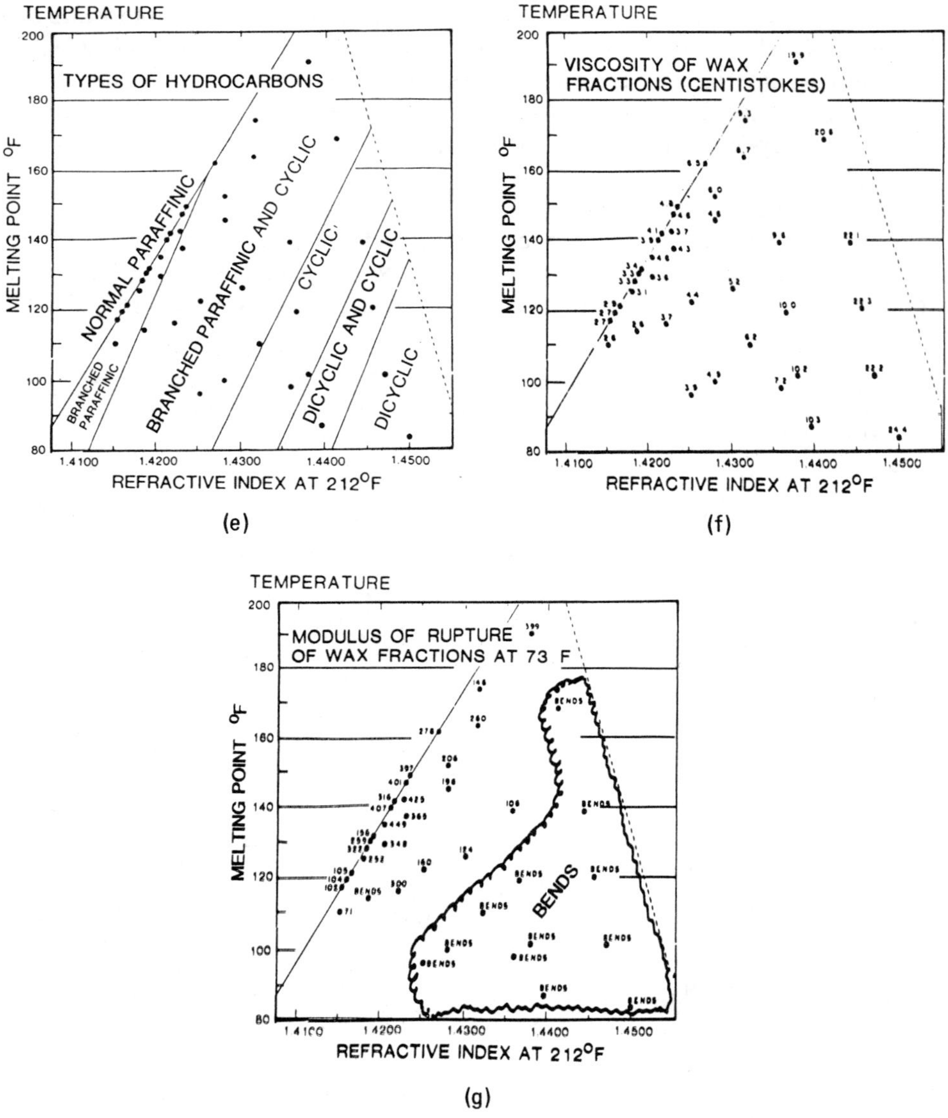

Fig. 8.17 — Various waxes are classified according to their refractive index when in the molten state and their melting points. (a) The area which the 'waxes' occupy compared with the hydrocarbon oils. (b) The average number of carbon atoms in the hydrocarbon chain. The melting point will depend on the structure of the hydrocarbon chain as shown in (c) which divides the wax triangle up into cyclic, dicyclic straight and branched chains (see also Figs. 6.7, 8.15 and 11.1. (d) The value of 'x' in the formula $C_n\,H_{(2n+x)}$ is an indication as to the number of double bonds or cyclic rings in the hydrocarbon chain (Fig. 8.15(c)). All these parameters go to produce (e) the viscosity of the wax and (f) the modulus of rupture of the wax fractions at 73°F. The cyclic and dicyclic waxes tend to be more flexible that the straight hydrocarbon chains. (Reproduced by permission from H. Bennett, Industrial Waxes, Vols. 1 and 2, Chemical Publishing Co., New York, 1975).

Density, viscosity, elasticity and plasticity and ball penetration are all variable quantities in the study of waxes and may be studied in this relationship with their refractive index.

8.9.4 General rules

(1) For a particular molecular weight *n*-paraffins have the highest melting point.

(2) For methylated compounds the divergence in their properties becomes more marked as the CH_3 unit moves towards the middle of the chain from the two extremes.

(3) Isomers have the highest viscosity in the melt.

(4) Side chains are responsible for the pronounced colloidal behaviour of certain high molecular weight iso-paraffins. Hence the side chains enhance the lyophilic character of the n-paraffins which have no affinity for solvents at room temperature (**lyophilic** means having a dispersed phase with a high affinity for the continuous phase: hence a lyophilic sol).

8.10 MICRO-CRYSTALLINE AND PARAFFINIC WAXES

Paraffinic waxes are formed by pressing a paraffin distillate from the distillation of crude oil. The Institute of Petroleum gives a definition of a **paraffinic** wax as: a distillate, of mainly *n*-paraffins, with a molecular weight of 225-450 and a setting point of 32–71°C (90-160°F).

8.10.1 Micro-crystalline waxes

These are sometimes called amorphous or petrolatum or petroleum ceresine waxes. They differ from paraffinic waxes in their degree of crystallinity. Micro-crystalline waxes are obtained from the solvent extraction of waxes and by crystallization from the paraffinic wax distillation residua.

Micro-crystalline waxes are characterized by a high molecular weight (between 450 and 1000) and a melting point of 150°F and above. They are soft, bendable and possess a high viscosity. Micro-crystalline waxes are a mixture of iso-paraffins and cycloalkanes, and the nearest naturally occurring micro-crystalline wax is the ozokerite (mineral/lignite). Micro-crystalline waxes contain an appreciable amount of the cyclic viscous oils which in the case of petroleum waxes are saturated (Fig. 8.15(c)).

Micro-crystalline waxes act as a binding, homogenizing and plasticizing agent in the formation of protective films and they give the films 'bendability' and promote a smaller crystallite or grain size in the n-paraffinic waxes and thus increase the density of the lattice structure. This promotes a greater adherence and coherence to the film and makes the film more moisture proof and therefore a better coating. They flow under compression (as opposed to fractu•ing). The use of microcrystalline waxes is manifested externally by a less lustrous and greasy film despite the oil content of the micro-crystalline waxes.

Note: an oil added to a Micro-crystalline wax increases its plasticity, but oil added to a paraffinic wax will make it crumbly. Micro-crystalline waxes are opalescent whereas paraffinic waxes are transparent.

Micro-crystalline waxes contract less when passing from the liquid to the solid phase than do paraffinic waxes (Kinsel and Philips, 1945; Bennett, 1975).

8.10.2 Oxidized micro-crystalline waxes

These are prepared from the residue of petroleum refining by dissolving the residue in methyl ethyl ketone (Section 11.7), chilling, filtering, and then refining. They are oxidized by air in the presence of a catalyst (cobalt or manganese soaps).

8.10.3 Petroleum waxes

Petrolatum waxes (mineral jelly) consist of a mixture of micro-crystalline wax and 10% mineral oil giving a white amorphous translucent plastic wax (see Table 8.8).

Table 8.8 — The effect of the addition of beeswax on the melting point of petroleum

Beeswax (%)	Melting point increase (°C)	
5 White	2	
10	5	
20	9	(optimum value between 10% and 20%)
5 Yellow	2	
10	10	
20	17	

Reproduced by permission from H. Bennet, Industrial Waxes, Vols. 1 and 2, Chemical Publishing Co., New York, 1975.

8.11 PAINT — WHAT IS IT?

A paint is an organic coating which performs three functions:

(1) It provides a decorative coating.

(2) It provides a barrier which isolates the metal from the environment to reduce corrosion.

(3) It ensures that the steel surface is equally oxygenated (if the steel surface has an equal replenishment of oxygen over the whole surface the 'corrosion battery' will be of zero volts and no corrosion will take place).

A paint is essentially a coloured pigment dispersed in a medium, which in paint technology is called a **binder**. The binder, or binding medium, forms a liquid on the metal surface which subsequently dries to form a solid film. The binder is the agent which determines the type of properties possessed by the paint. The sole function of the volatile component (thinners) is to control the viscosity for ease of manufacture and application. The viscosity controls the application thickness which can be obtained before 'runs' and 'sags' are obtained. Different types of paints require different 'thinners'. Automotive paints are essentially very different from household paints and the automotive 'primers' are dealt with in Chapter 9. The automotive primer paint is called an electrocoat and is put on by a process called electrophoresis or electrocoating. The pretreatment process, which prevents the undercutting of the paint film by the corrosion process, is discussed in Sections 6.18, Section 9.1 and in Section 8.19 as a coating process.

One coat of paint will not possess all the properties required for corrosion protection, and therefore it is necessary to have two- or three-coat systems. These are generally referred to as a primer, undercoat and top coat.

8.11.1 Primer

A primer's function is that of the first coat of paint and is put on to obtain good adhesion to the metal substrate. The primer usually has a high percentage of pigment in the binder, which means that it dries without a lustre. The primer can be pigmented with anodic metallic elements (e.g. zinc particles) or with corrosion inhibitors. Hydrophylic additions usually form a high percentage of additives. As the primer is the first coat of paint put onto a metal surface, wettability is needed so that surface irregularities are filled rather than bridged. For better adhesion, mechanical (abrasion by emery cloths) or chemical keying is used. It is important that these surface irregularities are not larger than the thickness of paint film to be applied, otherwise they will 'stick through' the paint film like 'rocks out of the sea'. Early paint breakdown will result (see the discussion of inorganic conversion coatings in Section 8.19 and of electrocoating in Chapter 9 to give a primer coat for motor vehicles).

8.11.2 Undercoat

The undercoat is the coat often missed out, but it does thicken the paint film, making the water and oxygen diffusion path longer. Like the primer coat, it is usually high in pigment content and like the primer dries to a matt finish.

8.11.3 Top coat

The top coat usually makes up for the matt finish of the primer. A top coat by itself does not have a good corrosion resistance, but does dry to give a bright lustre as well as giving the final colour to the article. The top coat has to possess stability to weathering, to be resistant to chemical staining (by bird droppings etc.) and to possess abrasion resistance.

It is impossible to apply a coat of paint and have it free of pin-hole defects (termed 'holidays' in the paint trade). These defects mean that the protection afforded by paints increases with film thickness. A paint film is only as good as its surface preparation. Paint films are best applied on dry warm days so that there is no possibility of sulphurous acid condensates forming on the metallic surface. These condensates produce early film breakdown as a result of the catalytic action of sulphates in the corrosion reaction. (Filter paper soaked in Potassium ferricyanide will turn blue in the presence of ferrous salts. This blue colour will indicate the areas of insufficient cleaning.)

The properties of a paint film are such that a paint film would not be used as a coating in many corrosive environments. For example, a paint film would not be considered for a submersible article in hydrochloric acid. A paint film is used in mildly corrosive atmospheric environments because they are **not impermeable** coatings. The properties of a paint film with regard to the passage of electrically charged ions and electrons are discussed in Section 8.12. The breakdown of paint films is discussed in Section 8.13.

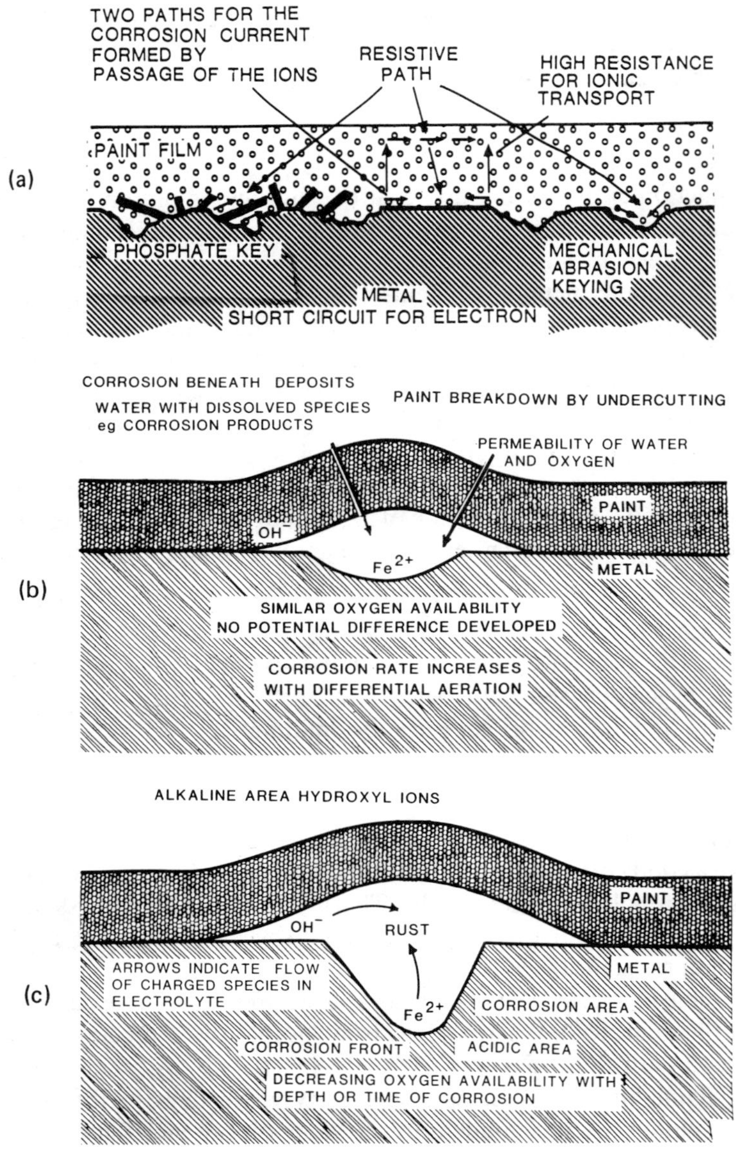

Fig. 8.18 — A paint film stops lateral corrosion and the passage of the anions and cations from each corrosion half-cell by providing an adherent film which increases the electrolyte resistance between anode and cathode. (a) The resistive path for a phosphate film, mechanical abrasion and a flat surface. (b) A paint film also provides similar oxygen replenishment to the whole surface and therefore no concentration cell is developed (Figs. 7.4 and 7.10). However, dirt particles which remain in contact with the paint film or non-adherent paint films cause an unequal concentration of oxygen and therefore provide the driving force for the corrosion of the substrate metal. Corrosion proceeds as osmosis (Fig. 8.19) causes more water to be pumped into the gaps where the paint film has lost its adherence to the substrate metal. (c) Once corrosion in initiated a difference in oxygen availability accelerates the corrosion rate. The metal is penetrated by the corrosion process as the corrosion front becomes further starved of oxygen. Both phosphate keying and abrasion are methods of increasing the diffusion path of the electric circuit in the liquid environment.

8.12 MECHANISM OF PAINT PROTECTION AND BREAKDOWN

A paint is generally considered to produce an organic coating which provides a barrier which isolates the metal from the environment and thus reduces corrosion. However, this is only a first approximation to the truth. The paint film acts as a '**semi-permeable**' membrane in that it allows only one kind of molecule to pass through. These are **uncharged** ions, such water as H_2O and oxygen. It is well established that a paint film will allow through seven times the amount of water required for an uncoated steel to corrode. This means that the diffusion of water through a paint film is not the rate determining step by which a paint film slows down the corrosion of steel. A paint film will allow the same amount of oxygen to permeate a paint film as is required for an uncoated steel to corrode. Hence, if the paint film is permeable to uncharged ions such as water and oxygen but impermeable to charged ions, i.e. those making up the de-icing salts (Na^+ and Cl^-, aggressive atmosphere (sulphates), or the products of the corrosion reaction (Fe^{2+} and OH^-: see Section 6.14 and Fig. 8.18), then the paint film will stop corrosion. The process of corrosion is stopped by arresting the lateral movement of those ions, which are the products of corrosion. These ions are prevented from spreading radially away from the micro-corrosion site at the metal–paint interface only if the paint film is adherent to the metal substrate. Therefore if there are any ionic contaminants, e.g. unwashed phosphoric acid (the phosphate film and electrocoats are subjected to at least two rinses Figs. 9.1, 15 and 17, and 29, 30, 31 and 34), ferrous sulphate from industrial condensates or chlorides from road de-icing salts, water can diffuse through the paint film to form an electrolyte at the metal–paint interface, which completes the corrosion cell (Section 7.6). Hence surface cleanliness and preparation are the most important aspects of preventing corrosion by the application of paint films. The most expensive paint is wasted if the surfaces contain non adherent impurities.

Therefore a paint film must have good adhesion to the substrate metal under dry, wet or corrosive conditions. This wet adhesion is complicated by the phenomenon of **osmosis**, which is now discussed.

8.13 OSMOSIS AND OSMOTIC PRESSURE (SOLUTION PRESSURE) — PAINT STRIPPING BY CORROSION

In order to understand the mechanism of paint breakdown and wet adhesion other elements of science have to be explained. This is the process of **osmosis** and the resultant **osmotic** pressure or solution pressure which can lead to paint blistering and the stripping of the paint film by corrosion. Osmotic pressure is often termed solution pressure and only exists across a semi-permeable membrane, such as a paint film, across which different concentrations of dissolved species exist. This arises from the fact that paint films are only impermeable to charged ions such as Fe^{2+}, OH^- and Cl^-, the former being the products of the corrosion reaction and the later being an ionic contaminant. In the case of corrosion, the dissolved species are the products of corrosion, the Fe^{2+} ions and the OH^- ions. Thus the rain water, without any dissolved species, will pass through the paint film by osmosis from a low concentration of dissolved species, i.e.on air side of the paint film, to a higher concentration of dissolved species which exists between the paint film and metal (Fig. 8.19).

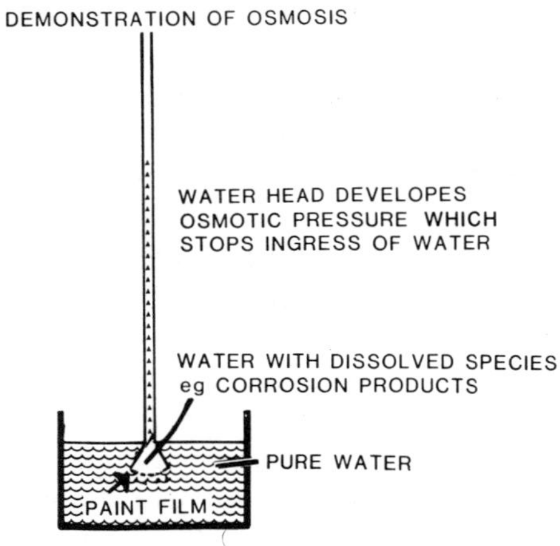

Fig. 8.19 — Diagrammatic view of the osmotic pressure which is generated by the diffusion of water through the semi-permeable membrane into a solution of sugar (in a laboratory demonstration cane sugar can be used). The water diffuses through the semi-permeable membrane so as to dilute the sugar solution and in such a direction that the pressure gradient is against the direction of diffusion. This pressure gradient is the osmotic pressure. Eventually the osmotic pressure will stop the water diffusing through the semi-permeable membrane. In the case of paint films the solution is an ionic solution and the continued presence of water in contact with the paint film will alter the glass transition temperature of the paint, making it more elastic so that the osmotic pressure separates and 'bursts' the paint off the substrate steel (Fig. 8.18(c)).

In doing so, the water creates an osmotic pressure (often called a 'solution pressure') beneath the paint film. This pressure tends to reduce this passage of water from the outside atmospheric side to the inside surface and beneath the paint film. However, this pressure also serves to blister and strip the paint film from the metal substrate, by an internal bursting pressure. This is rather like blowing up a balloon (the paint film) with water. This stripping of the paint film (Fig. 3.10) is further aided by the alkaline nature of the cathodic area (Section 7.9), which is plentifully supplied with oxygen as most paints are softened, by a process called **saponification**, by alkalis (see Sections 6.16, 7.9 and 8.16). The alkali formed is more able to creep and disbond the paint from the underlying metal at the cathode, and may attack the pigment of the primer paint. Adhesion failure of the paint often precedes catastrophic corrosion as shown in Plates 8.1 and 2.1(c).

Radioactive tracer techniques have shown that the permeation of ionic species increases with time of contact. Hence saline mud pockets will allow diffusion of the ionic species 'Cl^-' through paint film until the metal–paint surface is contaminated. When this happens, an electrolyte is formed at the metal–paint interface, and rapid corrosion can occur. The products of corrosion are more mobile in an electrolyte (Section 10.1). The contact of a paint film with water vapour produces internal expansion stresses within the paint film. This is demonstrated when shim steel is coated on one side only and arranged suspended over water as shown in Fig. 8.20.

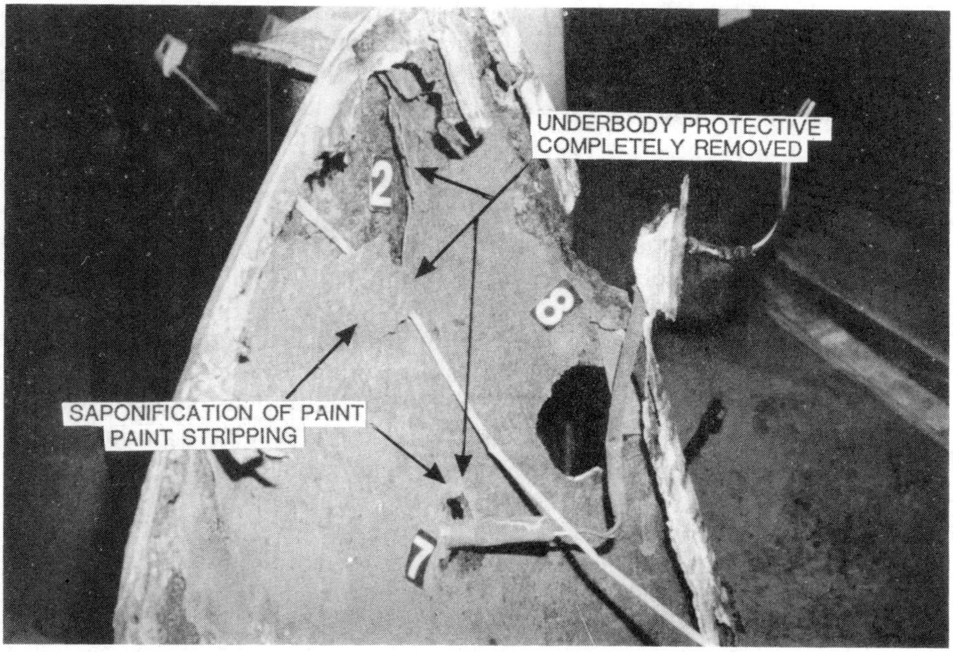

Plate 8.1 — The non-adhesion of paint is caused by the saponification of paint arising from the alkali formed in the cathodic reaction (see Fig. 3.10).

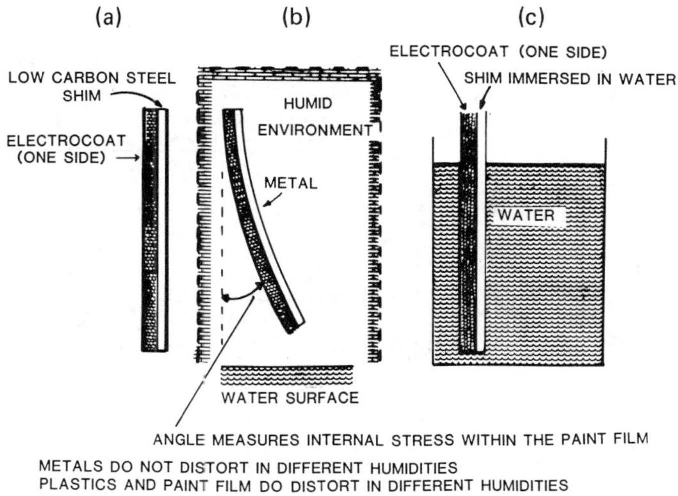

Fig. 8.20 — Two pieces of thin section steel shim are (a) coated on one side only and are then (b) suspended over water or (c) submersed in water. The results show that the piece suspended in a moist environment is bowed owing to the internal stresses developed within the paint film, while the shim immersed in water is not bent. There are several explanations to this phenomenon, but the most likely is the alteration in the paint's glass transition temperature caused by the uptake of water.

The same expansion is not produced when the shim steel and paint film are immersed in water. Hence most paint films are adversely affected by prolonged contact with water owing to the hygroscopic nature of the sodium chloride and the paint's function is degraded by prolonged contact with road de-icing salts.

8.14 SCAB AND FILIFORM CORROSION

Scab corrosion is where the paint film is damaged by external means, but the paint film still remains intact (coherent) within itself, although it is still non adherent to the substrate metal (Figs. 8.18(b) and 8.18(c)).

Filiform corrosion is characterized by thread-like filaments of the corrosive environment beneath the paint film. This corrosion arises as a result of the mechanism described in sections 8.12 and 8.13. The profile of the thread-like corrosion is shown in Fig. 8.21. The area along the axis and within the 'tube' is deficient in oxygen, but the area where the paint is still adherent to the metal substrate is not deficient in oxygen. This arises because oxygen has the least diffusion path when total diffusion path is considered (paint thickness + liquid path). The prolonged contact of the paint film with water or moisture tends to soften the protective coating as a result of the self-induced expansion Fig. 8.20). As a result of the paint film's physically absorbing water vapour, it can be levered off by the osmotic pressure developed by the diffusion of water through the semi-permeable paint membrane. Hence the corrosion path of filiform corrosion, is not manifested as a total surface phenomena but is restricted to these 'tubes' or 'threads' which are formed beneath the semi-non-adherent paint films. Another analogy is that these

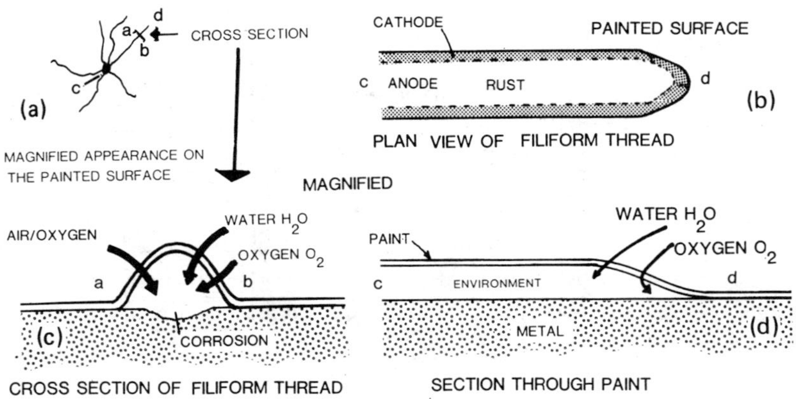

Fig. 8.21 — The diagram shows the various perspectives of filiform corrosion. (a) Filiform corrosion starts from non-adherent paint arising from small paint 'chips' of paint in contact with a second phase, be this dirt or crevices. The corrosion process spreads out from the central area in a thread-like spider's web manner which resembles a tunnel ((b) and (c)). The cathodic reaction takes place at (d) the interface between the paint and metal surface because the diffusion path of oxygen is the least at this point, while the anodic reaction takes place at (c) the centre of the tunnels. The diffusion path for oxygen is now the extra path within the liquid environment.

filaments are rather like an ant or mole warren except that none of the tubes cross and are restricted to two dimensions only at the metal–paint interface. The existence of filiform corrosion and the restriction of areas of corrosion to the very small 'tubes' which are invisible or unresolvable to the naked eye means that poorly prepared surfaces can have the start of filiform corrosion beneath the paint film. Thus what appears as thoroughly good adherent paintwork may contain the start of a corrosion cell. Therefore, in preparing a metal suffering from scab corrosion, it is wise to remove the old paint well back from the damaged areas, rather as one clears away wood rot well past the affected timbers. The problem with automotive paintwork is that most people are afraid to be drastic with the areas of scab and filiform corrosion and so rectification is often only temporary. In these cases the paint film soon fails.

8.15 STRESSES IN A PAINT FILM

Metals are dimensionally stable in a moist environment. Plastics, and other non-metallic compounds, are affected by the absorption of water vapour. This can be demonstrated by the following experiments. Thin strips of steel shim were electro-coated on one side and subjected to a moist environment. The metal–paint combination therefore acted as a 'bimetallic strip' and which bends to show the expansion of the paint film. From the curvature the stress within the paint film can be calculated.

8.16 PAINT SAPONIFICATION

Fig. 3.10 shows how the cathodic product (OH^- ions) undermined the paint film and caused the loss of adhesion of paint film from metal surface. This loss of adhesion resulted from the contact of the paint film with an aqueous environment over a long period of time. The science developed in Section 7.9 showed how the cathodic area becomes alkaline in nature, and this will affect paints if they are not 'alkali resistant'. It is already acknowledged that 'alkali-resistant' paints are required for both marine and cementitious conditions where high pH conditions exist. Alkali-resistant paint films, at the present time, are not acknowledged for the protection of automobiles in conditions which match marine conditions (Fig. 1.6(c)), perhaps because saponification is not shown up in the accelerated automotive tests (Section 10.22(1)) where the droplets of a salt spray are continually altering their location. Experiments have shown that the central area of a salt droplet (Fig. 2.6 and associated text) is anodic and the outer perimeter is cathodic. The salt spray cabinet (ASTM 117B) (Section 10.22) uses vertically hung mild steel panels and the salt droplets are continually running down the steel sample in such a manner that anodic areas are changed for cathodic areas as the drop runs over the surface. Therefore the cathodic environment, which is alkaline in nature, is continually changed for anodic sites (which are acidic in nature (Section 7.9). Hence the automotive paints under test in these accelerated salt spray conditions are not in continuously alkaline conditions. Therefore saponification does not appear to be a problem as it cannot occur if the anodic area is changed to cathodic with time (and vice versa). Plate 4.7(b) shows the paint stripping from the lower edge of the door where the clinch fold has been ground away. This internal crevice, with the associated high drainage hole (Plate 4.6) has resulted in the crevice's being kept full of stagnant water. The area is also one which has restricted drying. This causes loss of adhesion as the alkali creeps between the paint film and metal. In the presence of the alkali, the paint breaks down by a process of **hydrolysis**. This softening is accelerated when the paint films crack at the interface of the raw edge of the clinch fold. It is therefore worth stating that alkaline softening and stripping are only experienced where the paint film is kept continuously wet by poor ventilation or high drainage holes or by corrosion within a crevice. In normal surface conditions, such as flat, rapidly drying surfaces, saponification of the paint would not be seen from the frequent shower of road de-icing salts from the preceding vehicle. In this situation there is a good chance that a cathodic site will be changed for an anodic site on the 'next' set of droplets. The altering alkaline and acidic environments would therefore arrest the saponification of the paint film by destroying the akalinity beneath the paint film. The saponification reaction can be written as

$$\underset{\text{paint ester}}{R_1\text{–}COOR_2} + \underset{\text{water}}{H^+OH^-} \xrightarrow[\text{an alkaline medium}]{\text{saponification in}} \underset{\text{acid}}{R_1COOH} + \underset{\text{alcohol}}{R_2OH}$$

In this equation, R is any large organic hydrocarbon chain (Section see 11.7 *et seq.*). If the hydrolysis is carried out in an alkaline environment such as sodium hydroxide (NaOH), the sodium salt of the acid R_1–COOH is formed. Since the alkali salts of the higher acids are soaps, the alkaline hydrolysis reaction is known as **saponification** (derived from the Latin word meaning soap). Saponification to produce the acid R_1–COOH is more rapid in alkaline media than the hydrolysis in water of pH 7. (Bathroom soap is sodium stearate ($C_{17}H_{35}COO\,Na$).) ($R_1 = C_{17}H_{35}$–).

8.17 SUMMARY — ORGANIC COATINGS

Summarizing, therefore, a paint film is not a good 'barrier'. It contains holes or holidays and good coatings rely on there being a multi-coat system so that these holes are not coincident (in this respect the difference in thickness between internal cavity paint films and external paint films is shown in Fig. 3.5). The paint film only acts as a barrier to the electrolyte of the mildly corrosive environment by preventing the lateral movement of the anodic and cathodic corrosion products. A paint film allows water and oxygen to pass through in sufficient quantities that corrosion can proceed. A paint film delays the onset of corrosion by preventing the anodic and cathodic products formed from small local sites (micro-cells) from combining and therefore delays the onset of further corrosion. A paint film does allow the diffusion through it of charged (ionic) species if moist, intimate contact is maintained over a long period of time. When these ionic species reach the metallic surface a good electrolyte is formed which allows the products of corrosion from the micro-cells to spread laterally and radially outwards. Osmosis causes water to be pumped into the metal–paint interface, and the cathodic by-products destroy the adhesion of the primer paint film, allowing the corrosion cell to spread. The cathodic reduction of oxygen (eq. (7.8)) and the subsequent saponification of the primer paint film is the main cause of the loss of adhesion and lateral growth of these micro-cells to larger cells viewable to the unaided eye. Osmosis is the main reason that the non-adherent paint is full of water. This tends to confirm the common belief that a metal is porous and that the metal has 'sweated' water to produce the blister (Fig. 2.1(a); Plates 3.2 and 4.5(b)).

Corrosion performance can be improved by better paint technology, e.g. by making a paint film more resistant to alkalis, but a better way is by good design and regular washing (equivalent to paint maintenance) so that less reliance has to be sought from the paint coating. Regular washing is required to remove the long-term effects of prolonged contact with road de-icing salts. Corrosion protection by one-side zinc-coated steels ensures that a protective coating is applied to the inter-weld areas, except around the perimeter of the spot weld where vaporization of the zinc will have occurred. However, zinc is one of those metals which dissolves in an alkali medium as well as an acid medium (Fig. 7.18). Therefore **cathodic disbondment** of the paint films will be expected at cathodic sites on zinc-coated steels.

8.18 COMPARISON OF WAX AND ZINC COATINGS

The various parameters which are important to consider in a comparison of wax and zinc coatings are as follows:

8.18.1 Corrosion protection

Essentially there is a great difference in their protection mechanisms, and so a comparison can only be made under specific conditions.

8.18.2 Paint adhesion

Wax coatings are put on top of a paint film and so give a back-up protection to the paint film. Zinc coatings are often now used on one-side zinc-coated steel panels for those panels which have proved in earlier models to be more susceptible to corrosion or as a result of using thinner panels for lightness. Hence zinc is at the interface between the steel and paint systems and so for the specific case of corrosion beneath dirt particles (Fig. 2.6) the presence of zinc will give reduced adhesion of the paint system (plus waxes) to the metal. This arises because the stationary cathodic site will be alkaline and zinc dissolves in alkalis. On internal panels another reason for a reduction in adhesion is because the older pretreatment process is by spray which often does not treat the inner panels. Hence the electrocoating process is usually onto oily steel. In cases where the pretreatment process is inadequate, the zinc dissolves in the cathodic environment and the whole paint film will dis-bond (Fig. 3.10). Modern factory layouts now combine spray and dip pretreatment processes (Section 9.3).

8.18.3 Capillaries and inter-welds

Zinc coatings are put onto the steel sheet before the body panel is stamped out so are guaranteed to be present within the inter-weld crevice. This is not so with wax coatings whose liquids have to penetrate the inter-weld, often against gravity. In order to penetrate narrow capillaries the wax and its solvent must overcome their own inherent viscosity. The big plus for wax and oil coatings (if they penetrate the crevices), is that they will exclude water and road de-icing salts by virtue of their contact angle or water-disliking properties (compare Figs. 3.7 and 3.8: Sections 3.8 and 10.18). Zinc coatings, in contrast, will allow road de-icing salts to concentrate within the interwelds (Section 10.15) and allow the corrosion of the zinc to proceed. When the zinc (which is in contact with the liquid environment) is used up (Fig. 7.13) the steel may then corrode.

8.18.4 Water exclusion from interwelds

Wax coatings are far superior to zinc coatings. Wax coatings undergo an anomalous expansion (Fig. 8.16) at their transition temperature, which may give rise to cracking and shrinkage of the coating seen in Plate 4.16.

SUMMARY

(1) Differences in the replenishment of oxygen at a steel surface and dissimilar metal contacts provide ready-made ‘batteries’. These galvanic cells accelerate the corrosion of mild steel owing to the steel’s negative potential with respect to most other metals–alloys (Tables 7.1–7.3).

(2) Zinc is one of the metals which sacrificially protects steel at all states of aeration, i.e. it is more negative than steel throughout and dissolves to protect the underlying steel.

8.19 INORGANIC COATINGS

Inorganic coatings are used in the motor industry to provide corrosion protection. They are of two forms: (a) inorganic phosphate films produced on steel (Section 9.3, Plate 9.1); (b) inorganic oxides electrically produced on aluminium and its alloys.

Inorganic phosphate coatings are used to provide temporary corrosion protection on steel (in combination with class A oil protectives (Sections 6.18 and 8.8)) as well as a basis for paint coatings, where their function is two-fold.

(1) They must limit the area available for corrosion reactions, by preventing the lateral spread of corrosion beneath the paint film.

(2) They must increase the adhesion between the organic paint film and the metal in both dry and wet (aqueous) environments.

Inorganic phosphate coatings are attacked and dissolved by both strong acids and alkalis and they are sensitive to changes in pH. If a phosphate steel is galvanically connected to an unphosphatized steel plate in salt solution, then the phosphatized coating is anodic to the unphosphatized steel. However, after a period of time the polarity of the couple reverses and the unphosphatized coating becomes anodic and dissolves. This will have relevance to paint repair systems where no conversion coating is employed on bare steel ‘in-fills’. Atmospheric exposure tests show that unpainted phosphated steel sheets corrode at a faster rate than solvent-wiped unpainted steel sheets. It is postulated that it is the phosphate ions which accelerate the solution of the steel. This is important where conversion coatings are used but not thoroughly rinsed off with demineralized (or distilled) water to prevent the trapping of foreign ions beneath the paint coating. Compare the de-ionized water rinses in Figs. 9.1 and 9.3(b) with the mineral content of ordinary tap water (Table 10.2) and its conductivity (Table 10.1).

Inorganic phosphate coatings are discussed in greater detail in Chapter 9.

8.20 ANODIC OXIDE FILMS — MOTOR VEHICLE TRIM

Aluminium is a very high reactive metal and combines vigorously with oxygen to produce very high temperatures. This is most apparent by the white 'flashes' present in fireworks which owe their existence to this metal in the powdered form. However, aluminium in the form of aluminium alloys is used in motor vehicle construction and light-weight aircraft structures and domestically for the manufacture of cooking utensils where this rapid combination with oxygen is not apparent. This is due to the natural oxide film of 10 nm (0.01 μm) in thickness which protects the metal from further attack by acting as a durable, coherent and adherent barrier (Fig. 6.17).

Anodizing is the process whereby if aluminium is made the anode in a suitable electrolyte (Table 8.9), this air-formed film breaks down and a much thicker oxide film (about 50 nm (0.05 μm)) is formed on the surface. These are called **anodic oxide** films and are chemically applied to aluminium and its alloys by making metal the anode of a wide range of electrolytic baths and passing a current. Aluminium oxide is an extremely hard oxide, much stronger than its metal. The hardness and thickness of the anodic oxide film depends on the composition of the electrolytic bath (Table 8.9). A schematic view of the micrŏsctructure of the anodic oxide film is shown in Fig. 8.2.

Table 8.9 — Conditions for anodizing aluminium alloys in various electrolytes

Process	Acid (%)	Current density (A/dm^2)	Voltage (V)	Temperature (°C)	Time (min)	Film type
Chromic acid	3	0.4–0.6	v/t cycle	40	60	Grey–white
	10	0.4–0.6	v/t cycle	40	30	Opaque
Sulphuric acid	14	0.8–1.0	15	10–15	30–40	Hard translucent
	20	1.2–1.5	15	18–20	20	Normal translucent
Oxalic acid	5	0.5–1.5	30–60	20	30	Pale yellow
	5	1.5	50	35–40	60	Deep yellow
	5	1.5	60	35–40	60	Thick
	3+ 2% potassium oxalate	1.5	50	50	30–60	Thick, colourless

Reproduced by permission from D. R. Gabe, Principles of Metal Surface Treatment and Protection, Pergamon, Oxford, p. 116, Table 6.2.

Bright anodizing is a process which involves chemically polishing aluminium and its alloys, followed by the protection of this shiny surface by anodic oxide films. In this way, anodized aluminium trim has replaced chromium-plated steel trim. It has the advantage that the oxide coat is free on pin-holes.

Note: the oxide film is compressed of 'pores' which are capable of taking a dye (Fig. 8.2). It is estimated that there are 10^{12} pores in each square inch of surface. These pores do not adversely affect the corrosion protection afforded by the oxide film, although they can be 'sealed' by exposing the surface to steam. This facility is often made use of when aluminium articles are coloured, e.g. bronze-coloured aluminium doors for office blocks etc. Anodic oxide films also act a good substrates for paint systems.

Aluminium is beneath iron in the galvanic series (Table 7.1) and so will dissolve anodically to steel. However, the air-formed oxide film is a continuous, adherent, pin-hole-free and good electrical resistor and so prevents the dissolution of aluminium when used in contact with other aluminium alloys. The oxide film is stable in acidic solutions (pH 4.5–7), but in alkaline environments the oxide dissolves. Thus the high reflectivity of the polished aluminium product is maintained for a long period of time in industrial atmosphere exposures.

8.21 CONVERSION COATINGS

There are many preparations on the market which allow the motorist to repair rusting paint defects with the minimum of surface preparation. Rust is a hydrated oxide which can be written as follows:

ferrous hydroxide $Fe(OH)_2$ or $FeO{\cdot}H_2O$, (ferrous oxide hydrate)
ferrous ferrite Fe_3O_4 (magnetite) or $FeO{\cdot}Fe_2O_3$
ferric hydroxide $2Fe(OH)_3$ or $Fe_2O_3{\cdot}3H_2O$ (ferric oxide hydrate)
Section 6.13.7

↓ increasing ratio of O:Fe

These oxides are compounds with water and the idea behind most of the rust preparations is to turn the oxides into the black oxide magnetite.

Most of these depend on the beneficial effect conferred on corrosion by the tannate molecule (Section 8.7) or by fish oils. The benefits of these compounds were discovered very early when steel and leather articles were pulled out of disused drinking wells and found to be uncorroded. Fish oils are used based on the fact that steel did not rust at sea where fish were handled. These tannins (from leather manufacturers as a by-product) are added to phosphoric acid solutions and produce a black oxide on the surface of steel. The tannins are easily washed off steel and require some form of etch for a satisfactory adhesion.

Table 8.10 — Conversion coatings

Resin–pigment	Phosphator
7.2% proprietary vinylite	3.6% phosphoric acid
6.9% basic xinc chromate	3.2% water
1.1% talc	13.2% isopropanol
48.7% isopropanol	
16.1% butanol	

The resin and phosphator are mixed in the ratio 4:1. The coating structure is layered (after Van Eijnsbergen, I. F. H., Jahrbuch der Oberflach, 1958, **14**, 409) (Fig. 8.22) with the phosphate forming on a thin layer of substrate oxide. The chromate forms on the phosphate and the whole is sealed on the outside by the vinyl polymer (e.g. polyvinyl butyl).

Reproduced by permission from D. R. Gabe, Principles of Metal Surface Treatment and Protection, Pergamon, Oxford, p. 134.

Developments of these solutions were undertaken during World War II, when wash primers were developed for aluminium. These were found to be beneficial for steel and have since been produced industrially for the general public. The most popular conversion coatings are the single packs whose formulation may be shown in Table 8.10(a). Two-pack formulations are shown in Table 8.10(b) with the composite film shown in Fig. 8.22.

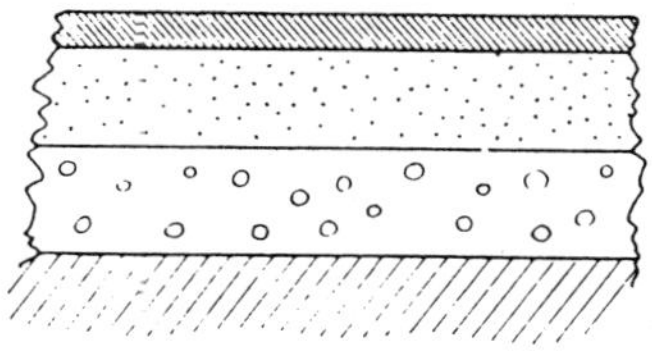

Fig. 8.22 — Schematic sectional structure of an etch primer coating. (After I. F. H. van Eijnsbergen, Jahrbuch der Oberflach, **14**, 1958, 409, reproduced by permission from D. R. Gabe, Principles of Metal Surface Treatment and Protection. Pergamon, Oxford, p. 134, Fig. 6.11.)

REFERENCES

Kinsel and Philips (1945). *Ind. Eng. Chem.*, **17**, 152.

Bennett, H. (1975). *Industrial Waxes, Vols. 1 and 2,* Chemical Publishing Co., New York.

9

Motor vehicle protection schemes

9.1 HISTORICAL INTRODUCTION TO PRETREATMENT PROCESSES

The role of a paint film in the automotive industry is to confer corrosion protection to the car body by arresting the cathodic or anodic reactions (sections 2.3 and 6.14). A paint film provides **equal oxygen availability** to the underlying metal, (Sections 2.4 and 8.11). To the layman, it does this by isolating the metal from the substrate, but this is not the whole story (Section 8.12). The paint film can provide some degree of electrical insulation but this is not always true as shown by the zinc-rich paints. A paint film must prevent the **lateral spread of corrosion** beneath the paint film. The paint film is chemically inert and must be stable in the environmental in which it is used. The visual effect of the paint film must be good so that the product looks good and will sell. Paint films act like permeable membranes and will allow a significant amount of uncharged species (e.g. water and oxygen) to diffuse through to the metal–paint interface. Therefore all the protective characteristics of the paint film depend on the paint film's being **adherent to the substrate metal** so as to prevent the lateral spread of the anodic and cathodic products. Paint films are impermeable to charged ions and the diffusion path between anode and cathode sites can be increased by mechanically roughening the surface. Therefore the industry attaches great importance to the preparation of the substrate metal, Plate 9.1 (Table 9.1).

9.2 THE PROBLEM OF PAINT ADHERENCE

The car body is spot welded together to form an integral rigid unit, complete with doors, boot and bonnet. This is termed a 'body in white'. The problem is now to make a paint film adhere to the metal surface. This is difficult for the following reasons.

Table 9.1 — Evolution of car body priming

Period	Body type	Primer type	Features
Pre-war	Body–chassis no box sections	Spray primer	All areas accessible
Post-war–1960	Welded unit construction; box sections to coat	Solvent-based dip primer	Messy, laborious and wasteful process; up to 20 bodies/hour production rate; much rectification; fire risk; health hazard
1960–1966	Welded unit construction	Water-based dip primer	Less hazardous for fire and health; other features similar
1967 onwards	Welded unit construction	Water-based anodic electropaints	Automated process; up to 60 bodies/hour production rate; no fire risk; less health hazard; uniform consistent coverage with little waste
1977 onwards	Welded unit construction	Water-based cathodic electropaints	Similar basic features to anodic electropaints

The evolution of the mass produced car body has taken 40 years. Originally the car body was a frame or chassis upon which all other panels were hung. The chassis gave way to the unitary method of construction where preformed panels were spot welded into rigid boxes which replaced the chassis. This gave rise to water and mud traps with an added problem of protecting the structure by paints.

(1) The steel sheet from the steel mills is protected from corrosion while in transit to the car body builder by a thin film of oil (class A temporary protective (Section 8.8)). In order to stamp the car panels from this sheet the car body panels must be covered with mill or pressing oils. These are used to lubricate the rolling and pressing operations to give a smooth flat 'polished' surface which will uniformly reflect the light from the painted surface. Any creases due to uneven stamping will be highlighted by the gloss paint. Hence when the car is spot welded together these pressing oils remain on the surface and within the interwelds.

(2) The bonding mechanism within the paint 'system' is chemical in nature and is completely different from the bond within the metal 'system'. There is no chemical bond between the paint and metal and therefore the corrosion process can just strip the paint film off the 'polished' steel.

(3) A pretreatment alkali degreasing process must be applied to remove the pressing oils. This must be followed by some 'process' which will provide a **key** for the paint film. This can be a chemical acid etch or a physical transformation process (e.g. abrasion by emery paper) which must be applied to roughen the surface (Fig. 8.18(a)).

(4) The process which applies the 'keying', must be cheap and be capable of application in mass production processes. (Figs. 9.1–9.3) Therefore a mechanical keying processes (rubbing down) must be discounted for mass production techniques (Table 9.1).

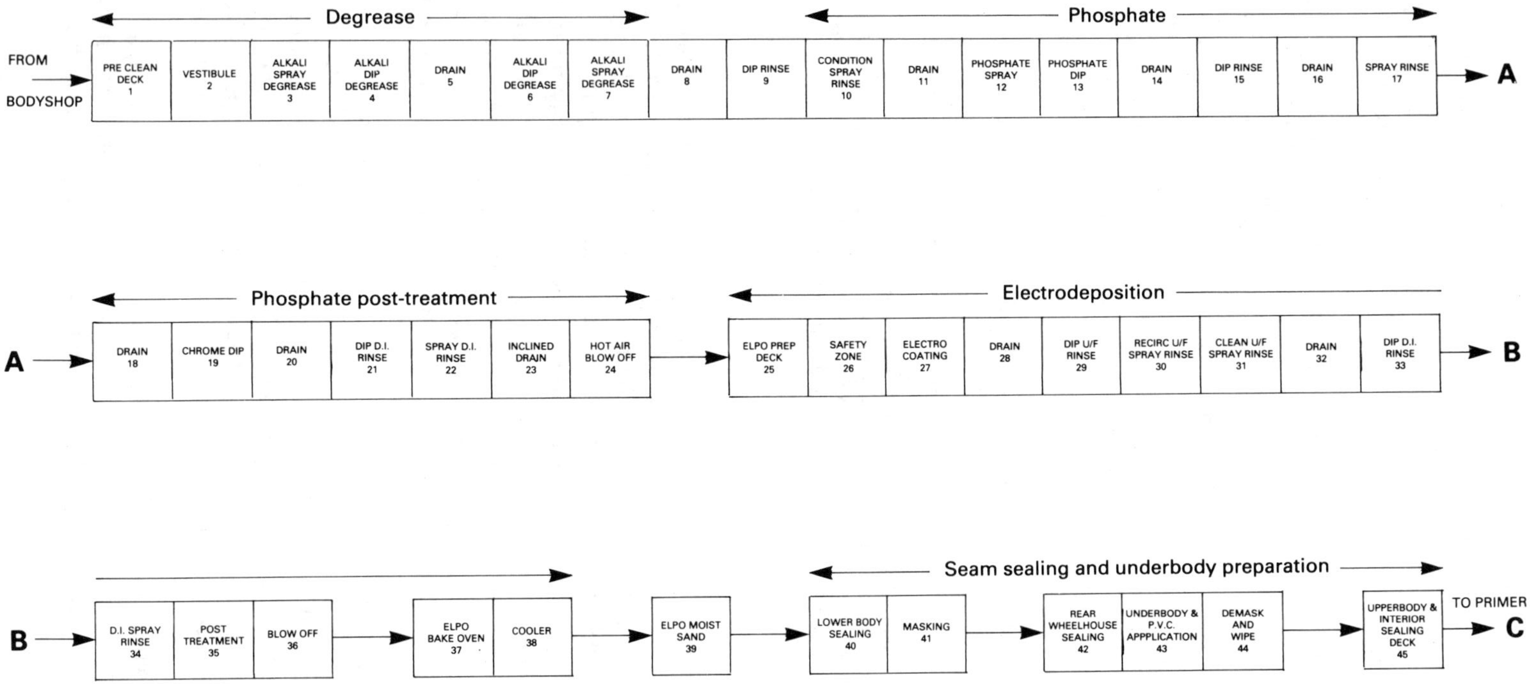

Abbreviations:

U/F Ultra filtration
RECIRC Recirculation
D.I. Deionised water
P.V.C. Poly vinyl chloride

Fig. 9.1 — The processes which are involved in the mass production pretreatment and painting of the automobile body. The degrease and phosphate spray are shown again in Fig. 9.3 and the electrocoating process in Fig. 9.4. Each one of these squares will be a large booth as shown in Fig. 9.3(b). ELPO, electroplating; U/F, Ultrafiltration; DI, de-ionized water; PVC, polyvinyl chloride.

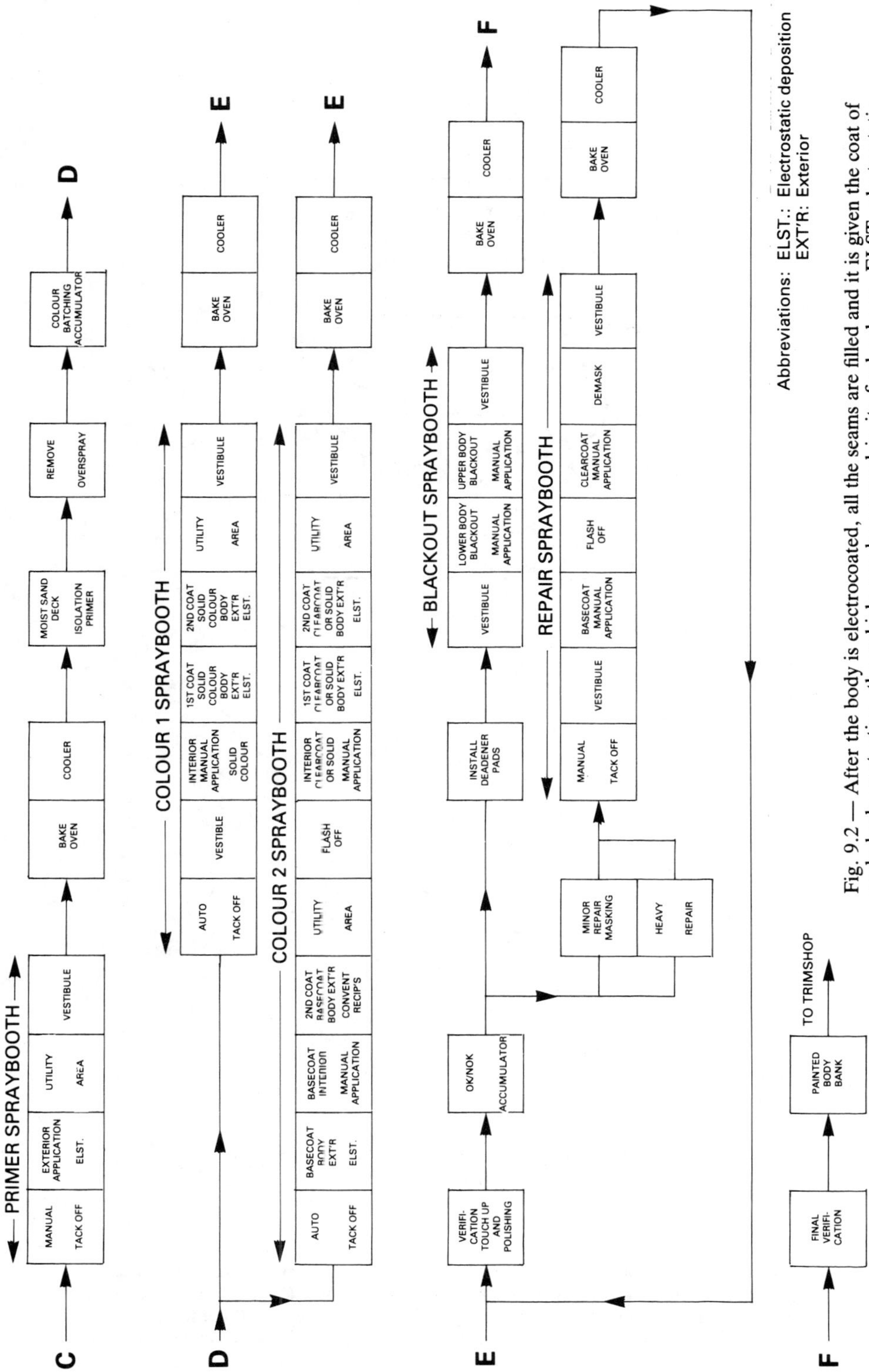

Abbreviations: ELST.: Electrostatic deposition
EXT'R: Exterior

Fig. 9.2 — After the body is electrocoated, all the seams are filled and it is given the coat of underbody protective, the vehicle can be sprayed in its final colours. ELST, electrostatic deposition; EXT'R, exterior.

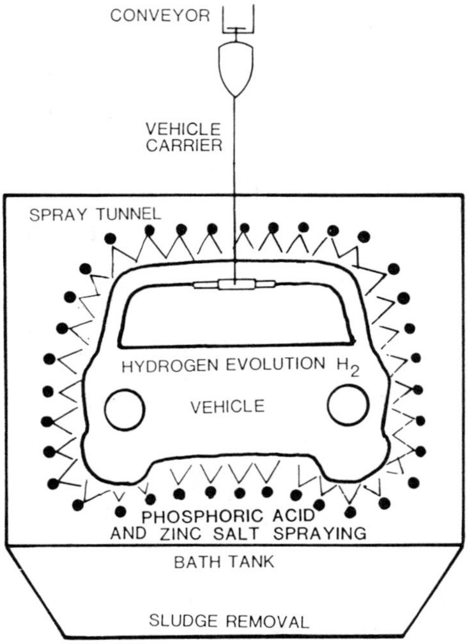

(a)

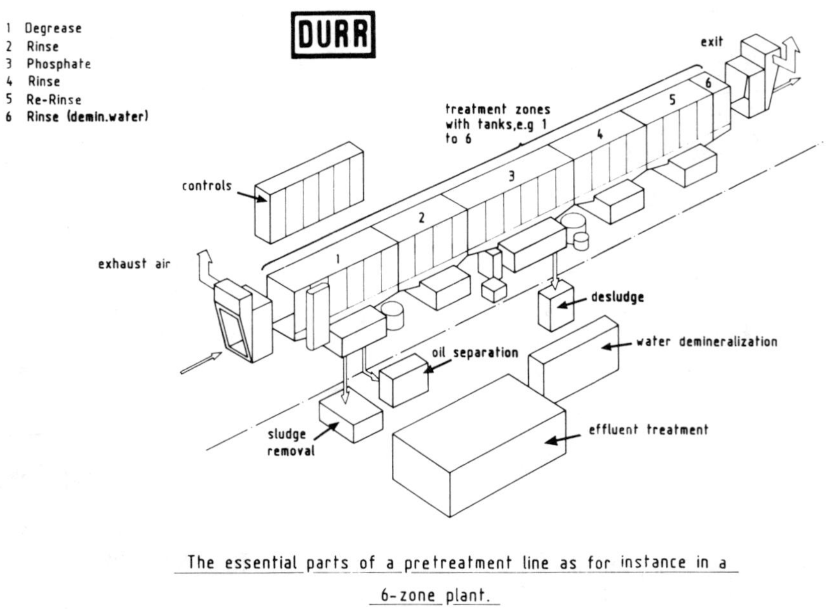

(b)

Fig. 9.3 — (a) The external application of the pretreatment salts by spraying. Spray and dip techniques are replacing the spray-only phosphate applications. (b) The total pretreatment process by spraying is shown diagrammatically in this figure. Note the three rinse tunnels to remove the residual acids. (Contributed by courtesy of DURR Ltd., Broxell Close, Warwick, CV34 5QF.)

9.3 THE PRETREATMENT PROCESS: CHEMICAL KEYING

The only cheap process is a chemical acid attack (called an etch or a pickle), which leaves the crystalline salt (Section 6.18 *et seq.*, eqn. (6.20)) of the acid and base metal (iron–steel) as an insoluble residue upon the surface, as shown in eqn. (6.22) (Plate 9.1). The introductory chemical science for the actual mechanism is discussed in Section 6.18.

This pretreatment 'salt' process is known as the complex zinc iron phosphate film, whose function is two-fold: (1) to give a good 'key' for the subsequent paint film and (2) to contain any corrosion due to stone chipping and thus to prevent the 'undercutting' of the paint film by corrosion, which has been described earlier (Section 8.11 *et seq.*). Therefore this salt restricts the lateral spread of corrosion due to movement of the cathodic hydroxyl ion OH^- between the metal and paint film which breaks the mechanical bond between the metal and paint. This undercutting corrosion is usually called either filiform corrosion (Section 8.14) if the corrosion spreads out like a spider's web beneath the paint film (Fig. 8.21) or scab corrosion where the paint blisters up from the metal surface (Figs. 8.18(b), 8.18(c) and 2.6).

It should be noted that it is more important to have good **adhesion under moist** conditions than under dry conditions owing to osmosis (Section 8.13). The phsophates can dissolve in water if the immersion times in water are prolonged.

The phosphate coating is applied by an acidic phosphoric acid spray (Fig. 9.3; eqns. (9.1) and (9.2)) in order to remove the hydrogen developed so that the metastable reaction (eqn. (9.2)) can occur and deposit the complex iron and zinc phosphate crystals, which is the keying mechanism. Where double salts are used, the iron (Fe) and zinc (Zn) are represented by the single letter (M). The reactions can be represented as follows:

$$M + 2H_3PO_4 \rightarrow M(H_2PO_4)_2 + H_2\uparrow \qquad (9.1)$$

i.e. the car body reacts with the acid to give a phosphate and hydrogen gas. Then

$$\underset{\text{soluble}}{3\,M(H_2PO_4)_2} \rightarrow \underset{\substack{\text{insoluble}\\ \text{'keying crystals'}}}{M_3(PO_4)_2} + 4H_3PO_4 \qquad (9.2)$$

i.e. the phosphate salt $M(H_2PO_4)_2$ formed in eqn. (9.1) is unstable and breaks down to produce a new salt $M_3(PO_4)_2$ and acid H_3PO_4. The keying crystals must be small so that they do not 'show' through the subsequent paint films like rocks out of the sea. Plate 9.1. The new salt produced acts as the **mechanical key** for the subsequent paint film. Unless a combined spray and immersion process is used, only the outer cosmetic panels are treated in the spray to the phosphate film and degreasing processes.

As a result of the Canadian Government Voluntary Code (Section 1.12) there has been a switch from spray application to dip and spray. In 1983, 58% of Japanese production was by the dip and spray application. The complicated industrial production line for the pretreatment process is shown schematically in Fig. 9.3(b).

9.4 HISTORICAL INTRODUCTION TO ELECTROCOATING

The first elementary work which has led to this modern technology was recorded in the 1890s when it was found that certain clays could be electrocoagulated by electricity. Nothing was done until the 1930s, when the process was used to deposit

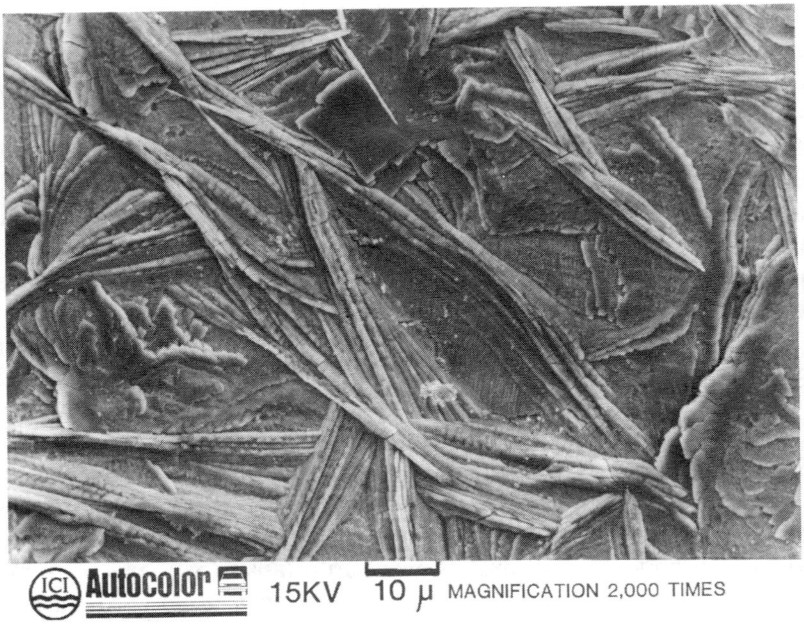

(a)

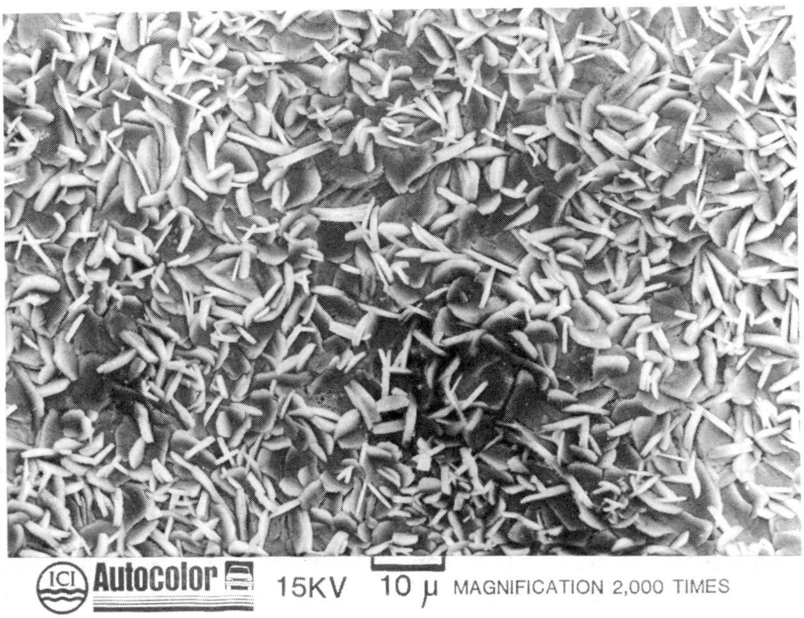

(b)

Plate 9.1 — Scanning electron micrographs from the surface phosphate deposits. (a) The older phosphate deposits which were larger and more open in their deposited sites as a result of the hydrogen evolved from the cathodic sites (equation 9.1). (b) The development within the pretreatment industry now produces small and denser deposits of phosphates. These smaller and denser phosphates reduce corrosion at the paint–metal interface and provide a good 'key' for the subsequent paint film.

rubber latex and to electrocoat the inside of food cans with a lacquer. Prior to the use of lacquer electrocoating on the food cans, the lacquer had to be placed on by total immersion and allowed to dry. Subsequent re-immersion was then undertaken in order to build up a sufficient film thickness. With lacquer electrocoating, this could be done in one operation as the deposited film acts as **its own thickness regulator**. The amount of lacquer deposited is dependent on the total current flowing, and the current flowing depends on how good an insulator the deposited lacquer happens to be. This is dependent on the total electrical resistance of the lacquer bath and deposited film. It is this principle which is used in the electrocoating of motor vehicle bodies to-day (Fig. 9.4).

9.5 SLIPPER DIP—YESTERDAY'S SYSTEM

The process of painting used by the motor industry at first was the 'slipper-dip' system, i.e. total immersion of the car body. This produced a paint film, just as putting a brush into a tin of household paint coats the brush. The disadvantage of this system, as with the household paint example, is that the workpiece was often marred by 'runs' and 'sags' in the paint surface. Removal of these defects was labour-intensive, because these runs had to be polished out by hand. This also meant that some of the protective coating was removed from the steel surface. There were other disadvantages, which result from the fact that the coating thickness depended on the viscosity. A more viscous paint would produce a thicker film. However, thicker paints would fill in the screw holes produced by the manufacturer for fixing points by surface tension effects. This made the insertion of screws difficult and the process had to be done 'blind'.

The slipper dip application of paint had a high tolerance of oily steel. The modern cathodic electrocoats are not as tolerant of oily substrates and require a particular form of phosphate deposition (see Plates 9.1(a) and 9.1(b)).

9.6 ELECTROCOATING

The electrocoating processes we use today are very similar to the electroplating of copper or nickel. Electropaints were formulated (1966–1976) so that the paint is a large singly charged unit of paint, called a macro-ion (denoted by $RCOO^-$), which is suspended in water by its charge (Fig. 9.5). This macromolecule can be made very large and still only have one unit charge. Hence it is much more electrically efficient than the electroplating of metals (Table 7.4; Section 7.2). Electrocoating processes are subdivided up into cathodic and anodic electrocoats (Figs. 9.4(c) and 9.4(b) respectively). There is no deposition of paint film in electrocoating without the passing of an electrical current. Hence a paint brush could be placed in the electrocoat medium, without a current flowing, and when the brush was removed there would be no paint on it. Electrocoating therefore requires that the article is an electrically conducting substance or has an **electrically conducting surface**.

9.7 ANODIC AND CATHODIC ELECTROCOATS

Anodic electrocoats (Fig. 9.5) were the first to be introduced. They are called anodic, because the workpiece (car body) is made anodic (positive) to the electroplating–

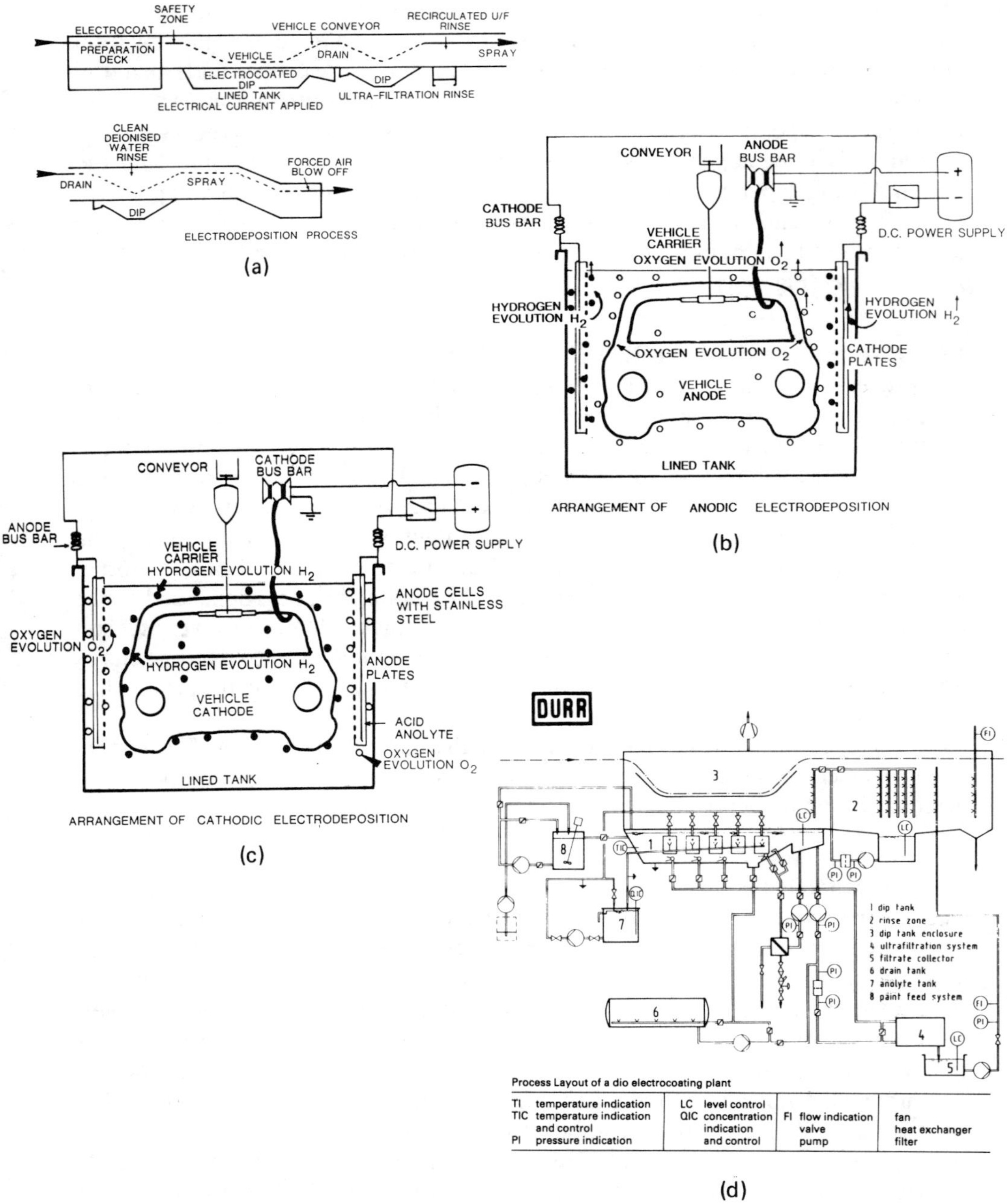

Fig. 9.4 — (a) An elementary diagram showing the electrocoating process by a process of total immersion. (b) The connections for the anodic process and the oxygen gases given off at the car body by the anodic process. The method by which the paint forms is discussed in Fig. 9.5. (c) The connections for the cathodic process and the hydrogen gases given off at the car body by the cathodic process. The method by which the paint forms is discussed in Fig. 9.6. (d) Schematic view of the electrocoating unit. This does not include many of the additional features which would make it a functional unit. Contributed by courtesy of DURR Ltd., Broxell Close, Warwick, CV34 5QF.)

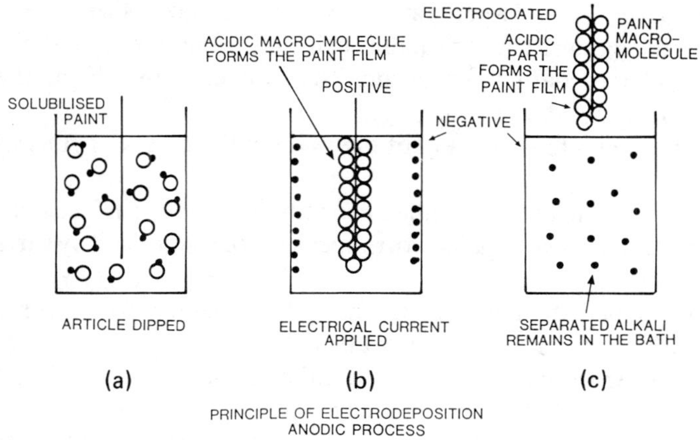

Fig. 9.5 — (a) The immersion of a ferrous article into the electrocoat would not cause any migration or deposition of the paint. The paint is formulated as a macromolecule RCOOX, which comprises an acid part $RCOO^-$ (○) and an alkali unit X^+ (●). (b) With the application of a positive potential and the flow of an electrical current within the aqueous paint solution, the acid macromolecule $RCOO^-$ (○) migrates towards the anode and is deposited as a powder by accepting a hydrogen ion. Switching off the current does not cause the paint to go back into solution as would a conventional paint left in its solvents would. The article can be removed and washed without losing the paint film as shown in (c). (c) The acid part of the macromolecule, now RCOOH, leaves the alkali part X^+ in the electrocoat bath. The addition of more paint and the neutralization of this alkali are discussed in Fig. 9.7.

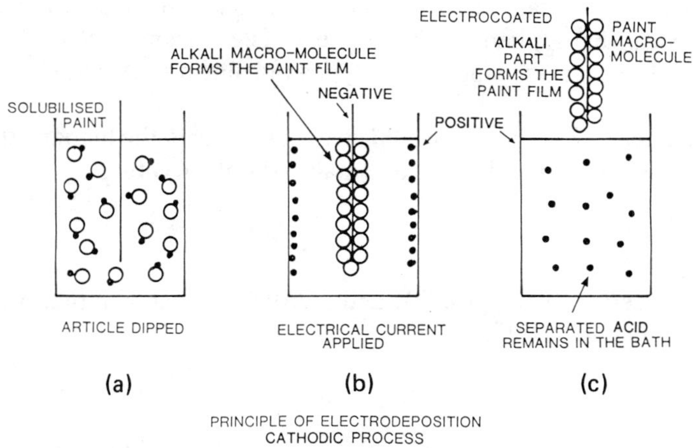

Fig. 9.6 — (a) The immersion of a ferrous article into the electrocoat would not cause any migration or deposition of the paint. The paint is formulated as a macromolecule $RNH_3^+-Y^-$, which comprises an acid part Y^- (●) and an alkali unit RNH_3^+- (○). (b) With the application of a negative potential and the flow of an electrical current within the aqueous paint solution, the alkali macromolecule RNH_3^+ (○) migrates towards the cathode and is deposited as a powder by accepting a hydroxyl ion which removes one hydrogen atom from the RNH_3^+ unit to produce RNH_2. Switching off the current does not cause the paint to go back into solution as would a conventional paint left in its solvents would. The article can be removed from the bath and washed without losing the paint film as shown in (c). (c) The alkali part of the macromolecule, now RNH_2 (neutral), leaves the acid part Y^- (●) in the electrocoat bath.

electrocoating bath. The 'first-generation' anodic electrocoats were very poor when compared with the modern cathodic paints (Fig. 9.6) which we use today. The speedy acceptance of electrocoating for automotive bodies arises from the following advantages over slipper dip.

(1) There is complete utilization of the paint with no sags or runs whatever the surface profile.

(2) Small screw holes were not filled in or coated over by surface tension (see the analogy of the paint brush (no paint worth speaking of is removed by surface tension) and Section 9.11).

(3) There is no need to manually flat the surface coating after deposition because the coating follows the surface uniformly (see Section 9.5).

(4) As the electrocoats are water borne, there is no fire hazard, as there is with conventional paints, and the thinners are cheap.

(5) The low viscosity of water means that the car body can be 'emptied' of paint quickly when it comes out of the electrocoat bath.

(6) As the paint is only deposited by the flow of an electrical current, the paint film does not wash off when water is used to wash the car body, nor is there any problem with **vapour phase washing** when the car bodies are stoved, as there was when the solvent-based slipper dip paints were stoved.

(7) The process can be fully automated, with the **'insulating' resistance** characteristics of the deposited paint acting as its own thickness regulator.

(8) Electrocoating could be applied to **already water-wet surfaces**, whereas this was impossible with the slipper dip paints whose solvents are not compatible with water. Therefore electrocoating saved a drying process.

(9) The electrocoat can only be applied to a conducting surface, and therefore only one coat can be applied by this process. The thickness of the electrocoat is limited by the electrical resistance of the deposited film in series with the resistance of the bath.

(10) Raw edges are coated with a **thicker** coat owing to the higher current density at these places which results from their small radii of curvature.

9.8 THE PAINT

In Section 6.16 (eqns. (6.9a), (6.9b), and (6.10)) it was shown that water could be written as splitting up as follows:

H_2O	$\rightleftharpoons$	H^+	+	OH^-	(9.3)
water	dissociates into	an acid part	and	an alkali part	(9.4)

This is also referred to as the electrolysis of water, and in this case the H^+ ions go to the anode, and the OH^- ions to the cathode. The acid part H^+ requires electrons and so, in the case of the paint system, will be drawn to the anode whereupon it will react with the paint macro-ion $RCOO^-$ to produce an insoluble compound as follows:

$RCOO^-$	+	H^+	$\rightarrow$	$RCOOH$	(9.5)
soluble				insoluble	

This equation says that the water-soluble paint, expressed as $RCOO^-$, plus the anodic product H^+, goes to a deposited paint (or insoluble compound) expressed as RCOOH.

These are called **anodic electrocoats**. As the anodic electrocoat requires free H^+ ions to form the insoluble paint, oxygen is liberated at the workpiece (Fig. 9.4(b)).

Anodic electrocoats have given way to the **cathodic electrocoats** in which the car body is made the cathode in a large 'electroplating bath' (Fig. 9.4(c)). In this case the OH^- goes to the cathode (negative) where it meets a differently formulated paint to give a deposited paint film as follows:

RNH_3^+	+	OH^-	$\rightarrow$	RNH_2	+	H_2O	(9.6)
soluble	and	cathodic product	goes to	deposited paint	and	water	(9.7)

The electrolysis of the paint causes the paint macro-ion to lose its charge whereupon the macro-ion is deposited as an insoluble powder (Section 6.13(3)) as shown above depending on whether an anodic (eqn. (9.5)) or a cathodic (eqn. (9.6)) system–paint is used.

This powder is then baked in order to form a coherent and adherent film. This is all the painting that internal and underbody panels are given. Plastisol coatings are used over the front wheel track and at metal overlaps.

The electrocoat therefore does not suffer sags and runs and indeed has other advantages. The paint, in the form of the powder, is an insulator and therefore the electrocoating process is a self-regulating process as regards its thickness. The thickness of the paint film on the car body in various places away from the opposite or counter-electrode is related to the electrolyte resistance of the electrocoat.

The thickness of film is inversely proportional to (1) the electrolyte resistances of the bath and the distance of the coated area from the opposite electrode plus, (2) the series resistance of any already-deposited paint film. In order therefore to reduce the electrical resistance of the bath to internal cavity members, large holes are introduced to increase the 'throw' of the painting process (Plate 2.5). Even with these large holes the internal box members do not get a very thick coat of electropaint and this factor, together with the poor corrosion resistance of the first generation of anodic electrocoats, led to the rapid growth of after-market rustproofing. The modern cathodic electrocoats do not give thick coatings in crevices as any DIY or garage repair will show, e.g. the interface behind door hinges is often completely denuded of paint (Plate 4.4).

The underbodies of the present generation of vehicle bodies are further protected by a PVC plastisol coating on top of the electrocoating, while the external body (uppermost panels) have primer surfacers and final gloss stove-enamelled coats.

9.9 DISADVANTAGES OF ELECTROCOATS

One big disadvantage of the electrocoat system has been the very thin electrocoats (or even the complete lack of paint) which are formed on the inter-welds and crevices, e.g. behind the door hinge. This is important for the following reasons.

(1) The inter-weld is an area which is denuded of any oxygen replenishment when the inter-weld contains water. This area is therefore anodic and suffers preferential attack.

(2) The inter-weld is denuded of electropaint owing to the high electrical resistance caused by the narrow cross-sectional area of the electrical path within the inter-weld.

(3) The corrosion mechanism produces an acidic environment within the inter-weld, which promotes the aggressive nature of the environment within the inter-weld. This can be seen by writing down the following chemical equations. The anodic reaction is

$$Fe \rightarrow Fe^{2+} + 2e^- \qquad (9.8)$$

i.e. steel dissolving in the crevice. Then the reaction of the Fe^{2+} ion with water is

$$2H_2O + Fe^{2+} \rightarrow Fe(OH)_2 + 2H^+ \qquad (9.9)$$

The $Fe(OH)_2$ is insoluble in excess of pure water. An excess of hydrogen ions (H^+) is the characteristic of an acid (Section 6.17 and 7.92).

In published tests carried out by the present author, it was reported that the inter-weld gap of 4 thou was the most corrosive environment owing to the possibility of the carry-over of the hygroscopic (phosphate) pretreatment salts. (Section 9.3).

9.10 THE REPLENISHMENT OF PAINT TO THE ELECTROCOAT BATH

When normal household paint or other conventional types of paint are removed from the paint pot, both solids and solvents are removed together. The paint removed when the article comes out of the electrocoat bath is the heavy macromolecule without the acid (cathodic; Fig. 9.6) or alkali (anodic; Fig. 9.5) part of the macromolecule. The latter ions, when combined with the paint macromolecule, cause the paint film to be soluble in water. When removed from the opposite ion the paint is insoluble and can be washed with ultrafiltrate (Section 9.11) without redissolving. This should be contrasted with a recently applied household paint which would readily wash off if the newly deposited paint film were sprayed with turpentine substitute. Hence the addition of the removed macromolecule causes some problems when the paint is made up. The added paint, which is unneutralized, needs to be solubilized with the opposite ions which are left in the bath at the counterelectrode to the car body.

9.10.1 Acid feed process (Fig. 9.7(a))

In this discussion anodic electrocoats only are considered: the same principles will apply to cathodic electrocoats except that, instead of supplying acid feed, alkali-deficient feed will be required (compare Figs. 9.5 and 9.6).

This acid feed (in anodic electrocoating) is very viscous because it contains the chemicals required for a priming paint (pigments, anti-oxidants resins and extenders) and must be carefully balanced with the remaining alkali left within the

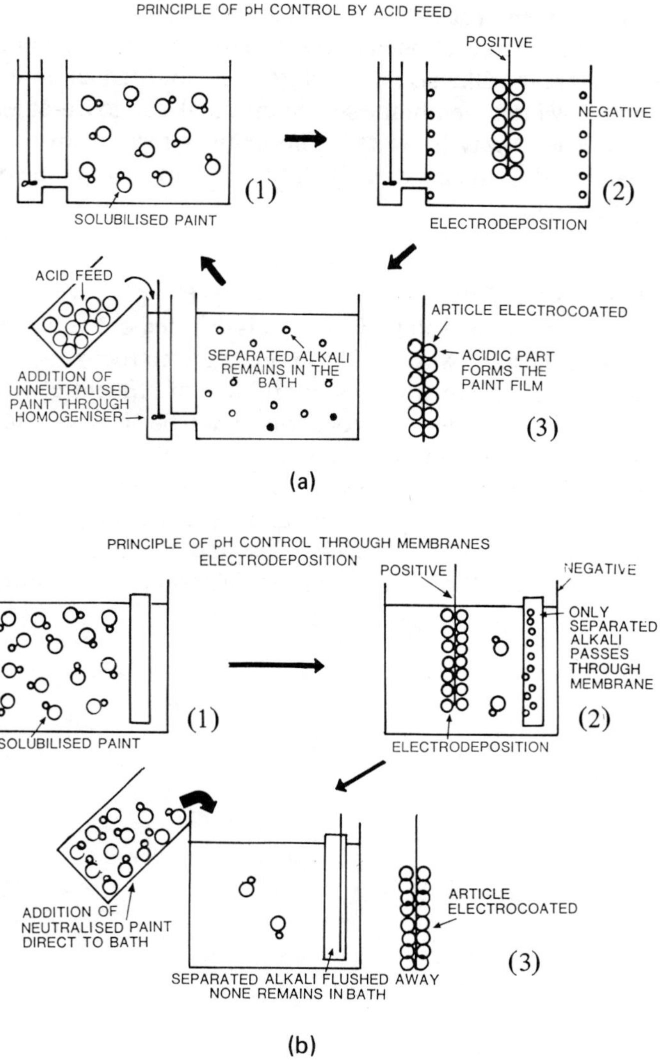

Fig. 9.7 — Two methods of replenishing the paint which is 'plated' out. The system is shown diagrammatically for the anodic electrocoating process (Fig. 9.5). Similar processes are available for the cathodic electrocoating process. (a) Addition of unneutralized paint $RCOO^-$ through a homogenizer as shown in (stage 3).(b) Use of dialysis membranes (stage 2) which take the alkali part (X^+) of the anodic electrocoat macromolecule RCOOX, which migrates to the cathode. This alkali is flushed away. The paint is then made up of fully neutralized electropaint RCOOX as shown in (stage 3).

bath to obtain neutrality. The disadvantage of this method of adding the deficient ions (in the unneutralized state) is the requirement for continued analysis of the electrocoat bath. This continued analysis is required as the deposited film is dependent on the bath composition.

9.10.2 Control by membranes

Another process has therefore been developed to remove the alkali. This is control by membranes. Here the alkali is removed from the counterelectrode to the car body by having a one-way filter membrane so that the alkali cannot re-enter the bath. This alkali is then washed away by water. The paint is topped up by the solubilized neutralized paint and so the ratio of alkali to acid parts is not adversely affected as shown in Fig. 9.7(b).

9.11 ULTRAFILTRATION

A very dilute solution of the solubilized paint is dragged out with the car body. This dragged-out paint must be washed off before the electrocoat is stoved. If it were washed off with water, there would be a problem of effluent control as this water could not be returned to the electrocoating bath because of the large volumes involved. Therefore a system is used where the water required for washing is 'extracted' from the electrocoat bath. This process is termed ultrafiltration. This rinsing liquid can then be returned to the bath, and the rinsing system becomes an enclosed system. In this respect no paint is lost and no (fresh) water is used. As the washing is all returned to the bath there is no disposal problem with the effluent.

10

Motor vehicle environment

10.1 CONDUCTIVITY OF WATER

Water is modelled as a covalently (Section 6.10) bonded molecule H_2O. In this form it will not conduct electricity and would be classed as an **insulator**. However this covalently bonded molecule is in equilibrium with a form where it forms ions as if the water molecule were formed of ions and thus ionically bonded (Sections 6.11 and 6.12.(7)). It is estimated that only one water unit (molecule) in every ten million is bonded in this ionic manner to give ions. Thus the following reaction is very biased towards the covalently bonded water molecule and is therefore indicated with the triple arrowhead:

$$\underset{\text{covalent bond}}{H_2O} \rightleftharpoons \underset{\text{ionic bond}}{(H^+)} + OH^- \qquad (10.11)$$

Hence this is often interpreted as stating that the water ionises or dissociates to form H^+ ions and OH^- ions and is given a constant K_w which expresses the degree of dissociation (Section 6.17).

Within a metal the current is carried by the electron cloud which must remain in the metal. In the liquid environment the current is carried by both positive and negative ions and for water this means that transport is by the H^+ ions and OH^- ions. The total current carried by the positive and negative ions must always equal the current carried by the electrons within the metals. Thus in pure water the corrosion of a metal will be limited by the small number of charge carriers in the aqueous environment. This is often expressed as a conductivity. Conductivity is the reciprocal of resistance. Conductivity is used to refer to the electrical properties of a liquid whilst **resistance** is used to refer to the electrical properties of a solid.

Both metals and liquids obey Ohm's Law:

$$V = i \times R$$

(where V is the voltage, i is the current, $C = 1/R$ is the conductivity and R is the resistance), which will apply to both the flow of the electronic charge in the metal as well as the ionic charge in water. The units of conductivity are mhos per centimetre, where mho is ohm written backwards to indicate that it is the reciprocal of the ohm (Ω).

However, the resistance of both metal and liquid conductors is proportional to their length l, and inversely proportional to their cross-sectional area A such that:

$$R = \frac{\rho \times l}{A}$$

where ρ is the specific resistance and has units of Ω cm. The reciprocal of the specific resistance is the specific conductance k with units of $\Omega^{-1}\ \mathrm{cm}^{-1}$:

$$\rho = \frac{R\,A}{l} = \frac{\Omega \times \mathrm{cm} \times \mathrm{cm}}{\mathrm{cm}} = \Omega\ \mathrm{cm}$$

$$k = 1/\rho$$

Thus the specific resistivity of an electrolyte is an intrinsic property of that electrolyte and does not depend on its dimensions. The values for several solutions are given in Table 10.1. The value of the resistivity varies depending on the purity of the water and the dissolved species. This transition of covalently bonded water to ionically bonded water is expressed as the ionization constant K_w for water. The value of this constant is temperature dependent, but at 25°C it is 1.01×10^{-14}. Pure water at 25°C contains an equal number of H^+ ions and OH^- ions and so the pH is the square root of the ionization constant K_w. If the H^+ ions exceed the OH^- ions the medium is acidic and is represented by a pH from 1 to 7. If the OH^- ions exceed the H^+ ions the medium is alkaline. At temperatures above 25°C the ionization constant of water is greater and so the pH of pure water is temperature dependent as shown in Table 6.10. Therefore in pure water at 25°C there are only 2×10^{-7} ($2 \times 10^{-7} =$ 2/10 000 000) charge carriers ($10^{-7}\ H^+$ and $10^{-7}\ OH^-$ ions). Thermal energy is not sufficient to produce a significant increase in charge carriers.This is a very small concentration and so pure water is a poor conductor of electricity.

Table 10.1 — The resistivity of water varies depending on its source. Approximate value of the resistivity (ohm centrimetres) are given. The effect of this resistivity has a marked effect on the corrosion profile in a given system as shown in Fig. 10.1

Pure water	20 000 000
Distilled water	500 000
Rain water	20 000
Tap water	1–5000
River water (brackish)	200
Sea water (coastal)	30
Open sea	20–25

Values from J. H. Morgan, Cathodic Protection, Leonard Hill Books, London, 1959.

Table 10.2 — Water from our drinking taps contains a number of impurities. A knowledge of the impurities is important when steel samples are coated with phosphate crystals and then washed with water. Chlorides and sulphates have a deleterious effect on the protection afforded by a paint coating

	Very soft lake water	Moderately soft surface water	Slightly hard lake water	Moderately hard river water	Hard borehole water (chalk formation)	Slightly hard borehole water containing sodium bicarbonate	Very hard underground water
pH value	6.3	6.8	7.4	7.5	7.1	8.3	7.1
Alkalinity to methyl orange ($CaCO_3$)	2	38	90	180	250	278	470
Total hardness ($CaCO_3$)	10	53	120	230	340	70	559
Calcium hardness ($CaCO_3$)	5	36	85	210	298	40	451
Sulphate (SO_4)	6	20	39	50	17	109	463
Chloride (Cl)	5	11	24	21	4	94	149
Silica (SiO_2)	trace	0.3	3	4	7	12	6
Dissolved solids	33	88	185	332	400 (also nitrate 51 mg/l NO_3)	620	1670

Reproduced by permission from L. L. Schrier, Corrosion, Vol. 1, Newnes Butterworth, London, 1977, Table 2.9.

Water, however, is not pure. It contains gases which dissolve in the water to produce charge carriers, which upsets the equilibrium between the H^+ ions and OH^- ions. Thus, when rain water falls through the atmosphere, it dissolves carbon dioxide and both sulphur dioxide and sulphur trioxide. These are acidic gases and dissolve in water to form a weak acid but the number of ions is significantly increased by this up-take. Assume that the pH becomes 1. This means that the H^+ ion concentration is 10^{-1} or 1/10. The OH^- ion concentration will be 10^{-13} or 1/10 000 000 000 000. Hence 10^{-1} is far greater than 10^{-13}, and the latter can be ignored when calculating the number of charge carriers present in the water. Hence such a solution will greatly assist the corrosion reaction in providing more charge carriers for the liquid environment.

The pH is also dependent on the geological strata over which the rain water has penetrated, and tap water is said to be either 'soft' or 'hard'. These are terms which refer to the amount of dissolved calcium and magnesium ions as shown in Table 10.2.

10.1.1 Dissolved gases and solids

The solubilities of oxygen, carbon dioxide and sulphur oxides in water are temperature dependent as shown in Table 10.3 and dependent on the salinity of the water.

Table 10.3 — The solubility (cubic metres per cubic metre of water under 1 standard atmosphere) of various gases in water will affect the resistivity and aggressiveness of the medium. The solubilities of the gases are also temperature dependent

	Temperature of water (°C)							
	0	10	15	20	30	40	50	60
Carbon dioxide	1.676	1.163	0.988	0.848	0.652	0.518	0.424	0.360
Nitrogen	0.0230	0.0183	0.0165	0.0152	0.0133	0.0119	0.0108	0.0100
Oxygen	0.047	0.037	0.033	0.030	0.026	0.022	0.020	0.019
Sulphur oxide	79.8	56.6	47.3	39.4	27.2	18.8	—	—

Data reproduced by permission from L. L. Schrier, Corrosion, Vol. 2, Newnes Butterworth, London, 1977, Table 21.19.

These oxides of carbon and sulphur are responsible for the industrial aggressive environments, while the dependence of the salinity on the solubility of oxygen is thought to account for the effect of salt concentration on corrosion rate for steel as shown in Figs 10.1 and 10.2.

Solids must ionize for the solid to be dissolved in water. Hence a salt, such as the road de-icing salts (NaCl), will produce ions (Na^+ and Cl^-) which will again act as charge carriers in the liquid environment without affecting the pH. The corrosion rates (Fig. 1.6), show the marked effect of the addition of these charge carriers, which increase the aggressive nature of the liquid medium. Note that the industrial atmospheres containing sulphur dioxides are more aggressive and this will be discussed in Section 10.6 under critical humidity.

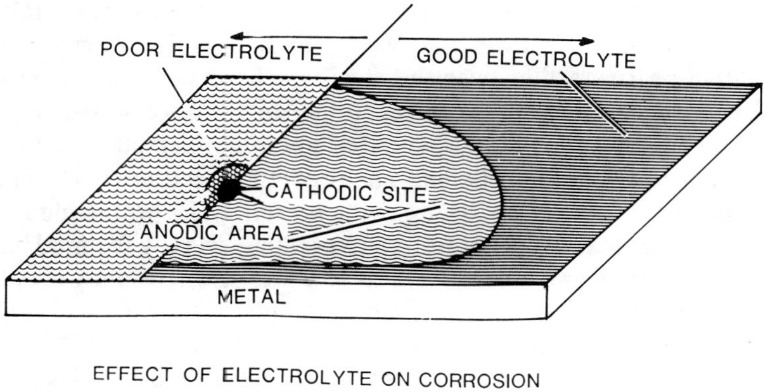

Fig. 10.1 — A schematic diagram indicating the effect of good and bad resistivity of the aqueous phase on the surface of mild steel. The conductivity of the electrolyte affects the extent of the degradation and corrosion of the anodic area. The anodic area will be small in poor electrolytes and extensive in good electrolytes.

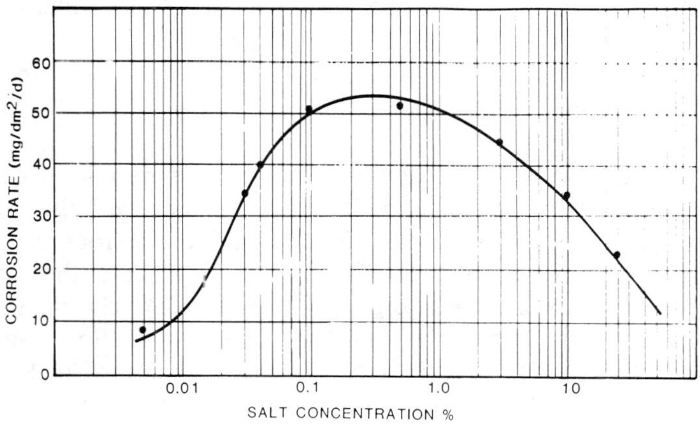

Fig. 10.2 — The effect of salt concentration on the corrosion rate of mild steel was studied by Bishop (1974) in the laboratory using intermittent salt spray at 5°C. The reason for the decrease in corrosion rate with the increase in salt concentration is due to the decrease in oxygen uptake with increase in salinity. (Redrawn with permission from R. R. Bishop, New Scientist, May16, 1974, p. 382, Fig. 1.)

10.2 ROAD DE-ICING SALTS: Their equivalence to sea water

Consider an ideal situation where a piece of steel can corrode in two media: pure water and sea water (Fig. 10.1). Both sea water and road de-icing salts contain sodium and calcium chlorides. The effect of the chloride ion upon the corrosion rate is immense. The mobility of the Fe^{2+} ion is very low in pure water, and it exists in an insignificantly small concentration. Scientists term this Fe^{2+} ion as being insoluble in water, and any ions produced by the corrosion process are taken out of solution in the liquid environment by eqns. (10.2) and (10.3) below. The Fe^{2+} ion is taken out of solution by the OH^- ions from the water. The 'insoluble' ferrous hydroxide $Fe(OH)_2$ initially forms a film over the steel which acts as a good barrier layer through which oxygen must diffuse to get to the steel surface so that the cathodic reaction can take place. The pH of a saturated solution of ferrous hydroxide is 10.5 and so the steel is also in an alkaline environment (Section 6.17). The rate of corrosion in alkaline media is low owing to the presence of a protecting passive film on the surface of the steel. Hence steel as a river piling keeping the mud banks from being eroded away by river traffic has a fairly low rate of corrosion. Crevices, where oxygen replenishment is difficult as a result of the physical geometry, have a much higher rate as this precipitation of the ferrous hydroxide removes OH^- ions from the water and thus unbalances the equilibrium shown in eqn. (10.1). This produces an acidic environment (Section 7.9). The reaction is represented as:

$$Fe^{2+} \quad + \quad 2OH^- \rightleftharpoons Fe(OH)_2 \qquad (10.2)$$
$$\text{product of anode} + \text{product of cathode} \rightarrow \text{iron hydroxide} \qquad (10.3)$$

The Fe^{2+} ion does, however, exist when in combination with the chloride ion Cl^- which is present in the road de-icing salts when they dissolve in water:

$$NaCl \quad \rightarrow \quad Na^+ + Cl^- \qquad (10.4)$$
$$\text{salt insoluble goes to or dissolves to produce ions}$$

The iron ion Fe^{2+} becomes more soluble and mobile in the presence of chlorides and so can move away from the metal–liquid interface and react with the hydroxyl ions OH^- away from the dissolving area. This prevents the ferrous hydroxide film $Fe(OH)_2$ from being deposited on the surface of the steel and thus there is no barrier film through which the oxygen has to diffuse. The presence of chlorides therefore is shown externally by the rust stains which accompany the first signs of metal penetration (Section 2.15). The road de-icing salts therefore make the liquid environment more aggressive by providing more mobility to the Fe^{2+} ions, and more charge carriers (Na^+ and Cl^- ions) in the liquid environment. Thus the corrosion reaction is accelerated. For the corrosion reaction to be accelerated, there must also be more access of the steel to oxygen and the effect of dissolved sodium chloride on the oxygen uptake will be considered. In order to sustain this increase in corrosion rate, the anions and cations (product of anode and cathode respectively) must combine (Section 7.7.1).

10.3 EFFECT OF DISSOLVED OXYGEN

Oxygen is removed from water by the following:

(i) corrosion of steel;
(ii) dissolved salts;
(iii) biological decay (this will not be considered as it does not affect motor vehicle corrosion);
(iv) increase in temperature.

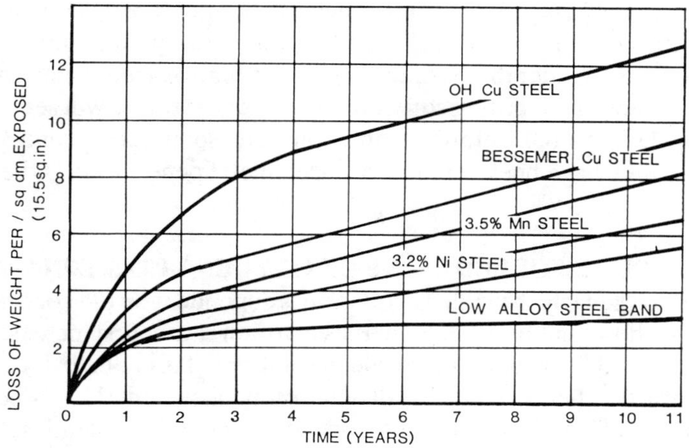

Fig. 10.3 — The corrosion rate of steels as a function of time, indicating some protective properties of the corrosion product. (After C. Larrabee, in H. H. Uhlig, Corrosion Handbook, Wiley, New York, p. 170.)

The corrosion of iron in oxygenated water is a parabolic curve as represented by Fig. 10.3. The initial corrosion rate may reach a figure of 100 mg dm^{-2} day^{-1}, which rapidly falls as the ferrous hydroxide is deposited on the steel surface. The steady state corrosion rate may be as low as 10–25 mg dm^{-2} day^{-1}. The corrosion reaction rate is controlled mainly by the diffusion of oxygen to the steel surface. This is shown in Fig. 10.2 and 10.3, with the comparable rates defined in millimetres per year, in Table 10.4. The corrosion rate reaches a maximum with high oxygen levels because the excess of oxygen required by the cathodic reaction is available to form a passive

Table 10.4 — The decrease in corrosion rate with increase in exposure period for mild steel, copper and aluminium immersed in sea water

Exposure time (months)	Average corrosion rate for period (mm/y)		
	Steel	Copper	Aluminium
1	0.33	—	—
2	0.25	—	—
3	0.19	—	—
6	0.15	—	—
12	0.13	0.034	0.004 3
24	0.11	0.019	0.002 1
48	0.11	0.018	0.001 7

Reproduced by permission of L. L. Schrier, Corrosion, Vol. 1, Newnes Butterworth, London, 1977, Table 2.15.

oxide film on the steel (Fig. 10.3). This effect is not often observed in automotive corrosion because most corrosion cells are formed by differences in the replenishment rates of oxygen. Under these conditions of very high oxygen replenishment, the metal is very susceptible to pitting corrosion, which once started will set up the local cells-batteries. Differential aeration conditions are soon set up and cannot be destroyed once they have been set up. Localized metal penetration is the result.

10.4 EFFECT OF DISSOLVED SALTS ON OXYGEN UPTAKE IN WATER

The amount of oxygen which water can dissolve is dependent on the concentration of dissolved salt as shown in Fig. 10.2. Here the corrosion rate at first increases with salt content and then at 3% of sodium chloride in solution it falls to below the corrosion rate for pure water when over 20% salt solution is dissolved. This is because the oxygen solubility falls with salt content. Indeed, many of the inland 'salt' lakes in America are devoid of life because of the low oxygen solubilities in these saline lakes. The initial rise in corrosion rate with salt content, shown in Fig. 10.2, is almost certainly due to the fact that the anodes and cathodes can be further away from each other when the medium contains salt than when the steel is in pure water. In pure water the anodes must be near to the cathodes owing to the poor conductivity (high resistivity; Table 10.1) of the medium. As the salt content rises, the conductivity rises and so the distance between the local electrodes may increase. Because of this increase in distance between anodes and cathodes, the protective Ferrous hydroxide $Fe(OH)_2$ is no longer deposited as the barrier layer.

10.5 HYGROSCOPIC SALTS, DELIQUESCENT SALTS

The air which we breath consists of oxygen (23% by weight), nitrogen (75% by weight), carbon dioxide (0.04%) and water vapour (0.7% by weight at 10°C). This water vapour is 'invisible' in the air and only shows its presence when the air becomes cold, whereupon the water condenses out as dew, fog or clouds. The amount of water taken in by the air is expressed as the relative humidity:

$$\text{relative humidity} = \frac{\text{amount of water vapour in the air}}{\text{amount of water vapour required to saturate the air}} \times 100$$

This is expressed as a percentage. A relative humidity of 30% means that the air is very dry, while a relative humidity of 95 % means that the air is very damp. Cold steel will reduce the temperature of the adjacent air and water will condense onto the metal, giving the concept of a metal 'sweating' (Section 2.1). Anyone wearing glasses will know that the same condensation or misting will occur when coming into a warm room from the outside cold environment. The warmer air in the room can contain more water vapour, but the colder layer of air adjacent to the glass in the spectacles, cannot contain so much water vapour and so it 'condenses out'. The same effect happens overnight and gives the dew in late evenings and early mornings, which condenses out on the car body and in crevices and on hygroscopic salts. A number of hygroscopic salts are shown in Table 10.5 with the relative humidity with which they are in equilibrium with saturated solutions. The main road de-icing salts are NaCl and $CaCl_2$.

Table 10.5 — Most salts are hygroscopic and attract water. Conversely, saturated solutions of salts have an equilibrium relative humidity. The table gives the relative humidity of the air in equilibrium with saturated solutions at 20°C. Some of these salts can be present in road de-icing salts. NaCl and $CaCl_2$ are the main contents of the road de-icing salts. Iron sulphate is formed on improperly prepared steel for painting in the winter months

Salt in solution	(%)	Salt in solution	(%)
$CuSO_4{\cdot}5H_2O$	98	NaCl	76
K_2SO_4	98	$CuCl_2{\cdot}2H_2O$	68
Na_2SO_4	93	$FeCl_2$	56
$Na_2CO_3{\cdot}10H_2O$	92	$NiCl_2$	54
$FeSO_4{\cdot}7H_2O$	92	$K_2CO_3{\cdot}2H_2O$	44
$ZnSO_4{\cdot}7H_2O$	90	$MgCl_2{\cdot}6H_2O$	34
$3CdSO_4{\cdot}8H_2O$	89	$CaCl_2{\cdot}6H_2O$	32
KCl	86	$ZnCl_2{\cdot}xH_2O$	10
$(NH_4)_2SO_4$	81	NH_4Cl	80

Reproduced by permission of L. L. Schrier, Corrosion, Vol. 1, Newnes Butterworth, London, 1977, Table 2.5.

A **hygroscopic salt** is a compound, a salt, which absorbs water from the atmosphere to produce an electrolyte. In some cases the compound will make a pool of water at its site and the material is said to be **deliquescent**. A hygroscopic salt therefore can be considered to be artificially increasing the relative humidity of its immediate environment. Thus the hygroscopic salt produces an environment which is wetted earlier as the relative humidity rises and remains wet for a longer period after the relative humidity falls. Hence the road de-icing salts NaCl and $CaCl_2$ will not start to 'dry out' until the relative humidity falls to 76% and 32% respectively (Table 10.5). The salt content of the sea can be carried by air inland over considerable distances from areas adjacent to the shore as shown in Table 10.6. There are many salts which are hygroscopic, but the main ones in the corrosion of the motor vehicle are phosphoric acid and its salts, which arise from the metal pretreatment process (Section 6.18 and Chpater 9) and the road de-icing salts.

The DIY 'rust cures' contain phosphoric acid. The excess or unreacted phosphoric acid from these preparations should be removed with distilled water (or demineralized water) in order to make sure that the hygroscopic contaminants are not trapped beneath the new paint films. Tap water is unsuitable as the water contains several dissolved salts (Table 10.2). Research has shown that sulphates are the main cause of paint breakdown by blistering. The sulphates may arise from industrial atmospheres (Fig. 1.6), which are not confined to places of large scale industry, but is also present in some 'pure' drinking water (Table 10.2). Research has

Table 10.6 — The salt content of air in various locations from the surf and the associated corrosion rates

Distance from surf	Salt content of air[a]	Rate of rusting (mm/year)[b]
50 yds	11.1	0.950
200 yd	3.1	0.380
400 yd	0.8	0.055
1300 yd	0.2	0.040
25 miles	—	0.048

a Salt content of the air was determined by exposing wet cloths and is expressed as mg NaCl d^{-1} $(100\ cm^2)^{-1}$.
b Specimens were ingot iron exposed for 1 year.
Reproduced by permission of L. L. Schrier, Corrosion, Vol. 1, Newnes Butterworth, London, 1977, Table 3.1.

also shown that it is much better to prepare steel work for painting in the summer months when there is less chance of water condensation from atmospheric water (and hence fewer sulphate contaminants) than in winter. This is due to the large amount of sulphur dioxide that is released and retained in the atmosphere during the winter period, (Fig. 10.4). Table 10.7 shows the impurities present in the air which we breathe. Table 10.8 shows the distribution of sulphur dioxide around five large industrial complexes in the United States.

The sulphites (from the gas SO_2), and sulphates (SO_3) are the main corrosion promoters as they produce hygroscopic salts (Table 10.5). These gases are produced by the burning of sulphur-containing fuels. Fig. 2.6 shows the effect of the road de-icing salts on the road aggregate and the water envelope associated with this atmospheric absorption of water vapour. The time of 'wetness' of the painted surface is complicated by drying and wetting cycles, but the presence of hygroscopic salts increases the period of wetness considerably.

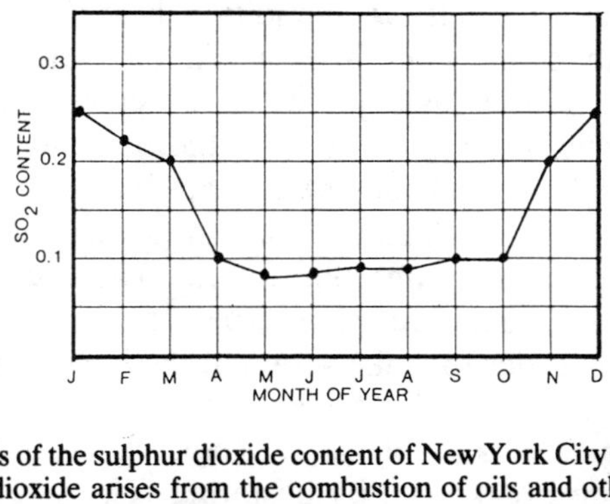

Fig. 10.4 — Analysis of the sulphur dioxide content of New York City air with time (month) of year. The sulphur dioxide arises from the combustion of oils and other industrial pollutants giving rise to 'acid rain'. Its concentration is a function of weather, cloud and rainfall. (After L. Greenburg and M. Jacobs, Industrial and Engineering Chemistry, reproduced with permission from H. H. Uhlig, Corrosion and Corrosion Control, Wiley, New York, p.170, Fig. 2.)

Table 10.7 — The concentration of the aggressive and deleterious gases during winter and summer (Fig. 10.4). These gases have a marked effect on the corrosion rate. The amount of sulphur dioxide is 1% of the amount of sulphur dioxide

Gas	Winter	Summer
Sulphur dioxide (industrial)	350	100
Sulphur dioxide (rural)	100	40
Hydrogen sulphide (industrial)	10.9	1.5
Hydrogen sulphide (rural)	1.7	0.5
Air-borne chlorides	8.2	2.7

Reproduced by permission of L. L. Schrier, Corrosion, Vol. 1, Newnes Butterworth, London, 1977, Table 2.33.

Table 10.8 — The sulphur dixoide content of the air varies with radius of location from American city centres

	Content (ppm) for the following distances (miles)					
City	0–5	5–10	10–15	15–20	20–25	25–30
Detroit	0.023	0.012	0.006	0.004	0.004	0.005
Philadelphia-Camden	0.030	0.018	0.016	0.021	0.012	0.012
Pittsburgh	0.060	0.030	0.015	0.018	0.009	0.010
St. Louis	0.111	0.048	0.029	0.020	0.018	0.014
Washington	0.003	0.001	0.001	0.001	0.001	0.002

After H. Meller, J. Alley and J. Sherrich, quoted by L. Greenburg and M. Jacobs, Industrial and Engineering Chemistry, **48,** (1956) 1517, and reproduced with permission from H. H. Uhlig, Corrosion and Corrosion Control, Wiley, New York, p. 171, Table 3.

The road de-icing salts are an impure form of the (table) salt which we sprinkle on our meals. This table salt sometimes becomes wet and will not run freely from the salt pot on very humid days. This table salt has been specially treated by the manufacturers to keep it 'free running' by preventing it from absorbing water vapour from the atmosphere. The impure road de-icing salts, therefore, absorb much more water than the treated normal table salts which are used domestically. The hygroscopic nature of sea salt is also shown by the difficulty of drying oneself on a cold day after a swim in the sea. The effect of keeping these salts on a painted surface for any length of time is shown in Fig. 3.10.

10.6 CRITICAL HUMIDITY

Water molecules are chemically bound to the surface of the metal by weak van der Waals bonds (Section 6.12). It is estimated that at a relative humidity of 55% there is a surface film some 15 molecules thick upon the surface of mild steel, and this increases to 90 molecular layers as the relative humidity rises to 100%. These surface-absorbed layers are capable of supporting the corrosion process above 60% relative humidity. This is very pronounced if the atmosphere contains sulphur dioxide (see Fig. 10.5, which illustrates this critical humidity). Relative humidity is also important in filiform corrosion (Section 8.14; Fig. 8.21) beneath enamelled steel as shown in Table 10.9.

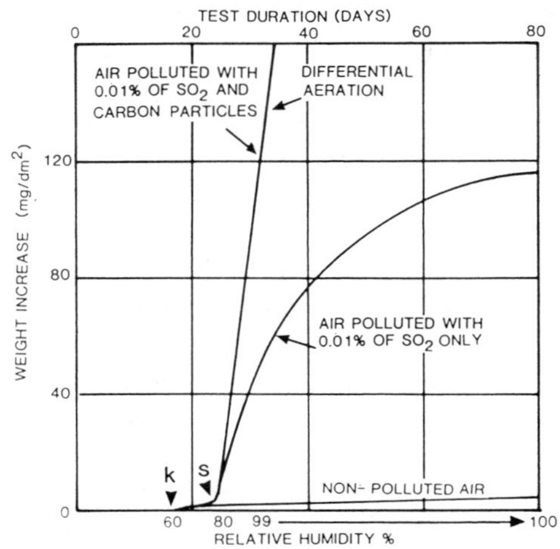

Fig. 10.5 — The corrosion rate of iron varies markedly with sulphur dioxide content, humidity and particulates, accelerating quickly at a critical humidity of over 70%. Vernon (1927) showed that below this critical humidity the corrosion rate was low. Particulates may accelerate the corrosion rate by producing a differential aeration cell at the point of contact with the metal substrate. In pure air, without the sulphur dioxide content, the corrosion rate was independent of humidity. In atmospheres containing SO_2, the corrosion rate is very dependent on humidity and particulates. Other workers have put forward a theory based on sulphuric acid regeneration to explain the catalytic action of this reaction by sulphuric dioxide. (After W. H. J. Vernon, Transactions of the Faraday Society, **23** (1) (1937) 113; with permission from L. L. Schrier, Corrosion, Vols. 1 and 2, Newnes Butterworth, London, 1977, p, 3.6, Fig. 3.1.)

Table 10.9 — Effect of relative humidity on filiform corrosion of enamelled steel. See also Table 10.10 and Fig. 8.21

Relative humidity (%)	Appearance
0–65	No corrosion
65–80	Very thin filaments
80–90	Wide corrosion filaments
93	Very wide corrosion filaments
95	Mostly blisters; scattered filiform corrosion
100	Blisters

Data from M. Van Loo, D. D. Laiderman and R. R. Bruhn, Corrosion **9** (1953), 2.

10.7 EFFECT OF CAPILLARY CONDENSATION

In the section dealing with surface tension, (Section 10.8), an analysis will show that the vapour pressure above a concave surface is less than the vapour pressure across a flat surface. Therefore, under equilibrium conditions, atmospheric water will condense out of the atmosphere into capillaries or crevices and thus produce a liquid environment where there should be dryness for that relative humidity. Table 10.10 gives the relative humidity for condensation for particular capillaries of different radii. Thus as the radius or 'crevice effect' decreases, so condensation can be obtained within the crevice or radius with drier air. The converse is also true; the capillary or crevice will remain wetter for long periods and in some locations the narrow capillaries will never dry out. This effect is very important when oxide films such as rust are overpainted as the porous nature of the oxide can trap water even in very dry atmospheres. This capillary effect, now applied to crevices, is also found for the larger increase in weight of steel panels exposed under various humidities and sulphur dioxide contents when particles of charcoal are deposited upon the surface of steel plates as shown in Fig. 10.5.

Table 10.10 — Different sizes of capillaries will attract water by offering a lower pressure over the concave surface. The smaller the capillary radii, the lower the pressure and therefore the lower the humidity for condensation to take place. This concept explains the presence of an electrolyte in relatively dry atmospheres in crevices and beneath painted surfaces and in incorrectly prepared surfaces for painting

Capillary radius (nm)	Relative humidity for condensation (%)
36	98
9.4	90
4.7	80
3.0	70
2.1	60
1.5	50

Reproduced by permission from L. L. Schrier, Corrosion, Vol. 1, Newnes Butterworth, London, 1977, Table 2.7.

10.8 SURFACE TENSION

The surfaces of liquids have unique properties because the liquid molecules are not 'fixed' in three-dimensional space. The internal water molecules are surrounded by, on the average, 12 nearest neighbours, whereas those at the surface are only surrounded by 7, 8 or 9 molecules. The molecules on the opposite side of the surface (in the vapour phase) are too far away to contribute anything to the attraction of the surface molecules (Fig. 10.6). Therefore the surface molecules have a resultant attraction force towards the bulk of the liquid, which manifests itself as the liquid's

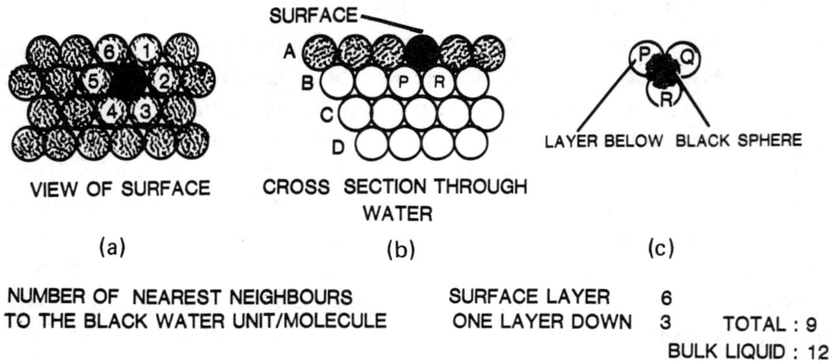

Fig. 10.6 — It is rather difficult to represent the three–dimensional structure of water in the liquid state. In this figure each water molecule (H_2O) is represented by a sphere. (a) A view of the surface molecules as seen by a fish. The black sphere has six nearest neighbours. (b) Discussion will be in terms of the layers A, B, C and D below the surface. (c) View of three molecules P, Q and R in layer B which are nearest to the black water molecule in layer A (a) and (b). There are therefore nine nearest neighbours to the black surface molecule, which would have another three neighbours making 12 nearest neighbours if it were below the surface and in layer C. This accounts for the resultant pull of the surface layers which explains the phenomenon of surface tension of water.

behaving as if it had a tight skin over the surface. The interaction of a liquid with a surface is a complex reaction of several interatomic forces. It is best studied by measuring the surface tension force T and the contact angle ϕ. Both of these factors are important because they dictate whether the liquid enters a capillary (depending on its contact angle ϕ) and how far it rises up that capillary (which depends on the magnitude of T and ϕ) according to the relationship:

$$h = \frac{2\mathrm{T} \cos \phi}{r\rho g}$$

where h is the height, r is the radius of capillary, g is the acceleration due to gravity, ρ is the density of the liquid, T is the surface tension of water or the liquid in question and cos ϕ is the cosine of the contact angle of the liquid with respect to the substrate.

Capillary rise is inversely proportional to the crevice gap or capillary radius r (Fig. 3.6). However, negative heights can be obtained by using hydrophobic substances. The contact angle is found from the tangent to the surface at the point where liquid and solid meet and is always the angle within the liquid. Cos ϕ is negative for contact angles of greater than 90°.

If a cross-section of a water droplet is examined on two surfaces 'A' (a painted surface) and 'B' (an oily surface) it appears that the liquid shape is completely different on the two surfaces, Fig. 3.6. In case A the attractive forces between the liquid molecules is less than the attraction between the liquid and substrate A and so the droplet will have a low profile as the liquid creeps over the surface. The contact angle ϕ will be small. In contrast, if the liquid–substrate B intermolecular attraction is less than the liquid–liquid attraction, then the liquid will have a high profile as the liquid is 'repelled' by the substrate and this is expressed as having a contact angle of greater than 90°. If these substrates now form capillaries, then the capillary formed from substrate A will allow the liquid to rise up the capillary, while that formed from the substrate B will exclude the liquid.

10.9 PROOF OF FORMULAE FOR SURFACE TENSION

The surface tension formula may be derived by two methods.

(1) Method 1: Using the concept of the surface tension as a force per unit length, the upward pull of the water–air meniscus in the capillary tube around the perimeter of the capillary ($2\pi r$) raises a known weight of water. When these forces are in equilibrium then

$$2\pi r \times T \cos \phi$$

is the resolved upwards force due to the surface tension which balances the weight of water mg which has been raised above the external surface. The mass of water is found from the volume of the water raised up the capillary:

volume V	$= \pi r^2 \times h$
density ρ	$= m/V$
mass of water	$= \pi r^2 \times \rho \times h$
weight of water	$= \pi r^2 \times \rho \times h \times g$

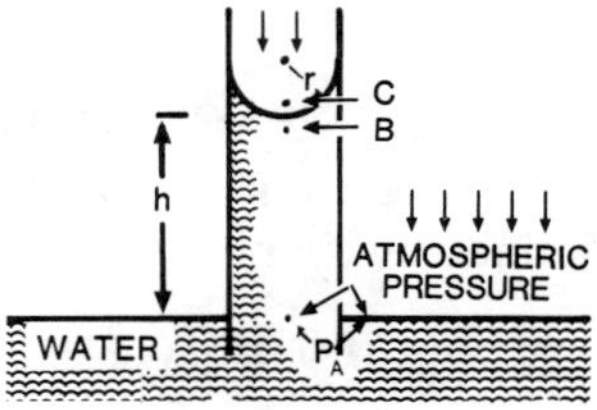

Hence

$$2T \cos \phi = hr\rho g$$

(2) Method 2: this uses the concept that the excess pressure within an air bubble within a liquid is $2T/r$ (at point C) and this must be equal to the excess pressure over a surface with an infinite radius (i.e. a flat surface). This is atmospheric pressure–minus the pressure due to the head of water pressure,

Pressure at C (P_c) is atmospheric pressure (P_a). Assume hemispherical profile.
Pressure at B (P_b) (just below the surface) is $P_a - h\rho g$.
Pressure over the curved surface is $2T/r$ (contact angle $\phi = 0$ for hemisphere).
At equilibrium $2T/r = P_c - P_b = P_a - (P_a - h\rho g)$.
Therefore $2T/r = h\rho g$ (contact angle = 0; $\cos \phi = 1$ for a hemisphere).

This result has very important consequences for water ingress into porous materials such as paint films or the narrow crevices discussed earlier.

(1) Water will condense more easily upon a concave surface than on a flat surface (discussed in greater detail below).
(2) Water will evaporate from a convex surface and condense upon a flat surface or concave surface.
(3) Once capillaries contain water it is very difficult to dry them out as their surface profile is always concave.
(4) Protective coatings (e.g. paints and chromium plating (Figs 8.3 and 8.4) contain micro-capillaries which take a long time to dry and act as 'preferential sites' for atmospheric condensation.

The atmospheric pressure p_A is made up of several partial pressures, i.e. the pressure due to oxygen p_{O_2}, the pressure due to nitrogen p_{N_2}, and the pressure due to water vapour p_{H_2O} and these **partial pressures** may be written as

$$p_A = p_{O_2} + p_{N_2} + p_{H_2O}$$

If the atmospheric pressure p_A is to be reduced then there is only one partial pressure which can be reduced (unless say oxygen or nitrogen is suddenly withdrawn or used up in the air) and that is the partial pressure of water vapour p_{H_2O}. This can only be reduced if the water condenses and, for water vapour to condense, the partial pressure of water vapour must equal the (SVP) of water saturated vapour pressure at that temperature. Hence the 'SVP' of water over the concave surface is very small and the water condenses on such a surface. Conversely, the 'SVP' over a convex surface is very high and so water evaporates from such a curved surface. This phenomenon is used in nature where plants have to survive in very dry climates, or where the rainfall is very low and so the plants have to rely on condensed water from the atmosphere for their water intake.

10.10 COMPACTION OF ROAD DIRT BY SURFACE TENSION

Sections 2.7 and 2.8 developed the concept that the the size of a spray droplet was defined by its momentum M, which is equal to its mass multiplied by its velocity, i.e.

$$M = m \times v$$

The mass in this relationship is the combined mass of the water and road dirt called the 'droplet' (Fig. 2.8). The magnitude of this momentum (mv) gives a parameter which characterizes how easy it is for that particle to alter direction. The smaller the quantity $M = mv$, the easier it is for the particle to follow the turbulent air around the moving vehicle. Nature also dictates, by the surface free energy theory, that there must be a nucleus upon which the water droplet can form or **nucleate**. Often in very dry spells, rain can be 'ordered' by 'seeding' a cloud with silver salts. These salts act as 'nuclei' for the atmospheric vapour to condense upon the nuclei and for a rain drop to grow of sufficient size to fall as rain.

Water, in small volumes, forms spheres (e.g. raindrops), which give it the minimum of surface energy. As the droplets become smaller, there comes a point where there are no internal molecules, only skin. In this form it would require a very high energy to form such a droplet. In practice the water droplet would have lower energy if it were a vapour. Another way of obtaining these very small droplets in nature is to use small dirt particles as the nucleus for formation. Nature uses this principle where particles of unburnt carbon act as nuclei for the water droplet for its nucleus. The resultant condensation became known as smog, before the smokeless fuel policy. The small particles of water in vehicle spray therefore use wear debris from tyres and road aggregate to act as the nucleus for micro-droplet formation.

Consider two water droplets as shown in Fig. 10.7. The water associated with the dirt particle can be distributed as follows:

(1) uniformly over the surface of the dirt particle; (2) as many discrete separate micro-droplet attached to the side of the dirt particle; (3) as a separate water droplet,

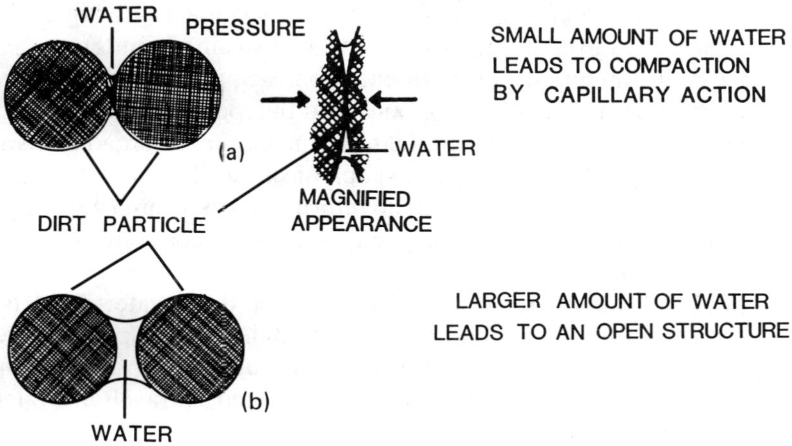

Fig. 10.7 — (a) Two dirt particles with a small amount of water will be attracted towards each other to gain the minimum surface area for the amount of water present. This will ensure that the dirt particles are in contact in three dimensions. This will be equivalent to the dirt particles being 'compacted' (better than using a steam roller). As the amount of water with each dirt particle increases, the minimum surface area for the associated water can only be obtained by pushing the dirt particles apart as shown in (b). This will not be 'self-compacting'. (Wet seaside sand will yield more water if the surface is vibrated. This water comes from the sand particles' becoming more closely packed on vibration, so that the extra water is thrown out of the 'structure'.)

completely separate from the two dirt particles. The only condition which gives the minimum surface area is condition (2). When two droplets as described in (2) come together, the dirt particle will be drawn together so that the water can form its minimum surface area. This is shown in Fig. 10.7.

Hence a water droplet with a lot of water will not be compacted by surface tension forces, whereas the dirt particles are artificially compacted more and more as the ratio of water to dirt decreases. Compaction and contraction also follow as the mud poultice dries. This comparison of the interaction of droplets with respect to their water:dirt ratio is fundamental to the micro-climates built up during the high speed road use of the vehicle. These dense compacted micro-climates are formed in several places. The important ones are those which are out of the direct line of vehicle wheel spray (Fig. 4.6 and Plates 2.1(b) and 2.1(d)). These areas also happen to be those which are hard to get clean with steam–high pressure spray cleaners.

10.11 CONCLUSIONS RELATING TO SURFACE TENSION

(1) The compaction of road dirt by the phenomenon of surface tension is important in the building-up of the aggressive micro-climate within the car body structure. This leads to the 'artificial compaction' of the mud–dirt poultices. This artificial compaction is the main reason why a soft fine spray is recommended for removing these mud compacts periodically. This form of washing can best be achieved by using a garden hose. Fierce jets of water will strip the protective coating off the car body as a result

of the long-term contact of de-icing salts and water in contact with organic protective coatings (Figs 8.18–8.21 and 3.10; see Section 4.13 steam cleaning).

(2) The small amount of water present upon each road dirt particle means that there is insufficient water to wash away and clear the road dirt which is deposited out of the wheel spray. This low water content also means that this road dirt is artificially more densely compacted by the phenomenon of surface tension into a high density mud pocket upon the (rear) wheel arch lip return which is an area out of direct water washes from the wheel, steam cleaning gun, or a garden hose, unless pointed out to the owner–operator.

(3) The slower the car travels, the larger will be these water droplets, and the higher the ratio of water to dirt which they can contain. At low speeds the rear wheel arch is only cleaned in the direct line (its trajectory) of wheel spray. The wheel spray from slowly moving vehicles has a high momentum and so travels in straighter lines and is not deviated by pressure eddies.

10.12 THE ROTATING WHEEL

The rotating wheel creates the spray which enters the car body by the mechanisms discussed in Chapters 2 and 5. This spray contains the road de-icing salts which act as an accelerator of the corrosion mechanism. In order to understand the mechanism of production of this corrosive micro-climate on internal surfaces of the car body panels and which are completely out of sight, the mechanics and velocities of trajectories from a rotating wheel will be discussed. This is difficult to show for readers who may not have had a mathematical background. Hence two explanations will be given.

A wheel of radius r, rotates at the angular velocity ω corresponding to the speed v of the vehicle. These are related by the formula

$$v = r\omega$$

(1) Method 1: A rolling wheel, whilst rotating about its axis, is also momentarily rotating about the point of contact with the road, thus giving a new radius of $2r$. Hence the velocity of the top of the wheel is $v = 2r \times \omega$, where $2r$ is the diameter of the wheel. This expression is twice as large as the tangential velocity of the tyre. This means that spray thrown off from the top of the tyre is thrown forward at twice the vehicle speed.

(2) Method 2: Let us assume that the vehicle speed is 60 mph (95 km/h). Now consider the wheel raised off the road surface and rotating; the axle is stationary with respect to the road. There will be a relative velocity of the wheel perimeter with respect to the road of 60 mph as shown in Fig. 10.8. For the car to travel at 60 mph (with respect to the road) the axle must travel at 60 mph and so the top of the wheel must have this speed added to its velocity as a result of its angular velocity. At the bottom of the tyre, the car's velocity will have to be subtracted. Hence water spray, and any of the items listed above, will be thrown off from the top of the rotating wheel at twice the vehicle speed (120 mph (190 km/h) with respect to the road). (See Sections 2.7 and 2.8). If this velocity is made with respect to the vehicle, then the undersides of the wheel arches are constantly bombarded with road dirt and low momentum spray at the vehicle speed. This happens over the entire circumference of

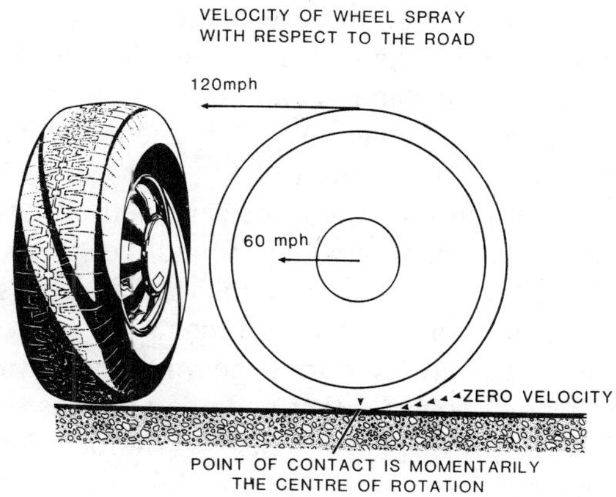

Fig. 10.8 — The easiest way to show that a rotating and travelling wheel throws spray off the top at twice the vehicle speed is to consider a bicycle wheel held just touching the surface of a water bath. The axle is stationary, but the wheel will shoot spray backwards. If the wheel travels forwards, then the point of rotation becomes momentarily the interface between the tyre and road. This means that the top of the wheel is travelling at twice the vehicle speed, because the centre of rotation is now the point of contact with the road. The spray which was shot backwards before is now lifted off the surface of the water or road and has a *zero* velocity with respect to the road or water bath.

the wheel arch including the points TUVDZYX in Fig. 2.9. Some areas, which are out of the direct trajectory of the wheel spray (W in Fig. 2.9, AB in Fig. 4.6), collect mud poultices which result from the micro-droplets when this 60 mph spray hits the car body or wheel arch. This micro-droplet spray is taken away in the pressure eddies beneath the vehicle body and deposited in areas out of the wheel trajectory spray.

One conclusion which immediately can be made from this analysis is that the motor industries' provision of partial wing–fender liners (types B, C and D in Fig. 2.9) is a waste of resources as far as anti-corrosion provision is concerned. Complete wing–fender liners (type A) must be supplied to counteract this build-up of the micro-climate on both front and rear wheel arches.

10.13 THE WINTER ENVIRONMENT

Winter motoring reflects six conditions.

(1) Fresh snow particles are thrown up by the wheels and consist of particles of ice, grit and salt which can abrade the painted surface under the wheel arch.

(2) Fresh snow which has started to thaw owing to the road de-icing salts can also be present. In this case the mush consists of ice crystals and a saline liquid which may permeate the matrix of ice crystals which act as a sponge for the salt water.

(3) There are dry road conditions where the road de-icing salts cover the surface of the road as a white powder. This powder is further pulverized by the traffic and as a result these particles may become suspended in the turbulent air around the car.

(4) Another condition is the wet road which arises through the melting of the snow due to a thaw or through rain upon the roads as outlined in (3) above.

(5) In moist but drying conditions, the winter salts dry out, and will be concentrated in the last remaining liquid to evaporate. This is usually at areas where drying is restricted. Examples include hemming folds of doors, inter-weld joints, metal overlaps, dirt particles as shown in Fig. 2.6, and in fact all those areas which produced a natural anodic site. See section 10.15.

(6) In moist foggy conditions, or when the car is left out overnight, the winter salts absorb water and make a pool of water around the salt particle (Figs 2.6 and 3.10). As a result of the hygroscopic nature of the road de-icing salts (Section 10.5), these areas remain wetter for longer periods of time.

(7) The temperature is usually low during the winter period and so the corrosion rate is low. Usually scientists consider that a temperature rise of 10°C doubles the rate of corrosion. However, this is not the case for motor vehicle corrosion as the increase in temperature will help to dry the corrosion areas. If the external crevice electrolyte is dried and eliminated, corrosion will stop (Section 2.6). Without this external electrolyte corrosion ceases (Section 7.18).

(8) The road de-icing salts contain other 'insoluble' aggregates which act as an abrasive to the underbody of the motor vehicle. Hence road dirt, ice and water abrade the underbody at the vehicle speed. These particles can be considered to be raised from the road surface by the tyre (and therefore to have zero velocity with respect to the road) and which then collide with the underbody at the vehicle speed.

10.14 ROAD DE-ICING SALTS

Road de-icing salts consist of naturally occurring sodium and calcium chlorides mixed with fine sands which also act as an abrasive owing to the relative speed of the car body and the road (Section 10.12). This mixture (Table 10.11), is spread onto the roads as a safety measure.

In some countries these salts (especially calcium chloride) are made use of to keep down the dust on minor (dirt) roads. They keep the dust down because they are

Table 10.11 — The road de-icing programme reflects society's expectation of life. The number of tons of salt used increases year by year. The amount of 'other aggregate also increases. These abrasives are responsible for the side impacts on doors in the turbulent eddy currents which exist down the vehicle body

Winter (year)	Tons of calcium chloride per ton of salt	Tons of abrasives per ton of salt	Salt	Calcium chloride	Abrasives
1969–1970	0.031	1.16	9 000 000	280 679	10 415 665
1973–1974	0.023	1.37	8 900 000	207 169	12 181 681
1978–1979	0.016	1.41	12 129 391	178 200	14 291 243

Reproduced with permission from R. Baboian, Economic and Environmental Factors in Automotive Corrosion by De-icing Salts, National Association of Corrosion Engineers, Houston, TX, 1981, p, 32, Tables 9 and 10.

hygroscopic and deliquescent (Section 10.5) thus absorbing the atmospheric water vapour at low humidities.

The reason that the chlorides calcium chloride and sodium chloride are used to keep the dust down on dirt tracks is shown in the Table 10.12. As the relative humidity in summer is round about 65%, it follows that the calcium chloride will absorb atmospheric water and therefore keep the dust down. If the relative humidity rises, then the dirt roads will appear wet as if it has rained. As salts deposit within the car body and out of sight, they will absorb water from the atmosphere throughout the year whenever the relative humidity is high enough. They will not 'dry' at the same time as a pure water droplet dries. They will therefore concentrate in the last areas to dry (Section 10.15).

Table 10.12 — Dried salts absorb water and make a pool of water at their deposition site. This is called deliquescence. The salts are 'dry' at humidities below the critical humidity

	Critical relative humidity for deliquescence (%)	
Temperature	NaCl	$CaCl_2$
25°C	76	30
10°C	76	41
0°C	—	45

Reproduced with permision from R. Baboian, The Automotive Environment, in Automotive Corrosion by De-icing Salts, National Association of Corrosion Engineers, Houston, TX, 1981, p. 7, Table 3.

The road de-icing salts melt the snow and ice because they depress the equilibrium temperature at which ice co-exists with its liquid (saline) state. If these saline additives are added to water, the temperature at which the water starts to freeze is lower than 0°C, just as the addition of tin to lead depressed the freezing point of lead solder (Fig. 6.31). Other freezing point depressants are anti-freeze for the vehicle radiator systems. The maximum freezing point decrease is concentration dependent. Thus the freezing points of water with soluble additions are usually shown in a phase diagram. One section of the phase diagram for both sodium chloride and calcium chloride is shown in Figs 10.9(a) and 10.9(b) respectively. This shows the maximum freezing point depression being − 21°C for sodium chloride and − 55°C for calcium chloride. The addition of sodium chloride to crushed ice cubes from the refrigerator at 0°C will actually produce this temperature of − 21°C. (This temperature depends on the composition of sodium chloride in the mix, and is correct for 23.31% sodium chloride (see Fig. 10 9(a)); − 55°C is for 30% calcium chloride (Fig. 10.9(b)). Hence it would appear that by the addition of the road de-icing salts to our roads we are making the roads and ice colder. In actual fact we do make them colder, but then the

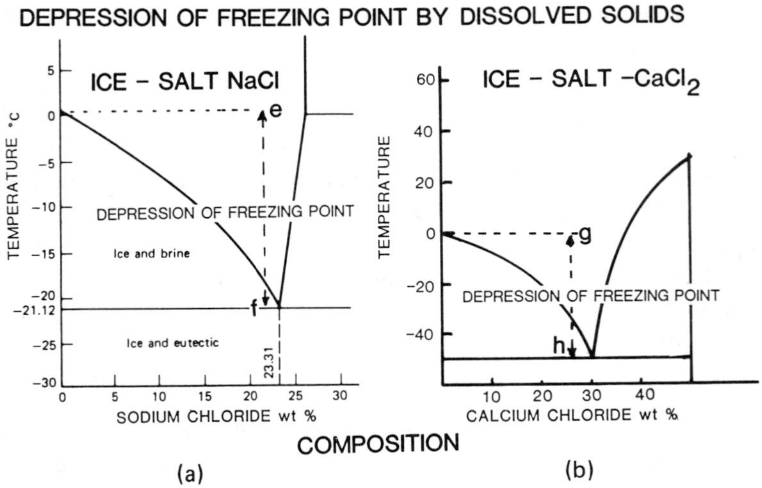

Fig. 10.9 — The effect of the road de-icing salts (a mixture of common salt NaCl and calcium chloride $CaCl_2$) is to depress the freezing point of a brine solution. The extent of the depressed freezing point is dependent on composition. The composition–freezing point relationship is shown by a 'phase diagram'. (a) The maximum depression of the freezing point (−21.12°C) is obtained with a 23.31% salt solution. (b) Calcium chloride is also used as a de-icer and also to lay the dust on 'dirt roads'. Calcium chloride is more efficient as a de-icer as the maximum depression of the freezing point is −51°C. (Redrawn with permission Kirk Othmer, Encyclopedia of Chemical Technology, Wiley, New York, Vol. 4, p. 433, Fig. 1, and Col. 21, p. 209, Fig. 2.)

road can take up heat from the surrounding atmosphere which may be at −5°C. More and more ice will melt in trying to keep the ice–salt mixture at −21°C until all the ice has melted. Hence the addition of road de-icing salts to our roads will only work if the ambient temperature is above −21°C. Calcium chloride must be used where the ambient temperature can be as low as −55°C.

10.15 CONCENTRATION OF THE ROAD DE-ICING SALTS WITHIN CREVICES

Throughout the text, and in the development of the corrosion mechanism, use has been made of the concept of the road de-icing salts' concentrating within the last areas to dry. Among the last areas to dry are (1) hemming folds or clinches (Fig. 4.4), (2) metal overlaps (Figs. 3.9 and 3.11), (3) at or between the motor vehicle panel and decorative brightwork, plastic guards or any attachment which forms a crevice, (4) behind the door–tailgate hinges, (5) beneath areas which have suffered 'Paint stripping' (Figs. 8.18 and 3.10; Section 8.13), (6) around the door–glass rubbers (Fig. 5.8) and (7) around the door returns, crevices and folds (Plates 4.7(a) and 4.7(b)).

The Figs 3.7 and 3.8 show diagrammatically what happens as water evaporates from a crevice which contains a very dilute solution of road de-icing salts within the crevice area. This is compared with a similar concentration of saline solution on a flat vertical panel (Fig. 3.7(b)). If the vehicle is stationary, then after, say 1 h of drying, the water profile would shrink as shown in Fig. 3.7(a). If the vehicle is now considered in motion, then a sudden 'bump' in the road can displace the droplet on the vertical panel to some other location. The only time that it would not be jolted into another position is if the water droplet were associated with a large particle of insoluble road aggregate or located within a crevice.

In contrast to this, the surface tension effects dictate that the saline solution remains within the crevice area, however rough the ride or 'bumping'. After further drying time, the water profile contracts until the road de-icing salt becomes concentrated in the crevice. A water droplet dissociated from any aggregate, however, may be displaced to any position, and does not end up in the same place where it can break down the protective paint film by prolonged contact with the paint film at one site (Fig. 3.10). The only stable place for the water droplet to reside without being displaced by mechanical means is in one the areas given above. These are the areas of severe corrosion, because not only are the road de-icing salts concentrated at these locations, but also the sites offer a difference in oxygen availability and an early condensation site by the crevice effect (Table 10.10). This latter fact aids the corrosion 'battery' (see Sections 2.4 and 7.9).

10.16 THE MECHANISM OF CONCENTRATION IN CREVICES

The road de-icing salts deposited upon the internal panels will, on a damp day (when the humidity is high), absorb water. The saline solutions on flat panels will become more mobile as the water and wax substrate repel each other. This mobile liquid will eventually be displaced to the lowest points in the section. These would in most cases be the inter-welds. On a (dry) day with a low relative humidity, the loss of water from the droplet would cause the salt to concentrate in the last remaining liquid to evaporate. This would be those areas where ventilation was poor and/or those areas which presented a small liquid surface to the air (e.g. clinch folds and inter-welds). Any saline solutions left on the flat panels will dry first. The last areas to dry will be the inter-welds and so these regions will suffer an increase in the concentration of road de-icing salts within the inter-welds. This effect will be more marked where the drainage holes are proud of the bottom of the clinch fold (of say the doors or the hatch-back door; Plates 2.3 and 4.6). (This is a very common design fault with most manufacturers).

10.17 TO COUNTERACT THIS PHENOMENON

To counteract this phenomenon it is necessary to wax or oil inject (Sections 3.5 and 8.8). The fluids must be thin enough to penetrate the narrow capillaries in order to exclude the solution of road de-icing salts (Figs. 3.6 and 3.8). The vehicle should be sprayed when dry (remembering that the capillaries and interwelds can be wet long after the flat panels have dried; Section 10.7). This is particularly important when the doors are sprayed.

It is important to appreciate that water, and solutions of road de-icing salts, can be excluded from the wax-coated capillary by means of a high contact angle ϕ. This means that the thickness of the coating does not play a part in this protection-exclusion mechanism but it does mean that the fluids must be thin enough to penetrate the narrow capillaries initially. See also Section 10.16.

10.18 WAX COATED CAPILLARIES AS THE ENVIRONMENT

The waxes are usually dispersed within a solvent composed of 'white spirits'. This liquid does not have a high surface tension (T is of the order of 0.026 N/m) and so the penetration of the medium (against gravity) is not good. When this value of the surface tension is compared with the value for water (T for water is 0.076 N/m) and placed in the equation which relates several physical factors together, a comparison can be obtained as to their ability to wick into the crevice. This expression is (Sections 3.6 and 10.8).

$$h = \frac{2T \cos \phi}{r\rho g}$$

Where h is the height of liquid in capillary, r is the radius of the capillary, ϕ is the contact angle which can be from 0° to 180°, T is the surface tension force, g is the acceleration due to gravity and ρ is the density.

Putting the values of T into this equation shows that the water will rise to nearly three times the height of the white spirit solution under the same conditions. The implication of this is that if water was present (in the inter-welds–clinches etc.) when the waves were injected, the water will fill the capillary in preference to the white spirit solvent which contains the waxes. When the water and solvents dry, the inter-welds would be left completely devoid of any protection. Hence the importance attached to the drying process in the rectification of rusting metal overlaps (Section 5.20).

10.19 THE MYTHS OF WATER-DISPLACING AGENTS IN WAX INJECTION

This capillary water discussed in Sections 5.20 and 10.18, within the crevice–inter-weld, cannot be displaced by the use of a water displacing agent in the wax. Indeed, a water-displacing agent will make sure that all the water in higher inter-welds, (and on the flat surfaces), is displaced to the lowest positions in the crevices (i.e. water within the clinch folds, surrounding a door), (Figs. 8.13 and 8.14). This is an area where it is certainly not advisable to have water filled inter-welds (see Section 2.4 and the general science behind this subject of corrosion in crevices).

10.20 THE IMPLICATION OF THESE POINTS

This means that in service the road de-icing salts, which dissolve in water, will be concentrated within the inter-weld capillaries which are barren of waxes. This would arise if the inter-welds were full of water at the time of injection or because the waxes did not creep into the inter-weld by virtue of their high viscosity, or the low magnitude of the surface tension T of their solvent. This concentration of road salts would arise as a result of the drying process, as explained in the Section 10.16.

10.21 SUMMARY AND CONCLUSIONS

(A) The most important point to appreciate in this discussion is that the fluids must be thin enough to penetrate the narrow capillaries to exclude water. Water has a low viscosity and a high surface tension force T and so will gain access to any capillary not coated by the waxes and produce crevice corrosion with the resultant high acidity (Section 7.9).

(B) Where the wax is giving an extra thickness layer to the electrodeposited paint for added corrosion resistance to the deposited salt particles, thickness will be important. To produce a thick film from a liquid on a vertical surface a liquid with a high viscosity is required. Thus there are two opposing requirements: a viscous medium does not give rise to capillary creep and a viscous medium does give rise to thicker films on vertical surfaces. In general there are more problems where corrosion is initiated at crevices, because of a lack of capillary creep than where initiation is due to rust-through on vertical surfaces. (This is due to the fact that flat surfaces are well ventilated and so dry a lot quicker than the unventilated inter-welds. Flat surfaces also have a thicker electrocoating film as a result of better current access. See Sections 3.2 and 9.1 *et seq*.)

(C) The scientific conclusion would be to go for the 'oil type' materials for both initial applications and re-treatment as they are non-hardening and will be taken into the capillary crevices when any water dries out. As they have this property, it should be borne in mind that water will also displace them. In general, therefore, the firm waxes, applied on a dry vehicle for preference, should be the standard first treatment and be followed by re-treatments by oil-based anti-rust compounds (should these be deemed necessary). If the vehicle is correctly designed and protected, it could be argued that these re-treatments would not be necessary.

10.22 THE MOTOR INDUSTRIES' ACCELERATED CORROSION TESTS

Over the years the raw products used in motor vehicle manufacture have undergone accelerated corrosion tests based on the traditional laboratory tests. These accelerated tests are numerous in number and some of them are as follows: (1) the ASTM 117 B salt spray test; (2) the cupric chloride accelerated acetic acid salt spray (CASS) test; (3) cyclic humidity tests, which are probably the most reliable; (4) the Corrodkote test, where ferric chloride is added to the acetic acid salt spray.

These tests are characterized by the fact that the conditions are precisely controlled so that a ranking order may be produced by any scientist, in any country, and the results will be reproducible. However, these accelerated tests, to accelerate the corrosive effects, are often remote from the conditions which will be met in practice. These accelerated tests do not therefore detect a mechanism of corrosion. Another series of tests can be used where the corrosive effects are much less and therefore a longer time is required to show up corrosion and these are the **field** or **service** tests. In these tests samples are exposed to the conditions which are met in practice. The Corrodkote test was the nearest service condition that was reached in that a corrosive acetic acid salt spray, modified with cupric chloride and ferric chloride, was splashed onto the specimen by rotating paddles and the soiled specimens were then exposed to a humid environment at high temperatures.

When it came to putting the finished motor car under an accelerated test cycle, the conditions chosen were not the ideal to simulate service conditions. The motor industries' accelerated corrosion tests bear no resemblance to the conditions which the vehicle meets on the roads. The accelerated tests conventionally used are to drive the vehicle through a salt ford or dip at low speeds (20 mph) and then to place the vehicle in a humidity chamber (100% relative humidity at 49°C) and allow the vehicle to 'sweat' for several hours. The disadvantages of this method has been given in Section 2.7 *et seq*. Other manufacturers have tested the vehicle by driving it over salt spray jets which have been concealed in the road surface, while another manufacturer has driven the vehicle over the beach when the tide has receeded. Other manufacturers have tested the vehicle by driving the vehicle with a saline solution contained in a pressurized can which was sprayed in the path of the vehicle. The vehicle therefore ran through this artificial spray. None of these methods will produce the small spray droplets required to form the low momentum micro-droplets which will so easily alter direction as a result of the pressure eddies around the vehicle body.

The large diameter spray from this 20 mph encounter with the salt 'dip' does not produce the (low momentum) fine particles of spray which will be 'self-compacting' on out of sight (or out of trajectory) ledges. Also, the high momentum (20 mph) droplets will not be drawn through the manufacturer's process drainage holes by the small pressure changes around the vehicle. This procedural test therefore leads to the acceptance of **bad design** faults which lead to premature failure by corrosion as described throughout this book. (Definition of bad design Section 2.5.1).

REFERENCES

Bishop, R. R. (1974), *New Scientist*, May 16.

Vernon, W. H. J. (1927), *Transactions of the Faraday Society*, **23** (1), 113.

11

Introduction to plastics

11.1 INTRODUCTION TO PLASTICS

Plastics are produced from crude oil, which occurs naturally beneath the earth's crust. Crude oil is a dark viscous liquid which is formed as a result of the bacterial decomposition of early (in geological time) marine life. Crude oil is a mixture of **hydrocarbons** which are a product of hydrogen (group 1 of the periodic table (Table 6.5; Section 6.4) and carbon (group 4 of the periodic table). These hydrocarbons are subdivided into **aliphatic** compounds (chain-like materials (Fig. 6.7; Section 11.2)) and **aromatic** compounds (ring-like compounds) based on benzene and its derivatives (Figs. 8.15 and 11.1; Section 11.3). The chemistry of carbon with hydrogen, nitrogen and oxygen is called **organic chemistry** to differentiate it from the **inorganic chemistry** which embraces the whole of the elements within the periodic table (including carbon, nitrogen and oxygen) and their reactions, with each other (e.g. $KClO_3$, potassium chlorate, and $NaNO_3$, sodium nitrate). Organic chemistry therefore includes the chemistry of animal and plant life (biochemistry). Organic chemistry is characterized by the covalent bond (Section 6.10) and the van der Waals bond. Ionic bonds may occur in organic chemistry as weak acids (e.g. vinegar is acetic acid, CH_3COOH). This ionizes (Section 6.13) as follows:

$$\underset{\text{covalent}}{CH_3COOH} \rightleftharpoons \underset{\text{ionic}}{CH_3COO^-} + \underset{\text{acid part}}{H^+}$$

The slight acidity (due to the 'excess' hydrogen ions) gives vinegar a 'nice' sharp taste to food and it is widely sprinkled on fish and chips (french fries).

11.2 ALIPHATIC COMPOUNDS

In aliphatic compounds the carbon and hydrogen combine to give long hydrocarbon chains with the general formula $C_nH_{(2n+x)}$ where n is the number of carbon atoms in the chain and for saturated hydrocarbons $x = 2$ (Figs. 6.7 and 6.16). As n increases so do the melting and boiling points of the hydrocarbon (see Table 11.1). **Saturated** cyclic hydrocarbons (like cyclohexane Fig. 8.15(c)) are aliphatic.

11.3 AROMATIC COMPOUNDS

In aromatic compounds the carbon and hydrogen are arranged in rings: the basic unit benzene C_6H_6 has one ring of six carbon atoms and six hydrogen atoms (Fig. 8.15(a)). Several of these rings can be joined together as shown in Fig. 11.1. These are usually referred to as 'cyclic' compounds where only one ring is involved or dicyclic if two rings are joined together. These **unsaturated** cyclic rings (i.e. containing double bonds) can be joined to aliphatic long chain groups, Fig. 11.5(d), without breaking the ring double bond. In writing complex chemical compounds, these hydrocarbon parts are further shortened by the terminology R_1, R_2 etc. (see Section 11.7). The mechanical properties of the hydrocarbons improve as the number of rings increases. See Section 11.7(4).

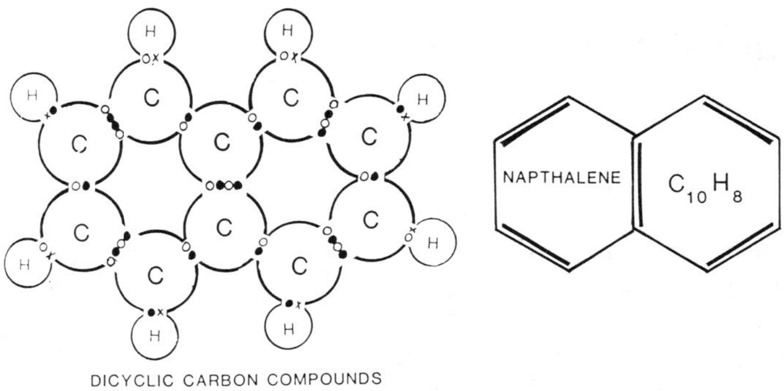

Fig. 11.1 — These are **aromatic** compounds based on the benzene ring, either singly Fig. 8.15(a) or as here with two rings.

11.4 FRACTIONAL DISTILLATION

The raw material for the plastics industry is the crude oil. This oil contains many hydrocarbons of differing number of C atoms and differing boiling points. The first requirement is to separate this oil into its constituent parts (by chain length and boiling point) by a process called **fractional distillation**. Fractional distillation produces the constituents shown in Table 11.1.

The boiling and melting points are dependent on the chain length. The boiling and melting points of a substance are the points where the thermal energy is capable of breaking the weak van der Waals bonds (Fig. 6.16). Thus the thermally dependent mechanical properties increase with increase in chain length.

11.5 NOMENCLATURE AND THERMOPLASTICS

The long-chain aliphatic hydrocarbons are usually referred to as **alkanes** when they are fully saturated (i.e. the hydrocarbons (Fig. 6.7) follow the formula $C_n\ H_{(2n+x)}$ where x is equal to 2 exactly. When x is not equal to 2 it means that either (a) the hydrocarbon is composed of cyclic compounds (Fig. 8.15(c) and 11.1) or (b) double bonds are present in the molecule (Fig. 11.2).

Table 11.1— Fractional distillation of crude oil

Fraction		Number of carbon atoms per molecule	Boiling point (°C)	Name and use
Gases & Volatile liquids		1–5	<40	Fuel gases, solvents
Examples	CH_4	1	−161	Natural gas (methane)
	C_2H_6	2	−90	Ethane
	C_3H_8	3	−42	Propane gas, bought in cylinders
	C_4H_{10}	4	0	Cigarette lighter fuel
	$C_{10}H_{22}$	10	174	Liquid solvent
Gasoline		5–10	40–190	Petrol, naptha etc.
Kerosine		11–14	190-260	Paraffin oil; jet fuel
Light fuel oils		15–19	260–330	Diesel oil (hence some waxing in cold weather)
Light lubricating oils		19–30	330–400	Medicinal paraffin oil to paraffin wax (now solids appear)
Bitumen		>30	>400	Heavy fuel oils to road asphalt

30 carbon atoms is a relatively short chain length: see section 11.21.

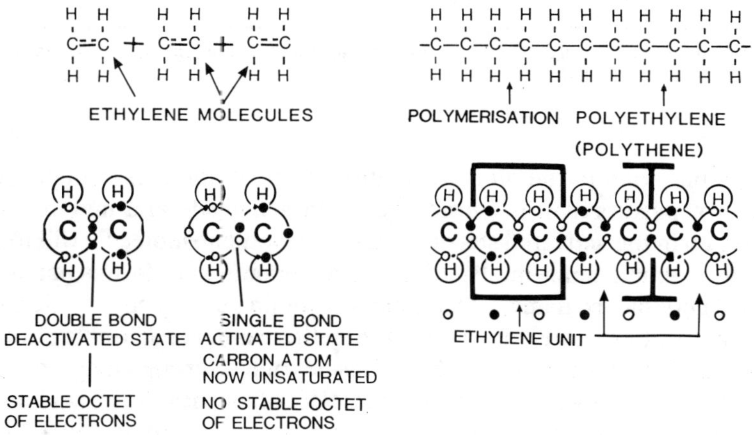

Fig. 11.2 — The structure of the **alkenes** is represented by the ethylene (ethene) molecule. This is an important molecule as the double bond can be broken and the ethene units can be joined up to form a polymer. This joining up by the breaking of the double bond is a characteristic of the **aliphatic** molecules (Section 11.2) and not of the **aromatic** double bond as in benzene. The majority of the thermoplastics (Section 11.8) can be considered to be formed from the $CH_2{=}CH_2$ unit, as shown in Fig. 11.5. This fundamental unit $CH_2{=}CH_2$ can be considered as $CH_2{=}X{-}C{-}Y$, where X and Y can be changed as shown in Fig. 11.5. In this form the $CH_2{=}CH{-}$ group is termed 'vinyl' and gives a number of important compounds.

The simplest hydrocarbon with a double bond in the molecule is ethene (C_2H_4), which can also be represented as $CH_2{=}CH_2$, the = sign representing the (covalent) sharing of four electrons between the carbon atoms. These straight-chained C_nH_{2n} molecules are given the general name **alkenes** to differentiate them from the fully saturated **alkanes** $C_nH_{(2n+2)}$. The formulae of some hydrocarbons are shown in Table 11.2.

Table 11.2 — Nomenclature

Alkanes (C_nH_{2n+2})		Alkenes (C_nH_{2n})		Unsaturated cyclic compounds	
Formula	Name	Formula	Name	Formula	Name
CH_4	Methane	—	—	C_6H_6	Benzene
C_2H_6	Ethane	C_2H_4	Ethene or ethylene	$C_{10}H_8$	Naphthalene
				THESE ARE AROMATIC	
C_3H_8	Propane	C_3H_6 or $CH_3{-}CH{=}CH_2$	Propylene[a]	Saturated Cyclic compounds	
C_4H_{10}	Butane	C_4H_8 or $CH_3{-}CH{=}CH{-}CH_3$ or $CH_2{=}C(CH_3){-}CH_3$	Butylene[a]	**THESE ARE ALIPHATIC** C_6H_{12} Cyclohexane C_3H_6 cyclopropane of propylane C_3H_6	

a Standard IUPAC nomenclature for these are propene and butene, but polypropylene has been in use too long for this to be appropriate here.

This double bond in the aliphatic hydrocarbon compounds is important in the technology of plastics as it is considered as an **active site** and may be chemically broken. When in the activated state (Fig. 11.2) the ethene molecule will join up with other separate ethene units–molecules (called **monomers**) to form a **polymer** (*mono* means one, *poly* means many). This chemical joining up (by the covalent bond), is called **polymerisation** (note: in polymerization there is never a double bond between the carbon and hydrogen, nor can the chains be joined through a hydrogen atom of the ethene). This polymerization produces a class of plastics which can be re-formed several times by heating and moulding to shape and cooling to 'freeze' them into the new shape. To these plastics a special name is given—**thermoplastics**. If the plastic molecule is bonded in three dimensions by the covalent bond, so that each hydrocarbon chain is linked through another carbon atom or a short chain (usually being produced from an aldehyde (Fig. 11.3)), then these resultant plastics cannot be reheated to reset them in another configuration. They can be heat set once and once only (see Fig. 11.3). These materials are called **thermosets**. The condensation reaction cross-links the carbon chains through the $-CH_2-$ group from the formaldehyde ($H{-}CHO$) as shown in Fig. 11.3(b).

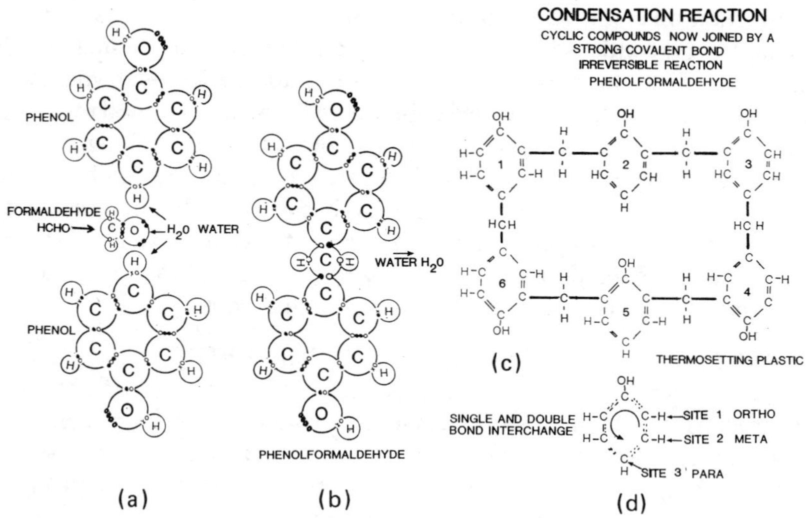

Fig. 11.3 — Fig. 6.16 showed how the scientist modelled the weak bond (van der Waals bond) which could be broken by thermal energy. (b) A stronger covalent bond can be produced which bonds the two phenol (a) chains together. This is termed a condensation reaction, which gives water (b) from the two hydrogens (from the cyclic hydrocarbon phenol m.pt. 43C) when combined with the oxygen from formaldehyde (gas b.pt. −21C). (c) Phenolformaldehyde is formed (NO m.pt. and used for ashtrays). These phenols, designated 1–6, can be further linked via more formaldehyde to other phenol units through the 'active sites' 1 and 3 (d) whose technical terms are **ortho** and **para** positions. Cross linking occurs in 3 dimensions (see also Fig. 11.8). Fig. 11.3(a) and (b) shows the electron configuration produced by this cross-linking. Fig. 11.4 shows that there is no glass transition temperature associated with this covalent cross linking bond. Aliphatic chains (even if they contain double bonds) **cannot** be cross linked with formaldehyde. This cross linking is particular to the 'double bond' of the **benzene** ring (d).

11.6 CRACKING, POLYMERIZATION AND CONDENSATION REACTIONS

11.6.1 Cracking

The fractional distillation process only separates the hydrocarbons into the components of the crude oil according to their boiling points. Often there are hydrocarbons in the crude fractions which are not as commercially saleable as some other hydrocarbons of shorter chain length. In order therefore to reduce the chain length of the hydrocarbon, a process called cracking is used to produce the more saleable hydrocarbon(s). Catalytic cracking therefore reduces the length of the carbon chain.

11.6.2 Polymerization

Polymerization is the reverse process of cracking. Here the length of the chain is increased to make a longer hydrocarbon chain molecule.

11.6.3 Condensation reaction

If the joining up of the organic chain involves the production of water, the reaction is called a condensation reaction (*n* indicates a large number of molecules):

$$\underset{\text{Glucose (carbohydrate)}}{nC_6H_{12}O_6} \rightarrow \underset{\text{cellulose}}{(C_6H_{10}O_5)_n} + \underset{\text{water}}{H_2O}$$

The transformation of glucose into starch (basis of food) and cellulose (plant tissue) is an example of a condensation reaction in plant life. Reactive adhesives (Section 11.26.2) make use of this condensation reaction to form a load-carrying medium.

Fig. 11.3 shows two hydrocarbon chains being covalently bonded together by the strong covalent bond with the production of water. This cross-linking is undertaken in the repair of punctures. The cross-linked patch or **vulcanized** patch will not come off, even if the temperature of the tyre rises, because it is bonded firmly with the strong covalent bond. A cycle patch makes use of the weak van der Waals bond which weakens as the temperature rises. This form of repair is therefore illegal on the motor vehicle tyre which becomes fairly hot at high speeds.

11.7 FUNCTIONAL GROUPS IN ORGANIC CHEMISTRY

Organic chemistry has the functional groups shown in Table 11.3 to the hydrocarbon. These characterize their chemistry and give rise to their nomenclature. Hence the gas CH_4 is called methane, the compound CH_3-OH is methano**l**† or methyl alcohol (the poisonous alcohol), $H-CHO$ is methana**l**† or formaldehyde (used to pickle parts of the human body for preservation), and CH_3-COOH is ethanoic acid† or acetic acid (vinegar). CH_3-Cl is chloromethane† or methyl chloride and CH_3-CHO acetaldehyde. The alcohol in beers, wines and spirits is ethanol† or ethyl alcohol C_2H_5-OH. As a point of interest the alcohol C_2H_5-OH oxidizes in the human body to produce acetaldehyde CH_3-CHO, which is responsible for the 'hang-over' if too much alcohol is consumed. Methyl alcohol CH_3-OH is added to ethyl alcohol (with a dye) (meths) which will oxidize to give formaldehyde, which is the liquid medical teaching schools use for preserving (pickling) organs. This reaction in the brain is very dangerous and results in a 'pickled' brain.

Table 11.3 — Functional groups

Group	Name	Group	Name
$-CH_3$ R[a]	Methyl	$-OH$	Alcohol
$-C_2H_5$ R[a]	Ethyl	$-OH$	Phenol when attached to a cyclic compound
$-NH_2$	Amine	$-Cl$	Chloride
$=CO$	Ketone	$-CHO$	Aldehyde
$-CN$	Nitrile	$-C_6H_5$ Ar[a]	Phenyl
$-COOH$	Carboxyl	$-O-$	Ether
$-C_nH_{(2n+1)}$	Alkyl	$RCO-OR_1$	Ester (see below text)

For the epoxy group. (Oxygen linked to two carbon atoms), see text.
Acidic character produced by the ionization of $-COOH$ to $-COO^- + H^+$.
a Standard IUPAC nomenclature and abbreviations.

These groups can be combined as follows: $CH_3-CO-C_2H_5$ is ethyl methyl ketone and $CH_3-O-C_2H_5$ is ethyl methyl ether and so forth. Hence organic chemistry builds up compounds by combining various 'units' together. The terminology of the compound arises from the combined fundamental units. Some compounds have popular usage terms which are a direct derivation of their chemical

† Standard IUPAC nomenclature and abbreviations.

name, e.g. polythene derived from either polyethene or polyethylene. Many of these are given in Fig. 11.2.

11.7.1 Functional groups in elastomeric chemistry

The rubber industry was well established before the plastics industry. The plastics industry was not a 'natural' product, although the raw products were naturally occurring. Considerable research and development work was required to exploit these distilled compounds which were a by-product of the petroleum industry. The rubber industry, in contrast, was based exclusively on the natural rubber latex which was tapped from the rubber tree. Initially these latexes were not perceived to be polymeric materials. To-day, the natural rubbers are used alongside the synthetic rubbers (often referred to as **elastomers**). Plastics, paint coatings and adhesives all use the 'basic' chemical units and associated chemical technology. If the various constituents making up adhesives are considered, the general pattern is as follows.

(1) **Butane** C_4H_{10} is lighter fuel: **Butadiene** is used to make synthetic rubber. The monomer is C_4H_6 or $CH_2{=}CH{-}CH{=}CH_2$ and the polymer is depicted as $(-CH_2{-}CH{=}CH{-}CH_2-)_n$ where n indicates many chains of $-CH_2{-}CH{=}CH{-}CH_2$.

(2) **Natural rubber** is a polymer of **isoprene** which is $CH_2{=}C(CH_3){-}CH{=}CH_2$. This is the monomer of rubber and so the scientist represents the polymer as the breaking of the two double bonds at either end as $(-CH_2{-}C(CH_3){=}CH{-}CH_2-)_n$ where n indicates many chains of $-CH_2{-}C(CH_3){=}CH{-}CH_2$ (See Fig. 11.2).

(3) **Neoprene** is made from the monomer **chloroprene**, Polychloroprene rubber is chains of

$$(-CH_2-\underset{\displaystyle Cl}{\underset{|}{C}}=CH-CH_2-)_n$$

(4) **Nitrile rubber** is an alternating copolymer of butadiene and acrylonitrile and consists of chains of

$$\underbrace{(-CH_2-CH=CH-CH_2}_{\text{butadiene}}\;\underbrace{-CH_2-\underset{\displaystyle CN}{\underset{|}{CH_2}}-)_n}_{\text{acrylonitrile}}$$

These **rubbers** are noted for their oil resistance and are extensively used for petrol–gasoline hoses. They are also used in the production of adhesives. A high impact plastic can be made by adding another copolymer (styrene Fig. 11.5(e)) to give **ABS**÷Acrylonitrile butadiene styrene. This styrene addition stiffens up the chain, Fig. 11.7.

(5) **Polysulphide** consists of the unit $-C-S-S-$. These have very good resistance to ozone and ultraviolet. They are used in the automotive and construction industries as sealants.

(6) **Polyurethane** consists of the fundamental unit $-O-CO-NH-$.

(7) **Polyurethane rubber** is chains of

$$(-R_1-O-CO-NH-R_2-NH-CO-O-R_3-)_n$$

R_1 and R_3 are aliphatic chains such as $-CH_3$ or $-C_2H_5$; R_3 is a cyclic compound such as C_6H_5-. The number of carbon atoms in R_1, R_2, and R_3 is dependent on the type of rubber and the physical properties required from the compound. The range of properties are shown below.

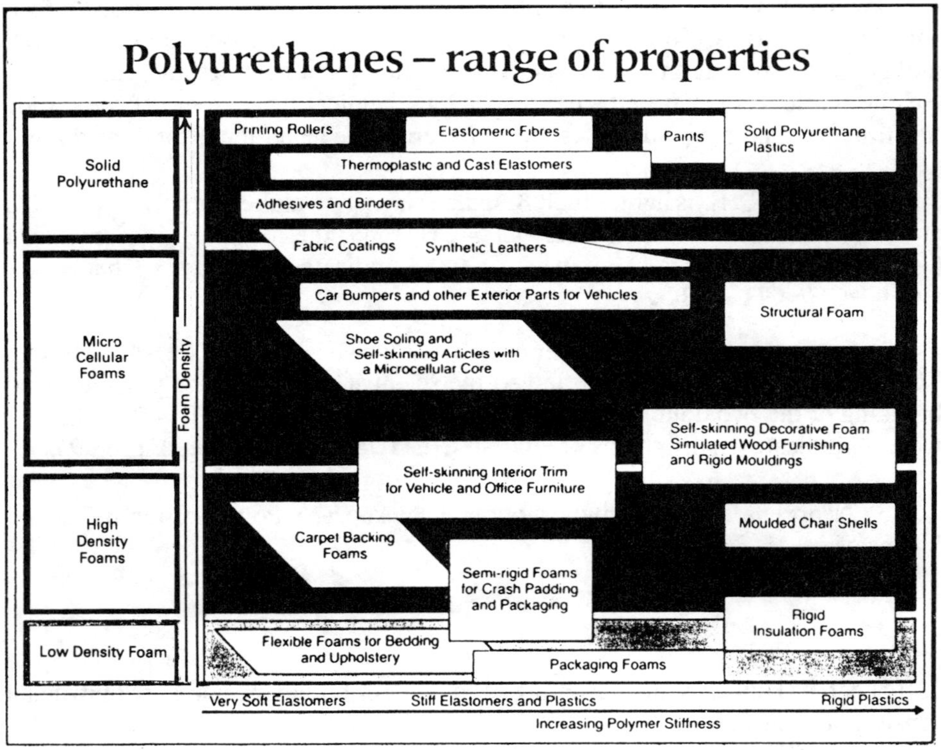

Acknowledgement: ICI Plastics

(8) **Nylon** is the generic term for the **polyamides**. The general polyamide consists of the unit $-CO-NH-$ joined up to hydrocarbon chains $-R_1-CO-NH-R_2-$. Thus nylon can be written as

$$(-NH-(CH_2)_6-NH-CO-(CH_2)_4-CO-)_n$$

or

$$(-NH-R_1-NH-CO-R_2-CO-)_n$$

The numbers of carbon atoms between each nitrogen atom pair govern the type of polyamide produced and the resultant properties, e.g. illustrated in nylon 6.4.

(9) The similarity between **polymethyl methacrylate** (PMMA) and **polyvinyl acetate** (PVA) can be seen by comparing the following representations of the two polymers: PVA is chains of

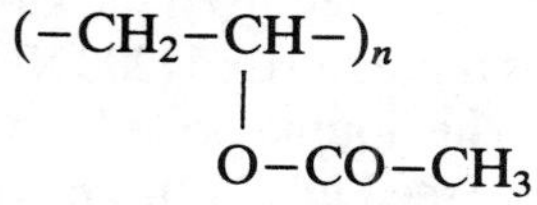

and PMMA is chains of

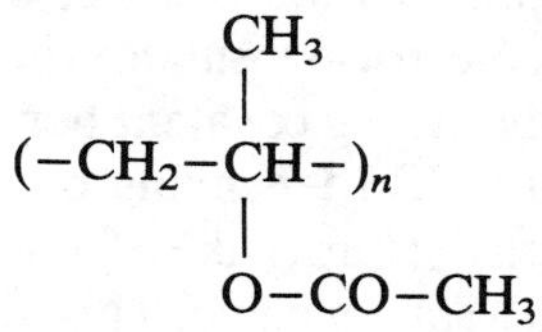

Hence copolymers may take on the names of the two polymers producing the copolymer. Ethylene–vinyl acetate copolymer would be chains of

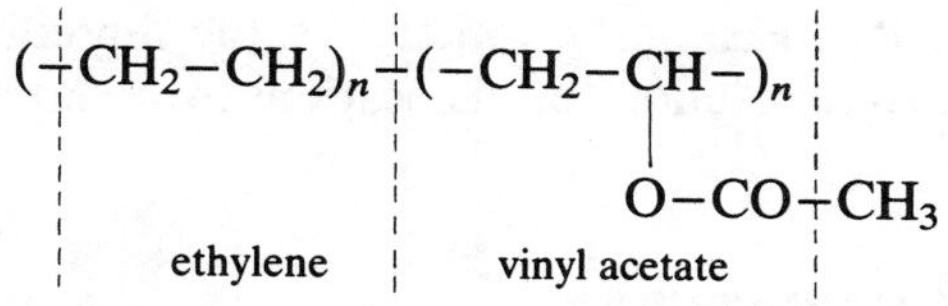

(10) An **ester** is the result of the reaction between an alcohol and an acid(equivalent to an acid plus base reaction). An ester is therefore the **alkyl** derivative of the acid. The fundamental unit of an ester is $-O-CO-$. In general terms it would be written $R_4-CO\text{-}OR_5$ where R_5 is from the alcohol and R_4 from the acid, in this case acetic acid. $CH_3-CO-OC_2H_5$ ($R_4=CH_3$; $R_5=C_2H_5$) is ethyl acetate. A **polyester** contains the characteristic $-O-CO-$ unit. A polyester is chains of these esters and they are written in general terms as

$$(-O-R_7-O-CO\text{-}R_8-CO-)_n$$

(11) **Epoxy resins** and **epoxides**: only when an oxygen atom is cyclically linked between two carbon atoms is the compound's name prefixed by the term **epoxy**. Thus ethylene oxide

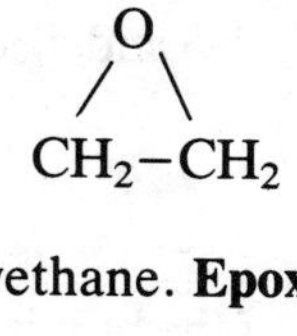

is epoxyethane. **Epoxide resins** are

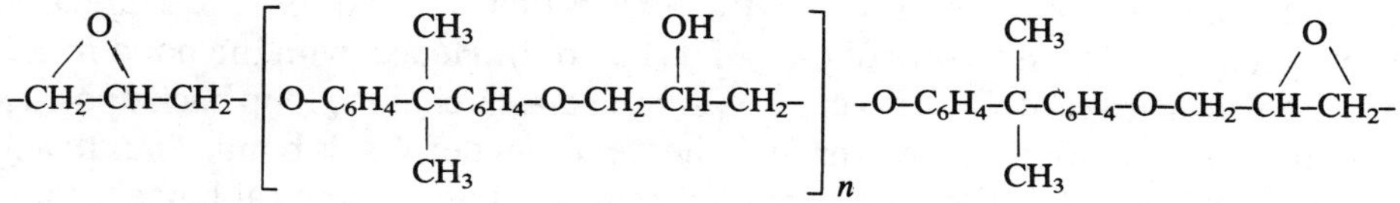

11.8 POLYMERS AND POLYMERIZATION

Polyethylene (PE) (polythene or, correctly, **poly(ethene)** is the simplest polymer to explain the fundamentals of the polymerization reaction and the properties of the class of materials which are called thermoplastics. PE is derived from the gas ethene (or, to use its older name, ethylene) C_2H_4 when the gas is pressurized (200 N/mm^2) at a temperature of 220°C in the presence of a catalyst. The double bond between the two carbon atoms of ethene is opened up and activated (see Fig. 11.2). The resulting product PE is a solid, stable, impermeable to water, unaffected by salt water, resistant to bacteria and acids. As a result of these properties, PE is used as a damp-proof membrane in the building industry. PE is an electrical insulator (in solid and molten states) because the electrons are fixed by the strong covalent bond into the molecule. PE has a melting point of between 70–90°C (depending on the chain length) and fairly readily catches fire. When PE burns, it burns like a 'super paraffin wax'. If the oxygen supply is restricted it burns with a smoky flame. This sooty flame arises from the fact that the hydrogen has 'first call' on any oxygen. PE is degraded by the ultraviolet radiation from the sun which turns PE products into rather brittle products. PE can be stabilized to ultraviolet radiation by the addition of carbon black. This is mixed into the molten PE plastic rather like the ingredients of a pudding are mixed. There are no solid solutions or dislocations involved in plastics (see Section 6.20 *et seq*).

11.9 GLASS TRANSITION TEMPERATURE

The hydrocarbon chain is held together through the carbon atoms by the strong covalent bond. A very weak 'residual' bond is left to 'bond' one chain to the adjacent chain. This is the van der Waals bond (Section 6.12). The glass transition temperature T_g is the temperature at which bond rotation and 'slip' between the weakly bonded molecular chains can take place as a result of thermal vibration. If the plastic is used at a temperature T_r (T_r is the temperature of use or room Temperature), and if T_r is below the glass transition temperature T_g, this relative motion is 'frozen out', and the polymer is fairly rigid. Hence the mechanical properties of the polymer are dependent on the difference between the glass transition temperature T_g and the temperature of use T_r as shown in Fig. 11.4: if $T_g = T_r$ the plastic is elastic or rubbery; if $T_g \ll T_r$ the plastic is rubbery or liquid; if $T_g \gg T_r$ the plastic is stiff, hard and may be brittle. The glass transition temperature increases with increase in the length of the hydrocarbon chain and the existence of, and length or size of, the side chains attached to the 'straight' hydrocarbon chain. These side chains give rise to a different naming of the hydrocarbons and this is shown in Figs. 11.4 and 11.5 and discussed in Section 11.7.

11.10 THE PHYSICAL STRUCTURE OF A PLASTIC

The plastic molecular chain is not a straight line. The direction of the bonds from the carbon atoms are such that they go to the corners of a tetrahedron. Their direction is shown clearly in the structure of diamond (Fig. 6.9). Hence, with the bond rotation around each carbon atom, it is very easy for the plastic chain to loop back upon itself and physically to bond to itself through the weak van der Waals bond. Thus to a first approximation each carbon atom can be considered as a 'chemical hinge' as it can rotate about this covalent bond. However, as a first principle, we shall consider them as semi-rigid 'matchsticks' in order to discuss the effect of thermal energy on the structure. The chemical hinge concept will somewhat modify the modelling of the plastic. Fig. 11.4 plots the specific volume (the reciprocal of the density) of the plastic with temperature as the plastic is slowly heated. Over the whole temperature range, the weak van der Waals bonds are broken at the temperature corresponding to their strength. (Note: the covalent intramolecular C–C or C–H bonds are not affected.) In order to gain an insight into the scientist's model of the physical structure of the plastic it is best to consider what happens to the plastic as it cooled down from the molten state. In the molten state, these matchsticks will have a lot of vibrational energy, which will allow them to rotate, but owing to their length there will be little translational movement or diffusion. The strongest van der Waals bond at this temperature will tend to align the matchsticks to get the best packing. If these matchsticks are tipped out onto the desk (which is analogous to a rapid or fast cool), there will be several sticks which cross each other, and there will be no obvious packing order. This lack of packing order is called **amorphous** and is the structure of the **'frozen liquid state'**. If the analogy is carried further and the plastic could have been slowly cooled (i.e. the matchsticks vibrated as they were tipped onto the desk and the desk was shaken when on the matchsticks were on the desk), they would have

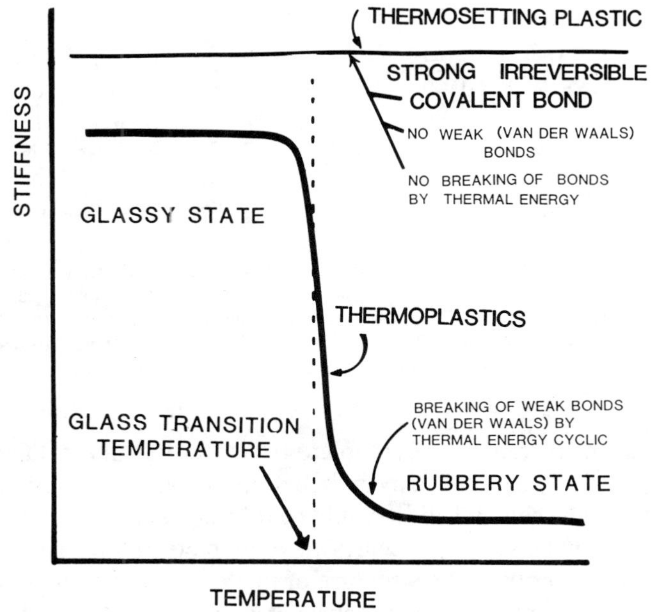

Fig. 11.4 — The scientific modelling of stiffness, as related to bond strengths as a function of temperature, shows that there is no glass transition temperature associated with covalently bonded cross-linking. Hence all (totally) covalently bonded structures are rigid. Elasticity or rubbery features are conferred on the plastic if the adjacent chains are bonded by a weak van der Waals bond (Figure 6.16) instead of the covalently bonded chains shown in Fig. 11.3. The extent of this elasticity or rubbery state is dependent on the glass transition temperature (see Fig. 11.5).

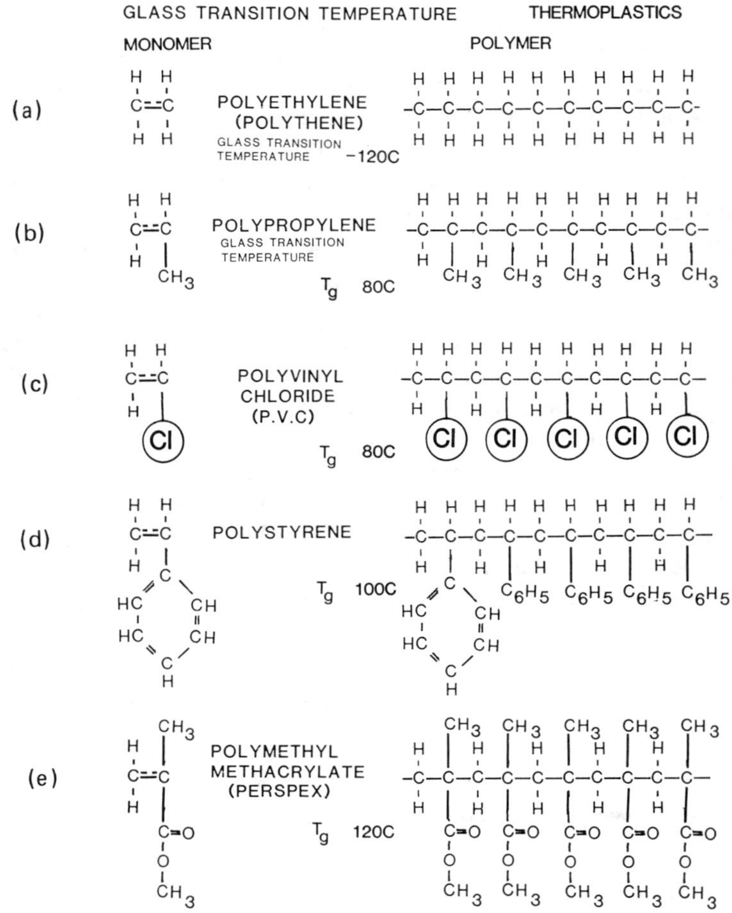

Fig. 11.5 — The glass transition temperature of various plastics which can be considered to be based on the ethylene unit as written as $CH_2{=}CH{-}Y$. The corresponding increase in the glass transition temperature is shown. This means a corresponding increase in rigidity of the plastic at room temperature. If the ethylene unit is considered as $CH_2{=}CX{-}Y$, then both X and Y can be changed (compare the structures of propylene and Perspex).

aligned themselves in regions of complete order, rather like soldiers standing to attention on parade. This is called a **crystalline region**, and if the ordered regions ran throughout the whole solid plastic it would be fully crystalline. Because of the size of these polymeric 'matchstick' molecules this complete re-arrangement to get full crystallinity is not possible. The structure of the 'crystalline' plastic is composed of regions of crystalline 'spheres', called **spherulites** within amorphous areas (regions of complete disorder). This is shown in Fig. 11.6.

Now, if these matchsticks are barbed (which is equivalent to the hydrocarbon's having a chemical side chain along its length), this re-arrangement to form the crystalline regions becomes *more* difficult. This means that organic thermoplastics with side chains are completely amorphous. Furthermore, these barbed chains cannot move easily relative to each other, and therefore have a higher T_g and

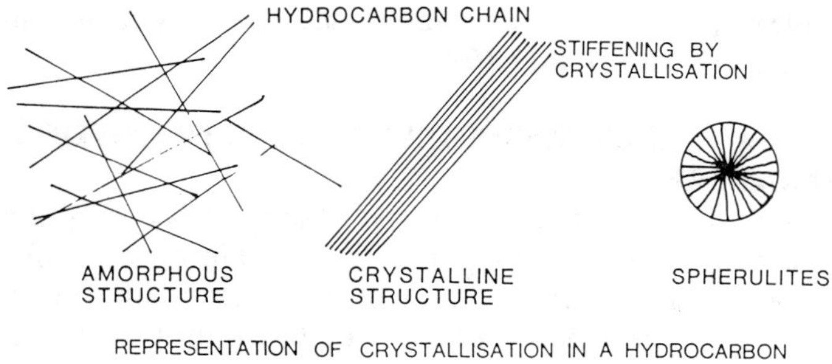

Fig. 11.6 — The concept of amorphous and crystalline structures is very important to the strength-related parameters of plastics. The small regions of crystallinity within the plastic are called spherulites. Note that there are no dislocations in a plastic, even in the crystalline regions. This means that plastics do not work harden.

stronger mechanical properties at all temperatures. The amorphous structure confers a 'glass-like transparency' to the 'frozen' plastic, while a crystalline structure tends to give opacity or milkiness to the plastic. If the matchsticks all align with each other, there is more chance that a stronger van der Waals bond will be formed. Hence one of the aims of the plastics technologist is to try to achieve some sort of crystallinity in the structure to give added strength. High density polyethylene (HDPE) buckets are stronger and more rigid than low density polyethylene (LDPE) buckets. Hence crystallinity (Fig. 11.7) is one of the many parameters which confer

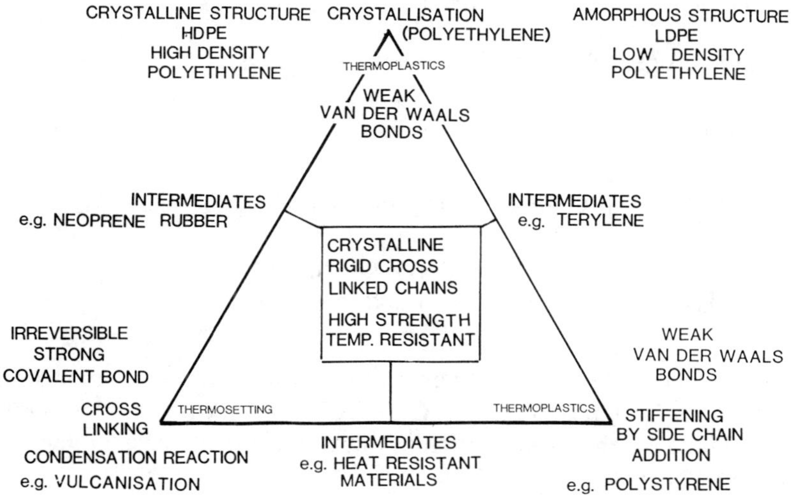

Fig. 11.7 — The properties of a plastic are scientific modelled in several compartments of this triangle. The triangular figure brings together the various models and relates the models to the ultimate strengths. The strength properties can be achieved by a mixture of the processes outlined. Hence a plastic with some cross-linking and some weak van der Waals bonds will be fairly stiff, pliable and heat resistant. However, it will soften on heating, but it would not be expected to lose its shape completely. A plastic completely cross-linked would remain rigid right up to the onset of chemical decomposition and the onset of charring.

strength to the plastic. Other methods of improving the strength are summarized in Fig. 11.7 and discussed in Section 11.16 *et seq*.

11.11 HEATING A PLASTIC THROUGH ITS GLASS TRANSITION TEMPERATURE

If a plastic is heated, then the weakest van der Waals bonds are the first to break. There will not be many of these and so there is no great increase in volume. As the temperature is increased still further, the next strongest bonds are broken. At a certain temperature, where the majority of bonds are broken, there will be an increase in volume. This temperature corresponds to the glass transition temperature T_g. Naturally this temperature will be higher if the hydrocarbon chain is longer. Above the glass transition temperature the number of bonds broken is smaller, but they are stronger than those broken at lower temperatures and so the volume increase is large. This is shown in Fig. 11.4.

11.12 STIFFENING BY SIDE CHAIN ADDITION

Side chain 'grafting' is one of the best methods for increasing the mechanical properties of the plastic and is shown in Figs. 11.5 and 11.7. This not only strengthens the plastic but raises the glass transition temperature. In order to consider this phenomenon several plastics have been chosen where the fundamental unit is the ethene molecule C_2H_4, which has the structure as follows:

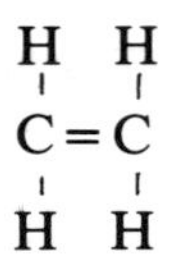

One of the hydrogen atoms can be replaced as follows:

replaced by	monomer	polymer
CH_3	propylene	polypropylene
Cl	vinyl chloride	polyvinyl chloride
C_6H_5	styrene	polystyrene

If propylene $CH_3{-}CH{=}CH_2$ is taken and one other hydrogen atom on the same hydrocarbon chain is replaced by $-CO-OCH_3$ we get polymethyl methacrylate (PMMA). Fig. 11.5 shows the structural units and the increase in glass transition temperature which results.

11.13 DEGREE OF CROSS-LINKING

If the plastic hydrocarbon is considered as a series of partial hinges with complete rotation at each carbon atom (Section 11.10) then it follows from this model that the plastic will be easily deformable. The readiness with which the plastic molecule can 'unhinge' and rotate to accommodate the deformation is dependent on the number and size of the side branches, e.g. whether they are $-H$, $-CH_3$, $-C_6H_5$, or $-CO-O-CH_3$ groups (Fig. 11.5). This rotation is further reduced if the plastic chain is cross-linked periodically throughout its length as shown in Fig. 11.3. If the accompanying hydrogen atoms are omitted, this can be depicted as shown in Fig. 11.8.

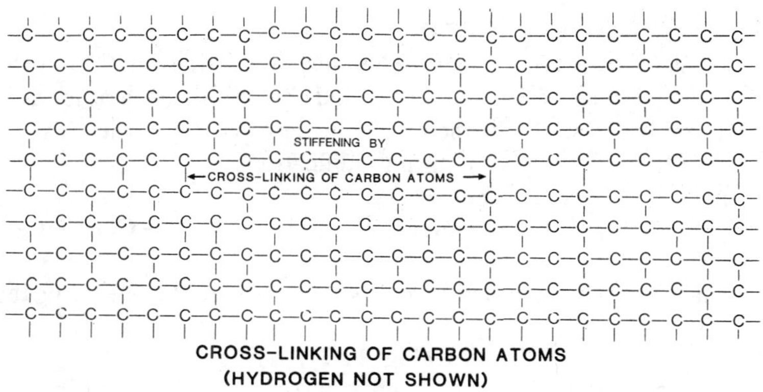

Fig. 11.8 — The scientist's model of a cross-linked structure is depicted as a three-dimensional array of covalently bonded carbon atoms. Note the similarity to diamond (Fig. 6.9(d)). In this diagram, for simplicity, the array is shown without the hydrogen atoms which are bonded to those carbon atoms which do not have four nearest covalently bonded neighbours. The actual compounds used in thermosetting reactions are a mixture of aliphatic and aromatic hydrocarbons. This produces a very complex three-dimensional cross-linking.

11.14 THERMOSETTING PLASTICS

Thermosetting plastics are formed when smaller or shorter hydrocarbon chains are irreversibly cross-linked into a colossal three-dimensional molecule by the joining up of the individual molecules into a thermally stable network. One method of achieving this cross-linking is by the condensation reaction shown in Fig. 11.3. This mechanism should be contrasted with the thermoplastic reaction which is thermally reversible.

The condensation reaction usually requires an **aldehyde** unit (ending in $-CHO$), e.g. $R-CHO$, which provides the oxygen and an $R-CH-$ group so that the two adjacent chains may be linked by the covalent bond. The oxygen removes the two hydrogen atoms as water so that the $R-CH$ group from the aldehyde can covalently bond into the adjacent chains, thereby producing the rigidly linked three-dimensional structure. Certain wood glues use formaldehyde $H-CHO$ ($R=H$) as the 'activator' in order to cross-link the glue and to form a load-bearing bond.

There are no van der Waals bonds of any consequence in a thermosetting unit ar so there is no glass transition temperature and no softening on heating. Thermo ting plastics can be heat set once and once only. They do not burn a temperatures because there are no short-chain hydrocarbons to fragm produce an inflammable vapour. (A flame is nothing more than burning g plastic does not form a vapour, because it is bonded in three-dimensions by strong bonds, it will not burn.) Thermosetting plastics just char, producing carbon. Examples of thermosetting plastics are phenol formaldehyde, melamine formaldehyde and urea formaldehyde. All contain the aldehyde derivative in their names. Other materials which use other mechanisms to cross-link are epoxy resins polyesters (unsaturated polyesters), polyurethane resins and polyurethane foam.

11.15 MECHANICAL STRENGTH

There are no dislocations in a plastic. The mechanical deformation of the plastic is modelled by the slipping of the weakly bonded chains pass each other as the van der Waals bonds are broken. Hence the mechanical properties of the plastic depend on one or more of the following structural modifications:

(a) number of side chains (Fig. 11.5);

(b) size of side chain and whether it is $-CH_3$, $-C_6H_5$, $-Cl$ or $-O-CO-CH_3$ (Fig. 11.5);

(c) the glass transition temperature T_g with respect to the operating temperature T_r (Fig. 11.5);

(d) the degree of crystallization (Fig. 11.7);

(e) the degree of cross-linking, which is particularly important in thermosetting adhesives–compounds (Fig. 11.8).

11.16 PLASTICIZERS, FILLERS AND STABILIZERS

In order to alter the properties of the plastics, certain additions may be made. These additives are called plasticizers. Plasticizers are usually a saturated long-chain paraffin (i.e. a hydrocarbon chain having no double bonds ($x=2$; Section 8.9)) of high melting point and low volatility (low saturated vapour pressure) whose function to produce a final product which is flexible while at the same time being dimensionally stable. PVC is used for the 'plastic' seat covering of motor vehicles, chairs and as an imitation of leather. PVC is also used for domestic window frames where it is termed 'UPVC'. PVC without a plasticizer would be hard and non-flexible without any 'leathery' feel. The flexibility is supplied by the plasticizer. Over a long period of time this plasticizer is driven out of the PVC owing to the high temperatures reached when a vehicle is left locked in the midday sun. These paraffinic substances are deposited on the windscreen, giving the waxy film on the inner window surfaces. The PVC seat material becomes 'harsher' and loses the ability to bend back on itself and so cracks. The PVC used for window frames does not contain a plasticizer. It is therefore termed unplasticized polyvinyl chloride (UPVC).

11.17 FILLERS

A filler is usually a cheap inorganic additive whose function is to 'pad out' the plastic and therefore to reduce the cost of the product when it is used in mouldings. Some of the cheapest fillers confer some other, additional, benefits. Thus carbon black produces stability to ultraviolet light, electrical conduction and colour. Other fillers are chalk, fuller's earth etc., which usually increase the hardness and modulus of the plastic as well as reducing costs.

11.18 STABILIZERS

A stabilizer is used to prevent degradation by ultraviolet light. Carbon black produces stability to the photodegradation caused by ultraviolet light.

11.19 OTHER ADDITIVES

Additives can be added which act as flame retardants (antimony trioxide). The charring discussed in Section 11.14 is a natural fire retardant, as carbon does not vaporize and therefore does not produce a flame.

11.20 DURABILITY OF PLASTICS

In general thermoplastics are combustible, whilst thermosetting materials are not. Thermosetting plastics will give off obnoxious vapours when heated and before charring. Thus thermoplastics are not very good materials to use unless the temperature environment is accurately known. Plastics do not rot or corrode and offer good resistance to moisture, some oils, acids and alkalis. Their durability is assessed in accelerated tests because their development has been so rapid. This rapid development means that there is not the 'in-service' testing available in the external environment. The laboratory accelerated testing shows that their anticipated lifetime should be lengthy. The major faults with plastics will be shown up with the accumulation of more severe and testing environments as well as different uses. Ultraviolet light, present in strong sunlight, is the major factor which can damage the structure of the plastic. Ultraviolet radiations are of short wavelength, and the energy associated with this short wavelength is just sufficient to initiate an oxidation reaction which destroys the strong covalent C–C bond which makes the plastic brittle.

11.21 PROPERTIES AND MOLECULAR WEIGHT

Plastics are a group of materials which have a wide range of mechanical properties. As scientists understand the mechanisms by which these properties are achieved, more use will be made of plastics in the field of automobile engineering. Plastics can be made rigid, springy (and hence development is being successfully undertaken to replace heavy steel springs by light-weight plastic units). Plastic springs (e.g. LITEFLEX (a trademark of General Motors)) have been used successfully on the 1981 Corvette vehicle. Thermoplastics are easily moulded at moderate temperatures, as a result of the ease with which the van der Waals bond may be broken by thermal energy. Facia boards and light housings are made from thermoplastics and the use of plastics is on the increase especially for easily moulded items (see Table 11.4).

11.21.1 Molecular weight

The length of the plastic molecule is usually determined by the molecular weight. The thermoplastic may have a molecular weight of 20 000–50 000. The length of the chain can be estimated by the fact that each carbon atom is associated with tw hydrogen units. Thus one $-CH_2-$ weighs 14 atomic mass units (a.m.u.) because atomic weight of carbon is 12 and of hydrogen is 1. Thus a molecular weight of 2 is equivalent to a chain with 1430 $-CH_2-$ units. This length of chain sh compared with those given in Table 11.1 from the fractional distillation of

Table 11.4

Plastics	Proportion of overall plastics use[a] (%)	Examples[b]
Polyurethane	36.7	Facias (semi-rigid foam); door handles; head rests
PVC	28.3	Headlinings; underseal
ABS	9.2	Vacuum-formed knee rolls; Parcel shelves
Phenolics	7.5	Small tough parts (knobs)
Polypropylene	7.1	Glove boxes; parcel shelves, bumpers
Polyamide (nylon)	2.5	Body panels; tyre cords; carpets; small gears[c]
Polymethyl methacrylate	2.4	Lens covers; acrylic paints
Polyethylene	1.4	Stone guards
Cellulose	1.4	Cellulose paints
Polycarbonate	0.9	
Others	2.5	PTFE door-locking mechanisms

a B. B. Hundy, personal communication.
b ICI Plastics.
c Not powertrain gear box, but window and wiper motor gears.

11.22 ADHESIVES—CHEMICAL WELDING

Adhesives are a group of chemical compounds which are capable of sticking two or more components to form a new complete entity. Some of these adhesives allow the bonding together of materials which are impossible, or impracticable to join by any other means. Adhesives are used to stick ceramics, Glass-reinforced plastics to metal, motor vehicle plastic trim to a glossy paint surface, and very thin non-weldable alloys to metals. Adhesives have their greatest use in the aerospace industries and as such adhesive technology is altering every day. Helicopter rotor blades are adhesively bonded, and some 80% of the aerodynamic surfaces of a jet liner's wing use adhesives. Adhesives have an advantage over conventional welding techniques in that they do not produce metallurgical changes within the structure. Welding produces a heat-affected zone (Section 6.32). This heat-affected zone may be detrimental to the strength of the product and is produced when the alloy steel passes through the austenite–ferrite transition temperature (Section 6.27). The use of adhesives requires no drilling, countersinking or other labour-intensive operations and so they are cheap and they are cosmetically attractive as they have good 'gap-filling' properties. The stresses associated with drilled holes for rivets or bolts are eliminated with the use of adhesives. Being of an amorphous structure, the adhesive bond often has a high fatigue life and high damping to vibration. For this reason the tail section of jet liners is adhesively bonded together. Metallic coupling of rubber to metal and the interface between brake linings and shoes are adhesively bonded, thus replacing the localized fixing provided by rivets. The metallastic bond between rubber and metal is through a thin film of copper which has to be plated onto the steel before the adhesive will bond to the metal state. It is thought that this takes place through the porosity of the electroplate and the copper-sulphur chemical bond which promotes adhesion. There is no problem of corrosion being produced by the adhesive components as there is by say the welding fluxes as is the case with soldering and brazing operations. Adhesives completely fill and bond the overlap to which they are applied and therefore they eliminate crevice corrosion. Adhesives can be very successfully used where two chemically dissimilar metals are joined together and a galvanic bimetallic couple could cause accelerated corrosion. The adhesive serves to form an electrically insulating layer between the two metals and also eliminates the crevice. Hence sealants or adhesives are used as inter-weld mastics. In contrast, however, sealants (Section 11.27) do not fill the crevice over which they are applied. In the manufacture of motor vehicle bodies, certain 'adhesives' are used which are performing the function of gap fillers. Certain inter-weld mastics are adhesive sealants.

One of the most difficult factors to achieve is the intimate contact of two surfaces which have different chemical bonds. Hence it is very difficult to get a paint to adhere to a metal, or to stick glass to metal, because the chemical bonds are so different. When it comes to bonding metals together, the time-old method of roughening up the surfaces to be stuck was always considered desirable. However, the surfaces of metals are already rough on the microscopic scale, and mountainous when compared with the dimensions of the interatomic bond of carbon (the main constituent of an adhesive). Polished steel has troughs and peaks differing in height by 0.0003 mm while the carbon–carbon bond is about 5×10^9 mm. The slip steps produced when dislocations leave the surfaces of rolled metals also produce an undulating surface. Some surfaces are extruded (with a clearly defined slip step), while adjacent surfaces can be intruded. An intrusion is a crystallographic 'valley'. The direction of the carbon bond is tetrahedral, (Fig. 6.9, illustrating the diamond structure, shows the natural bond direction). As the adhesive is a long hydrocarbon chain and very viscous, it follows that there can be difficulty for the adhesive entering the troughs between the ridges and making good contact over the whole surface area between the troughs and ridges rather than just sticking to and spanning the ridges rather like a suspension bridge spanning a valley. Hence the first fundamental for an adhesive must be that it is a thin low viscous liquid which 'somehow' has the property of stiffening up so that it can carry the high loads imposed upon it across the whole area between ridge and trough.

11.23 ADHESION

Adhesion is the ability of an adhesive to stick to a surface of the material being stuck. The material being stuck is termed the 'adherend'. This adhesion can be generated in several ways.

(a) **Specific bonding** takes place when the adhesive reacts chemically with the adherend. Hence there is a chemical bond formed. With some adhesives special cleaning techniques have to be used.

(b) **Mechanical interlocking** results when the adhesive penetrates the pores, small holes or scratches which act as a good 'key'. The joint is a physical bond (as opposed to the bond above which is chemical in nature).

(c) **van der Waals forces** are 'residual bonds' which operate over very short distances and are the bonds which hold most materials together, e.g. plastics (Fig. 6.16). They are considered weak bonds but they are the only bonds which can operate at the surfaces of all the different materials and chemical bonds which need to be stuck. As a result of the van der Waals short range, it follows that the adhesive and the adherend surface must be in intimate contact. This means that the surfaces to be stuck must be **clean** and be capable of being **wetted** by the adhesive solution. This means that the adhesive must see the substrate as a material with a low angle of contact (Section 3.6; Fig. 3.5).

11.24 COHESION

For an adhesive to work satisfactorily, it must possess good cohesion so that it remains intact when subjected to stress. The cohesive forces are a property of the van der Waals bond (Fig. 6.16) and the degree of cross-linking (Fig. 11.8) between the hydrocarbon chains and the number and size of the side chains (Figs. 11.5 and 11.7). These cohesive forces develop in the curing process and are a mixture of covalent and van der Waals bonds (Section 6.10).

11.25 DESIGN

The golden rule with adhesives is to maximize the surface area over which adhesion is to take place and to avoid peeling stresses where adhesives are very weak. Adhesives work best where the stresses that act on them are kept in shear so that the deformation forces are in the plane of the adhesive. The tensile strength of an adhesive is only passable in pure tension where the forces of deformation are at right angles to the plane of the adhesive.

Adhesives must be capable of spreading to make the maximum surface area of contact. In order to achieve this, the physical characteristics of the adhesive must be those of a material with a low glass transition temperature. The equivalent to a low glass transition temperature may be obtained by using solvent bases to carry the 'adhesive'. The adhesive must contain weak van der Waals bonds (Section 6.12), but these weak bonds must transform to strong bonds (in a short time) when the adhesion is required. However, they must not develop over a period of time so that the adhesive sets slowly in the container (shelf life is very important with adhesives). When the adhesive sets, it cross-links (Section 11.6) to form a material with a very high glass transition temperature T_g. Normally this would give the adhesive a brittle nature and the adhesive material would not perform its function under high loads. Therefore most adhesives are formulated as **co-polymers**. Co-polymers allow the chemist to combine brittle compounds (high T_g) with rubbery materials (low T_g) (Fig. 11.4).

The bond strength will depend on the thickness of adhesive, the type of solvent and the service environment (temperature and moisture content).

11.26 CLASSIFICATION OF ADHESIVES

There are several ways in which the scientist can satisfy the conditions required of adhesives: (a) the adhesive has to be applied as a liquid and (b) it has to stick to different substance with sometimes completely differing chemical bonds but (c) the adhesive in its final state has to be a 'solid' to carry the high loads. The methods used are as follows.

11.26.1 Solvent based adhesives

The 'adhesive' is carried in either an organic solvent or water, e.g. rubber compounds in an organic solvent which evaporates (or is absorbed in the adherend) leaving the rubber compounds to carry the load and to provide the 'stick'. These types of adhesives are often helped by leaving the applied surfaces exposed to the air to aid evaporation and then when 'tacky' the surfaces are brought together. If the adjacent surfaces were brought together before the solvent had evaporated, the stresses generated by the volume shrinkage (of some 70–80%) would 'self-destroy' the adhesive joint. The classes of adhesive are usually categorized by the solvent, i.e. water borne adhesive (WBA) and solvent borne adhesive (SBA).

11.26.2 Cross-linking

These adhesives are often referred to as **reactive adhesives**. The chemicals constituting the adhesive react with each other and produce a three-dimensional structural material (Fig. 11.3). This is the setting process which can be brought about by (a) heating (some cure at temperatures as low as 120°C and so are unstable at room temperatures and have to stored frozen: this type of adhesive is very attractive for the motor manufacturer as the adhesive can be cured at the same time as the paint coatings are cured), (b) a surface catalyst, (c) two reactive parts which are mixed prior to use, (d) reacting with the physically absorbed water on the surface (super glues) and (e) spreading a reactive component on one of the surfaces to be stuck, and the adhesive on the adjacent surface.

11.26.3 Thermal setting

Many 'adhesives' are thermoplastic (Section 11.5), melting to produce a liquid at elevated temperatures, and so may run into the rough surface and be easily spread. On cooling they revert to their 'solid state' and produce strong bonds. The highest temperature of heating is around 200°C. Above this temperature the carbon chain will start to 'char' and colour the adhesive.

11.26.4 Combination of methods

Some solvent-based adhesives have a reactive component in them in order to gain better adhesive strengths.

11.26.5 Conversion processes

Where the two adherends are completely different chemically, it is often not possible to use the same adhesive to stick both surfaces. Therefore chemical conversion processes are used to change one surface into a more suitable 'chemical state' so that the one adhesive will stick and adhere to both surfaces. Such chemical processes are the use of 'primers' (e.g. organo-silanes for glass, or anodizing (of aluminium) and acid etches by chromic acid for chlorinated plastics, polyethylene or polypropylene).

11.26.6 Oil absorbers

As steel used in the automotive industry is coated with an oil protection (Sections 3.4 and 8.8) and oils which lubricate the metal in the pressing operations, it is often necessary to formulate an adhesive which 'absorbs' the oil. There is a loss of bond strength by this process.

11.27 SEALANTS

A sealant is the same chemical compound as an adhesive except that it is formulated to fill gaps and so is not formulated to have a bond strength. Sealants are used widely in the motor vehicle production techniques, e.g. for open seams to exclude water and dust from the passenger compartment. The gap-filling properties are often achieved by the addition of fillers, e.g. chalk and clay (Section 11.17). There is a physical limit to the size of gap which can be filled, as the curing process seals off further activating species which can be atmospheric water vapour or oxygen. Bigger gaps also required a sealant with a good 'slump resistance', as it would be detrimental to find the gap-filling sealant sliding down the gap under gravitational forces.

11.28 MECHANICAL PROPERTIES

The mechanical properties of the hardened adhesive are dependent on the chemical constituents. If the chains of these chemical structures are capable of bending, rotation and slip (Figs. 11.4, 11.5, 11.7 and 11.8) then the adhesive will be rubbery. Hence if the adhesive has a low glass transition temperature it will be brittle (Fig. 11.4). The mechanical properties can be controlled by the addition of fillers which 'pad out' (Section 11.17) the adhesive and reduce the degree of elasticity (Fig. 11.9). Too much filler will make the adhesive brittle and crumbly. From these deductions, and an application of Fig. 11.4 and Section 11.7, it follows that an adhesive is very temperature sensitive as shown in Fig. 11.10. Hence special self-adhesive labels have to be used on freezer packs. Ordinary self-adhesive labels would fall off as their adhesive strength is very temperature dependent.

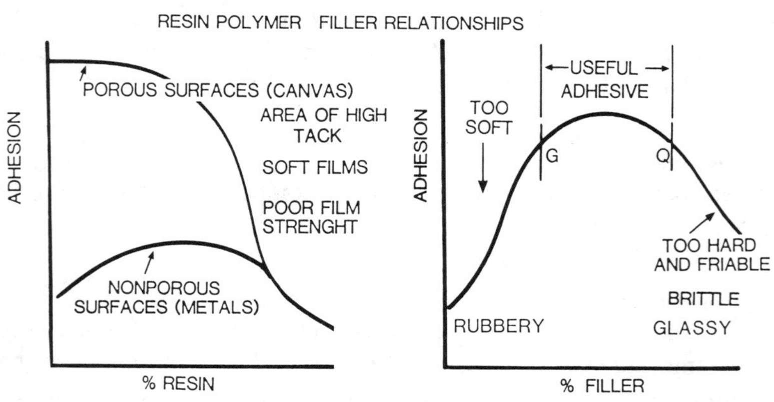

Fig. 11.9 — The adhesion obtained with adhesives is dependent on the percentage of resin and filler which are blended is neither too elastic nor too brittle. (Figure contributed by A. Hardy, Technical Manager, Bostik Ltd., Leicester.)

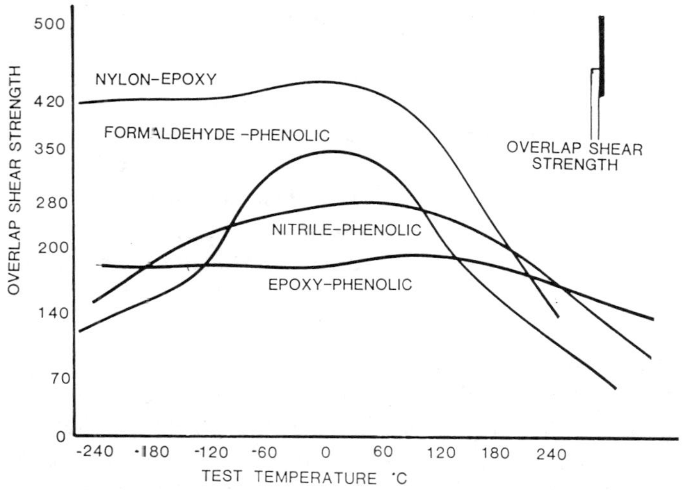

Fig. 11.10 — The strength of adhesives in overlap shear (as well as in peeling mode (not illustrated)) is dependent on the test temperature. (Figure contributed by A. Hardy, Technical Manager, Bostik Ltd., Leicester.)

11.29 TACKINESS

A property of adhesives (and sealants) is their 'tackiness'. Certain hydrocarbons have a natural affinity for a surface and are tacky to the touch. These naturally occurring materials (e.g. resins from trees, gum arabic) have been used for thousands of years ago as adhesives. These tacky materials are used in conjunction with modern polymers in order to obtain adhesion and strength. Other natural 'tacky' materials which can be used are bitumens (old style underseals), and tars. These have relatively short molecular chains and so have good 'tacky' adhesion.

Modern adhesives are a formulation where rubber can be reacted into the organic chain rather than blended into the adhesive. Materials such as perspex (polymethyl methacrylate (Fig. 11.5)) can be "graft" co-polymerized with rubber to produce a 'rubberized Perspex'. In this case the rubber forms a separate phase (particles of rubber about 0.1 μm in diameter) which are chemically bound into the acrylic hydrocarbon chain or phase. These adhesives provide for the applied stress to be uniformly distributed over the whole contact area with the ability to reduce stress concentration effects. This is achieved because the cracks within the harder phase are arrested by the 'rubber micro-sphere'.

11.30 EXAMPLES

The classification of adhesives is summarized in Table 11.5.

11.30.1 Windscreens

These have been sealed into the recess by mechanical means using rubber strip (Fig. 5.8). It is becoming more common to use polyurethane adhesives to bond windscreens into the opening and thus to increase the structural strength of the whole frame. Chemical type: adhesive polyurethane. Sealant: butyl-rubber-based mastic.

Table 11.5 — Classification of adhesives

Type	Example	Use
Water based	Polyvinyl acetate	Cracks in wall plaster
Solvent based	Rubber	Contact adhesives; cycle tube repairs
Cross-linking	Epoxy	Two-pack adhesives
Thermal setting		Adhesives which cure at the oven temperature: used to cure the electrocoat paint.

11.30.2 Bonnet stiffeners

Pads of mastic (high slump adhesive-sealant) are used to prevent the outer panel of the bonnet from vibrating against the inner bonnet stiffener. These mastic pads are heat cured when the car body goes through the paint stoving process. Chemical type: nitrile rubber–resin. This can also be used to seal the folded edges of door assemblies (Fig. 4.4(3)).

11.30.3 Anti-vibrational pads

Large panels are prevented from vibrating at speed by increasing the mass of the panel locally at the point where the maximum displacement will take place (Plate 2.2). These are applied after the electropaint has been cured.

11.30.4 Brake shoes

When worn, rivets score the brake drums. Adhesives eliminate this problem. In order to inform the driver when the shoes have become worn, warning lights are fitted which display the safe working thickness. Heat removal from the pads is improved with adhesive bonding.

11.30.5 Sealant compounds

Thermally cured sealants are applied to seal all the body joints against water ingress (note that they do not penetrate the crevice and so can form local sites of corrosion where or when water becomes trapped within the crevices). A typical curing cycle is during the stoving of external sprayed paint coats at 160°C for 30 min. Chemical type: PVC Plastisol.

11.30.6 Bonding felt to metal

Solvent-borne adhesives of the styrene butadiene rubber (Table 11.4) are used for this.

11.30.7 Prestik sealant contact adhesive

This material (which is replacing self tapping screws) is extruded into shape and applied in circles or rectangles to peelable paper. It is used for the fixing of rear lamp house units. Chemical type: polybutene.

11.30.8 PVC headliners

These are bonded to the steel body by solvent-based styrene butadiene rubber.

11.30.9 Epoxy adhesives

These are used for metal-to-metal bonding. The main advantage of the epoxy adhesive is its tolerance of oily surfaces, although clean (oil-free) surfaces will produce the best results. The adhesive is also capable of sticking some plastic thermosets to metals. Chemical type: epoxy resins.

11.30.10 Hyperbond graft co-polymer technology.

This involves a two-part adhesive and activator capable of bonding a wide range of materials metals plastics glass and wood. Chemical type: methacrylate.

ACKNOWLEDGEMENTS

The author wishes to acknowledge the help received from Mr. A. Hardy, Bostik Ltd., Leicester, and ICI Plastics, Motor Industry Group, Welwyn Garden City.

12

Concrete and durability

12.1 INTRODUCTION

Bridge and multistorey car park decks, made of concrete, become degraded by the road de-icing salts which are spread upon the roads to remove the black ice. A car park deck collapsed in 1984 in downtown Minneapolis as a result of corroded rebars (Plate 12.1). These reinforcing bars were corroded by the road de-icing salts' falling from the parked cars. No one was hurt but several cars were damaged when the 180 ft × 150 ft slab fell with its load of parked cars onto the ground level. This chapter explains the problems which arise with all cementitious materials when subject to aggressive soluble salts.

Concrete is a mixture of naturally occurring geological aggregates bonded together by a cementitious agent which 'sets' by reacting with water. This setting process is called 'hydration' and is not a dissolution process. The advantages of concrete are as follows.

(1) The constituents are relatively cheap and readily available in most countries of the world.

(2) Concrete can be made on site and can be cast in formwork. The formwork is cheap and can be readily erected.

12.2 PROPERTIES OF CEMENTITIOUS MATERIALS

Concrete has one serious limitation in that it is very weak in tension. This weakness is due to the many stress raisers which are present in the surface of the concrete matrix. Concrete does not possess dislocations (Section 6.21) and so there is no crack-blunting mechanism. When a crack is formed it is very easy to propagate it through the multiphase matrix. When concrete is used in compression, it is very strong and shows good engineering properties. Most concrete structures are therefore designed so that the material is used in compression only or with a reinforcement to improve the tensile properties. The reinforcement may be alkali-resisting glass, a polymer or

Plate 12.1 — Many of our structures today make use of reinforced concrete. Over a period of time, where concrete degradation has been extensively investigated, it has been shown that the physical concrete cover to embedded metals needs to be increased (Table 12.6) to protect the steel from the acid rain and aggressive chloride ions (Fig. 12.5). The quality of concrete (Section 12.14) is also important. This plate shows the result of rebar corrosion as a result of the ingress of chlorides from de-icing salts into the concrete car park deck. The upper deck, with parked cars, has fallen onto the cars beneath. Contributed by the *Minneapolis Star and Tribune*, 425 Portland Ave., Minneapolis, Minnesota 55488 and acknowledged with thanks.

steel wires–rods. The usual method of reinforcement is by steel which carries the total tensile stresses. Steel can be used in several ways: (a) the steel forms a rod which is unstressed (Fig. 12.1(a)); (b) the steel is a mesh which is unstressed; (c) the steel is in the form of a stranded wire which is initially unstressed whilst the concrete is in the unhardened state. When hardened the concrete is placed under compression. This can be undertaken by stressing the steel wires in tension against the outer concrete surfaces (Fig. 12.1(c)). There are two mechanisms which can be used: **post-tensioning** and **pre-tensioning**. These are shown in Figs. 12.1(c) and 12.1(b) respectively. The bulk of the hardened concrete is therefore placed in compression by the applied tensile forces placed in the encased steel wires (note the similarity to toughened glass: Section 6.45).

As a result of the very good tensile properties of steel compared with the compression strength of concrete, only a small fraction of the concrete section need be taken up by the steel reinforcement. Thus the relatively small cross-section of the steel reinforcement means that the steel must not suffer any localized form of corrosion (called 'pitting': Section 7.11; Fig. 7.14) which would increase the stress in

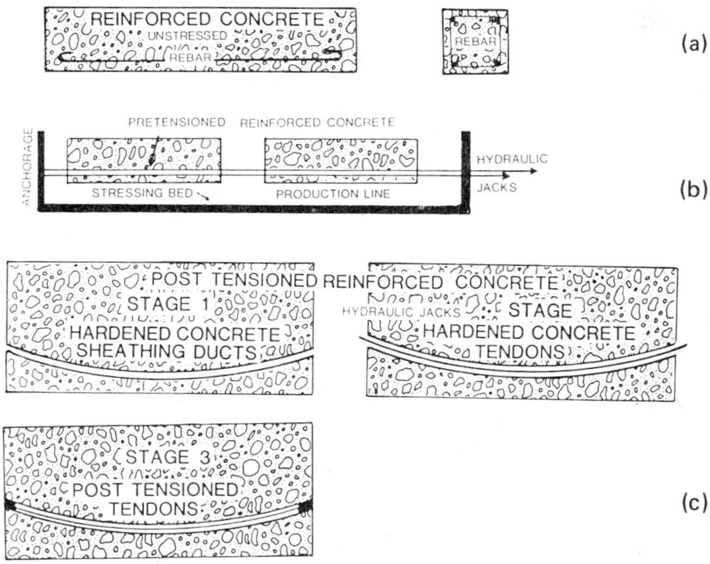

Fig. 12.1 — (a) Reinforced concrete is produced by casting concrete around the steel rebars. The concrete sets and hardens without being placed in compression. (b) Pretensioned concrete is produced by casting concrete around steel tendons which are stressed in tension. When the concrete has hardened, the tensile stresses are released. The reduction in stress allows the steel tendons to expand and so increases the friction between the steel and hardened concrete. This places the concrete in compression and the steel outside the concrete is then cut and each individual casting is separated. Note that the steel is not covered with concrete at the ends (c) Post-tensioned concrete is produced by casting concrete around the sheathing ducts which will house the steel tendons. The concrete sets and hardens without being placed under stress. After hardening, steel tendons are threaded into the sheathing ducts, and the steel tendons are anchored and stressed. These tendons are then grouted in order to keep the environment out of the sheathing ducts. The hardened concrete is now in compression.

one particular area. However, pitting is an unfortunate characteristic of steel when subjected to localized oxygen concentration cells.

In the hydrating and setting state, concrete produces an alkali (calcium hydroxide $Ca(OH)_2$) which causes pitting in glass fibres, but this same alkali produces chemical protection (Fig. 7.15) upon steel and therefore pitting corrosion does not exist in ideally constituted reinforced concrete.

However, there are specific circumstances where the corrosion of the steel reinforcement does occur. This corrosion is often localized and leads to loss of cross-section and in some cases to brittleness because the hydrogen reaction is the cathodic reaction:

$$H^+ + H^+ + 2e^- \rightarrow H_2$$

The hydrogen causes two side effects: (a) hydrogen reduces the strength of the steel; (b) hydrogen reduces the strain at fracture. The combination of these two effects reduces the energy absorbed by the steel. This embrittlement of the steel is termed **hydrogen embrittlement** of the steel. The corrosion of the steel produces an increase in volume which fractures the concrete. This reduces the beneficial function of the physical barrier on the steel. Usually the loss of the physical barrier is shown up by large lumps spalling off the concrete structure. A large number of cases of steel reinforcement corrosion have been examined and the factors which give rise to corrosion are fairly well established.

12.3 CODES OF PRACTICE AND CORROSION

If steel is immersed and covered to a depth given by the relevant code of practice in a good quality, well-compacted, concrete, it should not corrode within the lifetime of the structure. (The codes of practice duscussed will be for normal engineering construction of concrete structures. The Department of Transport sometimes modifies these codes. Where this is known, the reader's attention will be drawn to this modification. In all cases the Department of Transport relevant codes should be used for road construction.) In order to understand why this is so, it is necessary to investigate some of the properties of the 'cement'.

12.4 CONSTITUENTS OF CEMENT

In order to write down abbreviated chemical equations for the hydration of cement particles another short-hand notation has to be used. This is often referred to as the 'cement chemists notation', which is as follows (see Table 12.1).

The compounds are all 'broken down' into their two constituent oxides (acid and alkali oxides) and the chemical oxide compound is then abbreviated, e.g. calcium carbonate is $CaCO_3 = CaO.CO_2 = C\overline{C}$ where $C = CaO$ and $\overline{C} = CO_2$. Hence the abbreviations for the oxides have to be learnt, and this is much easier if the normal chemical abbreviations are recognized.

12.5 MANUFACTURE OF CEMENT

The manufacture of cement is very energy intensive (Fig. 1.11). The raw materials from which cement is made are cheap and naturally occurring. These are chalk

Table 12.1 — Cement chemists' notation

Oxide	Chemical	Abbreviation
Calcium oxide	CaO	C
Carbon dioxide	CO_2	$\overline{C}$
Magnesium oxide	MgO	M
Sodium oxide	Na_2O	N
Iron oxide	Fe_2O_3	F
Aluminium oxide	Al_2O_3	A
Silicon dioxide	SiO_2	S
Sulphur trioxide	SO_3	$\overline{S}$
Water	H_2O	H
Calcium silicate hydrate	Very complex	CSH

There is no fixed formula for the hydrate which is formed.

(limestone) and shale or clay. These two minerals are ground and heated to about 1400°C. At this temperature they break down, with the limestone changing into the alkali oxide CaO which reacts with the heated clay to form a 'clinker' which consists of four main compounds. These new chemical compounds are expressed in the cement chemist's notation as follows (see Table 12.2): C_3S; C_2S; C_3A; C_4AF. These are the constituents of the clinker which react with water (called hydration) and form the 'adhesive' of the cement. This clinker must be ground to produce the smallest possible particle of cement so that the surface area to mass ratio is high. This is due to the fact that the clinker does not dissolve in water — it hydrates. This grinding process is energy intensive as the particle size needs to be about 74 μm in diameter.

Table 12.2 — Compounds in the cement clinker (Shirley, 1980)

Abbreviations	Chemical	Normal chemical notation
C_3S	Tri-calcium silicate	$3(CaO)\cdot SiO_2$ (= Ca_3SiO_5)
C_2S	Di-calcium silicate	$2(CaO)\cdot SiO_2$ (= Ca_2SiO_4)
C_3A	Tri-calcium aluminate	$3(CaO)\cdot Al_2O_3$ (= $Ca_3Al_2O_6$)
C_4AF	Tetra-calcium alumino ferrite	$4(CaO)\cdot Al_2O_3\cdot Fe_2O_3$ (= $Ca_4Al_2O_{10}$ =Ca_2AlFeO_5)

12.6 STRENGTH–TIME RELATIONSHIP

The hardening of concrete is a process called 'ageing'. The mechanism of ageing of concrete is by hydration which produces an increase in strength with time providing that the concrete is always maintained wet. Once the concrete dries, the hydration process is stopped and it cannot be restarted. Hence once concrete has 'dried' it will not continue to harden even if subsequently wetted. Therefore in sunny conditions, or in hot dry climates, provision must be made to keep the concrete wet for as long as possible. Each of the constituents present in the cement adhesive hydrates and hardens at a different rate. The most rapid increase in hardness is produced by the constituent C_3S (tricalcium silicate). The rate at which each constituent of cement hydrates and hardens is shown in Fig. 12.2. As a result of this increase in strength with time, the scientist considers the strength of concrete to be that at 28 days. The hardening of concrete with respect to time is shown in Table 12.3.

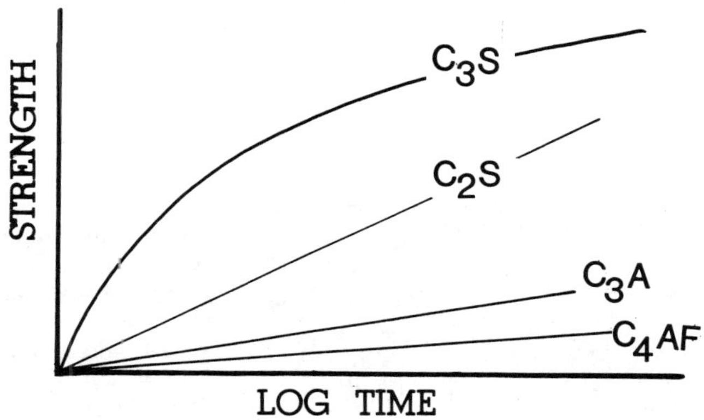

Fig. 12.2 — The hardening rate of the components of cement with time.

Table 12.3 — Hardening of concrete with time

Age (days)	Fraction of 28 day compressive strengths OPC	RHPC
3	0.40	0.55
7	0.65	0.75
28	1.00	1.00
90	1.20	1.15
360	1.35	1.20

12.7 TYPES OF CEMENT AND THEIR COMPOSITIONS

Each of the new chemical compounds formed in the cement clinker has different properties. These different properties are used to make different kinds of cement. Table 12.4 and Fig. 12.3 give the functions of each type of cement together with their compositions and relate the four compounds present in cement to their function in the hydration process.

Table 12.4 — Constituents of cement

Constituent	Percentage of constituents in the following cements			
	SRPC	OPC	LHPC	RHPC
C_3S	43	49	30	55
C_2S	37	25	46	16
C_4AF	14	9	13	9
C_3A	2	11	5	11
Other	4	6	6	9

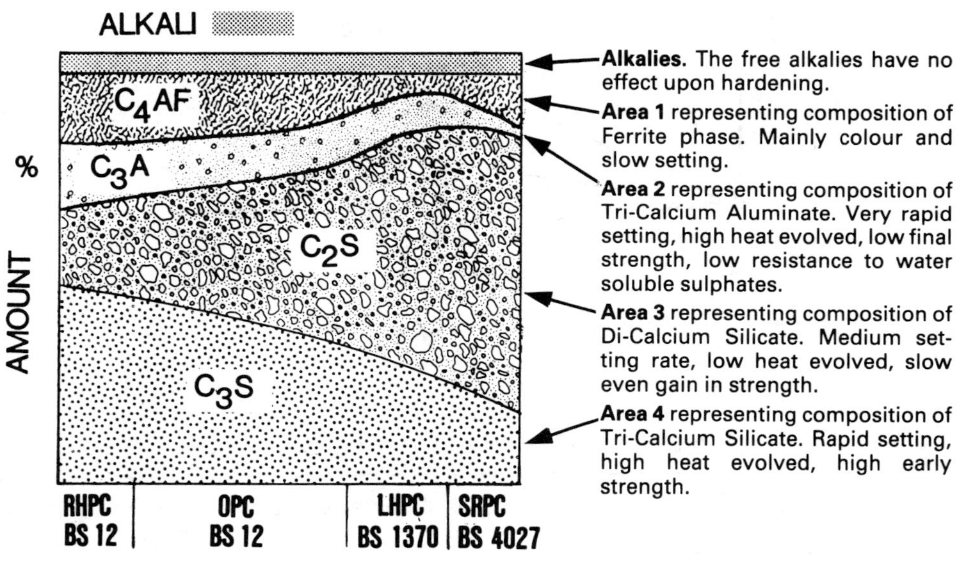

Fig. 12.3 — The relative composition of various cements and the constituents function in cement clinker.

12.8 THE HYDRATION OF THE CEMENT CLINKER

Fig. 12.4(a) represents how the cement particles hydrate. The hydration reaction gives out heat. Such a reaction is termed **exothermic** and therefore the casting of large structures (e.g. dams) should contain 'plumbs' and larger aggregates to absorb some of the heat evolved. Hence the grading of the aggregates making up the concrete structure is important. The larger irregular (gravel) or aggregates, termed coarse aggregates, have a void space of about 50%. This void space must be filled with the (sand) or fine aggregates which contain about 33% void space. This void space within

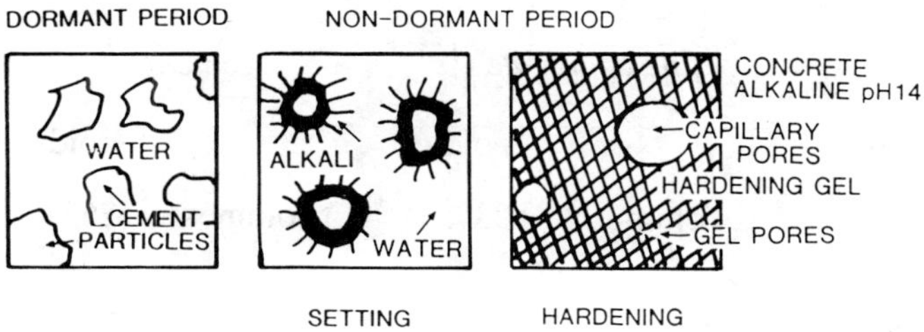

(a)

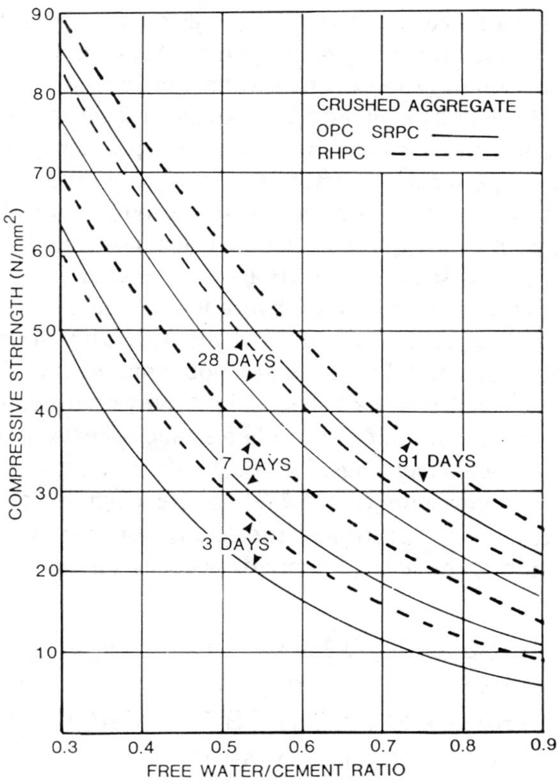

(b)

Fig. 12.4 — (a) Hydration of a cement particle is a surface interaction which produces CSH hydrate fibres by a process of osmosis. (b) strength as a function of water : cement ratio, cement type and age.

the fine aggregate is filled by the cement and water. Hence a traditional concrete mix will contain the following:

coarse aggregate		*fine aggregate*		*cement*		*water*
	Voids		Voids		Voids	
4 volumes	50%	2 volumes	33%	2/3 volumes	26%	0.5 volumes
In practice						
4 volumes	50%	2 volumes	33%	1 volume	26%	0.5 volumes

However some of the fine aggregate will 'jack up' some of the coarse aggregate and so extra cement has to be added. The ratio of the weight of water to the weight of cement is called the **water:cement** ratio. The void space between the cement–fine–coarse aggregates is filled by the cement hydrate fibres CSH. The strength of the concrete is dependent on time and the free water:cement ratio (Fig. 12.4(b)).

The hydration proceeds by a phenomenon known as osmosis (Section 8.13). The surface of the ground clinker reacts with the water, producing a semi-permeable membrane. This semi-permeable membrane allows only water molecules through the membrane and the resulting osmotic pressure produced within the membrane fractures this semi-permeable film shooting the calcium silicate hydrate (CSH) into the surrounding water. These fibres of CSH are the mechanism by which the cement paste stiffens up. The other result of hydration is the production of calcium hydroxide $Ca(OH)_2$. The presence of calcium hydroxide is important because it, and the free alkalis NaOH and KOH (sodium and potassium hydroxides), provides the alkaline environment which confers a chemical passivity on the embedded steel (the term chemical passivity means that a very adherent and coherent oxide film is formed conferring passivity on the steel (Figs. 7.8 and 7.15(a)) rather like the oxide film of aluminium does on aluminium (Fig. 6.17)). Some cement constituents do not produce this alkali, e.g. the constituent tricalcium aluminate C_3A (see equations below). The hydration of the four compounds of cement is expressed as follows:

$$\underset{\text{Clinker}}{2(3CaO\cdot SiO_2)} + \underset{\text{Water}}{6H_2O} \rightarrow \underset{\text{hydrate}}{3CaO\cdot 2SiO_2\cdot 3H_2O} + \underset{\text{alkali}}{3Ca(OH)_2} + \underset{\text{heat evolved}}{X\ \text{J}} \quad (12.1)$$

$$2(2CaO\cdot SiO_2) + 4H_2O \rightarrow 3CaO\cdot 2SiO_2\cdot 3H_2O + Ca(OH)_2 + 2X\ \text{J} \quad (12.2)$$

$$3CaO\cdot Al_2O_3 + 6H_2O \rightarrow 3CaO\cdot Al_2O_3\cdot 6H_2O + 3X\ \text{J} \quad (12.3)$$

The following should be noted: (1) the heat liberated is represented in terms of X J (X joules) (Section 1.18; Table 1.12(b)); (2) tricalcium aluminate does not hydrate to produce the alkali $Ca(OH)_2$; (3) the differing amounts of alkali produced and therefore the difference in the chemical passivity afforded to embedded steel. The hydration product is a **gel**, which has an amorphous or a non-crystalline structure (and therefore little strength). This is responsible for the setting process. The setting process is accompanied by a change in flow characteristics or stiffness characteristics.

To improve on strength, the gel must form a compound in the crystalline state. This is the hardening process. The ions responsible for the promotion of the crystalline state are the alkali ions Ca^{2+} and Na^{+} (see Section 12.23). The covalently bonded silicates (Fig. 6.10) and aluminates do not favour the formation of the crystalline state. The final hardening of the cementitious matrix produces holes (of size about 3 nm) (1 nm is the same as 10^{-9} m) in the gel (termed **gel pores**) and this means that the water within them behaves as a very viscous fluid, and this accounts for the durability of concrete products in hostile environments. A high water:cement ratio produces larger capillaries (diameter $\approx$ 10 nm) and diffusion of hostile species is increased, leading to a reduction in durability. This is very important with regard to carbonation (see later). These are called **capillary pores** and arise from the evaporation of the unreacted water within the matrix. The minimum water:cement ratio required for complete hydration is 0.28. Anything more than this will produce the porosity, which is attributable to capillary pores. The addition of water is therefore an 'undesirable' workability agent. Hence concrete can be made with too much water. The porosity (Section 12.10) allows the ingress of aggressive ions with the result that embedded steels corrode. This could be the reason for the car park collapse shown in Plate 12.1.

12.9 WORKABILITY, HEAT EVOLVED AND FLASH SETTING

One of the most important properties of concrete is its castability, which is measured by its workability. This is determined by producing a cone of concrete and measuring the slump or slide of the concrete cone against its mould cone. The **slump** is measured in millimetres. The slump test is relatively easy to carry out and so would be undertaken at the site where the motorway bridge is being built. The VEBE test, which is a laboratory based test, is another alternative and is used for batches of concrete which do not show a slump. The VEBE test is the time taken (in seconds) for a cone of concrete to be vibrated horizontal. The **compacting factor** is another workability test which measures relative bulk densities when loosely compacted concrete is fully compacted. These tests are undertaken at the time of mixing.

12.9.1 Flash setting

Flash setting is where the workability of the concrete is altered by the rapid hydration of tri-calcium aluminate caused by: (a) its rapid hydration and (b) its exothermic reaction (see Table 12.5) which raises the bulk temperature and further increases the

Table 12.5 — Heats evolved from the cement constituents (Verbeck and Foster, 1950)

Constituent	Heat evolution (J/g) at the following ages					
	3 days	7 days	28 days	90 days	360 days	6.5 years
C_3S	243	222	377	436	490	490
C_2S	50	42	105	176	226	222
C_3A	888	1559	1378	1303	1169	1374
C_4AF	289	494	494	410	377	465

rate of hydration and decreases the setting time. As the component tricalcium aluminate continues to give out more heat with time of setting, it can cause severe thermal strains within large structures. In order to get over this problem certain additions are made to the clinker at the grinding stage which overcome the problem. Gypsum ($CaSO_4 \cdot 2H_2O$) (about 3-5%) is added at the grinding stage to retard the hydration of the C_3A component so that this long-term heat evolution is avoided, and the initial setting time is controlled. The compound tricalcium aluminate is important for three reasons.

(1) Its presence in cement is reported to aid in the protection of embedded metals as it acts as a 'sponge' for chlorides and so helps to protect the embedded metals from the adverse effects of chlorides (see Section 12.13).

(2) The hydration product of the tricalcium aluminate is unstable in the presence of water-soluble sulphates (see Section 12.16). Hence, for sulphate-resisting Portland cement (SRPC), clays are chosen where the amount of alumina (Al_2O_3) is minimal so as to reduce the amount of C_3A in the resulting cement.

(3) It evolves heat over a long period. When this evolution of heat stops, the thermal strains introduced on cooling can be severe and cause cracking within the concrete.

Hence it would seem a contradiction to add gypsum (itself a calcium sulphate) to cement at the grinding stage of manufacture in order to overcome the problem of flash setting when sulphates are one of the causes of concrete or cementitious degradation. The retardation reaction forms the compound Ettringite $3CaO \cdot Al_2O_3 \cdot 3CaSO_4 31\text{–}32H_2O$. This compound will dissolve in the presence of CaO, and water without damage to form a more stable monosulphate. This large molecule (Ettringite) does severe damage, but fortunately the damage occurs within days of mixing and the damage can be repaired by the subsequent hydration of the cement which occurs after 7 days of mixing provided that there is water present. Hence sulphate attack is a term used to describe the effect of sulphates on cementitious materials (concrete, mortar etc.) when the mortar–concrete is older (and drier). In this case there is no further hydration possible and so the 'self repair' discussed above can no longer take place. Hence when the concrete becomes dry all hydration ceases and there is no hydration to repair the damage caused by the volume expansion which results from the formation of the compound Ettringite. This Ettringite is no longer in an environment of CaO and water and so cannot dissolve (see Section 12.15). Note: gypsum is not added to cement at the grinding stage to reduce the possibility of sulphate attack.

12.10 WATER:CEMENT RATIO OF CONCRETE

The water:cement ratio has a profound effect on the strength of the resulting concrete (or mortar). The strength decreases with increase of water as shown in Fig. 12.4(b). This is explained by the more open structure of the hydrating clinker as a water:cement of 0.40 produces 30% porosity compared with the 50% porosity produced by a water:cement ratio of 0.70.

12.11 CARBONATION

Carbonation is the loss of the alkalinity of the concrete matrix owing to the ingress of the acidic atmospheric oxides which are present in an industrial atmosphere. These acidic oxides are carbon dioxide (CO_2), sulphur dioxide (SO_2) and sulphur trioxide (SO_3).

The carbonated areas change from an alkaline region, of pH 12 (which are chemically protective towards steel) to a region of pH 9, which does not confer any chemical protection even though the physical protection is still present (Fig. 7.15). The greater the depth of concrete cover on top of the embedded steel, the longer this alkaline nature adjacent to the steel survives in the industrial atmosphere. This is the reason for the changes in the codes of practice over the years as shown in Tables 12.6 and 12.7.

Table 12.6 — Carbonation of concrete

Code of practice (year)	Cover specified (mm)	Life span expected with respect to to 1984 specification
1948	25	15 years
1957	37	30 years
1969	40	37 years
1972	50	60 years
1984[a]	60	100 years

The ingress of acidic gases obeys the square root law: depth of carbonation $x = (Dt)^{1/2}$ where D = coefficient of diffusion and t = time.
[a] BS5400, Table 13, Part 4, 1984.

Table 12.7 — Carbonation time (years) for various depths of cover and water: cement ratios (Beckett, 1983)

Water:cement ratio	Time (years) for carbon dioxide diffusion through the following depths of cover					
	5 mm	10 mm	15 mm	20 mm	25 mm	30 mm
0.45	19	75	>100	>100	>100	>100
0.50	6	25	56	99	>100	>100
0.55	3	12	27	49	76	>100
0.60	1.8	7	16	29	45	65
0.65	1.5	6	13	23	36	52
0.70	1.2	5	11	19	30	43

Ordinary Portland cement (no additives), sand and gravel.

Hence the amount of water mixed in with the concrete affects the rate at which carbonation occurs. The water : cement ratio (Section 12.8) has a profound effect on both the strength of the resulting concrete–mortar and its porosity (Fig. 12.4 and Table 12.7). It is this porosity which allows the ingress of the acidic gases into the concrete. The structure of the concrete matrix in the fully hardened state is that of a cementitious matrix composing the gel and 'holes' (of size about 3 nm) in the gel. These are termed **gel pores**, and they account for the durability of concrete products with a low water : cement ratio in hostile environments. A high water : cement ratio produces larger capillaries (diameter $\approx$ 10 nm) and the diffusion of hostile species is increased, leading to a reduction in the protection afforded to embedded metals. These larger pores are called **capillary pores** and they arise from the evaporation of the unreacted water within the matrix. The minimum water : cement ratio required for complete hydration is 0.28. Anything more than this will produce the larger capillary pores. This is why the water : cement ratios are reduced as the severity of the environment increases (Tables 12.8(c), 12.8(d) and 12.9).

Another factor which affects the durability of steel within the inert aggregate–cement matrix is the presence of chlorides. The addition of chlorides (in the form of calcium chloride $CaCl_2$) to concrete as an accelerator (Table 12.8(a)) has been banned, but chlorides can diffuse into the concrete from such environments as road de-icing salts on concrete bridge decks, motorway fly-overs, multistorey car parks and structures which are subjected to marine environments. The presence of chlorides increases the rate at which the corrosion process takes place and is dependent on the pH of the medium as shown in Fig. 12.5.

Table 12.8(a) — Limits of chloride content of concrete

Type or use of concrete	Maximum total chloride content expressed as a percentage of chloride ion by mass of cement (inclusive of p.f.a.[a] or g.g.b.f.s. when used)
Prestressed concrete Heat-cured concrete containing embedded metal	0.1
Concrete made with cement complying with BS 4027 or BS 4248	0.2
Concrete containing embedded metal and made with cement complying with BS 12, BS 146, BS 1370 or BS 4246 or combinations with g.g.b.f.s. or p.f.a.	0.4

Note: this table and Tables 12(b)–12(e) and 12.9 are reproduced by permission of the British Standards Institution from Structural Use of Concrete, BS 8110, 1985. Complete copies of these tables can be obtained from the British Standards Institution, Linford Wood, Milton Keynes MK14 6LE, to which reference should be made before concrete specifications are drawn up.
Reproduced by permission from BS 8110, Part 6, Table 6.4.
[a]p.f.a., pulverized fuel ash; g.g.b.f.s., ground and granulated blast furnace slag.

Table 12.8(b) — Exposure conditions

Environment	Exposure conditions
(A) Mild	Concrete surfaces protected against weather or aggressive conditions
(B) Moderate	Concrete surfaces sheltered from severe rain or freezing whilst wet Concrete subject to condensation Concrete surfaces continuously under water Concrete in contact with non-aggressive soil (see class 1 of BS 8110, Table 6.1 or Table 12.9) Note: for aggresive soil conditions see BS 8110, Section 6.2.3.3
(C) Severe	Concrete surfaces exposed to severe rain, alternate wetting and drying or occasional freezing or severe condensation
(D) Very severe	Concrete surfaces exposed to sea water spray, de-icing salts (directly or indirectly), corrosive fumes or severe freezing conditions whilst wet
(E) Extreme	Concrete surfáces exposed to abrasive action, e.g. sea water carrying solids or flowing water with pH ≤4.5 or machinery or vehicles
(F) Freezing and thawing and de-icing salts	Where freezing and thawing actions under wet conditions exist, enhanced durability will be obtained by the use of air-entrained concrete

Reproduced by permission from BS 8110, Part 3, Table 3.2.

Table 12.8(c) — Nominal cover to all reinforcement (including links) to meet durability requirements

	Nominal cover (mm)				
Conditions of exposure (see Table 12.8(b) (F))					
Mild	25	20	20[a]	20[a]	20[a]
Moderate	—	35	30	25	20
Severe	—	—	40	30	25
Very severe	—	—	50[b]	40[b]	30
Extreme	—	—	—	60[b]	50
Properties					
Maximum free water:cement ratio[c]	0.65	0.60	0.55	0.50	0.45
Minimum cement content[d] (kg/m^3)	275	300	325	350	400
Lowest grade of concrete	C30	C35	C40	C45	C50

Reproduced by permission from BS 8110, Part 3, Table 3.4. This table relates to normal-weight aggregate of 20 mm nominal maximum size. For different sizes see Table 12.8(e).

[a] These covers may be reduced to 15 mm provided that the nominal maximum size of aggregate does not exceed 15 mm.

[b] Where concrete is subject to freezing while wet, air entrainment should be used (see Table 12.8(b) (F)).

[c] See Table 12.8(d).

[d] See Table 12.8(e).

Table 12.8(d) — Nominal cover to all steel (including links) to meet durability requirements

	Nominal cover of prestressed concrete (mm)			
Conditions of exposure (see Table 12.8(b) (F))				
Mild	20	20[a]	20[a]	20[a]
Moderate	—	30	25	20
Severe	—	40	30	25
Very severe	—	50[b]	40[b]	30
Extreme	—	—	60[b]	50
Properties				
Maximim free water:cement ratio	0.60	0.55	0.50	0.45
Minimum cement content (kg/m^3)	300	325	350	400
Lowest grade of concrete	C35	C40	C45	C50

Reproduced by permission fron BS 8110, Part 4, Table 4.8. This table relates to normal-weight aggregate of 20 mm nominal maximum size.

[a] These covers may be reduced to 15 mm provided that the nominal maximum size of aggregate does not exceed 15 mm.

[b] Where concrete is subject to freezing while wet, air entrainment should be used (see Table 12.8(b) (F)).

Table 12.8(e) — Adjustments to minimum cement contents for aggregates other than 20 mm nominal maximum size

Nominal maximum aggregate size (mm)	Adjustments to minimum cement contents in Tables 12.8(c) and 12.9 (kg/m^3)
10	+40
14	+20
20	0
40	−30

Reproduced by permission from BS 8110, Part 6, Table 6.3.
In no case should the cement content be less than 275 kg/m^3 for reinforced concrete or 300 kg/m^3 for prestressed concrete. See Table 12.8(d).

Table 12.9 — Concrete exposed to sulphate attack

Class	Concentration of sulphates expressed as SO_3: In soil: Total SO_3 (%)	In soil: SO_3 in 2:1 water:soil extract (g/l)	In groundwater (g/l)		Dense, fully compacted concrete made with 20 mm nominal maximum size aggregates complying with BS 882 or BS 1047: Cement[a] content not less than (kg/m^3)	Free water:cement[a] ratio not more than
1	<0.2	<1.0	<0.3	All cements listed in BS 8110, Section 6.1.2.1 BS 12 cements combined with p.f.a.t.[b] BS 12 cements combined with g.g.b.f.s.[b]	—	—
2	0.2–0.5	1.0–1.9	0.3–1.2	All cements listed in BS 8110, Section 6.1.2.1 BS 12 cements combined with p.f.a. BS 12 cements combined with g.g.b.f.s.	330	0.50
				BS 12 cements combined with minimum 25% or maximum 40% p.f.a.[c] BS 12 cements combined with minimum 70% or maximum 90% g.g.b.f.s.	310	0.55
				BS 4027 cements (SRPC) BS 4248 cements (SSC)	280	0.55
3	0.5–1.0	1.9–3.1	1.2–2.5	BS 12 cements combined with minimum 25% or maximum 40% p.f.a.[b] BS 12 cements combined with minimum 70% or maximum 90% g.g.b.f.s.	380	0.45
				BS 4027 cements (SRPC) BS 4248 cements (SSC)	330	0.50
4	1.0–2.0	3.1–5.6	2.5–5.0	BS 4027 cements (SRPC) BS 4248 cements (SSC)	370	0.45
5	>2	>5.6	>5.0	BS 4027 cements (SRPC) and BS 4248 cements (SSC) with adequate protective coating (see BS 8110, Section 6.2.3.3)	370	0.45

Reproduced by permission from BS 8110, Part 6, Table 6.1.
Abbreviations: See Table 12.8(a).
[a] Inclusive of p.f.a. and g.g.b.f.s. content.
[b] For reinforced concrete see BS 8110, Section 3.3.5; for plain concrete see BS 8110, Section 6.2.4.2.
[c] Values expressed as percentage by mass of total content of cement, p.f.a. and g.g.b.f.s.
See BS 8110, Section 6.2.4.3 for adjustments to the mix proportions.
Within the limits given in this table, the use of p.f.a. or g.g.b.f.s. in combination with sulphate-resisting Portland cement (SRPC) will not give lower sulphate resistance than combinations with cements to BS 12.
If much of the sulphate is present as low solubility calcium sulphate, analysis on the basis of a 2:1 water extract may permit a lower site classification than that obtained from the extraction of total SO_3. Reference should be made to BRE Current Paper 2/79 for methods of analysis, and to BRE Digests 250 and 222 for interpretation in relation to natural soils, and fills respectively.

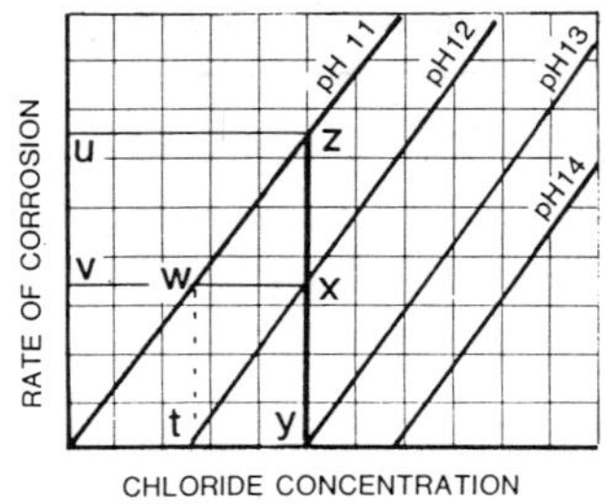

Fig. 12.5 — The effect of corrosion rate on chloride concentration for different pH environments. For a concentration y of chlorides, there will be a corrosion rate of v in a concrete of pH 12 or u if the pH is 11.

12.12 POTENTIAL-pH TABLE

If steel is immersed in a good quality, well-compacted, chloride-free concrete, to a depth given by the relevant codes of practice, (Table 12.8) it will not corrode within the lifetime of the structure.

A scientist called Pourbaix first examined the relationship of pH and potential and given in Fig. 7.15 (Section 7.10). To-day this electrode potential is being used to examine non-destructively for rebar corrosion within bridge decks and motorway fly-overs. This technique is called potential mapping, and is shown in Fig. 12.6.

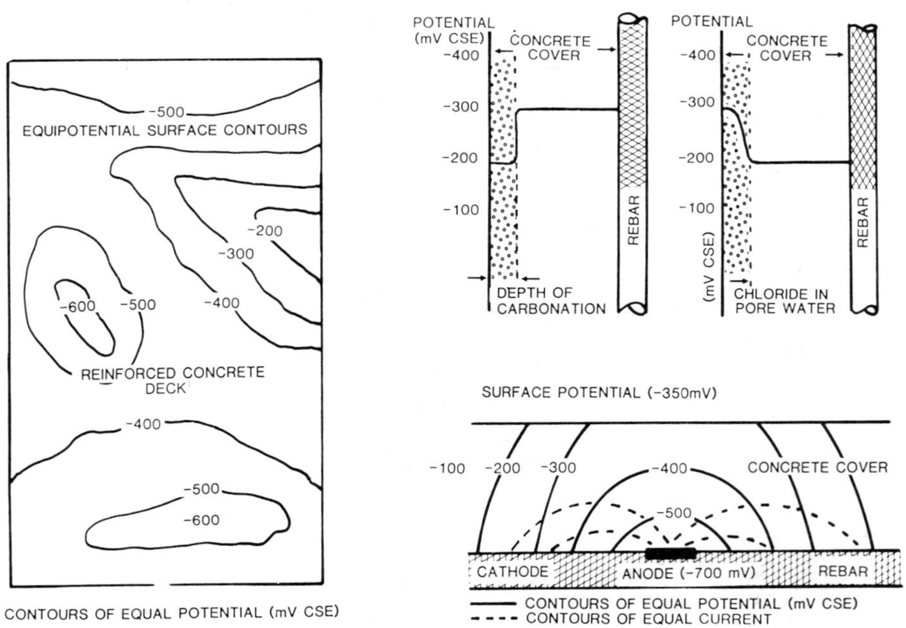

Fig. 12.6 — Potential mapping of the rebars within a bridge deck showing the areas most likely to be corroding. (After D. E. Tonini and S. W. Dean (eds.), Chloride Corrosion of Steel in Concrete, STP 629, ASTM, Philadelphia, PA, 1987, and R. D. Browne, M. P. Geoghegan, and A. F. Baker, in A. P. Crane (ed.), Corrosion of Reinforcement in Concrete Construction, Ellis Horwood, Chichester, Chap. 13, Figs. 3 and 4.)

12.13 THE EFFECT OF CHLORIDES

The effect of chlorides on this Pourbaix diagram is enormous and has far-reaching effects on the protection of embedded steel. The addition of chlorides shifts the whole of the line UKFH to the right (Fig. 7.15) so that the point W now enters the corrosion region. This means that the steel will corrode whether or not the concrete is carbonated or alkaline in nature (Fig. 12.5). The calcium aluminates and the aluminoferrites are compounds which react with the chlorides and remove them from the depassivating role. This, however, only happens if the concentration of chlorides is low. This fact explains why the effect of chlorides on the corrosion of embedded metals is dependent on the cement content of the mix. When the concrete carbonates, these chlorides are liberated from the alumino complexes and are again free to act as depassivators.

12.14 QUALITY OF CONCRETE

From the above, it can be seen that the 'Quality' of concrete has a big part to play in the corrosion or otherwise of reinforced concrete. The quality of concrete is determined by the (1) the water:cement ratio, (2) cement type and content, (3) aggregate grading, (4) the degree of compaction and (5) the ratio of aggregate to cement. Hence concrete behaves as both a physical and a chemical passivator or 'barrier' to embedded metals. As a physical barrier to the ingress of oxygen and water vapour it is very poor, owing to the micro-voids produced by the hydration mechanism. The chemical protection of the steel, by the passivation, is very efficient as it can only be destroyed by the ingress of chlorides and carbon dioxide and sulphur

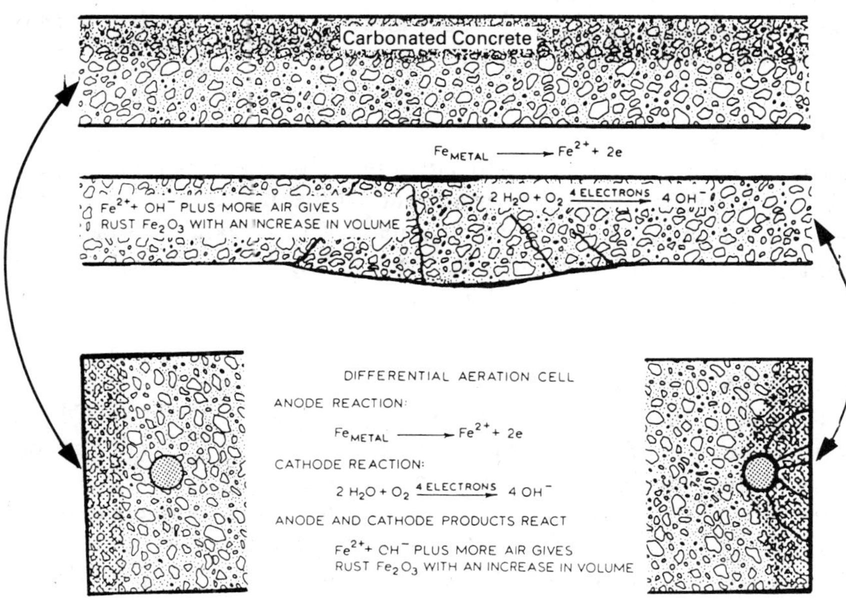

Fig. 12.7 — Diagrammatic view of the corrosion mechanism responsible for the degradation of the rebar.

dioxide. Both the gases may reduce the alkalinity of the concrete, and therefore the maintenance of the chemical passivation relies on the **physical** nature of the concrete barrier. Breakdown occurs when the physical barrier thickness is insufficient to prevent the diffusion of these gases to the steel. The corrosion mechanism is indicated diagrammatically in Fig. 12.7.

12.15 SULPHATE ATTACK

When concrete (or mortar) is in contact with soluble sulphates, degradation of the cementitious agent occurs unless a sulphate-resisting cement is used. This degradation is called **sulphate attack**. Sulphate attack is essentially manifested as a scaling or 'powdering' of the concrete due to expansive forces produced by a metastable compound Ettringite. It is manifested in hardened and aged (dried) cementitious products. The conditions necessary for sulphate attack are (1) the cementitious constituent tricalcium aluminate, (2) Water, (3) water-soluble sulphates and (4) time.

The attack of the cementitious paste only takes place if all four conditions are present. Therefore the degradation of cementitious agencies can be arrested by the absence of any one item.

12.16 CAUSE OF SULPHATE ATTACK

A constituent within the cementitious adhesive transforms in the presence of water and the sulphate ion, and given plenty of time, into a metastable compound which has chemically bound 31–32 molecules of water within the molecule. This compound is called ETTRINGITE (Section 12.9). The compound responsible for this reaction (called sulphate attack) is the compound tricalcium aluminate. The expansion resulting from the uptake of these 32 molecules of water is often given as between 0.2% and 2.0%.

12.17 RATE OF ATTACK

The rate of attack depends on the porosity of the sulphate-containing medium, the solubility of the sulphates and the permeability of the mortar or concrete product, together with the amount of tricalcium aluminate. The compound tricalcium aluminate was the absorber of the chloride ions (Section 12.9.1(1)) and would prevent the depassivation of the steel in the alkaline matrix.

12.18 BRICKS AND THE MORTAR BEDS

Bricks are made from clay by a process of firing to about 950–1100°C. Clays often contain sulphates and this firing temperature is not high enough to decompose the soluble sulphates. The firing process produces many capillaries within the brick where the chemically bound water in the clay has to evaporate at these high temperatures. Therefore most (domestic household) bricks are very porous and are capable of dissolving large amounts of water (rain) which then transports the soluble sulphates to the mortar bed where they may react with the tricalcium aluminate. The greater the amount of water-soluble sulphates and the greater the length of time that the sulphates are in contact with the cement matrix, the greater the degree of degradation. Engineering bricks and blue bricks are fired to a much higher temperature and a process called sintering takes place. Sintering closes up the capillaries within the bricks and hence they are used for damp-proof courses and road and railway bridges.

12.19 EFFECT OF RICHNESS OF MIX

As the necessary constituent for sulphate attack is tricalcium aluminate, which is present in the cement, one would have expected that the richer the mortar was in cement, and therefore the richer in the component tricalcium aluminate, the greater would be the degree of sulphate attack. In practice, however, a rich mix of cement : lime : sand of say 1 : 0.5 : 4.5) (17% cement) is much more resistant to sulphate attack than a more porous 1 : 1 : 6 mix (12.5% cement). The actual scientific explanation for this anomaly is not fully clear, but it is obvious that continuous sulphate ion movement from the porous bricks to the mortar bed will require some compatibility in transfer coefficients.

12.20 SULPHATE RESISTING MORTARS

Sulphate-resisting mortars may be made by three methods.

(1) Richer mixes e.g. a 1 : 0–0.25 : 3 (22% cement content), can be specified.

(2) Sulphate-resisting Portland cement (SRPC) can be used. SRPC is manufactured with the compound responsible for sulphate attack, tricalcium aluminate, kept to below 3.5% (Table 12.4).

(3) High alumina cement, which does not contain the compound C_3A, can be specified.

12.21 SULPHATE ATTACK OF CONCRETE

Concrete, like the brickwork mentioned above, also withstands the presence of sulphates, if the concrete is made from a rich mix. The guide for sulphate resistance is given in Table 12.9.

12.22 RESISTANCE TO INDUSTRIAL POLLUTANTS

Cement is a 'calcium alkali adhesive' and will react with any chemical which reacts with calcium hydroxide $Ca(OH)_2$ or calcium carbonate. Hence it is 'attacked' by acids, fats and some oils and carbon dioxide, which leads to **carbonation**. This latter point is important as the alkalinity of the concrete is altered, which reduces the chemical passivity afforded to embedded metals. (See earlier sections 12.11 and 12.13).

12.23 ALKALI-SILICA ATTACK (CONCRETE CANCER)

This form of degradation has been found in car park decks. It is not due to the chloride ion. Some aggregates react with the alkali (NaOH, KOH) constituent of the cement and form a gel and this swells by imbibing (absorbing) water. The expansive forces cause cracking and disruption of the concrete. This is overcome by making sure that the total alkali content is less than 0.6%. The total alkali content is expressed as sodium oxide equivalents (Na_2O(%) + (62/94)K_2O(%)). Calcium hydroxide $Ca(OH)_2$ is a much weaker alkali and does not cause alkali–silica attack.

12.24 RESISTANCE TO SEA WATER AND CHLORIDES

Concrete is resistant to sea water and abrasion by shingle. Chlorides do not affect the durability of concrete. Chlorides do diffuse through the concrete and affect embedded metals. If the porosity is low and increased cover to embedded metals is specified and used, this is usually sufficient.

12.25 RESISTANCE TO FIRE

Concrete is resistant to fire as a result of the combined effects of (1) poor thermal conductivity, (2) endothermic reaction on the loss of water from the concrete matrix and (3) the thermal expansion accompanying the increase in temperature being offset by the contraction experienced on the loss of water.

12.26 RESISTANCE TO FROST

The fine capillaries and gel pores may allow water to penetrate the concrete mass. The number and diameter of these capillaries are increased with increase in the water: cement ratio. Damage is shown first at corners and as small cracks around coarse aggregates. Added resistance to frost attack may be achieved by the introduction of an air entraining agent.

REFERENCES

Brown, R. D., Geoghegan, M. P., and Baker, A. C. Corrosion of Reinforcement in Concrete Construction. Ed. A. P. Crane. Chapter 13. Publishers, Ellis Horwood, Chichester.

Beckett, D. (1983) Concrete, February, 16.

Shirley, D. E. (1980) Introduction to Concrete. Cement and Concrete Association, p. 3, Table 1.

Verbeck, G. J., and Forster, C. W. (1950) Proceedings of the American Society for Testing and Materials, **50**, 1235–1257.

Index